pm

D1286837

WITHDRAWN

ALGEBRA AND ANALYSIS
PROBLEMS AND SOLUTIONS

ALGEBRA AND ANALYSIS

PROBLEMS AND SOLUTIONS

G. LEFORT

Faculty of Science, Paris

Translated by SCRIPTA TECHNICA, INC.

Translation Editor

BERNARD R. GELBAUM
University of California, Irvine

1966

W. B. SAUNDERS COMPANY
PHILADELPHIA

THIS BOOK WAS ORIGINALLY PUBLISHED BY
DUNOD, PARIS, 1964, UNDER THE TITLE
ALGÈBRE ET ANALYSE. EXERCICES

Publishers:

NORTH-HOLLAND PUBLISHING COMPANY—AMSTERDAM

Sole distributors for U.S.A. and Canada:
W. B. SAUNDERS COMPANY — PHILADELPHIA

LIBRARY OF CONGRESS CATALOG CARD NUMBER
66—23603

PRINTED IN BELGIUM

Editor's Foreword

A glance at the Table of Contents of this volume will convince the reader that he will have been exposed to the full spectrum of undergraduate mathematics after he has completed the book. The exercises are neatly partitioned into eight sections, each dealing with a well-defined and important area of algebra and/or analysis.

Each section is begun with a summary discussion of the material to be treated. The concepts, vocabulary, notations, etc., are laid out and thereupon the material is subdivided into chapters, each of which consists of a block of exercises followed by a block of their solutions.

All the work is carefully graded and leads from the rather elementary to the quite sophisticated.

This collection should prove to be a most valuable asset for all students of mathematics.

BERNARD R. GELBAUM
University of California, Irvine

Preface

The assimilation of the course in general mathematics presents serious difficulties for many students. The importance given to reasoning and the axio-
matic construction of the theories give an impression of complete novelty and a
total break with prior teaching. Their usual thought processes are poorly
adapted to the problems treated and the geometric intuition already acquired is
completely unusable. Furthermore, the definitions presented often seem quite
arbitrary and for the student they are merely the rules of a subtle game which
one must follow even though he has no firm interest in them.

How, then, shall we orient his work in order to surmount as rapidly as possible
these inevitable difficulties, reestablish the connection between this new mathematics and the mathematics that he has learned in school, and recreate an
intuition, that is, a sufficient comprehension of the theories so that the solution
of a problem is suggested before the logical chain of reasoning is completed?

In learning a chapter in which new definitions are introduced, the student
ought to seek from his previous knowledge particular examples of mathematical
objects that verify the given axioms. This will enable him to understand better
the meaning of these axioms and the conclusions that can be drawn from them
and it will facilitate the liaison between his prior knowledge and the course that
he is taking. At the same time, he needs to distinguish in these examples between
that which relates to the theory studied and the special properties associated
with the particular case in question. Thus, the vectors of elementary geometric
space provide an example of a three-dimensional vector space over R. But this
space is also a Euclidean space in which the distance is defined with the aid of an
inner product, which introduces properties that one does not encounter in other
examples of vector spaces.

But above all, it is essential to accompany the study of a chapter in a course
with the solution of simple exercises illustrating the material in order to familiarize oneself with the definitions and the salient properties. It is only in this way
that the lifeless language of the mathematical statements is transformed into an
effective tool for the study of subsequent problems. These exercises should

(1) lead the student into the habit of carefully verifying that all the axioms
of a definition or that all the hypotheses of a theorem are verified and show how
the labor involved in this verification can sometimes be simplified;

(2) provide simple examples illustrating the theories studied and show how the
results of the course can be used;

(3) initiate new theoretical applications that will gradually show the student how to overcome the difficulties of abstract reasoning.

Finally, the student who is aware of the new difficulties encountered should not neglect calculating techniques (study of the more common functions, evaluation of derivatives and their primitives, solution of differential equations, computation with matrices, actual solution of linear equations, determination of eigenvalues, etc.). Certainly, the solution of a theoretical problem assumes at least partial mastery of the techniques involved; otherwise, the details of a proof will themselves cause so many difficulties that it will be impossible to have an overall picture of the problems.

It was with a view to these different requirements that I assembled the exercises in this book and I emphasize the fact that, from the very beginning of the academic year, the students can find in it exercises corresponding to those portions of the course that they are currently studying. In fact, the only information necessary for beginning the exercises in a chapter is already theoretically given. One difficulty no doubt remains: The different parts of the course are not treated in an immutable order. The order followed was approximately that of the *Cours de Mathématiques Générales* by C. Pisot and M. Zamansky. However, I believe that the examples chosen are sufficiently simple and that the indications for solutions are sufficiently complete for the students who have followed a course arranged in a different order to be able to work the majority of the exercises when they have completed the corresponding study of the theory.

Thus, this collection of exercises is primarily a permanent working instrument designed for use throughout the year. However, it can also be used during a pre-examination review. In the latter case, the student will be interested first of all in those exercises dealing with the portions of the course that have appeared most difficult to him. He will then choose from the collection of statements preferably those exercises touching on several questions of a theoretical rather than a technical problem since these texts are more allied than others with the examination proofs. As a rule, these exercises appear at the end of the different chapters or parts of the collection. The high school candidates will also use it profitably in reviewing in this way the essential concepts of algebra and analysis before their examination.

The users of this book and, in particular, individual students, who, for various reasons, cannot appear regularly in courses and exercises of general mathematics will tell whether I have achieved the purpose that I set for myself, namely, the facilitation of their work by aiding them to assimilate more rapidly the new concepts and methods that so often surprise them at the beginning of the year and that sometimes discourage them.

Practical suggestions
for the use of this collection of exercises

(1) The student who uses an exercise book by himself is his own instructor and taskmaster. Therefore, in comparing his work with the solutions given, he needs to use the utmost care to see that he has not omitted a step in the reasoning and he needs to be very exacting with himself. Failure to make this check very carefully will considerably reduce the profit of the exercise. It is for this reason that I have explicitly given the solution of all the exercises though I allowed myself to go through the calculations more rapidly than the reasoning behind them.

(2) The exercises given are relatively short (especially the first few in each chapter) so as to keep the solution of each of them from being too great a labor. For some of the exercises, I use the results of a preceding exercise. When I do, I note the fact in the beginning.

(3) In the presentation of the solutions, the different questions involved are rather sharply separated. Therefore, it is easy to find the solution of a question that one does not know how to handle or to verify the result of a calculation without going through the study of the exercise.

(4) The algebraic material has been placed entirely ahead of the portion dealing with analysis. Therefore, it is possible to familiarize oneself with the algebraic properties involved before proceeding to the study of analysis and to use these properties in that study. However, the majority of the exercises in analysis assume a knowledge only of fundamental algebraic concepts. Therefore, it is quite possible to treat the study of algebra and analysis simultaneously as is done in the majority of such courses.

(5) The collection has been divided into eight parts, some of these in turn have been divided into chapters. Therefore, it is easy to find the exercises dealing with any one particular part of the course. I have further facilitated this research for the readers of the Pisot and Zamansky book referred to in the preface by specifying what chapters of that book correspond to the different portions of the present exercise book.

(6) At the head of each of these parts, I give the theorems that are frequently used in the proofs and I indicate the way in which they should be used. Here, it is not a matter of a summary of the course but rather of an extra feature designed for applications. This material emphatically does not relieve the student of the need to refer to his course.

(7) In the proofs, I often recall the statements of the theorems used (or simply their name when there is no danger of ambiguity). Furthermore, for the benefit of those who have the book by Pisot and Zamansky, I indicate references to theorems in that book with the letters PZ in order to enable them to locate the theorem in the overall theory and to check its proof if need be *.

(8) Certain classical theorems that do not constitute part of the general mathematics course are given in the form of exercises. In fact, I thought it would be interesting to have the reader derive certain important results, as he will then be somewhat familiar with them when he encounters them in subsequent work. However, it will not be worthwhile for him to try to remember these now.

* For the convenience of readers who do not have access to the Pisot and Zamansky book, a special table has been provided on page X. This table will aid the reader to locate the theorem in question in one or more of the following books:

[1] Apostol, T.M., *Mathematical Analysis: A Modern Approach to Advanced Calculus*, Addison-Wesley, Reading, 1957
[2] Cater, F., *Lectures on Real and Complex Vector Spaces*, Saunders, Philadelphia, 1966
[3] Hafstrom, J., *Introduction to Analysis and Abstract Algebra*, Saunders, Philadelphia, 1967
[4] Halmos, P. R., *Finite-dimensional Vector Spaces*, Van Nostrand, Princeton, 1958
[5] Herstein, I. N., *Topics in Algebra*, Blaisdell, Waltham, 1963
[6] Tenenbaum, M. and Pollard, H., *Ordinary Differential Equations*, Harper, New York, 1963

The first column in the table on page X lists the page number in this book where reference is made to Pisot and Zamansky. The second column indicates one or more of the texts listed above as a recommended alternative reference.

As an additional convenience to the reader, the editors have included a special glossary on page 581. This glossary lists and defines many terms whose meaning may not be readily apparent from the text.

TABLE OF ALTERNATIVE REFERENCES

See page ix for explanation and bibliographical data

Page	Reference	Page	Reference
1	2, 3, 4	279	1
13	2, 4	281	1
25	3, 5	288	1
27	3, 5	293	1
29	2, 3, 4	300	1
39	2, 3, 4, 5	301	1
45	3, 5	309	1
49	3, 5	321	1
54	5	348	1
57	5	351	1
87	2, 4	362	1
90	2, 4	373	2, 4
95	3, 5	392	1
96	2, 4	400	1
97	2, 4	412	1
109	2, 4	415	1
110	2, 4	435	1
137	2, 4	436	1
139	2, 3, 4, 5	455	1
144	2, 4	459	1
166	2, 4	468	1
191	3, 5	475	1
194	2, 4	491	1
198	1	494	1
200	2, 4	500	1
211	1	503	1
213	1	504	1
242	1	509	1
246	1	515	1
251	1	517	1
255	1	531	6
265	1	551	6
275	1	568	6
277	1	576	6
278	1		

Contents

Part 1

GENERAL CONCEPTS AND BASIC ALGEBRAIC STRUCTURES

Exercises on the definitions and results of Book I and
of chapters I, II, and III of Book II of
Cours de Mathématiques Générales
by C. Pisot and M. Zamansky

Introduction

1. *Supplementary concepts regarding functions*

DEFINITIONS

A function or mapping f of E into F is said to be

injective if $$x \neq x' \Rightarrow f(x) \neq f(x')$$

in other words, if $\forall y \in F$, the equation $f(x) = y$ has at most one solution;

surjective if $$f(E) = F$$

in other words if $\forall y \in F$, the equation $f(x) = y$ has at least one solution;
bijective if it is injective and surjective, in other words, if $\forall y \in F$, the equation $f(x) = y$ has one and only one solution. In this case, an inverse function exists and is defined on F (we say also that the mapping is *one-to-one* or that f is a *bijection*).

REMARKS

(1) An injective mapping f of E into F is a bijective mapping of E onto $f(E)$ (since it is obviously a surjective mapping onto $f(E)$).
(2) For a function to have an inverse function, it is necessary and sufficient that it be injective. The inverse function is then a bijective mapping of $f(E)$ onto E.
(3) A composite function of two injective functions is injective; a composite function of two surjective functions is surjective; a composite function of two bijective functions is bijective.

2. *Remarks on the procedure for solving certain algebraic problems*

(1) To show that a certain subset Γ of a group G is a subgroup

We use multiplicative notation for the group operation and we denote by x^{-1} the inverse of an element x.

First method: Show that

$$\left\{ \begin{array}{l} \text{the neutral element } e \text{ belongs to } \Gamma, \\ x\in\Gamma \Rightarrow x^{-1}\in\Gamma, \\ x\in\Gamma \text{ and } y\in\Gamma \Rightarrow xy\in\Gamma. \end{array} \right.$$

We know also that the group operation (which is closed in Γ by virtue of the third condition) is associative in Γ since it is associative in G.

Thus, Γ is a group and hence is a subgroup of G.

Second method: The three conditions above may be replaced with verification of the single condition

$$u\in\Gamma \quad \text{and} \quad v\in\Gamma \Rightarrow uv^{-1}\in\Gamma.$$

This is true because

$$v = u \quad \Rightarrow uu^{-1} = e\in\Gamma,$$
$$u = e \quad \Rightarrow v^{-1}\in\Gamma \text{ if } v\in\Gamma,$$
$$u = x \quad \text{and} \quad v = y^{-1} \Rightarrow xy\in\Gamma \text{ if } x\in\Gamma \quad \text{and} \quad y\in\Gamma.$$

(2) To show that a certain subset B of a ring A is a subring

We need to verify by one of the methods of (1) that B is an additive subgroup of A and that the product of two elements of B is also an element of B. These conditions are sufficient since the multiplication properties (associativity, distributivity with respect to addition) are satisfied in A and hence in B.

In sum, it will be sufficient to show that

$$x\in B \quad \text{and} \quad y\in B \quad \Rightarrow \quad x - y\in B \quad \text{and} \quad xy\in B.$$

(3) To show that a certain subset F of a vector space E is a vector subspace

We need to show by one of the methods of (1) that F is an additive subgroup of E and that λx belongs to F whenever x belongs to F and λ belongs to the scalar field K. As in (2), these conditions are sufficient since the properties of multiplication by a scalar are satisfied in E and hence in F.

In sum, it will be sufficient to show that

$$x\in F \quad \text{and} \quad y\in F \Rightarrow x - y\in F,$$
$$x\in F \quad \text{and} \quad \lambda\in K \Rightarrow \lambda x\in F.$$

(4) To show that a certain set E on which a binary operation is defined is a group, a ring, or a vector space

The first method consists in verifying the defining axioms of the structure in question. This method, however, is sometimes rather long.

If we know that the set E is a subset of a set $\mathscr{E}$ that has a group (or ring or vector space) structure, it will be sufficient to show that E is a subgroup (or a subring or a vector subspace) of $\mathscr{E}$, and the methods of (1), (2), or (3) yield the result rather rapidly.

Finally, one can sometimes show that E is isomorphic to a set E' that we already know possesses the structure in question. This gives the result immediately. (For the definition of an isomorphism, see PZ, Book I, Chapter III, part 3, § 2.)

Exercises

101

If two subsets E and F of a set A satisfy the two relations

$$E \cup F = A, \qquad E \cap F = \emptyset,$$

show that E and F are the complements of each other in A.

102

Suppose that E and F are two subsets of a set A. Prove the following:

(1) $$\mathop{C}_{A}(E \cap F) = \left(\mathop{C}_{A}E\right) \cup \left(\mathop{C}_{A}F\right),$$

(2) $$\mathop{C}_{A}(E \cup F) = \left(\mathop{C}_{A}E\right) \cap \left(\mathop{C}_{A}F\right).$$

(3) The four sets

$$E \cap F, \qquad \mathop{C}_{E}(E \cap F), \qquad \mathop{C}_{F}(E \cap F), \qquad \text{and} \qquad \mathop{C}_{A}(E \cup F)$$

constitute a partition of A.

103

Let A denote a set and let E denote a subset of A. The mapping e of A onto the set consisting of the two numbers 0 and 1 defined by

$$e(x) = 0 \quad \text{if} \quad x \notin E,$$
$$e(x) = 1 \quad \text{if} \quad x \in E$$

is called the *characteristic function* of the subset E.

(1) Suppose that e and f are the characteristic functions of two subsets E and F of A. What are the sets whose characteristic functions are $1-e$, ef, $e+f-ef$?

(2) Use the results of (1) to show that

$$(E \cap F) \cap G = E \cap (F \cap G)$$
$$(E \cap F) \cup G = (E \cup G) \cap (F \cup G)$$
$$\mathop{C}(E \cap F) = (\mathop{C}E) \cup (\mathop{C}F).$$

104

Let E and F denote two finite sets of m and n elements respectively. Show that there are n^m mappings of E into F. Evaluate the number of subsets of E by using the concept of characteristic function defined in exercise 103.

105

Consider the injective mappings f of E into F, where E and F have m and n elements respectively.

(1) Show that for injective mappings f to exist, it is necessary and sufficient that $m \leq n$.

(2) Show by induction on m that the number of injective mappings of E into F is

$$n(n-1)\cdots(n-m+1).$$

(3) Show that if the image sets $f(E)$ and $f'(E)$ are identical, the mapping $f^{-1} \circ f'$ is a permutation of E.

(4) What is the number of m-element subsets of F.

(5) Show that this number is the coefficient of x^m in the expansion of the polynomial $(x+1)^n$.

106

$(\mathscr{F})$ is the set of functions mapping a set E into itself. For two functions f and g belonging to $(\mathscr{F})$, we denote their composite function by $f \circ g$.

(1) Show that this operation is associative.

(2) Determine the regular elements for this operation. (An element f is said to be *regular* if the equation $f \circ g = f \circ h$ implies $g = h$.)

107

An example of a non total ordering

(1) In the set N of natural numbers, we shall write $a \prec b$ if a divides b. Show that this defines an ordering relation Ω which, however, does not totally order N.

(2) In this question, the terms 'upper bound', 'lower bound', 'maximal element', and 'minimal element' relate to the ordering Ω. Let E denote a finite subset of N. Show that the set of upper bounds of E has a minimal element and that the set of lower bounds of E has a maximal element.

108

We say that a relation $\mathcal{R}$ defined on a set E is a quasi-ordering if it satisfies the two following conditions:

$$(a\mathcal{R}b \quad \text{and} \quad b\mathcal{R}c) \Rightarrow a\mathcal{R}c$$
$$a\mathcal{R}a.$$

(1) Show that $(a\mathcal{R}b$ and $b\mathcal{R}a)$ defines an equivalence relation R' between a and b. (The corresponding equivalence classes will be denoted by A, B, ...)

(2) In the quotient set $\mathcal{E} = E/\mathcal{R}'$ (the set of equivalence classes), we shall say that A precedes B (which fact we denote by $A \prec B$) if there exist elements $a \in A$ and $b \in B$ such that $a\mathcal{R}b$.

Show that the definition is independent of the particular elements a and b chosen and that it defines an ordering in $\mathcal{E}$.

(3) Let E denote the collection of finite subsets a of the set N of natural numbers. Let us define a relation $\mathcal{R}$ by $a\mathcal{R}b$ if the sum of the elements of a does not exceed that of the sum of the elements of b.

Show that $\mathcal{R}$ is a quasi-ordering. Define the associated equivalence relation $\mathcal{R}'$ and the equivalence classes. Show that the equivalence classes are finite sets. Use this result to show that the collection of finite subsets of N is denumerable.

109

Consider a set E of four elements e, i, j, and k and a binary operation under which this set is closed, the table for which is as follows:

	e	i	j	k
e	e	i	j	k
i	i	j	k	e
j	j	k	e	i
k	k	e	i	j

(The product of two elements appears at the intersection of the row containing the first element and the column containing the second.)

Show that we have defined an abelian group on E.

Find the subgroups of E.

110

Let E be a given set and let G be a group of bijective functions of E onto itself, the group operation being the composition of functions.

We say that a subset F' of E is equivalent to a subset F of E if there exists a function f of G such that $f(F)=F'$. (If E is the three-dimensional space of elementary geometry and G is the group of displacements, these equivalent subsets are congruent geometric figures.)

(1) Show that this 'equivalence' of subsets of E is indeed an equivalence relation (to be denoted by $F' \sim F$).

(2) Let E denote a finite set and let G denote the set of permutations of E. Show that G is a group and give a direct definition of equivalence of two subsets of E.

111

An equivalence relation defined on a set E is said to be compatible with a closed binary operation denoted by T if

$$a \sim a' \quad \text{and} \quad b \sim b' \quad \Rightarrow \quad a \mathsf{T} b \sim a' \mathsf{T} b'.$$

Show that we can then define a binary operation on the set F of equivalence classes by taking as the result of the operation on two classes A and B the class to which $a \mathsf{T} b$ belongs, where a and b are two arbitrary elements of A and B respectively.

Show that, if E is a group, so is F.

112

Determination of equivalence relations that are compatible with the structure of an abelian group G or a commutative ring A

(See exercise 111 for the definition of compatibility of an equivalence relation with a closed binary operation.)

The group operation for the group G will be denoted additively.

(1) The class Ω to which the neutral element 0 belongs is a subgroup.

(2) Conversely, every subgroup Γ defines an equivalence relation that is compatible with the structure of the group G if we take $x \sim y$ for all x and y such that $x - y \in \Gamma$.

(3) If G is the additive group Z of integers (positive, zero, and negative), its only subgroups are the sets Γ_p of multiples of an integer p.

(4) An equivalence relation that is compatible with the structure of a commutative ring A is defined by

$$x \sim y \quad \text{if} \quad x - y \in \mathscr{I},$$

where $\mathscr{I}$ is an ideal of the ring A.

If A is the ring of integers, the ideals $\mathscr{I}$ are the sets Γ_p defined in (3).

113

In the set E of pairs of real numbers (a, b), with the first, a, nonzero, we define the product of two elements by

$$(a, b) (a', b') = (aa', ab' + b).$$

(1) Show that E is a noncommutative group.

(2) Show that the elements $(a, 0)$ constitute a subgroup F of E that is isomorphic to the multiplicative group of nonzero real numbers. (For the definition of isomorphism, see PZ, Book I, Chapter III, part 3, § 2.)

(3) Define a mapping φ of E into R by

$$\varphi[(a, b)] = a.$$

Show that

$$\varphi[(a, b) (a', b')] = \varphi[(a, b)]\varphi[(a', b')].$$

Show that $\varphi^{-1}(1)$, that is, the set of elements of E whose image under φ is 1, is a subgroup of E that is isomorphic to the additive group of real numbers.

114

The concept of group homomorphism

Let G and Γ denote two groups and let e and ε denote their respective neutral elements (identities). Denote the group operations in G and Γ multiplicatively.

Then, a mapping f of G into Γ is a *homomorphism* of G into Γ if, for all elements a and b of G,

$$f(ab) = f(a)f(b).$$

EXAMPLES:

(a) The function φ in (3) of exercise 113.

(b) The mapping $x \rightarrow x^2$ of the multiplicative group of nonzero rational numbers into itself.

(1) Show that $f(e) = \varepsilon$ and that $f(a^{-1}) = [f(a)]^{-1}$.

(2) Show that $f^{-1}(\varepsilon)$ is a subgroup of G and that $f(G)$ is a subgroup of Γ.

(3) Verify the results of (2) for example (b) above.

(4) Show that an isomorphism between G and Γ is a bijective homomorphism.

115

(A) The symbol $[x]$ denotes the greatest integer not exceeding x, where x is a real number. Let a denote a given irrational number. Let f denote the mapping of the set $\mathbf{Z}$ of integers into the interval $[0,1]$ defined by

$$f(q) = aq - [aq].$$

Show that f is injective and that if two numbers y_1 and y_2 belong to the set $f(\mathbf{Z})$, so does the number $|y_2 - y_1|$.

Show that, for any $\varepsilon > 0$, at least one element (and hence infinitely many) of $f(\mathbf{Z})$ belongs to the interval $[0, \varepsilon]$. From the preceding result, show that there exist infinitely many rational numbers p/q satisfying the two inequalities

$$q > 0 \qquad\qquad \left| a - \frac{p}{q} \right| \leqslant \frac{\varepsilon}{q}.$$

(B) *In this part of the exercise, we shall use the results of exercises 111, 112, and 114.*

Show that the relation

$$x \sim y \qquad \text{whenever } x - y \text{ is an integer}$$

is an equivalence relation on the set $\mathbf{R}$ of real numbers and that it is compatible with the structure of the additive group of real numbers. Thus, the set of equivalence classes is an additive group T (known as the group of real numbers modulo 1).

Denote by F the mapping of $\mathbf{Z}$ into T that assigns to every integer q the class $F(q)$ of the real number $f(q)$ defined in (A). Show that $F(\mathbf{Z})$ is a subgroup of T that is isomorphic to $\mathbf{Z}$.

116

Let $\mathcal{L}$ denote the set of linear functions f with real coefficients that are defined for all real x by $f(x) = ax + b$ (where a and b are real constants).

We define an addition and a multiplication on $\mathcal{L}$ by

$$(f + g)(x) = f(x) + g(x),$$
$$(fg)(x) = (f \circ g)(x) = f[g(x)].$$

(1) Show that $\mathcal{L}$ is an abelian group under addition.
(2) Show that this multiplication is associative but not commutative.
(3) Is $\mathcal{L}$ a ring?

117

Let E denote the set of real numbers of the form $m + n\sqrt{3}$, where m and n are integers.

(1) Show that E is a subring of the ring of real numbers.
(2) Let us define an equivalence relation on E by

$$m + n\sqrt{3} \sim m' + n'\sqrt{3}$$

if $m' - m$ and $n' - n$ are multiples of 2. How many numbers are there in the quotient space A? Show that we can define on A a ring structure with the sum and product of two classes taken as the class of the sum and product respectively of any two elements in these classes. (One may either prove this directly or use the results of exercises 111 and 112.) Construct a multiplication table for A. Is the ring A a field?

(3) Show that A contains an ideal $\mathscr{I}$ consisting of two elements.

118

Let ω denote an irrational root of a quadratic equation with rational coefficients

$$\omega^2 = a\omega + b.$$

(1) Show that the set K of real numbers $A(\omega)$, where A denotes an arbitrary polynomial with rational coefficients, is a ring. Show that all numbers belonging to K can be obtained from polynomials A of degrees 1 and 0.

(2) Show that every number in K is rational or else a root of a quadratic equation with rational coefficients.

(3) Show that K is a field.

119

Let A denote a finite ring without divisors of 0 (that is, such that the equation $xy = 0$ implies that either x or y is 0).

(1) Show that every nonzero element of A is regular on the right and on the left and that the equations

$$ax = b \quad \text{and} \quad x'a = b$$

have unique solutions x and x' if a is not 0. (One can show that the mapping $x \to ax$ is bijective.)

(2) Let a denote a particular element of A and let u_0 denote the solution of the equation $au_0 = a$. Show that u_0 is a neutral element with respect to multiplication on the right. Show that there exists an element v_0 that is neutral with respect to multiplication on the left. Then show, by observing the product $v_0 u_0$, that $u_0 = v_0$.

(3) Show that A is a field.

(4) Exhibit a ring without divisors of 0 and not a field.

120

Examples of vector spaces

(1) Show that the set R of real numbers is a vector space over the field Q of

rational numbers. (The product of a real number by a rational number is taken as the product of two numbers in the usual sense.)

(2) Show that the set of mappings of the interval $[0, 1]$ into R that assume only finitely many distinct values constitutes a vector space Φ over R. (One can use the following result: The mappings of A into R constitute a vector space over R. See PZ, Book II, Chapter III, part 3, § 1.)

(3) Show that the set of functions on $[0, 1]$ into R that are continuous in the complement of a denumerable set is a vector space E over R. (Use the fact that, if two functions f and g are continuous at x_0, so is the function $\lambda f + \mu g$.)

(4) Show that the set of rational n-tuples $(x_1, x_2, ..., x_n)$ that satisfy a system of equations with rational coefficients

$$(S) \qquad \sum_j a_{ij} x_j = 0, \qquad \text{where} \qquad i = 1, 2, ..., p$$

constitutes a vector space Σ over the field Q of rational numbers. Here, the sum of two sets and the product of a set by a rational number λ are defined by

$$(x_1, x_2, ..., x_n) + (x'_1, x'_2, ..., x'_n) = (x_1 + x'_1, x_2 + x'_2, ..., x_n + x'_n),$$
$$\lambda(x_1, x_2, ..., x_n) = (\lambda x_1, \lambda x_2, ..., \lambda x_n).$$

121

(1) The set $\mathscr{F}$ of mappings of $[-1, 1]$ into the field R of real numbers has a commutative ring structure if we define the sum and product of two mappings f and g by

$$(f + g)(x) = f(x) + g(x),$$
$$(fg)(x) = f(x)g(x).$$

Show which of the following sets are subrings of $\mathscr{F}$ and which are ideals in $\mathscr{F}$.
(a) the set $\mathscr{P}$ of polynomials;
(b) the set $\mathscr{P}_n$ of polynomials of degree not exceeding n;
(c) the set $\mathscr{Q}_n$ of polynomials of degree n;
(d) the set $\mathscr{N}$ of mappings that assume the value 0 at the point O;
(e) the set $\mathscr{U}$ of mappings that assume the value 1 at the point O.

(2) The set $\mathscr{F}$ also has a vector space structure on R if we define the product of a function f by a real number λ by

$$(\lambda f)(x) = \lambda f(x).$$

Show which of the subsets $\mathscr{P}$, $\mathscr{P}_n$, $\mathscr{Q}_n$, $\mathscr{N}$, and $\mathscr{U}$ are vector subspaces.

101

The first relation $E \cup F = A$ implies that every element x of A that does not belong to E, that is, that belongs to $\complement E$, belongs to F: $\complement E$ is contained in F.

The second relation implies that, if x belongs to F, then x does not belong to E and hence does belong to $\complement E$, so that F is contained in $\complement E$

$$\complement E \subset F \quad \text{and} \quad F \subset \complement E \Rightarrow F = \complement E$$

(and also, equivalently, $E = \complement F$).

102

First question:

$$x \in E \cap F \quad \Leftrightarrow \quad x \in E \quad \text{and} \quad x \in F.$$

Expressing the negation of these properties, we have

$$x \in \complement_A (E \cap F) \quad \Leftrightarrow \quad x \notin E \quad \text{or} \quad x \notin F.$$

(The conjunction 'or' means that x satisfies at least one of the two properties indicated [it may or may not satisfy both of them].) In other words,

$$x \in \complement_A (E \cap F) \Leftrightarrow (x \in \complement_A E \quad \text{or} \quad x \in \complement_A F) \Leftrightarrow x \in \left(\complement_A E\right) \cup \left(\complement_A F\right).$$

Second question:

We can reason as in (1) or we can use the result of (1). If

$$E' = \complement_A E \quad \text{and} \quad F' = \complement_A F,$$

we know that

$$\complement_A E' = E \quad \text{and} \quad \complement_A F' = F.$$

If we apply the result of (1) to E' and F', we have

$$\complement_A (E' \cap F') = \left(\complement_A E'\right) \cup \left(\complement_A F'\right) = E \cup F$$

Taking the complements, we have

$$E' \cap F' = \left(\underset{A}{\complement}E\right) \cap \left(\underset{A}{\complement}F\right) = \underset{A}{\complement}(E \cup F).$$

Third question:

By using the results of (2), we can define the four sets in question by

$$x \in E \cap F \quad \Leftrightarrow \quad x \in E \quad \text{and} \quad x \in F,$$

$$x \in \underset{E}{\complement}(E \cap F) \quad \Leftrightarrow \quad x \in E \quad \text{and} \quad x \notin F,$$

$$x \in \underset{F}{\complement}(E \cap F) \quad \Leftrightarrow \quad x \notin E \quad \text{and} \quad x \in F,$$

$$x \in \underset{A}{\complement}(E \cup F) \quad \Leftrightarrow \quad x \notin E \quad \text{and} \quad x \notin F.$$

We can easily see that any two of these sets are disjoint (since one of the conditions defining any one of them is contradicted by one of the conditions defining any other) and that their union is the set A (since any element of A necessarily satisfies one or the other of the four pairs of conditions).

The following diagram represents the four sets, where A is the plane R^2, the shaded area is $\underset{A}{\complement}(E \cup F)$, and the dotted area is $E \cap F$.

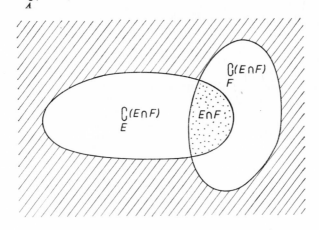

103

First question:

$$1 - e(x) = \begin{cases} 1 & \text{if} \quad x \notin E, \quad \text{that is, if} \quad x \in \complement E, \\ 0 & \text{if} \quad x \in E, \quad \text{that is, if} \quad x \notin \complement E. \end{cases}$$

The difference $1 - e$ is the characteristic function of $\complement E$.

$$(ef)(x) = e(x)f(x) = \begin{cases} 1 & \text{if} \quad x \in E \cap F, \\ 0 & \text{if} \quad x \notin E \cap F, \end{cases}$$

since at least one of the factors is 0 in the second case. The product ef is the characteristic function of $E \cap F$.

$$\varphi(x) = e(x) + f(x) - e(x)f(x)$$
$$= e(x) + f(x)[1 - e(x)] = f(x) + e(x)[1 - f(x)].$$

If e or f assumes the value 1, so does φ. If $e(x)$ and $f(x)$ are 0, so is $\varphi(x)$:

$$\varphi(x) = \begin{cases} 0 & \text{if} & x \notin E & \text{and} & x \notin F, \\ 1 & \text{if} & x \in E & \text{or} & x \in F. \end{cases}$$

Thus, φ is the characteristic function of $E \cup F$.

Second question:

To verify that two sets are identical, it is sufficient to show that their characteristic functions are identical, that is, that they take the same value for every x.

Let us denote by e, f, and g the characteristic functions of the sets E, F, and G. Thus, we need to verify with the aid of the results of (1) that

$$[e(x)f(x)]g(x) = e(x)[f(x)g(x)]$$

(associativity of the product in R), that

$$g(x) + e(x)f(x) - e(x)f(x)g(x)$$
$$= [e(x) + g(x) - e(x)g(x)][f(x) + g(x) - f(x)g(x)]$$

(which is immediate if we note that $g^2(x) = g(x)$ since $g(x)$ is equal to either 0 or 1), and that

$$1 - e(x)f(x) = [1 - e(x)] + [1 - f(x)] - [1 - e(x)][1 - f(x)].$$

104

Reasoning by induction on the number m of elements of E.

If $m = 1$, the image of the unique element of E can be any one of the n elements of F, so that there are n mappings.

Assume that the result is established for a set of $m - 1$ elements. Consider the set E as the union of a set E_1 of $m - 1$ elements and a set E_2 with a single element. A mapping f of E into F is defined by its restrictions f_1 and f_2 to the two sets E_1 and E_2; also, to two distinct pairs (f_1, f_2) and (f_1', f_2'), there correspond distinct functions f and f'. The set of mappings f_1 of E_1 into F has, by hypothesis, n^{m-1} elements and the set of mappings f_2 of E_2 into F has n elements. The number of elements of the product set, that is, of the set of functions f is therefore

$$n^{m-1} \times n = n^m.$$

A subset of E can be defined by its characteristic function, that is, by a mapping of E into the set $[0, 1]$, which has two elements. The number of subsets of E (including the empty set and the set E itself) is the number of mappings of E into a two-element set, which is 2^m.

105

First question:

The application f is injective; the number of elements of $f(E)$ is equal to the number of elements of E, namely m. Since $f(E)$ is contained in F, it is necessary that $m \leq n$. The condition is sufficient because, for a set Φ of m elements contained in F, there exists a bijective mapping of E into Φ, that is, an injective mapping of E into F.

Second question:

The result is obvious if $m = 1$. Assume it established for a set of $m - 1$ elements.

A set E of m elements is the union of a set E_1 of $m - 1$ elements and a set E_2 consisting of a single element x_0. The restriction f_1 of f to E_1 is injective: The number of such functions is, by hypothesis, $n(n-1)\cdots(n-m+2)$. Once the function f_1 is chosen, the image $f(x_0)$ of the last element of E can be taken arbitrarily in $\underset{F}{\complement} f_1(E_1)$, that is, in a set with $n - (m-1)$ elements. Therefore, to every function f_1 there correspond $n - (m-1)$ distinct functions f, and to distinct functions f_1 there correspond disjoint sets of functions f. The number of injective functions of E into F is therefore

$$n(n-1)\cdots(n-m+2)(n-m+1).$$

Third question:

The inverse mapping f^{-1} of f is defined in $f(E)$ since f is injective; f^{-1} is also injective. The composite mapping $f^{-1} \circ f'$ therefore maps E into E and it too is injective. The set $(f^{-1} \circ f')(E)$ has m elements and hence coincides with E. The mapping $f^{-1} \circ f'$ is a bijective mapping of E onto itself, that is, it is a permutation p of E.

Equivalent result: $f' = f \circ p$.

Fourth question:

Every subset Φ of F that has exactly m elements is the image of E under an injective mapping. More precisely, if f is one of the mappings such that $f(E) = \Phi$, the other such mappings f' are of the form $f \circ p$, where p is a permutation of E. Therefore, there are $m!$ mappings f the image of E under which is Φ.

(We recall that there are $m!$ permutations of E.) The number of injective mappings of E into F is $\binom{n}{m} \times m!$, where

$$\binom{n}{m} = \frac{n(n-1)\ldots(n-m+1)}{m!}.$$

Fifth question:

In the product $(x+1)^n = (x+1)(x+1)\cdots(x+1)$, one can identify a term by indicating the factors in which one takes the term x, that is, by considering a subset Φ of the set of the first n integers. The terms in x^m are those for which Φ has exactly m elements. The number of these, that is, the coefficient of x^m is $\binom{n}{m}$.

106

First question:

To show that

$$(f \circ g) \circ h = f \circ (g \circ h),$$

it will be sufficient to show that the images of an element of E under these two functions are identical. Let x be an element of E and define $u = h(x)$ and $v = g(u)$:

$$[(f \circ g) \circ h](x) = [f \circ g](h(x)) = (f \circ g)(u) = f[g(u)] = f(v),$$
$$(g \circ h)(x) = g[h(x)] = g(u) = v,$$
$$[f \circ (g \circ h)](x) = f[(g \circ h)(x)] = f(v).$$

Second question:

For f to be regular on the left, it is necessary and sufficient that it be injective. If f is injective and if $f \circ g = f \circ g'$, then, for every x,

$$f[g(x)] = f[g'(x)] \Rightarrow g(x) = g'(x) \Rightarrow g = g'.$$

If f is not injective, there exist two distinct elements a and b whose images $f(a)$ and $f(b)$ are identical. If, for every x in E, we have $g(x) = a$ and $g'(x) = b$, then $f \circ g = f \circ g'$. The mapping f is not regular on the left.

Similarly, for f to be regular on the right, it is necessary and sufficient that it be surjective. If f is surjective, then, for every x in E, there exists a u such that $f(u) = x$. Then, if $g \circ f = g' \circ f$, we have $g(x) = g'(x)$ for every x.

If f is not surjective but if the restrictions of g and g' to $f(E)$ are identical, we have $g \circ f = g' \circ f$. But g and g' can be different.

Finally, for f to be regular (that is, on the left and on the right), it is necessary and sufficient that it be bijective.

107

First question:

Reflexivity: $a \prec a$ since a divides a.

Transitivity: $a \prec b$ and $b \prec c \Rightarrow a \prec c$.
This is true because

$$b = aq \qquad \text{and} \qquad c = bq' \Rightarrow c = a(qq').$$

Antisymmetry: $a \prec b$ and $b \prec a \Rightarrow b = a$.
This is true because, if a divides b, then $a \le b$ and, if b divides a, then $b \le a$, which together imply that $a = b$.

However, Ω is not a total ordering since neither of the two relations $6 \prec 10$ and $10 \prec 6$ is satisfied.

Second question:

The upper bounds of E are the common multiples of the elements of E. We know that they are all multiples of one of these, namely, the least common multiple of the set. This is the least upper bound of E in the sense of the ordering relation Ω.

Similarly, the lower bounds of E are the common divisors of the elements of E. These are all divisors of the greatest common divisor, which is the greatest lower bound in the sense of the ordering relation Ω.

108

First question:

Reflexivity:

$$a \mathcal{R}' a$$

as a consequence of property (2) of $\mathcal{R}$.

Symmetry:

$$a \mathcal{R}' b \Rightarrow b \mathcal{R}' a$$

since a and b play the same role in the definition of $\mathcal{R}'$.

Transitivity:

$$(a \mathcal{R}' b \qquad \text{and} \qquad b \mathcal{R}' c) \Rightarrow a \mathcal{R}' c.$$

This is true because

$$a\mathscr{R}'b \;\Rightarrow\; a\mathscr{R}b \quad \text{and} \quad b\mathscr{R}a,$$
$$b\mathscr{R}'c \;\Rightarrow\; b\mathscr{R}c \quad \text{and} \quad c\mathscr{R}b,$$
$$\left.\begin{array}{l} a\mathscr{R}b \text{ and } b\mathscr{R}c \;\Rightarrow\; a\mathscr{R}c \\ c\mathscr{R}b \text{ and } b\mathscr{R}a \;\Rightarrow\; c\mathscr{R}a \end{array}\right\} \Rightarrow a\mathscr{R}'c.$$

Second question:

If a' and b' are elements of A and B, we know that

$$a'\mathscr{R}'a \;\Rightarrow\; a'\mathscr{R}a,$$
$$b'\mathscr{R}'b \;\Rightarrow\; b\mathscr{R}b',$$

and, by hypothesis,

$$a\mathscr{R}b.$$

The transitivity of $\mathscr{R}$ thus implies that $a'\mathscr{R}b'$: The relation $\mathscr{R}$ is verified for all pairs (a', b') if it is verified for any particular pair (a, b).

The relation thus defined on the equivalence classes is obviously *reflexive* and *transitive*. Specifically,

$$A \prec A$$

since we know that $a\mathscr{R}a$ for any $a \in A$, and

$$A \prec B \quad \text{and} \quad B \prec C \;\Rightarrow\; A \prec C$$

since we know that, for any a, b, and c in A, B, and C respectively,

$$a\mathscr{R}b \quad \text{and} \quad b\mathscr{R}c, \quad \text{so that} \quad a\mathscr{R}c.$$

Let us now show that this relation is also *antisymmetric*. Suppose that $A \prec B$ and $B \prec A$.

If a and b are elements of A and B respectively, they satisfy the relations $a\mathscr{R}b$ and $b\mathscr{R}a$; that is, by definition, we have $a\mathscr{R}'b$, so that a and b belong to the same equivalence class. Therefore, A and B are the same set.

Third question:

One can verify immediately that $\mathscr{R}$ is a quasi-ordering. If $K(a)$ is the sum of the elements of a, the relation $a\mathscr{R}'b$ is defined by

$$K(a) \leqslant K(b) \quad \text{and} \quad K(b) \leqslant K(a), \quad \text{so that} \quad K(a) = K(b).$$

The set of subsets a of E such that $K(a)$ has a particular value h is an equivalence class, which we denote by A_h. There are only finitely many of these subsets a since the number of elements of a is at most equal to h and each of these elements is equal to 1, 2,..., or h (the number of subsets a is at most h^h). All the equivalence classes are finite sets.

To number the finite subsets of the set of positive integers, we can number successively those belonging to A_1, A_2, A_3, ... Since each class contains a finite number of elements, this operation can be carried out without any difficulty. Since every element belongs to one class, all the elements are thus numbered.

109

The product of an element multiplied on the right or on the left by e is equal to that element; that is, e is the neutral element. The table for the operation is symmetric about the principal diagonal. Thus, reversal of the order of the factors does not change the result: the operation is commutative.

Associativity of the operation

One can use the commutativity to reduce the number of necessary verifications.

For products of the form $(\alpha\beta)\gamma$, where the three factors are distinct, we need to show that the result is independent of the permutation chosen for i, j, and k. For this, we need only note the following products from the table:

$$(ij)k = j, \qquad (jk)i = j, \qquad (ki)j = j.$$

The products $\alpha(\beta\gamma)$ are equal to the preceding ones since they are equal to $(\beta\gamma)\alpha$ (because of the commutativity).

For products of the form $(\alpha\beta)\gamma$ that contain two equal factors, one can verify that

$$(ii)j = e = i(ij), \qquad (ii)k = i = i(ik)$$

and similarly for

$$(jj)i, \qquad (jj)k, \qquad (kk)i, \qquad \text{and} \qquad (kk)j.$$

Finally, if all three factors α, β, and γ are equal or if at least one of them is equal to e, the result is obvious.

Symmetry of an element. The relations $ik = e$ and $jj = e$ show that every element has an inverse. The inverse of i is k and vice versa, and j is its own inverse.

E is indeed an abelian group since we have defined an associative and commutative operation possessing a neutral element and since each element has an inverse.

A *subgroup G* contains the neutral element e. For any element that it contains, it also contains the image of that element. Furthermore, it contains the result of the group operation performed on any two elements of the subgroup.

If G contains i, it also contains k, which is symmetric to i, and it contains the element $j = ii$. Thus it is identical to E. If G contains k, it contains i also and is again identical to E.

Finally, the set $\{e, j\}$ is a subgroup of E since j is its own image and since

$$ee = e, \qquad ej = j, \qquad \text{and} \qquad jj = e.$$

110

First question:

The neutral element of the group G is the identity mapping e of E into itself:

$$e(F) = F \Rightarrow F \sim F.$$

If $F' \sim F$, there exists by definition a function f of G such that $f(F) = F'$. The inverse function f^{-1} of f is the (group) inverse of f in G and f^{-1} is also a bijection of E onto E.

$$F = f^{-1}(F') \Rightarrow F \sim F'.$$

If $F'' \sim F'$ and $F' \sim F$, there exist functions f and g such that

$$F'' = g(F') \qquad \text{and} \qquad F' = f(F).$$

The composite function $g \circ f$ is also an element of G and

$$(g \circ f)(F) = F'' \Rightarrow F'' \sim F.$$

Thus, the three properties of the definition of an equivalence relation are satisfied.

Second question:

A permutation p is a bijective mapping of E onto itself. The identity mapping is a permutation; the inverse of a permutation is a permutation; finally, since the composite of two bijective mappings is a bijective mapping, the composite of two permutations of E is a permutation of E. Since we know (see exercise 106) that the composition of functions is an associative operation, we have established that the set G of permutations of a set is a group.

If the equation $F' = p(F)$ holds, it follows that F and F' must have the same number of elements since p is injective (that is, two distinct elements of F have distinct images).

Conversely, if F and F' have the same number of elements, there exists a permutation p of E such that $p(F) = F'$. We can define such a permutation explicitly by listing on one line the elements of F followed by those of $\complement F$ and then listing on the line below it the elements of F' followed by those of $\complement F'$. Since F and F', and consequently $\complement F$ and $\complement F'$ have the same number of elements, the permutation of E that we obtain by assigning to each element of E the element written below it maps F onto F'.

$$\overbrace{a_{i_1}\ldots a_{i_k}}^{F} \qquad \overbrace{a_{i_{k+1}}\ldots a_{i_n}}^{\complement F}$$

$$\underbrace{a_{i_1'}\ldots a_{i_k'}}_{F'} \qquad \underbrace{a_{i_{k+1}'}\ldots a_{i_n'}}_{\complement F'}$$

Thus, for two subsets of E to be equivalent, it is necessary and sufficient that they have the same number of elements.

111

(1) *It is necessary that the definition of the operation be meaningful,* that is, that the class of $a\mathsf{T}b$ depend only on A and B and not on the choice of a and b in A and B. This condition is assumed satisfied. That is,

$$a' \sim a \qquad \text{and} \qquad b' \sim b \qquad \text{imply} \qquad a'\mathsf{T}b' \sim a\mathsf{T}b.$$

The class of $a'\mathsf{T}b'$ is the class of $a\mathsf{T}b$.

(2) *The operation defined on F has the properties of the original operation:*
If the first operation is associative, so is the second. Consider three classes A, B, and C containing three elements a, b, and c respectively:

$$(a\mathsf{T}b)\mathsf{T}c = a\mathsf{T}(b\mathsf{T}c) \Rightarrow (A\mathsf{T}B)\mathsf{T}C = A\mathsf{T}(B\mathsf{T}C).$$

If ω is the neutral element in E, the class Ω of ω is the neutral element in F:

$$\omega\mathsf{T}a = a\mathsf{T}\omega \Rightarrow \Omega\mathsf{T}A = A\mathsf{T}\Omega.$$

Finally, if a^{-1} is the inverse of a, the class A' of a^{-1} is the inverse of the class A of a:

$$a\mathsf{T}a^{-1} = a^{-1}\mathsf{T}a = \omega \Rightarrow A\mathsf{T}A' = A'\mathsf{T}A = \Omega.$$

Thus, the operation defined on F meets the requirements of a group structure.

112

First question:

If $x\in\Omega$, then $x\sim 0$ by definition. If $x\in\Omega$ and $y\in\Omega$, then $x-y\sim 0-0=0$, since the equivalence relation is compatible with addition. Therefore, $x-y\in\Omega$ and Ω is a subgroup of G.

Second question:

The three defining properties of an equivalence relation are satisfied because (by virtue of the fact that Γ is a subgroup) we have: first

$$x - x = 0 \in \Gamma \Rightarrow x \sim x;$$

second,

$$x - y \in \Gamma \Rightarrow y - x = -(x - y) \in \Gamma,$$

so that

$$x \sim y \Rightarrow y \sim x;$$

third,

$$x - y \in \Gamma \quad \text{and} \quad y - z \in \Gamma \Rightarrow x - z = (x - y) + (y - z) \in \Gamma,$$

so that

$$x \sim y \quad \text{and} \quad y \sim z \Rightarrow x \sim z.$$

The equivalence relation thus defined is compatible with the group operation:

$$x' \sim x \quad \text{and} \quad y' \sim y \Rightarrow x' - x \in \Gamma \quad \text{and} \quad y' - y \in \Gamma \Rightarrow$$
$$x' + y' - (x + y) = (x' - x) + (y' - y) \in \Gamma \Rightarrow x' + y' \sim x + y.$$

Third question:

Let us denote by Γ^+ the set of positive elements of Γ and let us denote the set of their inverses by Γ^- (so that Γ is the union of $\Gamma^+, \{0\}$, and Γ^-). The set Γ^+ has a smallest element p.

The products obtained by multiplying positive integers by p belong to Γ and hence to Γ^+. Specifically, mp is the sum of m terms (each equal to p) and hence is an element of the group. To any element x belonging to Γ^+ we assign the remainder r resulting from dividing it by p:

$$r = x - mp.$$

This r, being the difference of two elements in Γ, is itself an element of Γ. Since it is less than p, it cannot belong to Γ^+. Therefore, r is 0. The elements of Γ^+ are multiples of p and, therefore, so are the elements of Γ^-. Thus, Γ consists of the set Γ_p of the multiples of p.

Fourth question:

The equivalence relation must be compatible with the additive abelian group structure of A. Thus, it is defined by a subgroup $\mathscr{I}$ and $x \sim y$ if $x - y \in \mathscr{I}$.

The equivalence must be compatible with multiplication; that is,

$$x \sim x' \quad \text{and} \quad y \sim y' \Rightarrow xy \sim x'y'.$$

In particular, if $y \in \mathscr{I}$, it follows that $y \sim 0$ and hence, for any element $x \in A$,

$$xy \sim x0 = 0, \qquad \text{that is,} \qquad xy \in \mathscr{I}.$$

In other words, $x\mathscr{I}$ has to be contained in $\mathscr{I}$. Thus, the defining requirements of an ideal are satisfied.

Conversely, if $\mathscr{I}$ is an ideal, the equivalence relation is compatible with multiplication in the ring. If $x' \sim x$ and $y' \sim y$, then, by definition, $x' - x \in \mathscr{I}$ and $y' - y \in \mathscr{I}$. Thus, the products $y(x' - x)$ and $x'(y' - y)$ belong to $\mathscr{I}$ and hence

$$x'y' - xy = x'(y' - y) + y(x' - x);$$
$$x' \sim x \qquad \text{and} \qquad y' \sim y \quad \Rightarrow \quad x'y' \sim xy.$$

If A is the ring of integers, its additive subgroups are the sets Γ_p of multiples of an integer p. These sets Γ_p are obviously ideals and they are the only ideals in **Z**.

The only quotient rings in **Z** are therefore the ones studied in PZ, Chapter III, part I, § 3, in the passage dealing with fields of characteristic p.

113

First question:

The operation is associative:

$$[(a, b)(a', b')](a'', b'') = (aa'a'', aa'b'' + ab' + b),$$
$$(a, b)[(a', b')(a'', b'')] = (a, b)(a'a'', a'b'' + b')$$
$$= (aa'a'', aa'b'' + ab' + b).$$

It is not commutative, as is shown by the facts that

$$(2, 0) (1, 1) = (2, 2) \qquad \text{but} \qquad (1, 1) (2, 0) = (2, 1).$$

(That a property does not always hold is often shown by exhibiting a counter-example.)

The element $(1, 0)$ *is the neutral element, and the inverse of* (a,b) *is* $(1/a, -b/a)$ since

$$(a, b) \left(\frac{1}{a}, -\frac{b}{a} \right)_{1} = \left(\frac{1\frac{a}{a}}{a}, -\frac{b}{a} \right) (a, b) = (1, 0).$$

(Since the multiplication is not commutative, we needed to verify the result by multiplying on both the right and the left.)

Second question:

The mapping $a \rightarrow (a,0)$ is a bijection of the set of nonzero real numbers onto

F. The product of two real numbers is mapped into the product of the corresponding elements of *F*:

$$aa' \rightarrow (aa', 0) = (a, 0)(a', 0).$$

Thus, *F* is a subgroup of *E*. (The image $(1, 0)$ of 1 is the neutral element, and the image of $1/a$ is the element $(1/a, 0) = (a, 0)^{-1}$.) This subgroup *F* is isomorphic to the multiplicative group of nonzero real numbers.

Third question:

$$\varphi[(a, b)(a', b')] = \varphi[(aa', ab' + b)] = aa',$$
$$\varphi[a, b]\,\varphi[a'b'] = aa'.$$

The set $\varphi^{-1}(1)$ is the set of elements of the form $(1, b)$. To show that it is a subgroup of *E*, one need only show that $(1, b)(1, b')^{-1}$ is an element of $\varphi^{-1}(1)$:

$$(1, b)(1, b')^{-1} = (1, b)(1, -b') = (1, b - b').$$

The mapping ψ that assigns b to $(1, b)$ is a bijective mapping of $\varphi^{-1}(1)$ onto **R**. Furthermore,

$$\psi[(1, b)(1, b')] = b + b' = [(1, b)] + \psi[(1, b')].$$

Thus, ψ is an isomorphism between $\varphi^{-1}(1)$ and the additive group of real numbers.

114

First question:

$$ae = a \Rightarrow f(ae) = f(a)f(e) = f(a).$$

When we multiply on the left by the inverse $[f(a)]^{-1}$ of $f(a)$ in Γ, we obtain

$$f(e) = \varepsilon.$$

Similarly, since $a^{-1}a = aa^{-1} = e$, we conclude that

$$\varepsilon = f(e) = f(a^{-1}a) = f(a^{-1})f(a) = f(aa^{-1}) = f(a)f(a^{-1}),$$

which proves that $f(a^{-1})$ is the inverse $[f(a)]^{-1}$ of $f(a)$ in Γ.

Second question:

Let us take two elements a and b in $f^{-1}(\varepsilon)$:

$$f(ab) = f(a)f(b) = \varepsilon\varepsilon = \varepsilon \Rightarrow ab \in f^{-1}(\varepsilon),$$
$$f(b^{-1}) = [f(b)]^{-1} = \varepsilon^{-1} = \varepsilon \Rightarrow b^{-1} \in f^{-1}(\varepsilon),$$
$$f(e) = \varepsilon \Rightarrow e \in f^{-1}(\varepsilon).$$

This proves that $f^{-1}(\varepsilon)$ is a subgroup of *G*.

If α and β are elements of $f(G)$, it follows from the definition that there exist elements a and b of G such that

$$\alpha = f(a) \qquad \text{and} \qquad \beta = f(b);$$

therefore,

$$\alpha\beta^{-1} = f(a)[f(b)]^{-1} = f(a)f(b^{-1}) = f(ab^{-1}).$$

The image $\alpha\beta^{-1}$ of ab^{-1} also belongs to $f(G)$, which, consequently, is a subgroup of Γ.

Third question:

The set $f^{-1}(1)$ consists of the two elements -1 and 1. It is indeed a subgroup of the multiplicative group of rational numbers. The image of the group is the set of squares of rational numbers, which is also a subgroup.

Fourth question:

In the definition of isomorphism, the only extra condition is that the function f be a one-to-one (that is, bijective) mapping of G onto Γ (see PZ, Book I, Chapter III, part 3, § 2).

115

(A) Suppose that $f(q) = f(q')$, that is $aq - [aq] = aq' - [aq']$ or

$$a(q' - q) = [aq'] - [aq].$$

It is impossible for $q' - q$ to be nonzero because a would then be a rational number (since $[aq']$ and $[aq]$ are integers). The condition $f(q) = f(q')$ thus implies that $q = q'$. Thus, f is injective.

Suppose that $y_1 < y_2$. Let us express the fact that they are elements of $f(Z)$:

$$y_1 = aq_1 - [aq_1], \qquad y_2 = aq_2 - [aq_2];$$
$$y_2 - y_1 = a(q_2 - q_1) - ([aq_2] - [aq_1]).$$

The difference $y_2 - y_1$ lies between 0 and 1. Thus, the integer $[aq_2] - [aq_1]$ is thus the greatest integer not exceeding $a(q_2 - q_1)$; that is, it is $[a(q_2 - q_1)]$. We therefore have

$$y_2 - y_1 = a(q_2 - q_1) - [a(q_2 - q_1)] = f(q_2 - q_1).$$

In accordance with the preceding results, if the interval $[0, \varepsilon]$ contained no element of $f(Z)$, the difference between two elements of $f(Z)$ would exceed ε and the set $f(Z)$ contained in the interval $[0, 1]$ would have only finitely many elements, which is impossible since Z is infinite and f is injective.

We note that $f(q)$ is nonzero since a is irrational. Therefore, one may construct an infinite sequence of elements of $f(q)$ belonging to $[0, \varepsilon]$ as follows: We know that $[0, \varepsilon]$ contains at least one element y of $f(Z)$. Suppose that we know n such elements: $y_1,..., y_n$. Let us take a number ε_n that is less than any of these y_i (which is possible since none of them is zero). We know the interval $[0, \varepsilon_n]$ contains at least one element y of $f(Z)$ that, by virtue of the construction, is different from the preceding ones.

Let us denote by q an integer such that $f(q) \in [0, \varepsilon]$ and let us denote by p the integer $[aq]$:

$$0 < aq - p \leqslant \varepsilon \Rightarrow \left| a - \frac{p}{q} \right| \leqslant \frac{\varepsilon}{|q|}.$$

By replacing p and q by $-p$ and $-q$ if necessary, we obtain the desired result.

(B) The integers constitute a subgroup Z of the additive abelian group R. We also know (cf. exercise 112) that the relation $x - y \in Z$ is an equivalence relation that is compatible with the group operation in R. Therefore, the quotient set has a group structure (cf. exercise 111).

The mapping F is injective. If $q \neq q'$, it follows from the injectiveness of f that $f(q) - f(q') \neq 0$. Since $0 < f(q) < 1$ and $0 < f(q') < 1$, we have $|f(q) - f(q')| < 1$. Therefore, $f(q)$ and $f(q')$ are not equivalent and the classes $F(q)$ and $F(q')$ are different. Therefore, F is injective.

F is a homomorphism of Z into T (for the definition and properties of a homomorphism, see exercise 114):

$$F(q) + F(q') = \text{class of } f(q) + \text{class of } f(q') = \text{class of } [f(q) + f(q')]$$

by virtue of the definition of the operation in T, so that

$$f(q) + f(q') \sim a(q + q') \sim f(q + q').$$

Thus, the class of $f(q) + f(q')$ is the class of $f(q + q')$; that is,

$$F(q) + F(q') = F(q + q').$$

Thus, the results of exercise 114 yield the following: $F(Z)$ is a subgroup of T; furthermore, the function F is injective; therefore, F is a bijection of Z onto $F(Z)$; F is a bijective homomorphism of Z and $F(Z)$; thus, it is an isomorphism.

116

First question:

The set $\mathscr{L}$ is certainly closed with respect to addition:

$$f(x) = ax + b \quad \text{and} \quad g(x) = a'x + b' \Rightarrow (f + g)(x) = (a + a')x + (b + b').$$

To show that $\mathscr{L}$ is an additive abelian group, two procedures are possible: On the one hand, we can use the representation of f in terms of the pair of real numbers (a, b), in which case the verification of the group properties is immediate. On the other hand, for the particular case of $\mathscr{L}$, one can perform the proof shown in PZ, Book II, Chapter III, part 3, § 1.

Second question:

One can easily verify that the set $\mathscr{L}$ is closed with respect to multiplication:

$$f(y) = ay + b \quad \text{and} \quad g(x) = a'x + b' \Rightarrow (fg)(x) = aa'x + ab' + b.$$

To study the properties of this multiplication, we can again follow two courses: First, we can use the representation of f with the aid of the pair of real numbers (a, b). Then, the multiplication in $\mathscr{L}$ is defined as in exercise 113, and one may refer to that exercise for the proof of the necessary properties. On the other hand, just as in exercise 106, we may show that the composition of functions is associative and we can give an example showing that it is not commutative. For example, take $f(x) = x - 2$ and $g(x) = 2x + 1$.

Third question:

$\mathscr{L}$ is an additive abelian group. The multiplication defined is associative. For $\mathscr{L}$ to be a ring, it is necessary and sufficient that the multiplication be distributive with respect to addition.

The multiplication is not distributive on the right:

$$[f(g+g')](x) = f[(g+g')(x)] = f[g(x) + g'(x)],$$
$$[fg + fg'](x) = f[g(x)] + f[g'(x)].$$

These two expressions are not equal if b is nonzero. This can easily be verified by taking, for example, $g(x) = g'(x) = x$ and $f(x) = ax + b$:

$$f[g(x) + g'(x)] = 2ax + b, \qquad f[g(x)] + f[g'(x)] = 2ax + 2b.$$

Thus, $\mathscr{L}$ is not a ring.

On the other hand, we can easily verify that the multiplication is distributive on the left, that is, that

$$(f + f')g = fg + f'g.$$

This shows clearly that in giving proofs *concerning a noncommutative operation*, one must not forget that the *properties of operating on the right and on the left need to be considered separately.*

117

First question:

We need to show that E is an additive subgroup of R (one can verify that the difference between two numbers in E is a number in E) and that E is closed with respect to the multiplication defined:

$$m+n\sqrt{3}-(m'+n'\sqrt{3}) = (m-m')+(n-n')\sqrt{3},$$
$$(m+n\sqrt{3})(m'+n'\sqrt{3}) = mm'+3nn'+(mn'+nm')\sqrt{3},$$

where

$$(m-m'),\quad (n-n'),\quad (mm'+3nn'),\quad (mn'+nm')$$

are indeed integers.

Second question:

We can use the result of exercise 112, (4). To show that the set A is a ring, note that the equivalence relation is defined by

$$(m+n\sqrt{3})-(m'+n'\sqrt{3})\in\mathscr{I},$$

where $\mathscr{I}$ is the set of numbers in E whose coefficients m and n are multiples of 2. Thus, we know that it is necessary and sufficient that $\mathscr{I}$ be an ideal, that is, that the following two conditions be satisfied:

$\mathscr{I}$ is an additive subgroup;
the product of an element of $\mathscr{I}$ and an arbitrary element of the ring is again an element of $\mathscr{I}$.

These are easily verified.

One may also proceed directly by verifying the following three assertions:
(1) the relation is indeed an equivalence relation;
(2) the sum and the product of two classes are actually defined; that is, the class of the sum or product of two elements is dependent only on the classes of these elements (see remarks below);
(3) the operations thus defined define a ring on the quotient set A. The last is obvious by virtue of the definition of the operations on the classes: The properties of the operations on E are equally valid in A.

To show, for example, that the class of the product of two elements u and v in E depends only on the classes of u and v, we need to show that

$$u' \sim u \quad \text{and} \quad v' \sim v \;\Rightarrow\; u'v' \sim uv.$$

Since the multiplication is commutative, to do this we need to show only that

$$u' \sim u \ \Rightarrow \ u'v \sim uv.$$

This is true because, when this result is established, it implies that

$$v' \sim v \Rightarrow v'u' \sim vu' \qquad \text{or} \qquad u'v' \sim u'v,$$

and, by combining the two results, we have

$$u' \sim u \qquad \text{and} \qquad v' \sim v \ \Rightarrow \ uv \sim u'v \sim u'v'.$$

The verification of the first result is immediate:

$$u'v - uv = (u' - u)v.$$

By hypothesis, the coefficients of $u' - u$ are multiples of 2; hence, so are the coefficients of $(u' - u)v$ (one need only use the formulas of (1)).

The elements of A are equivalence classes. Eeach class contains one and only one number of the form $m + n\sqrt{3}$ with m and n equal to 0 or 1 and hence can be defined by that number. The set A thus contains the four elements represented by

$$0 \qquad 1 \qquad \sqrt{3} \qquad 1 + \sqrt{3}.$$

The multiplication table, which is symmetric about the diagonal since the multiplication is commutative, is as follows:

	0	1	$\sqrt{3}$	$1 + \sqrt{3}$
0	0	0	0	0
1	0	1	$\sqrt{3}$	$1 + \sqrt{3}$
$\sqrt{3}$	0	$\sqrt{3}$	1	$1 + \sqrt{3}$
$1 + \sqrt{3}$	0	$1 + \sqrt{3}$	$1 + \sqrt{3}$	0

A is not a field. The product of $1 + \sqrt{3}$ by itself is 0; $1 + \sqrt{3}$ has no inverse, as one can see from the table.

Third question:

A subset B of A is an ideal if B is an additive subgroup and if the product of an element of B by an arbitrary element of A belongs to A.

If B contains 1 or $\sqrt{3}$, it contains all the elements of A since these elements can be obtained by multiplying 1 or $\sqrt{3}$ by the elements of A. This case is not of interest to us.

However, the set B consisting of 0 and $1 + \sqrt{3}$ is obviously an ideal: It is an additive subgroup since $(1 + \sqrt{3}) + (1 + \sqrt{3}) = 0$.

The table shows that the product of $1+\sqrt{3}$ by an element of A is equal to 0 or $1+\sqrt{3}$ and hence is also an element of B.

118

First question:

The difference and the product of two polynomials with rational coefficients are polynomials with rational coefficients. The difference and the product of two numbers in K are therefore numbers in K, so that K is a subring of the commutative ring of real numbers.

We need to show that, for any polynomial A, there exists a first-degree polynomial A_1 such that $A(\omega)=A_1(\omega)$. Since the coefficients of A are rational numbers, we need to show that every integral power ω^n of ω is equal to a first-degree polynomial in ω with rational coefficients. We prove this by induction on n. The result is obvious for $n=1$. Let us suppose that

$$\omega^n = a_n\omega + b_n,$$

where a_n and b_n are rational numbers. Therefore,

$$\omega^{n+1} = a_n\omega^2 + b_n = a_n a\omega + a_n b + b_n.$$

Second question:

We know that

$$x \in K \Rightarrow x = p\omega + q,$$

where p and q are rational numbers. If $p=0$, x is rational. If $p \neq 0$, then $\omega = (x-q)/p$ and x is a root of the equation

$$\left(\frac{x-q}{p}\right)^2 = a\frac{x-q}{p} + b$$

or

$$x^2 = (ap+2q)x + bp^2 - apq - q^2,$$

which is an equation with rational coefficients (since a, b, p, and q are rational numbers) with nonvanishing constant term (since otherwise x and ω would be rational).

Third question:

We have seen that K is a commutative ring. The number 1 belongs to K and is therefore the unit element of K. Since the ring K has a unity, it will be a field, and hence commutative, if the inverse of every nonzero element x in K is an element of K.

If x is rational, x^{-1} is rational and belongs to K. If x is not rational, it is the root of a quadratic equation with rational coefficients:

$$x^2 = \alpha x + \beta, \qquad \text{where} \qquad \beta \neq 0,$$

$$\frac{1}{x} = \frac{x - \alpha}{\beta} = \frac{p\omega + q - \alpha}{\beta} \qquad \text{if} \qquad x = p\omega + q.$$

Thus, $1/x$ is a polynomial in ω with rational coefficients and hence is an element of K.

119

First question:

Suppose that a is a nonzero element of A. Then,

$$ax' = ax \Leftrightarrow a(x' - x) = 0 \Leftrightarrow x' - x = 0.$$

(For the product $a(x' - x)$ to be zero, it is necessary that one of the factors be zero.) Thus, a is regular on the left. Similarly, one can show that a is regular on the right:

$$x'a = xa \Leftrightarrow (x' - x)a = 0 \Leftrightarrow x' - x = 0.$$

The mapping $x \to ax$ of A into A is injective by virtue of what was said above. Since A is a finite set, this mapping is bijective (the image set having as many elements as A and hence coinciding with A). In other words, the equation $ax = b$ has a unique solution.

The proof regarding the equation $xa' = b$ is similar (the mapping $x \to xa'$ is injective).

Second question:

From (1), the equation $au_0 = a$ has one solution. If b is an arbitrary element of A, we know that the equation $b = x'a$ has a solution. Therefore,

$$bu_0 = (x'a)u_0 = x'(au_0) = x'a = b.$$

The element u_0 is a neutral element on the right.

Similarly, we define v_0 as the solution of the equation $v_0 a = v_0$ and use the fact that b can be written ax:

$$v_0 b = v_0(ax) = (v_0 a)x = ax = b.$$

Finally,

$$v_0 u_0 = v_0$$

since u_0 is an element that is neutral on the right, and

$$v_0 u_0 = u_0$$

since v_0 is an element that is neutral on the left. Therefore $u_0 = v_0$ and this element is the neutral element (that is, it is neutral on the right and on the left).

Third question:

We know that A is a ring with a unit element. To show that it is a field, we need only show that every nonzero element has an inverse. The equations $ax = u_0$ and $x'a = u_0$ have unique solutions α and α', and

$$\alpha' = \alpha' u_0 = \alpha'(a\alpha) = (\alpha'a)\alpha = u_0\alpha = \alpha.$$

The element α, being the right inverse and the left inverse of a, is the inverse of a.

Fourth question:

The integers constitute a ring without divisors of zero. It is clearly an infinite ring since the answer to the preceding question shows that every finite ring without divisors of zero is a field.

120

First example:

R is an additive abelian group. On the other hand, the multiplication of a real number x by a rational number λ verifies the defining axioms of a vector space:

$$\lambda(x + y) = \lambda x + \lambda y,$$
$$(\lambda + \mu)x = \lambda x + \mu x,$$

since multiplication in R is distributive with respect to addition;

$$\lambda(\mu x) = (\lambda\mu)x$$

since multiplication in R is associative; finally,

$$1x = x$$

since the rational number 1 is the unit element of the field R.

Second example:

The set Φ is a subset of the vector space $\mathscr{F}$ over R of mappings of $[0, 1]$ into R. To show that Φ is a vector space over R, let us show that it is a vector subspace of $\mathscr{F}$. That is, that it is an additive subgroup that is closed under multiplication by a member of **R**. (We do not need to verify the axioms concerning multiplication by a member of **R** in the case of Φ since we already know that they are satisfied throughout $\mathscr{F}$.)

If f and g are functions of Φ, they assume a finite number of values and the difference $f-g$ assumes a finite number of values. This difference belongs to Φ, which is therefore a subgroup of the additive group $\mathscr{F}$.

Similarly, if f assumes only a finite number of values, λf assumes only a finite number of values and belongs to Φ. Thus, Φ is closed under multiplication by a member of **R**.

Third example:

The outline of the proof is the same as in (2) since E is again a subset of the vector space $\mathscr{F}$. If f and g are functions of E, f must be continuous in $\complement D_1$ and g in $\complement D_2$, where D_1 and D_2 are denumerable sets. The functions f and g are both continuous in

$$(\complement D_1) \cap (\complement D_2) = \complement(D_1 \cup D_2)$$

(by virtue of exercise 2) and $f-g$ is continuous on that set.

The set $D_1 \cup D_2$, being the union of two denumerable sets is a denumerable set. (We can put the elements of D_1 in correspondence with the even integers and those of D_2 that do not belong to D_1 in correspondence with the odd integers.) The function $f-g$, which is continuous in the complement of a denumerable set, belongs to E. This proves that E is a subgroup of the additive group $\mathscr{F}$.

Finally, the functions f and λf are continuous at the same values of their arguments. If f is continuous in the complement of a denumerable set, so is λf. The set E contains λf if it contains f. Thus, it is closed under multiplication by a member of **R**.

Since E is an additive subgroup of $\mathscr{F}$ that is closed under multiplication by a real number, it is a vector subspace of $\mathscr{F}$ and hence is a vector space on **R**.

Fourth example:

Let us denote by $\{x_j\}$ the n-tuple $(x_1, x_2,..., x_n)$, that is, the solution of the system (S) and let us denote by $\{0\}$ the n-tuple $(0, 0, ..., 0)$.

Let us verify first that the sum of two solutions $\{x_j\}$ and $x'_j\}$ and the product of a solution by a rational number are elements of Σ, that is, solutions of (S):

$$\sum_j a_{ij}(x_j + x'_j) = \sum_j a_{ij}x_j + \sum_j a_{ij}x'_j = 0,$$

$$\sum_j a_{ij}(\lambda x_j) = \lambda\sum_j a_{ij}x_j = 0.$$

Let us now show that the addition defined satisfies the requirements of an abelian group operation.

(1) *Associativity:*

$$[\{x_j\}+\{x_j'\}]+\{x_j''\} = \{x_j+x_j'+x_j''\} = \{x_j\}+[\{x_j'\}+\{x_j''\}].$$

(2) *Neutral element:* The set $\{0\}$ is a solution of (S) and is the neutral element with respect to addition.

(3) *Existence of an inverse:* If $\{x_j\}$ is a solution of (S), so is $\{-x_j\}$ and

$$\{x_j\}+\{-x_j\} = \{0\}.$$

(4) *Commutativity:*

$$\{x_j\}+\{x_j'\} = \{x_j+x_j'\} = \{x_j'\}+\{x_j\}.$$

It remains for us to verify the axioms regarding multiplication by a rational number.

(1) *Distributivity with respect to addition in Σ:*

$$\lambda[\{x_j\}+\{x_j'\}] = \lambda\{x_j+x_j'\} = \{\lambda x_j+\lambda x_j'\} = \lambda\{x_j\}+\lambda\{x_j'\}.$$

(2) *Distributivity with respect to addition in Q:*

$$(\lambda+\mu)\{x_j\} = \{(\lambda+\mu)x_j\} = \{\lambda x_j+\mu x_j\}$$
$$= \{\lambda x_j\}+\{\mu x_j\} = \lambda\{x_j\}+\mu\{x_j\}.$$

(3) *Associativity:*

$$\lambda[\mu\{x_j\}] = \lambda\{\mu x_j\} = \{\lambda\mu x_j\} = (\lambda\mu)\{x_j\}.$$

(4) *Multiplication by the unit element:*

$$1\{x_j\} = \{1x_j\} = \{x_j\}.$$

121

First question:

A set $\mathscr{E}$ is a subring if it satisfies the conditions

$$f\in\mathscr{E} \quad \text{and} \quad g\in\mathscr{E} \Rightarrow f-g\in\mathscr{E} \quad \text{and} \quad fg\in\mathscr{E}$$

(see introduction, § 2, ɪɪ). These conditions are satisfied by $\mathscr{P}$ and $\mathscr{N}$ but they are not satisfied by $\mathscr{P}_n$ or Q_n. The product of two polynomials of degree n is a polynomial of degree $2n$.

$$\mathscr{U} \quad : \quad f(0) = 1 \quad \text{and} \quad g(0) = 1 \Rightarrow (f-g)(0) = 0.$$

A subring $\mathscr{B}$ of a commutative ring $\mathscr{A}$ is an ideal if it satisfies the condition

$$f\in\mathscr{B} \Rightarrow \varphi f\in\mathscr{B} \quad \forall\,\varphi\in\mathscr{A}.$$

Thus, $\mathcal{N}$ is an ideal:

$$f(0) = 0 \;\Rightarrow\; (\varphi f)(0) = \varphi(0)f(0) = 0 \qquad \forall\,\varphi \in \mathcal{F}.$$

$\mathcal{P}$ is not an ideal: a polynomial function has a finite number of zeros. If we multiply it by a function φ with infinitely many zeros, we obtain a function that is not a polynomial function. For example, we may take

$$\varphi(0) = 0 \qquad \text{and} \qquad \varphi(t) = t \sin(1/t) \qquad \text{if} \qquad t \neq 0.$$

Second question:

A set $\mathcal{E}$ is a vector subspace if it is an additive subgroup and if it is closed under multiplication by a real number (cf. introduction, § 2, III).

$\mathcal{P}$ and $\mathcal{N}$ satisfy these conditions. They are additive groups since they are subrings and they are closed under multiplication by a real number.

$\mathcal{U}$ is not an additive subgroup (by virtue of the study made in the solutions of the first question) and hence is not a vector subspace.

$\mathcal{P}_n$ is a vector subspace: a linear combination of two polynomials of degree not exceeding n is a polynomial of degree not exceeding n.

Q_n is not a vector subspace since it does not contain the neutral element of $\mathcal{F}$, that is, the zero function.

Part 2

COMPLEX NUMBERS, POLYNOMIALS, AND RATIONAL FRACTIONS

Exercises dealing with the definitions and results of
Chapters IV, V, and VI of Book II of
Mathématiques Générales of C. Pisot and M. Zamansky.
The results of the second part of Chapter IV of Book IV
are also assumed known.

Review

Ideals of the ring of polynomials

Every ideal of polynomials is a principal ideal; that is, it consists of multiples of a single polynomial.

APPLICATIONS. 1) The sums of arbitrary multiples of given polynomials $A_1,...,A_n$ constitute an ideal. The polynomial of which they are all multiples is the greatest common divisor D of $A_1,...,A_n$.

(2) The polynomial D belongs to the preceding ideal. Therefore, there exist polynomials $U_1, U_2,...,U_n$ such that

$$D = A_1 U_1 + A_2 U_2 + \cdots + A_n U_n.$$

In particular, if $n=2$, there exists a unique pair U_1, U_2 such that

$$\deg U_1 < \deg A_2 - \deg D, \qquad \deg U_2 < \deg A_1 - \deg D.$$

(3) The common multiples of $A_1,...,A_n$ constitute an ideal. The polynomial of which they are all multiples is the smallest common multiple of $A_1,...,A_n$.

Lagrange's polynomials

For any set of $2n$ numbers x_i and y_i (where $i=1, 2,...,n$), there exists a unique polynomial A of degree not exceeding $n-1$ such that

$$A(x_i) = y_i$$

for every i. It is defined by

$$A(x) = \sum_{i=1}^{n} y_i \frac{\prod_{j \neq i} (x - x_j)}{\prod_{j \neq i} (x_i - x_j)}.$$

Complex numbers

(1) The complex numbers constitute a field.

(2) The mapping $z \to \bar{z}$ that assigns to every complex number its conjugate is an isomorphism of $\mathscr{C}$ onto itself; that is,

$$\overline{(x + y)} = \bar{x} + \bar{y}, \qquad \overline{(-x)} = -\bar{x},$$

$$\overline{(xy)} = \bar{x}\bar{y}, \qquad \overline{\left(\frac{1}{x}\right)} = \frac{1}{\bar{x}}.$$

(3) The complex numbers of unit absolute value constitute a multiplicative group isomorphic to the additive group of real numbers modulo 2π. This result is proven in analysis with the aid of the properties of the complex exponential function.

In what follows, we shall accept the following properties: The argument of a complex number z of unit absolute value is a real number θ defined modulo 2π, and we set

$$z = e^{i\theta} = \cos\theta + i\sin\theta,$$

$$e^{i\theta} e^{i\theta'} = e^{i(\theta+\theta')}, \qquad e^{-i\theta} = \frac{1}{e^{i\theta}} = \overline{(e^{i\theta})},$$

$$e^{i\theta} + e^{-i\theta} = 2\cos\theta, \qquad e^{i\theta} - e^{-i\theta} = 2i\sin\theta.$$

(4) Every complex number z can be defined by

$$z = re^{i\theta},$$

where r is a nonnegative real number.

The use of complex numbers in geometry

The vector space C over the field of reals is isomorphic to R^2. The points of a plane can be represented by complex numbers. The point assigned to a complex number is called the image of that number.

In particular, to every function f that maps a complex number z into a complex number z', there corresponds a geometric transformation between the images M and M' of these numbers.

A mapping of the form $z' = z+a$ is a translation;
A mapping of the form $z' = az$ is a dilatation (of the distance between the point and the origin O);
A mapping of the form $z' = \bar{z}$ is a symmetry about the real axis;
A mapping of the form $z' = 1/z$ is a dilatation and symmetry about the real axis.

Properties of polynomials with complex coefficients

(1) A polynomial has at least one zero (d'Alembert). From this, one can deduce that every polynomial can be decomposed in a unique fashion into a product of first-degree factors (prime factors).

(2) The elementary symmetric functions of the zeros x_i of a polynomial

$$\sum_{p=0}^{n} a_p x^p$$

are expressible as ratios (possibly with change of sign) of the coefficients of that polynomial:

$$\sum_{i_1,\ldots,\,i_k} x_{i_1} \cdots\cdots x_{i_k} = (-1)^k \frac{a_{n-k}}{a_n}.$$

Decomposition of a rational function into partial fractions with first-degree denominators

(1) Such a decomposition is unique. This fact is frequently used to obtain the properties of the decomposition. For example, if the fraction has real coefficients, the decomposition includes with each term the complex conjugate term. Also, if the fraction is an even function, then, for every term in the decomposition, the decomposition also includes the term obtained by replacing x with $-x$; an analogous assertion holds if the fraction is an odd function.

(2) If x_0 is a simple pole of the rational function A/B, the coefficient λ in the term $\lambda/(x-x_0)$ is

$$\lambda = \frac{A(x_0)}{B'(x_0)},$$

where B' is defined by $B(x) = (x-x_0)B'(x)$.

(3) Suppose that the denominator B of the rational function A/B is of the form $(x-x_0)^\alpha B_1$, where B is a polynomial prime to $x-x_0$ (so that x_0 is a pole of order α of A/B). Then the part of the partial-fraction decomposition corresponding to the factor $(x-x_0)^\alpha$ consists of a sum of fractions of the form $\lambda_i/(x-x_0)^i$, for $i = 1,\ldots,\alpha$, where λ_i is the coefficient of z^{-i} in the expansion of $A(x_0+z)/B_1(x_0+z)$ in powers of z.

(4) Once a certain number of coefficients is calculated, it is sometimes simpler to complete the calculations by assigning to the variable suitably chosen particular values. Then, the unknown coefficients are the solution of a system of linear equations (this is the method of undetermined coefficients).

Exercises

201

Let r_1, r_2, and r_3 denote the remainders resulting from dividing a polynomial A by $x-x_1$, $x-x_2$, and $x-x_3$ respectively. Determine the remainder R resulting from dividing A by the product $(x-x_1)(x-x_2)(x-x_3)$.

202

The polynomials considered in this exercise belong to the ring $R(x)$. Determine the polynomial A of minimum degree such that A is divisible by x^2+1 and $A-1$ is divisible by x^3+1.

Generalization: For given polynomials B and C, can we find a polynomial A such that A is a multiple of B and $A-1$ is a multiple of C?

203

Let a and b denote two relatively prime polynomials. Show that the greatest common divisor of a and c, where c is a third polynomial, is equal to the greatest common divisor of a and bc. What result do we obtain in the special cases in which

(1) a divides the product bc,
(2) a is prime to c?

204

Suppose that a and b are two given real numbers. Show that there exist unique polynomials P and Q such that

$$\begin{cases} \deg P < 2n, \qquad \deg Q < 2n, \\ (x-a)^{2n} P(x) + (x-b)^{2n} Q(x) = 1. \end{cases}$$

Then show that

$$Q(x) = P(a+b-x), \qquad P(x) = Q(a+b-x).$$

Will these last equations hold if we relax the conditions on the degrees of P and Q?

205

Let A denote a polynomial in two indeterminates x and y with complex coefficients.

Suppose that $A(x, y) = A(y, x)$ and that A is divisible by $x-y$, that is, that there exists a polynomial Q such that

$$A(x, y) = (x-y) Q(x, y).$$

Show that A is divisible by $(x-y)^2$.

206

Let us define an equivalence relation on the ring $R(u)$ of polynomials with real coefficients by

$$p \sim q \Leftrightarrow p-q \text{ is a multiple of } u^2 - 1.$$

(1) Show that this equivalence relation is compatible with the operations in $R(u)$ and define a ring structure on the quotient set E. (One may use exercise 112 of Part 1 or consult PZ, Book II, Chapter V, part 1.)

(2) Each equivalence class can be represented by the polynomial of degree 1 or 0 that it contains. Write the expression for the sum and the product of two elements of E, that is, the sum and product of two classes. Show that E possesses divisors of zero.

(3) Show what elements of E have inverses. State the condition under which the equation $AX + B = 0$ has a unique solution, where A and B are given elements and X is an unknown element of E.

(4) Show that, if $D_1 D_2 = 0$, the second-degree equation

$$(X - D_1)(X - D_2) = 0$$

has at least four solutions.

207

Evaluate the roots of the equation

$$x^8 + x^4 + 1 = 0$$

(1) by solving the system

$$x^4 = u, \qquad u^2 + u + 1 = 0,$$

(2) by decomposing the polynomial $x^8 + x^4 + 1$ into a product of second-degree factors with real coefficients.

208

Evaluate $\cos \frac{1}{5}\pi$
  (1) by using the equation $z^5 + 1 = 0$,
  (2) by using de Moivre's formula to express $\cos 5\alpha$ as a function of $\cos \alpha$ and showing that the equation thus obtained has multiple roots. Could this fact have been predicted?

209

Binomial equations

Consider the equation

$$(E) \qquad (z-a)^n = u(z-b)^n,$$

where a, b, and u are given complex numbers.
  (1) Show that the above equation is equivalent to the system

$$\frac{z-a}{z-b} = x, \qquad\qquad x^n = u.$$

Find expressions for the roots when $a=j$, $b=j^2$, $u=1$, and $n=6$. Check the result obtained by expanding both sides of the equation and finding the roots directly.
  (2) Show that the image points M, in the complex plane, of the roots z of equation (E) lie on a circle or a straight line according to the value of $|u|$. If $n \geq 3$, how can we choose a, b, and u so that the roots will all be real?

210

We recall that the nth roots of unity are the numbers

$$z_k = e^{2k\pi i/n} = \cos 2k\pi/n + i \sin 2k\pi/n \qquad (k = 0,1,...,n-1).$$

  (1) Show that the nth-roots of unity constitute a group Γ that is isomorphic to the additive group Z_n of integers modulo n (that is, the quotient group of Z defined by the relation $x \sim y$ whenever $x - y$ is a multiple of n).
  (2) The powers of an element z_k of Γ constitute a subgroup Γ_k. How may we choose k in such a way that Γ_k will be identical to Γ?

211

Direct study of the nth roots of unity

One may study the solutions of the equation

$$z^n - 1 = 0$$

without using the expression for the absolute value and argument (though we suppose it known that the equation has n distinct roots).

(1) Show that the set G_n of nth roots of unity is a multiplicative group.

(2) Let z denote an nth root of unity. Show that the set of integers r such that $z^r = 1$ consists of all the multiples of a divisor r_0 of n (the expression z^{-r} is defined as the inverse of z^r). From this result, show that the multiplicative group of powers of z is isomorphic to the additive group Z_{r_0} of integers modulo r_0.

(3) Let n denote a prime. Show that G_n is generated by successive powers of an arbitrary element not equal to 1.

(4) Suppose that p is a prime and that $n = p^\alpha$. Show that there are exactly $p^{\alpha-1}$ elements z in G_n that do not generate the group.

(5) Let n_1 and n_2 denote relatively prime numbers. Show that the groups G_{n_1} and G_{n_2} have only the unit element in common. From this result, show that, if the powers of u and v generate G_{n_1} and G_{n_2} respectively, then the powers of the product uv generate the group $G_{n_1 \times n_2}$.

(6) Show that the group G_n can be generated by successive powers of certain of its elements (one may use the decomposition of n into prime factors) and that it is isomorphic to the additive group Z_n of integers modulo n.

212

Let a_p, where $p = -n$, $-n+1,\ldots,n-2$, $n-1$, denote $2n$ complex numbers. Use these numbers to define $2n$ complex numbers b_q, where $q = -n$, $-n+1,\ldots,$ $n-2$, $n-1$, by

$$b_q = \frac{1}{\sqrt{2n}} \sum_{p=-n}^{n-1} a_p e^{ipq\pi/n}.$$

(1) Show that

$$a_p = \frac{1}{\sqrt{2n}} \sum_{q=-n}^{n-1} b_q e^{-ipq\pi/n}.$$

(2) Evaluate the b_q when $a_p = e^{i\alpha p}$, where α is a given real number.

(3) Evaluate the b_q when

$$a_p = \cos \alpha p = \frac{1}{2}(e^{i\alpha p} + e^{-i\alpha p}).$$

Write the equation expressing a_{-n} as a function of the b_q and show that

$$\cot x = \frac{1}{2n} \sin \frac{x}{n} \sum_{q=-n}^{n-1} \frac{1}{\cos q\pi/n - \cos x/n}.$$

213

A homographic function f, undecomposed into partial fractions, is defined on the extended complex plane C' (that is, the complex plane C with the element ∞ added) by

$$f(z) = \frac{az+b}{cz+d} \text{ if } z \text{ is finite and other than } -\frac{d}{c},$$

$$f(-\frac{d}{c}) = \infty, \qquad f(\infty) = \frac{a}{c},$$

where a, b, c, and d are finite complex numbers such that $c \neq 0$ and $ad - bc \neq 0$, or by

$$f(z) = \frac{az+b}{d} \qquad \text{if} \qquad z \text{ is a finite complex number,}$$

$$f(\infty) = \infty,$$

where a, b, and d are finite complex numbers such that $ad \neq 0$.

We recall that the composition of functions defines a group operation in the set G of undecomposed homographic functions.

(1) Show that the functions belonging to G are bijections of $\mathscr{C}'$ onto itself.

(2) Show that the only function in G that maps more than two elements of $\mathscr{C}'$ into themselves is the identity mapping.

(3) Show that, for any three distinct complex numbers z_1, z_2, and z_3, there exists one and only one function φ in G such that

$$\varphi(z_1) = 0, \qquad \varphi(z_2) = \infty, \qquad \varphi(z_3) = 1.$$

(4) Use the functions defined in (3) to show that

$$\frac{Z_3 - Z_1}{Z_3 - Z_2} : \frac{Z_4 - Z_1}{Z_4 - Z_2} = \frac{z_3 - z_1}{z_3 - z_2} : \frac{z_4 - z_1}{z_4 - z_2},$$

where z_1, z_2, z_3, and z_4 are mapped by a function f in G into Z_1, Z_2, Z_3, and Z_4.

(5) Show that every function g that satisfies this property for every choice of z_1, z_2, z_3, and z_4 is a function of G. Show that there exists one and only one function h in G that maps three given distinct complex numbers $z_1, z_2,$ and z_3 into three given distinct complex numbers $Z_1, Z_2,$ and Z_3.

214

A necessary and sufficient condition for three points in the complex plane A, B, and C (corresponding to complex numbers, a, b, and c) to form an equilateral triangle is that

$$a^2 + b^2 + c^2 - ab - bc - ca = 0.$$

215

(1) If points m and M in the complex plane correspond to each other in an inversion about the circle of radius $\sqrt{k}$ (centered at the origin), the complex numbers corresponding to them satisfy the relation

$$z\bar{Z} = k, \qquad \text{where } \bar{Z} \text{ is the conjugate of } Z.$$

(2) Suppose that the medians of a triangle whose vertices are M_1, M_2, and M_3, are continued so as to intersect the circle circumscribed around that triangle at points P_1, P_2, and P_3 respectively. Let z_1, z_2, z_3, u_1, u_2, and u_3 denote the complex numbers corresponding to M_1, M_2, M_3, P_1, P_2, and P_3 and denote by z the center of gravity G of the triangle. Show that

$$3z - (z_1 + z_2 + z_3) = 0,$$

$$\frac{1}{z - u_1} + \frac{1}{z - u_2} + \frac{1}{z - u_3} = 0.$$

(3) Show that, for three given points, there exist in general two triangles whose medians (extended) intersect the circumscribed circle at those points.

216

Let A, B, and C denote three points on a plane and let a, b, and c denote the complex numbers corresponding to them when an xy system of rectangular coordinate axes is drawn in the plane. Define

$$u = a + bj + cj^2, \qquad v = a + bj^2 + cj,$$

where j denotes the number $e^{2i\pi/3}$, which is one of the cube roots of unity.

(1) How are u and v transformed if we perform on the points A, B, and C a translation, a rotation about the origin O, or a homothetic transformation relative to the origin O? (One can see in PZ, Book IV, Chapter IV, part 2 how the corresponding complex numbers are transformed in these three cases.) Show that if the numbers u/v and u'/v' corresponding to points A, B, and C on the one hand and A', B', and C' on the other are equal, then the triangles ABC and $A'B'C'$ are similar.

(2) Characterize the triangles ABC for which one of the two numbers, u or v, is zero.

(3) Show that a necessary and sufficient condition for AB to be equal to AC is that either v be 0 or the ratio u/v be real. What is a necessary and sufficient condition for the triangle ABC to be isosceles?

217

Consider transformations T of the following form: A point m corresponding to a complex number z is mapped into a point M corresponding to a complex number Z according to

$$Z = \frac{az + b}{cz + d},$$

where a, b, c, and d are given complex numbers such that $ad - bc \neq 0$.

Let us accept as proven the facts that the transformations T constitute a group Γ, where the group operation is composition of transformations, and that the group Γ is isomorphic to the group G of undecomposed homographic functions. The group operation in Γ will be written multiplicatively.

(1) What is the image of the real axis in the transformation T_0 defined by

$$Z = \frac{z - i}{z + i}.$$

(2) Show that the transformations T that conserve the real axis (called the R-transformations) can be defined by homographic functions with real coefficients. Show that the R-transformations constitute a subgroup Γ_1 of Γ.

(3) Show that the transformations T that conserve the circle C of radius 1 with center at O can be represented in the form

$$T = T_0 R T_0^{-1},$$

where R is an arbitrary transformation in Γ_1. Show that these transformations constitute a subgroup Γ_2 of Γ.

(4) Give an expression for the homographic function t that defines a transformation T in Γ_2 with the aid of the coefficients a, b, c, and d of the function r that defines R (as in (3)).

Show that the most general function t can be put in the form

$$t(z) = \alpha \frac{z - z_0}{1 - z\bar{z}_0},$$

where α is a complex number of unit absolute value and $\bar{z}_0$ is the conjugate of z_0. Show that, conversely, every function of this type defines a transformation belonging to Γ_2.

218

Evaluate the products

$$P = \prod_{k=1}^{n-1} (e^{2ik\pi/n} - 1),$$

$$S = \prod_{k=1}^{n-1} \sin \frac{k\pi}{n}.$$

219

(1) Let A denote a third-degree polynomial with complex coefficients. Assume that there exist complex numbers λ and μ such that

$$A(\lambda z + \mu) = A(z).$$

Show that λ is equal to j or j^2 and that

$$A(z) = a(z - z_0)^3 + d,$$

where a, z_0, and d are complex numbers.

(2) Suppose now that A is an nth-degree polynomial with complex coefficients and that there exist complex numbers λ and μ such that

$$A(\lambda z + \mu) = A(z).$$

Show that, if A has no divisors that remain invariant when we replace z with $\lambda z + \mu$, then

$$A(z) = a(z - z_0)^n + d.$$

What must the values of λ and μ be?

(3) Find expressions for all 6th-degree polynomials that remain invariant when z is replaced by $\lambda z + \mu$.

220

Suppose that

$$(E) \qquad x^3 + px + q = 0,$$

where p and q are real numbers. Consider the following three cases:

(1) (E) has a double root.
(2) (E) has a complex root with real part equal to a given real number a.
(3) (E) has a complex root of given absolute value r.

In each of these three cases, state the conditions that p and q must verify and show how the roots can be evaluated explicitly. (One may use the relationships between the coefficients and the zeros of a polynomial.)

221

Let x_1, x_2, and x_3 denote the roots of the equation

$$(E) \qquad x^3 + px + q = 0.$$

(1) Exhibit a polynomial A whose zeros are the squares of the roots of equation (E).

(2) Use the expressions for the symmetric functions of the roots of (E) in terms of p and q to show that the square of the difference of any two roots of (E) can be expressed as a function of p, q, and the third root.

(3) Exhibit a polynomial B whose zeros are the squares of the differences of the roots of (E).

222

Determine λ such that the polynomial

$$A(x) = x^4 - x^3 + \lambda x^2 + 6x - 4$$

will have two zeros x_1 and x_2 the product of which is equal to 2. (One may express the symmetric functions of the zeros of A in terms of the sum and product of x_1 and x_2 on the one hand and in terms of the other two zeros on the other.)

223

Decompose the rational function R defined by

$$R(x) = \frac{x^4 + 3x^2 + 1}{(x + 1)^2(x^2 + x + 1)}$$

into partial fractions.

224

(1) Let A denote an nth-degree polynomial with complex coefficients. Let $x_1, x_2, ..., x_n$ denote the zeros of A (which may or may not be distinct). Let A' denote the derivative of A. Show that the poles of the fraction A'/A are all simple and that the partial-fraction decomposition of A'/A can be represented in the form

$$\frac{A'(x)}{A(x)} = \sum_{i=1}^{n} \frac{1}{x - x_i},$$

with the understanding that a zero of A of order α appears α times in the summation. Evaluate the sums

$$\sum_{i=1}^{n} \frac{1}{x_i - a}, \quad \sum_{i=1}^{n} \frac{1}{(x_i - a)^2},$$

where a is a given complex number.

(2) Use the preceding results to evaluate the sum

$$S = \sum_{i=1}^{4} \frac{2x_i^2 + 1}{(x_i^2 - 1)^2} \, ,$$

where the numbers x_i are the zeros of the polynomial $x^4 - x + 1$.

225

Decompose

$$R(x) = \frac{1}{x^{2q} + 1}$$

into partial fractions. Add the terms representing conjugate poles to obtain a decomposition into fractions with real coefficients.

226

In the following, all the polynomials are assumed to belong to the ring $R(x)$.

(1) Suppose that the real part of the rational function A/B is zero and that the poles x_i are real and distinct. Show that a necessary and sufficient condition for the zeros of A to be real and distinct and to separate the poles is that all the coefficients λ_i of the elements $1/(x-x_i)$ in the partial-fraction decomposition of A/B have the same sign.

(2) If the zeros of the polynomials P and Q are all real (though not necessarily distinct) and if k is a positive number, show that the zeros of the polynomial $P'Q + kPQ'$ are all real (where P' and Q' are the derivatives of P and Q). To solve this problem, one may use the results of exercise 224.

Solutions

201

The remainder resulting from division of A by $x - x_i$ is a constant r_i, which is equal to $A(x_i)$ since

$$A(x) = (x - x_i)Q_i(x) + r_i.$$

The polynomial R is a polynomial of degree not exceeding 2 such that

$$A(x) = (x - x_1)(x - x_2)(x - x_3) Q(x) + R(x),$$

so that

$$R(x_i) = A(x_i) = r_i \qquad (i = 1, 2, 3).$$

Study of Lagrange's polynomials (see PZ, Book II, Chapter IV, part 5) then enables us to write

$$R(x) = r_1 \frac{(x - x_2)(x - x_3)}{(x_1 - x_2)(x_1 - x_3)} + r_2 \frac{(x - x_3)(x - x_1)}{(x_2 - x_3)(x_2 - x_1)} + r_3 \frac{(x - x_1)(x - x_2)}{(x_3 - x_1)(x_3 - x_2)}.$$

202

Our problem is to find the polynomial A of lowest degree that satisfies the equations

$$A = (x^2 + 1)U = (x^3 + 1)V + 1$$

for suitably chosen polynomials U and V. We have

$$(x^2 + 1)U - (x^3 + 1)V = 1.$$

The polynomials $x^2 + 1$ and $x^3 + 1$ are relatively prime. Therefore, this last equation means that U and V are the required polynomials of Bézout's theorem, which asserts their existence. To find the polynomials of minimum degree that can be used for U and V, we first proceed as if we were finding the greatest common divisor of $x^2 + 1$ and $x^3 + 1$:

$$x^3 + 1 = x(x^2 + 1) - (x - 1),$$
$$x^2 + 1 = (x + 1)(x - 1) + 2.$$

Combining these equations, we get

$$2 = x^2 + 1 - (x + 1)[x(x^2 + 1) - (x^3 + 1)].$$

Grouping terms, dividing by 2, and equating the coefficients of x^2+1 and x^3+1 with those in the equation for U and V above, we get

$$U = -\tfrac{1}{2}(x^2+x-1), \qquad V = -\tfrac{1}{2}(x+1),$$

so that

$$A = -\tfrac{1}{2}(x^2+1)(x^2+x-1).$$

The generalization is immediate:

$$A = BU = CV+1.$$

For the equation $BU-CV = 1$ to hold, it is necessary and sufficient that U and V be relatively prime (Bézout's theorem).

203

Consider the ideals $\mathscr{I}_1$ and $\mathscr{I}_2$ formed respectively by the polynomials $ua+vc$ and $xa+ybc$, where u, v, x, and y are arbitrary polynomials.

$$xa+ybc = xa+(yb)c \;\Rightarrow\; \mathscr{I}_2 \subset \mathscr{I}_1.$$

Since a and b are relatively prime, there exist polynomials α and β such that $a\alpha+b\beta=1$ and

$$ua+vc = ua+v(a\alpha+b\beta)c = (u+v\alpha c)a+(v\beta)bc.$$

Every polynomial in $\mathscr{I}_1$ belongs to $\mathscr{I}_2$, so that $\mathscr{I}_1 \subset \mathscr{I}_2$.

The two ideals $\mathscr{I}_1$ and $\mathscr{I}_2$ are identical: $\mathscr{I}_1$ is the ideal consisting of the multiples of the greatest common divisor of a and c, and $\mathscr{I}_2$ is the ideal consisting of the multiples of the greatest common divisor of a and bc; these two greatest common divisors are equal.

If a divides bc, the greatest common divisor of a and bc is equal to a, and the greatest common divisor of a and c is also equal to a; therefore, a divides c.

If a is prime to c, the two greatest common divisors are equal to 1. If a polynomial is prime to two factors, it is prime to their product.

204

The polynomials $(x-a)^{2n}$ and $(x-b)^{2n}$ are relatively prime. Therefore, we know from Bézout's theorem that there exist unique polynomials P and Q satisfying the given conditions.

If we set $x=a+b-u$, we transform the given equation into

$$(b-u)^{2n}P(a+b-u)+(a-u)^{2n}Q(a+b-u) = 1.$$

The polynomials P_1 and Q_1 defined by

$$P_1(u) = Q(a+b-u) \qquad \text{and} \qquad Q_1(u) = P(a+b-u)$$

are of degree less than $2n$ and constitute another pair satisfying the given conditions. Since only one such pair exists,

$$P_1(u) = Q(a+b-u) = P(u), \qquad Q_1(u) = P(a+b-u) = Q(u).$$

If R and S are two polynomials satisfying the equation

$$(x-a)^{2n}R(x) + (x-b)^{2n}S(x) = 1,$$
$$(x-a)^{2n}[R(x)-P(x)] + (x-b)^{2n}[S(x)-Q(x)] = 1.$$

$(x-a)^{2n}$ must divide the product $(x-b)^{2n}[S(x)-Q(x)]$. Since it is prime to $(x-b)^{2n}$, it must divide the second factor, so that

$$S(x) = Q(x)+(x-a)^{2n}q(x) \qquad \text{and} \qquad R(x) = P(x)-(x-b)^{2n}q(x),$$

where q is an arbitrary polynomial. Thus,

$$S(a+b-u) = Q(a+b-u)+(b-u)^{2n}q(a+b-u)$$
$$= P(u)+(u-b)^{2n}q(a+b-u).$$

For the equation $S(a+b-u) = R(u)$, it is necessary and sufficient that $q(a+b-u) = -q(u)$. This relation is not always satisfied since q is an arbitrary polynomial.

205

By hypothesis,

$$A(x, y) = A(y, x)$$

or

$$(x-y)Q(x, y) = (y-x)Q(y, x).$$

If $x-y \neq 0$,

$$Q(x, y)+Q(y, x) = 0.$$

This equation, which is valid for $x \neq y$, is also valid when $x=y$. If we assign to y a numerical value y_i, we see that the polynomials $Q(x, y_i)+Q(y_i, x)$ vanish identically. Now, the coefficients of the different powers of y are linear combinations of these polynomials. (One may use, for example, Lagrange's interpolation formula.) These coefficients are identically zero and the polynomial $Q(x, y)+Q(y, x)$ is the zero polynomial.

Let us now consider Q as a polynomial $\bar{Q}$ in x whose coefficients are polynomials in y. If x assumes the value y, the polynomial $\bar{Q}$ vanishes and is hence divisible by $x-y$. The coefficients of the quotient obtained by dividing $\bar{Q}$ by $x-y$ are polynomial functions of the coefficients of $\bar{Q}$ and of $-y$ (this last because the coefficient of x in the expression $x-y$ is 1). Therefore, the coefficients of the quotient are polynomials in y and the quotient is a polynomial in the two indeterminates x and y. Let us denote it by $R(x, y)$:

$$Q(x, y) = (x-y) R(x, y) \Rightarrow A(x, y) = (x-y)^2 R(x, y).$$

206

First question:

The set of multiples of the polynomial $u^2 - 1$ is an ideal in $R(u)$. Thus, we know that the quotient set has a ring structure (cf. exercise 112 of Part 2, question 4).

We might also study the question directly as is done in PZ: The additional hypothesis that the polynomial is prime in $K(x)$ plays no role with regard to the proof of the properties required here.

Second question:

If a is a polynomial belonging to the class A in question, it is equivalent to the remainder r resulting from dividing a by $u^2 - 1$ since $a - r$ is a multiple of $u^2 - 1$. The class A contains the polynomial r, the degree of which is either 0 or 1. Every other polynomial in A can be obtained by adding to r a multiple of $u^2 - 1$ and is of degree at least equal to 2. Let us therefore identify A and the polynomial r.

Let us now take $A = a_1 + a_2 u$ and $B = b_1 + b_2 u$, where a_1, a_2, b_1 and b_2 are arbitrary real numbers. Then,

$$A + B = a_1 + b_1 + (a_2 + b_2)u$$
$$AB = a_1 b_1 + a_2 b_2 + (a_1 b_2 + a_2 b_1)u$$

(we replace $a_2 b_2 u^2$ with the equivalent polynomial $a_2 b_2$). In particular,

$$(1 - u)(1 + u) = 0.$$

Neither of the factors on the left is zero: Thus, each is a divisor of 0

Third question:

$$(a_1 + a_2 u)(a_1 - a_2 u) = a_1^2 - a_2^2.$$

If $a_1^2 - a_2^2 \neq 0$, then

$$\frac{a_1}{a_1^2 - a_2^2} - \frac{a_2}{a_1^2 - a_2^2} u$$

is the inverse of $a_1 + a_2 u$.

If $a_1^2 - a_2^2 = 0$, then $a_1 + a_2 u$ is a divisor of 0 and has no inverse.

If A has an inverse, the equation $AX + B = 0$ has a unique solution since

$$AX + B = 0 \Leftrightarrow A^{-1}(AX + B) = 0 \Leftrightarrow X = -A^{-1}B.$$

If A is a divisor of zero, there exists an element A' such that $AA'=0$. If the equation has a solution X_0, it has infinitely many. This is true because, if P is an arbitrary element of E,

$$A(X_0+PA') = AX_0+PAA' = AX_0.$$

Therefore, X_0+PA' is a solution if X_0 is a solution. (We need only take for P a real number r. This is true because every element P in E can be written in the form $r+sA$, where r and s are two real numbers. Then, $PA'=rA'$.)

Fourth question:

The equation given has roots D_1 and D_2. Expanding the left-hand member and using the fact that $D_1D_2=0$, we get

$$X^2 - X(D_1+D_2) = 0.$$

Thus, it also has the roots $X=0$ and $X=D_1+D_2$.

207

First question:

$$u_1 = j, \qquad u_2 = j^2, \qquad \text{where} \qquad j = \frac{-1+i\sqrt{3}}{2} \qquad \text{and} \qquad j^3 = 1.$$

The system is replaced with the two equations

$$x^4 = j, \qquad x^4 = j^2.$$

Since $j^4 = j^3 \times j = j$, it follows that j is a fourth root of j. The other roots are obtained by multiplying j by the fourth roots of unity:

$$x_1 = j = \frac{-1+i\sqrt{3}}{2}, \qquad x_2 = ij = \frac{-\sqrt{3}-i}{2},$$

$$x_3 = -x_1 = \frac{1-i\sqrt{3}}{2}, \qquad x_4 = -ij = \frac{\sqrt{3}+i}{2}.$$

The roots of the equation $x^4=j^2$ can be obtained in the same way. They are $j^2, ij^2, -j^2,$ and $-ij^2$. We might also note that, if $x^4=j$, the conjugate $\bar{x}$ of x satisfies the equation $\bar{x}^4=\bar{j}= j^2$. Therefore, the roots are $\bar{x}_1, \bar{x}_2, \bar{x}_3,$ and $\bar{x}_4$.

Second question:

$$x^8 +x^4 +1 = (x^4 +1)^2 -x^4 = (x^4 -x^2 +1)(x^4 +x^2 +1),$$
$$x^4 -x^2 +1 = (x^2 +1)^2 -3x^2 = (x^2 -x\sqrt{3}+1)(x^2 +x\sqrt{3}+1),$$
$$x^4 +x^2 +1 = (x^2 +1)^2 -x^2 = (x^2 -x+1)(x^2 +x+1).$$

Finally,

$$(x^8 + x^4 + 1) = (x^2 - x\sqrt{3} + 1)(x^2 + x\sqrt{3} + 1)(x^2 - x + 1)(x^2 + x + 1).$$

Thus, the problem of solving the original equation is reduced to that of solving four quadratic equations. By using classical formulas, we obtain, from these four equations, x_4 and $\bar{x}_4$, x_2 and $\bar{x}_2$, x_3 and $\bar{x}_3$, and x_1 and $\bar{x}_1$.

208

First question:

The roots of the equation $z^5 = -1 = e^{i\pi}$ are the numbers

$$z_k = e^{i(2k+1)\pi/5} \qquad (k = 0, 1, 2, 3, 4).$$

With the exception of $z_2 = -1$, these numbers are roots of the equation

$$\frac{z^5 + 1}{z + 1} = z^4 - z^3 + z^2 - z + 1 = z^2 \left[z^2 + \frac{1}{z^2} - \left(z + \frac{1}{z} \right) + 1 \right] = 0.$$

The classical method of solving equations of this form leads us to replace this equation with the system

$$\begin{cases} u = z + \dfrac{1}{z}, \\ u^2 - u - 1 = 0. \end{cases}$$

But then $z = e^{i\alpha}$ is equivalent to $u = e^{i\alpha} + e^{i\alpha} = 2\cos\alpha$. Therefore, the roots of the last equation are $2\cos\frac{1}{5}\pi$ and $2\cos\frac{3}{5}\pi$. Since $0 < \frac{1}{5}\pi < \frac{1}{2}\pi < \frac{3}{5}\pi < \pi$, we have

$$2\cos\tfrac{1}{5}\pi = 1 + \sqrt{5} > 0, \qquad 2\cos\tfrac{3}{5}\pi = 1 - \sqrt{5} < 0.$$

Second question:

$$\cos 5\alpha = \cos^5\alpha - 10\cos^3\alpha\sin^2\alpha + 5\cos\alpha\sin^4\alpha$$
$$= 16\cos^5\alpha - 20\cos^3\alpha + 5\cos\alpha.$$

If $\alpha = \frac{1}{5}\pi$, then $\cos 5\alpha = \cos\pi = -1$ and $x = \cos\frac{1}{5}\pi$ is a root of the equation

$$16x^5 - 20x^3 + 5x + 1 = 0.$$

The above shows that $\cos\frac{3}{5}\pi$ and $\cos\pi = -1$ are also roots of this equation. Therefore, the left-hand side is divisible by $x + 1$:

$$16x^5 - 20x^3 + 5x + 1 = (x+1)(16x^4 - 16x^3 - 4x^2 + 4x + 1)$$
$$= (x+1)(4x^2 - 2x - 1)^2.$$

The zeros of $4x^2-2x-1$ are $\cos \frac{1}{5}\pi$ and $\cos \frac{3}{5}\pi$. In particular, $\cos \frac{1}{5}\pi$ is the positive zero, namely $\frac{1}{4}(1+\sqrt{5})$.

The equation in $\cos \alpha$, obtained by writing $\cos 5\alpha = \cos 5\alpha_0$, has the five roots of the form $\cos (\alpha_0 + 2k\pi/5)$, where k assumes the values 0, 1, 2, 3, and 4.

In the particular case of $\alpha_0 = \pi/5$, these roots are

$$\cos \tfrac{1}{5}\pi, \cos \tfrac{3}{5}\pi, \cos\tfrac{5}{5}\pi = -1, \qquad \cos \tfrac{7}{5}\pi = \cos \tfrac{3}{5}\pi, \qquad \cos \tfrac{9}{5}\pi = \cos \tfrac{1}{5}\pi,$$

the roots $\cos \frac{1}{5}\pi$ and $\cos \frac{3}{5}\pi$ being double roots.

209

First question:

b is not a root of the equation. Therefore, we may take for a new unknown

$$\frac{z-a}{z-b} = x.$$

Then, x is a solution of the equation $x^n = u$. Now, solution of the equation amounts to finding the nth roots x_k of the number u. To each root x_k there corresponds one and only one value of z except in the case of $x_k = 1$, that is, $u = 1$. In this last case, we find only $n-1$ values of z. In fact, the original equation is then of degree $n-1$. We note finally that the values obtained for z are all distinct since z is a homographic function of x.

The equation $x^6 = 1$ has six roots, of which one is 1. We find five values of z corresponding to the other five roots:

$$\begin{cases} x_1 = -j^2 \\ z_1 = -2 \end{cases} \quad \begin{cases} x^2 = j \\ z_2 = -1 \end{cases} \quad \begin{cases} x_3 = -1 \\ z_3 = -\tfrac{1}{2} \end{cases} \quad \begin{cases} x_4 = j^2 \\ z_4 = 0 \end{cases} \quad \begin{cases} x^5 = -j \\ z^5 = 1 \end{cases}.$$

If we carry out the operations in the equation and collect terms of like power, we obtain

$$2z^5 + 5z^4 - 5z^2 - 2z = z(z^2-1)(2z^2+5z+2) = 0$$

and we again find the roots obtained above.

Second question:

Let us denote by A, B, and M the images of a, b, and z:

$$\frac{MA}{MB} = \left| \frac{z-a}{z-b} \right| = |x| = \sqrt[n]{|u|}.$$

If $|u| \neq 1$, the points M lie on the circle C, which is the locus of points the ratio of the distances of which from A and B is equal to $\sqrt[n]{|u|}$.

If $|u| = 1$, the points M lie on the perpendicular bisector of AB.

If the roots are all real, their images M lie on the x-axis. Since the roots are all distinct, their images cannot all lie on both the x-axis and on a circle if $n \geq 3$. A necessary and sufficient condition for all the roots to be real is that the x-axis be the perpendicular bisector of AB, that is, that $|u| = 1$ and a and b be conjugates.

210

First question:

The complex numbers of absolute value 1 constitute a multiplicative group. To show that Γ is a subgroup, we need only show that
  (a) Γ contains the unit element: $z_0 = 1$;
  (b) Γ contains the inverse of each of its elements:

$$z_k z_{n-k} = 1;$$

  (c) Γ contains the product of any two of its elements:

$$z_h z_k = \begin{cases} z_{h+k} & \text{if} \quad h+k < n \\ z_{h+k-n} & \text{if} \quad h+k \geq n. \end{cases}$$

The mapping $k \to z_k$ is a bijective mapping of Z_n onto Γ (we identify an element of Z_n, that is, an equivalence class, with that integer k lying between 0 and n that it contains). This bijective mapping is a group isomorphism by virtue of (c) since, in Z_n, the sum $(h) + (k)$ is the class of $h + k$, that is, $(h+k)$ if $h+k < n$ and $(h+k-n)$ if $h+k \geq n$.

Second question:

Let us show, just as in (1), that Γ_k is a subgroup:
  (a) Γ_k contains $z_k^n = 1$;
  (b) the image of z_k^p is z_k^{n-p} (we may suppose that $p < n$);
  (c) $z_k^p \times z_k^{p'} = z_k^{p+p'}$.
For Γ_k to be the group Γ, it is necessary and sufficient that n elements of Γ_k be distinct, for example, that 1 (since $z_k^n = 1$), $z_k, \ldots, z_k^{n-1}$ be distinct. If any two of them are equal, their quotient, which is a power of z_k, is equal to 1. Therefore, it is necessary and sufficient that none of these numbers be equal to 1.

$$z_k^p = e^{2 p k \pi i / n} = 1 \Leftrightarrow kp/n \text{ is an integer.}$$

If n is prime to k and divides kp, it divides p. Here, this is impossible since $p \leq n-1$.

If n and k have a common divisor d, the ratio n/d is an integer and $(k/n) \times (n/d) = (k/d)$ is an integer. Then, the equation $z_k^p = 1$ has solutions (in particular, $d = n/d$).

A necessary and sufficient condition for Γ_k to be identical to Γ is that k be prime to n. (If n is a prime, k is arbitrary.)

211

First question:

It will be sufficient to verify that the quotient z'/z of two elements of G_n is an element of G_n:

$$\left(\frac{z'}{z}\right)^n = \frac{z'^n}{z^n} = 1.$$

Second question:

The set E of integers r such that $z^r = 1$ is a subgroup of the additive group Z:

$$r \in E \quad \text{and} \quad r' \in E \implies r' - r \in E \quad \text{since} \quad z^{r'-r} = z^{r'}/z^r = 1.$$

The subgroups of Z consist of the multiples of an integer r_0. (See Part 1, exercise 112, question (3).) By definition, n belongs to E; r_0 is a divisor of n.

We know that the relation $r' - r \in E$ defines an equivalence in Z and that the quotient set Z_{r_0} has a group structure (cf. exercise 112 of Part 1). The mapping that assigns to each equivalence class the number z^p, where p is the integer $< r_0$ and contained in that equivalence class, is an isomorphism between Z_{r_0} and the multiplicative group of powers of z.

(a) This mapping is bijective.

(b) It conserves the operation

$$z^p z^q = \begin{cases} z^{p+q} & \text{if} \quad p+q < r_0, \\ z^{p+q-r_0} & \text{if} \quad p+q \geqslant r_0. \end{cases}$$

Third question:

If z is an element of G_n other than 1, the number r_0, being a divisor of n, must now be equal to n. The group of powers of z has n distinct elements and may be identified with G_n.

Fourth question:

The number r_0 associated with an element z of Γ_n is a divisor of p^α. Thus, it is a power of p, let us say, p^β. If the powers of z do not generate Γ_n, then r_0 is less than p^α and it divides $p^{\alpha-1}$. Thus, z is an element of $G_{p^{\alpha-1}}$, and this group has $p^{\alpha-1}$ elements. Conversely, it is obvious that no element of $G_{p^{\alpha-1}}$ generates the group G_{p^α}.

Fifth question:

If z belongs to G_{n_1} and G_{n_2}, the number r_0 assigned to z is a divisor of n_1 and n_2. Thus, $r_0 = z = 1$.

Since $(uv)^r = 1$, we conclude that $u^r = v^{-r}$. The number $u^r = v^{-r}$ is an element of both G_{n_1} and G_{n_2}. It is equal to 1, and r is a multiple of n_1 (the smallest integer such that $u^r = 1$) and of n_2 (the smallest integer such that $v^r = 1$). Thus, r is a multiple of $n_1 \times n_2$.

The powers of uv with exponent less than $n_1 \times n_2$ are all distinct. They are obviously elements of $G_{n_1 \times n_2}$. Thus, they generate that group.

Sixth question:

PROOF BY INDUCTION ON THE NUMBER h OF FACTORS PRIME TO n

The result is true if $h = 1$: We saw in (4) that $p^\alpha - p^{\alpha-1}$ elements of G_{p^α} generate the group.

Suppose that the result holds for $h-1$ factors. Consider the product

$$n = p_1^{\alpha_1} \cdots p_h^{\alpha_h} = n_1 \times n_2,$$

where we set

$$n_1 = p_1^{\alpha_1} \quad \text{and} \quad n_2 = p_2^{\alpha_2} \cdots p_h^{\alpha_h}.$$

Here, n_1 and n_2 are relatively prime. By hypothesis, G_{n_1} and G_{n_2} are generated by the powers of certain of their elements. Therefore, the result is also valid for $G_{n_1 \times n_2} = G_n$ by virtue of (5).

If the powers of z generate G_n, the smallest integer r such that $z^r = 1$ is n. Under these conditions, the group of powers of z, that is, G_n, is isomorphic to the group Z_n of integers modulo n (the result obtained in (2)).

212

First question:

In the expression

$$s_p = \sqrt{2n} \sum_{q=-n}^{n-1} b_q e^{-ipq\pi/n},$$

let us replace b_q by its defining expression as a function of the a_r (we now use the subscript r instead of p, which we are already using elsewhere in the above expression):

$$s_p = \sum_{q=-n}^{n-1} \left(\sum_{r=-n}^{n-1} a_r e^{irq\pi/n} \right) e^{-ipq\pi/n}.$$

This s_p is a sum of $4n^2$ terms. The commutativity and associativity of addition in $\mathscr{C}$ permits us to change the order of the summations:

$$s_p = \sum_{r=-n}^{n-1} a_r \left(\sum_{q=-n}^{n-1} e^{i(r-p)q\pi/n} \right).$$

The coefficient of a_r, which we denote by A_r, defines a geometric progression with ratio

$$z = e^{i(r-p)\pi/n}$$

and first term z^{-n} (relative to $q = -n$) and last term z^{n-1}. If the ratio z is not equal to 1, that is, if $r \neq p$, we know that

$$A_r = \frac{z \times z^{n-1} - z^{-n}}{z-1} = \frac{z^n - z^{-n}}{z-1} = 0,$$

since

$$z^{2n} = 1.$$

Finally, all the terms in A_p are equal to 1 and their sum A_p has the value $2n$. Therefore,

$$s_p = 2na_p = \sqrt{2n} \sum_{q=-n}^{n-1} b_q e^{-ipq\pi/n}.$$

Division by $2n$ yields the formula desired.

Second question:

$$a_p = e^{i\alpha p} \Rightarrow b_q = \frac{1}{\sqrt{2n}} \sum_{p=-n}^{n-1} e^{ip(\alpha+q\pi/n)}.$$

The expression $\sqrt{2n}b_q$ also defines a geometric progression. The ratio u is $e^{i(\alpha+q\pi/n)}$, the first term is u^{-n}, and the last term is u^{n-1}:

$$\sqrt{2n}\, b_q = \frac{u^n - u^{-n}}{u-1},$$

$$u^n = e^{i(n\alpha+\pi q)} = e^{iq\pi}e^{in\alpha},$$

$$u^{-n} = e^{-iq\pi}e^{-in\alpha} = e^{iq\pi}e^{-in\alpha},$$

$$u^n - u^{-n} = e^{iq\pi}(e^{in\alpha} - e^{-in\alpha}) = (-1)^{|q|}2i\sin n\alpha,$$

$$u - 1 = e^{i(\alpha+q\pi/n)} - 1 = e^{\frac{1}{2}i(\alpha+q\pi/n)}\left[e^{\frac{1}{2}i(\alpha+q\pi/n)} - e^{-\frac{1}{2}i(\alpha+q\pi/n)}\right]$$

$$= e^{\frac{1}{2}i(\alpha+q\pi/n)}\, 2i\sin\left(\tfrac{1}{2}\alpha + q\pi/2n\right),$$

$$\sqrt{2n}\, b_q = (-1)^{|q|}e^{-i(\frac{1}{2}\alpha+q\pi/2n)}\, \frac{\sin n\alpha}{\sin\left(\tfrac{1}{2}\alpha + q\pi/2n\right)}.$$

Third question:

The b_q depend linearly on the a_p. For $a_p = \cos \alpha p$, the value of b_q is the arithmetic mean of the values of b_q corresponding to $a_p = e^{i\alpha p}$ and $a_p = e^{-i\alpha p}$. Let us denote by b_q' and b_q'' these last two values. This b_q' is the value determined in (2) and b_q'' is derived from b_q' by switching α and $-\alpha$ (not i and $-$i since the coefficients of the sums are complex numbers that would then be replaced with their conjugates):

$$\sqrt{2n}\, b_q' = (-1)^{|q|} \sin n\alpha \, [\cot (\tfrac{1}{2}\alpha + q\pi/2n) - i],$$

$$\sqrt{2n}\, b_q'' = (-1)^{|q|} \sin n\alpha \, [\cot (\tfrac{1}{2}\alpha - q\pi/2n) + i],$$

$$\sqrt{2n}\, b_q = \tfrac{1}{2}(-1)^{|q|} \sin n\alpha \, [\cot (\tfrac{1}{2}\alpha + q\pi/2n) + \cot (\tfrac{1}{2}\alpha - q\pi/2n)]$$

$$= \tfrac{1}{2}(-1)^{|q|} \frac{\sin n\alpha \sin \alpha}{\sin (\tfrac{1}{2}\alpha + q\pi/2n) \sin (\tfrac{1}{2}\alpha - q\pi/2n)}$$

$$= (-1)^{|q|} \frac{\sin n\alpha \sin \alpha}{\cos q\pi/n - \cos \alpha}.$$

Conversely, $a_{-n} = \cos n\alpha$ is expressed as a function of the b_q as follows:

$$\cos n\alpha = \frac{\sin n\alpha \sin \alpha}{2n} \sum_{q=-n}^{n-1} \frac{(-1)^{|q|}\, e^{iq\pi}}{\cos q\pi/n - \cos \alpha}$$

$$= \frac{\sin n\alpha \sin \alpha}{2n} \sum_{q=-n}^{n-1} \frac{1}{\cos q\pi/n - \cos}.$$

The substitution $x = n\alpha$ then yields

$$\cot x = \frac{1}{2n} \sin \frac{x}{n} \sum_{q=-n}^{n-1} \frac{1}{\cos q\pi/n - \cos x/n}.$$

213

First question:

We verify that the equation $Z = f(z)$ has one and only one solution z for every Z in $\mathscr{C}'$.

For $c \neq 0$,

$$z = \begin{cases} \dfrac{dZ - b}{a - cZ} & \text{for} \quad Z \text{ a finite complex number} \neq \dfrac{a}{c}, \\[2ex] \infty & \text{for} \quad Z = \dfrac{a}{c}, \\[2ex] -\dfrac{d}{c} & \text{for} \quad Z = \infty. \end{cases}$$

For $c = 0$,

$$z = \begin{cases} \dfrac{dZ - b}{a} & \text{for} \quad Z \text{ a finite complex number,} \\[2em] \infty & \text{for} \quad Z = \infty. \end{cases}$$

Second question:

Let us determine the solutions of the equation $z = f(z)$. If $c \neq 0$, the element ∞ is not a solution. The solutions are the complex solutions z of the equation

$$cz^2 + (d - a)z - b = 0,$$

which has at most two distinct solutions.

If $c = 0$, then ∞ is a solution. The complex solutions are given by

$$z = \frac{az + b}{d} \Leftrightarrow z(a - d) + b = 0.$$

This equation has at most one finite solution and at most two solutions in the extended complex plane except in the case of $a - d = b = 0$. Then, the function f is the identity mapping.

Third question:

$$\varphi(z_1) = 0 \Leftrightarrow az + b = a(z - z_1),$$

$$\varphi(z_2) = \infty \Leftrightarrow cz + d = c(z - z_2),$$

$$\varphi(z_3) = \frac{a}{c} \frac{z_3 - z_1}{z_3 - z_2} = 1 \Rightarrow \frac{a}{c} = \frac{z_3 - z_2}{z_3 - z_1},$$

$$\varphi(z) = \frac{z - z_1}{z - z_2} \times \frac{z_3 - z_2}{z_3 - z_1} = \frac{z - z_1}{z - z_2} : \frac{z_3 - z_1}{z_3 - z_2}.$$

Fourth question:

If φ and ψ are functions of G that map z_1, z_2, and z_3 and Z_1, Z_2, and Z_3 into $0, \infty$, and 1 respectively, the function $\psi \circ f \circ \varphi^{-1}$ maps $0, \infty$, and 1 into themselves. This is a function in G since G is a group. It has more than two invariant elements. It is the identity function and therefore

$$\psi \circ f = \varphi.$$

If we express the fact that the images of z_4 under these two functions are the same, we obtain

$$\frac{Z_4 - Z_1}{Z_4 - Z_2} : \frac{Z_3 - Z_1}{Z_3 - Z_2} = \frac{z_4 - z_1}{z_4 - z_2} : \frac{z_3 - z_1}{z_3 - z_2},$$

which is equivalent to the equation desired.

Fifth question:

Consider three fixed values z_1, z_2, and z_3 and a variable z. Let us set

$$Z_1 = g(z_1), \qquad Z_2 = g(z_2), \qquad Z_3 = g(z_3), \qquad Z = g(z).$$

By hypothesis,

$$\frac{Z - Z_1}{Z - Z_2} \cdot \frac{Z_3 - Z_1}{Z_3 - Z_2} = \frac{z - z_1}{z - z_2} \cdot \frac{z_3 - z_1}{z_3 - z_2}.$$

In the notation of (4),

$$\psi(Z) = (\psi \circ g)(z) = \varphi(z).$$

Since ψ is bijective, this equation implies that

$$g = \psi^{-1} \circ \varphi.$$

Since the function g is a composite of two functions in G, it is a function in G itself.

The preceding result is valid in particular for functions in G. The desired function h satisfies

$$(\psi \circ h)(z) = \varphi(z).$$

Therefore, it is the function $\psi^{-1} \circ \varphi$.

214

For the triangle ABC to be equilateral, it is necessary and sufficient that

$$AB = AC \qquad \text{and} \qquad (AB, AC) = (Ox, AC) - (Ox, AB) = \pm\tfrac{1}{3}\pi \bmod 2\pi.$$

If we express these quantities in terms of the numbers a, b, and c, this condition becomes

$$|b - a| = |c - a| \qquad \text{and} \qquad \arg(c - a) - \arg(b - a) = \pm\tfrac{1}{3}\pi \bmod 2\pi$$

$$c - a = (b - a)e^{i\pi/3} \qquad \text{or} \qquad c - a = (b - a)e^{-i\pi/3}.$$

Therefore, the desired necessary and sufficient condition is

$$[c - a - (b - a)e^{i\pi/3}][c - a - (b - a)e^{-i\pi/3}] = 0.$$

If we carry out the multiplication, we obtain

$$a^2 + b^2 + c^2 - ab - bc - ca = 0.$$

215

First question:

If $k > 0$, the relation $z\bar{Z} = k$ is equivalent to

$$Om \cdot OM = k \quad \text{and} \quad (Ox, Om) = (Ox, OM) \bmod 2\pi.$$

The points m and M correspond to each other in an inversion about the circle of radius $\sqrt{k}$.

A similar result remains valid if $k < 0$. (The inversion can be decomposed into an inversion about the circle of radius $\sqrt{|k|} = \sqrt{-k}$ and a reflection about the origin.)

Second question:

The center of gravity G of the triangle $M_1 M_2 M_3$ can be defined by the vector relation

$$M_1 G = \tfrac{2}{3} M_1 I = \tfrac{2}{3} \tfrac{1}{2}(M_1 M_2 + M_1 M_3).$$

Using only position vectors, we have

$$3OG = OM_1 + OM_2 + OM_3.$$

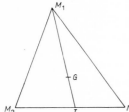

(The above equation is most rapidly obtained by starting with a study of the center of gravity.) This vector equation is equivalent to

(1) $$3z = z_1 + z_2 + z_3.$$

We know that the lengths GP_1, GM_1, GP_2, GM_2, GP_3, and GM_3 satisfy the relations

$$GP_1 \cdot GM_1 = GP_2 \cdot GM_2 = GP_3 \cdot GM_3.$$

(From geometry, if two lines through a point G intersect a circle at points M_1 and P_1 for the first line and M_2 and P_2 for the second, the products $GM_1 \cdot GP_1$ and $GM_2 \cdot GP_2$ of the distances are equal. This is true whether G is inside, on, or outside the circle. We might note that an inversion of the type described is referred to geometrically as an 'inversion of power k'.) This in turn means that the points P_1, P_2, and P_3 on the one hand and the points M_1, M_2, and M_3 on the other correspond to each other in an inversion about a circle centered at G and of radius

$$\sqrt{GP_1 \cdot GM_1} = \sqrt{GP_2 \cdot GM_2} = \sqrt{GP_3 \cdot GM_3} \equiv \sqrt{k}.$$

If the origin is placed at G, the complex numbers represented by the six points are $u_1 - z$, $u_2 - z$, $u_3 - z$, $z_1 - z$, $z_2 - z$, and $z_3 - z$ respectively. We know that

$$(u_\alpha - z)\overline{(z_\alpha - z)} = k.$$

Equation (1) can be written

$$(z-z_1)+(z-z_2)+(z-z_3) = 0$$

or

$$(z-z_1)+(z-z_2)+(z-z_3) = 0$$
$$\overline{(z-z_1)}+\overline{(z-z_2)}+\overline{(z-z_3)} = 0.$$

Therefore, it implies that

$$\frac{1}{z-u_1} + \frac{1}{z-u_2} + \frac{1}{z-u_3} = 0.$$

Third question:

Conversely, if the complex numbers u_1, u_2, u_3, and z, corresponding to the four points P_1, P_2, P_3, and G, satisfy eq. (2), there exists a triangle whose medians meet at G and intersect the circumscribed circle at the points P_1, P_2, and P_3.

The lines GP_1, GP_2, and GP_3 intersect the circle $P_1P_2P_3$ at three points M_1, M_2, and M_3 corresponding to the complex numbers z_1, z_2, and z_3. We then know that

$$(z-u_i)\overline{(z-z_i)} = k \text{ (real)} \qquad (i = 1, 2, 3).$$

Therefore,

$$\sum_i \overline{(z - z_i)} = 0 \qquad \text{and} \qquad \sum_i (z - z_i) = 0.$$

The point G is the center of gravity of the triangle $M_1M_2M_3$.

If u_1, u_2, and u_3 denote the complex numbers corresponding to the given points, the complex numbers corresponding to the centers of gravity of the desired triangles are the roots of equation (2). This equation is of degree 2 in z and has two solutions (distinct or not).

There is one obvious exceptional case: If the three given points are on a single line, the problem posed has no solution.

216

First question:

If we perform a translation on the points A, B, and C, the complex numbers a, b, and c corresponding to them are replaced by

$$a_1 = a+t, \qquad b_1 = b+t, \qquad c_1 = c+t,$$
$$u_1 = a+bj+cj^2 +t(1+j+j^2) = u,$$
$$v_1 = a+bj^2 +cj +t(1+j^2 +j) = v.$$

(One can easily verify that $1+j+j^2 = 0$.)

If we perform a rotation about the origin O through an angle α, the complex numbers corresponding to the images of the points A, B, and C will be

$$a_2 = ae^{i\alpha}, \qquad b_2 = be^{i\alpha}, \qquad c_2 = ce^{i\alpha},$$

so that

$$u_2 = ue^{i\alpha}, \qquad v_2 = ve^{i\alpha}.$$

If we perform a dilatation of ratio k, we get

$$a_3 = ka, \qquad b_3 = kb, \qquad c_3 = kc, \qquad u_3 = ku, \qquad v_3 = kv.$$

In the plane, a similarity transformation, which is a composite of the above transformations, leaves the ratio u/v invariant. In an arbitrary displacement, the quantities $|u|$ and $|v|$ are also invariant.

On the other hand, we note that, when we perform successively a translation, a rotation, and a dilatation, we can transform a triangle ABC into a triangle DEO, where O is the coordinate origin and E is a point chosen arbitrarily on the x-axis. The ratio u/v relative to the triangle DEO is equal to the ratio u/v relative to the triangle ABC.

Let us now compare the triangles DEO and $D'EO$, representing the images of ABC and $A'B'C'$ under the transformation. For these, the ratios u/v and u'/v' are equal. If d, d', and e are the complex numbers corresponding to the points D, D', and E,

$$\frac{d + ej}{d + ej^2} = \frac{d' + ej}{d' + ej^2} \quad \Rightarrow \quad d = d'.$$

The two triangles DEO and $D'EO$ are the same. The triangles ABC and $A'B'C'$, being both similar to DEO, are similar to each other.

Second question:

Let us express the conditions $u = 0$ and $v = 0$ in connection with the triangle $OB'C'$ derived from ABC by the translation AO (which does not change u and v):

$$u = 0 \Rightarrow b' + c'j = 0,$$
$$v = 0 \Rightarrow b' + c'j^2 = 0.$$

The point B' is the image of the point C' under a rotation through an angle $\pm \frac{1}{3}\pi$. Thus, the triangles $OB'C'$ and ABC are equilateral.

Third question:

By means of a translation and a rotation, the triangle ABC can be transformed into a triangle $A'B'C'$, where the y-axis is the perpendicular bisector of $B'C'$.

Then, the complex numbers corresponding to B' and C' are the real numbers b' and $-b'$. Therefore, a necessary and sufficient condition for $A'B'$ to be equal to $A'C'$ is that the complex number a' corresponding to A' be a purely imaginary number:

$$u' = a' + b'(j - j^2) = a' + ib'\sqrt{3},$$
$$v' = a' + b'(j^2 - j) = a' - ib'\sqrt{3}.$$

If v' is zero, then $a' = ib'\sqrt{3}$ is purely imaginary. Otherwise, the ratio u'/v' is well defined and is real if and only if a' is purely imaginary. If $v' = 0$, then $v = 0$; if $v' \neq 0$, then $v \neq 0$ and $u'/v' = u/v$. Therefore, a necessary and sufficient condition for AB to be equal to AC is that v vanish or that the ratio u/v be real.

The preceding study shows that a necessary and sufficient condition for BA to be equal to BC is that either

$$b + cj^2 + aj = jv = 0$$

or

$$\frac{b + cj + aj^2}{b + cj^2 + aj} = \frac{j^2 u}{jv} = \text{a real number.}$$

The result is analogous for $CA = CB$.

A necessary and sufficient condition for the triangle ABC to be isosceles is that either v be equal to 0 or that one of the ratios u/v, ju/v, j^2u/v be real, that is, that u^3/v^3 be a real number.

217

First question:

If z is real, $z - i$ and $z + i$ are conjugates and

$$|Z| = \left| \frac{z - i}{z + i} \right| = 1.$$

Conversely, setting $Z = e^{i\varphi}$, we have

$$\frac{z - i}{z + i} = e^{i\varphi} \quad \Leftrightarrow \quad z = i\frac{e^{i\varphi} + 1}{1 - e^{i\varphi}} = -\cot\frac{\varphi}{2} \quad \text{if} \quad e^{i\varphi} \neq 1.$$

The image of the real axis is the circle of radius 1 with center at the origin and with the point corresponding to the number 1 removed.

Second question:

The homographic function that defines the transformation assigns to a real number z a real number Z. Let us denote by z_1 a real number and let us denote by Z_1 the corresponding value. The function $Z - Z_1$ is a homographic function that vanishes at $z = z_1$. Therefore,

$$Z - Z_1 = k(z - z_1) \qquad \text{or} \qquad Z - Z_1 = k\frac{z - z_1}{z - z_2}.$$

In the first case, the number $k = (Z - Z_1)/(z - z_1)$ is the quotient of two real numbers if z is real; k is a real number.

In the second case, the ratio

$$\frac{z - z_1}{Z - Z_1} = \frac{z}{k} - \frac{z_2}{k}$$

is real for real values of z. If we assign to z the value 0, we see that z_2/k is a real number; if we assign to z the value 1, we see that $1/k$ is a real number.

In both cases, the homographic function obtained has all real coefficients.

We know that the set of homographic transformations with real coefficients constitutes a subgroup G_1 of the group G. The isomorphism between G and Γ obviously implies that the set Γ_1, which is the image of G_1, is a subgroup of Γ.

Third question:

If the transformation T conserves the circle C of radius 1 with center at the origin, the transformation $T_0^{-1}TT_0$ conserves the real axis $x'x$. Specifically, the images of $x'x$ under these transformations are

$$x'x \xrightarrow{T_0} C \xrightarrow{T} C \xrightarrow{T_0^{-1}} x'x.$$

The transformation $T_0^{-1}TT_0$ is an R transformation and therefore

$$T = T_0 R T_0^{-1}.$$

Conversely, a transformation of this type conserves the circle C:

$$C \xrightarrow{T_0^{-1}} x'x \xrightarrow{R} x'x \xrightarrow{T_0} C.$$

To show that these transformations constitute a subgroup Γ_2 of Γ, we need only show that $T'T^{-1}$ belongs to Γ_2 if T' and T belong to Γ_2.

$$T = T_0 R T_0^{-1} \implies T^{-1} = T_0 R^{-1} T_0^{-1},$$
$$T' = T_0 R' T_0^{-1},$$
$$T'T^{-1} = (T_0 R' T_0^{-1})(T_0 R^{-1} T_0^{-1}) = T_0 R'(T_0^{-1} T_0) R^{-1} T_0^{-1}$$
$$= T_0 (R'R^{-1}) T_0^{-1}.$$

Since Γ_1 is a group, $R'R^{-1}$ is a transformation in Γ_1, and hence $T'T^{-1}$ is a transformation in Γ_2.

(One might also reason directly regarding the transformations T by showing that T^{-1}, and hence $T'T^{-1}$, conserves the circle C.)

Fourth question:

If R is defined by

$$Z = \frac{az + b}{cz + d},$$

where a, b, c, and d are real numbers, then T is defined by the function t:

$$Z = \frac{[c - b + i(a + d)]z + b + c + i(a - d)}{[-(b + c) + i(a - d)]z + b - c + i(a + d)},$$

where a, b, c, and d are arbitrary real numbers.

Let us set

$$\lambda = c - b + i(a + d), \qquad \mu = b + c + i(a - d),$$

$$\alpha = -\frac{\lambda}{\bar{\lambda}}, \qquad\qquad z_0 = -\frac{\mu}{\lambda}.$$

Then,

$$Z = \frac{\lambda z + \mu}{-\bar{\mu}z - \bar{\lambda}} = -\frac{\lambda}{\bar{\lambda}} \frac{z + \mu/\lambda}{(\bar{\mu}/\bar{\lambda})z + 1} = \alpha \frac{z - z_0}{1 - \bar{z}_0 z}.$$

(Since α is the quotient $\lambda/\bar{\lambda}$ of two complex conjugates, its absolute value is 1.)

Since a, b, c, and d are arbitrary real numbers, λ and μ are arbitrary complex numbers, α is an arbitrary complex number of unit absolute value, and z_0 is arbitrary. Therefore, if α and z_0 are given, we can determine a, b, c, and d in such a way that the corresponding function t will be precisely defined by

$$t(z) = \alpha \frac{z - z_0}{1 - \bar{z}_0 z}.$$

This form characterizes the functions t.

218

The numbers $e^{2ik\pi/n}$ such that $1 \leqslant k \leqslant n - 1$ constitute the set of all nth roots of unity except 1 itself. Therefore, they are the zeros of the polynomial A defined by

$$A(x) = \frac{x^n - 1}{x - 1} = 1 + x + \cdots + x^{n-1}.$$

The decomposition of A into prime factors is

$$A(x) = \prod_{1}^{n-1} (x - e^{2ik\pi/n}).$$

Therefore,

$$n = A(1) = (-1)^{n-1} \prod_{1}^{n-1} (e^{2ik\pi/n} - 1) = (-1)^{n-1} P,$$

$$e^{2ik\pi/n} - 1 = e^{ik\pi/n}(e^{ik\pi/n} - e^{-ik\pi/n}) = e^{ik\pi/n} 2i \sin k\pi/n,$$

$$P = \prod_{1}^{n-1} (2i e^{ik\pi/n} \sin k\pi/n) = 2^{n-1} i^{n-1} S \prod_{1}^{n-1} e^{ik\pi/n}.$$

The argument of a product is the sum of the arguments of the factors. Therefore, the argument of the product

$$\prod_{1}^{n-1} e^{ik\pi/n}$$

is

$$\sum_{1}^{n-1} \frac{k\pi}{n} = \frac{\pi}{n}\left(\sum_{1}^{n-1} k\right) = \frac{\pi}{n} \frac{(n-1)n}{2} = \frac{(n-1)\pi}{2},$$

$$\prod_{1}^{n-1} e^{ik\pi/n} = e^{i(n-1)\pi/2} = i^{n-1},$$

$$P = 2^{n-1} i^{n-1} S i^{n-1} = 2^{n-1} S(-1)^{n-1},$$

$$S = (-1)^{n-1} \frac{P}{2^{n-1}} = \frac{n}{2^{n-1}}.$$

219

First question:

The number λ cannot be equal to 1 since the inequality $\mu \neq 0$ implies

$$A(z+\mu) \neq A(z).$$

(The polynomial A would assume every value infinitely many times if $A(z+\mu)$ were equal to $A(z)$.) Therefore, the relation $Z = \lambda z + \mu$ can be written

$$Z - z_0 = \lambda(z - z_0),$$

where z_0 denotes the number $\mu/(1-\lambda)$.

If B is the polynomial defined by $B(u) = A(z_0 + u)$, it remains unchanged when we replace u with λu:

$$B(u) = au^3 + bu^2 + cu + d,$$

$$B(\lambda u) = \lambda^3 au^3 + \lambda^2 bu^2 + \lambda cu + d.$$

For these polynomials to be identical, it is necessary and sufficient that

$$a(\lambda^3 - 1) = 0, \qquad b(\lambda^2 - 1) = 0, \qquad c(\lambda - 1) = 0;$$
$$a \neq 0, \text{ since } B \text{ of degree } 3 \quad \Rightarrow \quad \lambda^3 - 1 = 0.$$

The number λ cannot be equal to 1. Therefore, it is equal either to j or to j^2. Therefore, the other two equations imply that $b = c = 0$.

$$B(u) = au^3 + d,$$
$$A(z) = B(z - z_0) = a(z - z_0)^3 + d.$$

Second question:

Reasoning identical to that in (1) shows that the polynomial B defined as above by $B(u) = A(z_0 + u)$ remains unchanged when we replace u with λu.
   If

$$B(u) = \sum_0^n b_k x^k,$$

setting $B(u)$ equal to $B(\lambda u)$ yields the conditions

$$(\lambda^k - 1) b_k = 0 \qquad (k = 0, 1, ..., n).$$

Since b_n is nonzero, λ is an nth root of unity.

FIRST CASE: $\lambda^k - 1 \neq 0$ if $0 < k < n$. All the coefficients b_k are zero with the exception of b_0 and b_n:

$$A(z) = B(z - z_0) = b_n(z - z_0)^n + b_0.$$

In this case,

$$\lambda = e^{2ik\pi/n},$$

where the integer k is prime to n (see exercise 210) and $\mu = z_0(1 - \lambda)$.

SECOND CASE: The lowest power of λ that is equal to 1 has an exponent $p < n$. We know that the integers k such that $\lambda^k = 1$ are multiples of p (in particular, this is the case with n).
   Therefore, the coefficients b_k are equal to 0 if k is not a multiple of p and they are arbitrary if k is a multiple of p. B is a polynomial in u^p. If we decompose this polynomial into prime factors, we obtain

$$B(u) = b_n \prod_1^m (u^p - u_i),$$

where the u_i are arbitrary complex numbers

$$A(z) = b_n \prod_1^m [(z - z_0)^p - u_i].$$

Each of the factors $u^p - u_i$ remains unchanged when we replace u by λu. Therefore, each of the factors of A is unchanged when we replace z by

$$\lambda z + z_0(1 - \lambda).$$

Third question:

If $n = 6$, the roots of unity (in addition to 1) are

$$-1, j, j^2, -j, -j^2.$$

We need only apply the results of (2). We keep the notation of (2).

If $\lambda = -1$, $p = 2$ and A is the product of three polynomials in $(z - z_0)^2$:

$$A(z) = a[(z - z_0)^2 - u_1][(z - z_0)^2 - u_2][(z - z_0)^2 - u_3].$$

If λ is equal to j or j^2, $p = 3$ and A is the product of two polynomials of the type defined in (1):

$$A(z) = a[(z - z_0)^3 - u_1][(z - z_0)^3 - u_2].$$

Finally, for $-j$ and $-j^2$, p is equal to 6:

$$A(z) = a(z - z_0)^6 + d.$$

220

First question:

(E) has a double root if the polynomial $x^3 + px + q$ and its derivative $3x^2 + p$ have a zero in common.

In accordance with the division algorithm, let us write the first of these polynomials in the form

$$x^3 + px + q = \tfrac{1}{3}x(3x^2 + p) + \tfrac{2}{3}px + q.$$

At the common zero of the two polynomials, the remainder must vanish. Therefore, this common zero is equal to $-3q/2p$. We then substitute this value into one of the polynomials, for example, $3x^2 + p$ and set that polynomial equal to 0.

$$3\left(-\frac{3q}{2p}\right)^2 + p = 0 \quad \Rightarrow \quad 4p^3 + 27q^2 = 0.$$

(These calculations are not valid if $p = 0$; the result however does remain valid.)

The double root is $-3q/2p$. The third root can be obtained by expressing the fact that the sum of the three roots is zero, that is, that the coefficient of x^2 is zero:

$$2\left(-\frac{3q}{2p}\right) + x_3 = 0 \quad \Rightarrow \quad x_3 = \frac{3q}{p}.$$

Second question:

The coefficients of the polynomial $x^3 + px + q$ are real. If the complex number x_1 is a root of (E), its conjugate $\bar{x}_1$ is also a root of (E). If x_3 is the third root, then

$$x_1 + \bar{x}_1 + x_3 = 0 \Rightarrow x_3 = -(x_1 + \bar{x}_1) = -2\mathrm{Re}(x_1).$$

For (E) to have a complex root with real part equal to a, it is necessary and sufficient that $-2a$ be a root of (E) and that the other two roots be complex numbers. Therefore,

$$(-2a)^3 + p(-2a) + q = -8a^3 - 2ap + q = 0.$$

In this condition is satisfied, then

$$x^3 + px + q = (x + 2a)(x^2 - 2ax + p + 4a^2).$$

The other two roots of (E) are the roots of a quadratic equation

$$x^2 - 2ax + p + 4a^2 = 0.$$

These roots will be complex if $a^2 - (p + 4a^2) < 0$. The desired conditions are

$$q = 2ap + 8a^3, \qquad p > -3a^2.$$

Third question:

If $q = 0$, the nonzero roots of (E) are the solutions of the equation

$$x^2 + p = 0.$$

These are complex numbers of absolute value r if $p = r^2$.

Suppose now that q is nonzero. Let us write the expression for the product of the roots of (E):

$$x_1 \bar{x}_1 x_3 = |x_1|^2 x_3 = -q.$$

For (E) to have a complex root of absolute value equal to r, it is necessary and sufficient that $-q/r^2$ be a root of (E) and that the other two roots be complex numbers. Thus,

$$\left(-\frac{q}{r^2}\right)^3 + p\left(-\frac{q}{r^2}\right) + q = 0$$

or

$$q^2 + pr^4 - r^6 = 0.$$

If this condition is satisfied, then

$$x^3 + px + q = \left(x + \frac{q}{r^2}\right)\left(x^2 - \frac{q}{r^2}x + r^2\right).$$

The other two roots of (E) are roots of a quadratic equation:

$$x^2 - \frac{q}{r^2}x + r^2 = 0.$$

These will be complex if

$$\frac{q^2}{r^4} - 4r^2 < 0.$$

Thus, the conditions sought are

$$q^2 + pr^4 - r^6 = 0, \qquad q^2 < 4r^6.$$

(The condition obtained when $q = 0$ is a particular case of the general formulas.)

221

First question:

The equation whose roots are squares of the roots of (E) is obtained by substituting the first of the following equations into the second:

$$u = x^2,$$
$$x^3 + px + q = 0.$$

We then have the equivalent system

$$u = x^2,$$
$$x(u + p) + q = 0,$$

so that the desired equation is

$$q^2 = u(u + p)^2.$$

We may take

$$A(u) = u(u + p)^2 - q^2.$$

We might also have calculated the symmetric functions of x_1^2, x_2^2, and x_3^2 with the aid of the symmetric functions of x_1, x_2, and x_3. For example,

$$x_1^2 + x_2^2 + x_3^2 = (x_1 + x_2 + x_3)^2 - 2(x_1x_2 + x_2x_3 + x_3x_1) = -2p,$$

and similarly for the other two.

Second question:

$$(x_1 - x_2)^2 = (x_1 + x_2)^2 - 4x_1x_2,$$
$$x_1 + x_2 + x_3 = 0 \quad \Rightarrow \quad x_1 + x_2 = -x_3,$$
$$x_1x_2 + x_2x_3 + x_3x_1 = p \quad \Rightarrow \quad x_1x_2 = p - x_3(x_1 + x_2) = p + x_3^2.$$

Finally,
$$(x_1 - x_2)^2 = x_3^2 - 4(p + x_3^2) = -(4p + 3x_3^2).$$

Third question:

The formula in (2) shows that the zeros of the polynomial B are the numbers $-(4p + 3x_i^2)$. Therefore, the polynomial B can be derived from the polynomial A whose zeros are the x_i^2 by making the change of variable

$$v = -(4p + 3u) \Leftrightarrow u = -\frac{v + 4p}{3},$$

$$B(v) = A\left(-\frac{v + 4p}{3}\right) = -\frac{v + 4p}{3}\left(\frac{v + p}{3}\right)^2 - q^2,$$

$$B(v) = -\frac{1}{27}(v^3 + 6pv^2 + 9p^2v + 4p^3 + 27q^2).$$

The polynomial B, like the polynomial A, is defined only up to a constant factor. Therefore, we may drop the factor $-1/27$.

222

Let us set

$$\begin{array}{ll} x_1 + x_2 = s, & x_1 x_2 = p = 2, \\ x_3 + x_4 = u, & x_3 x_4 = v. \end{array}$$

We shall calculate the symmetric functions of x_1, x_2, x_3, and x_4 in terms of s, p, u, and v on the one hand and in terms of the coefficients of the polynomial on the other:

$$\begin{aligned} \Sigma x_i &= s + u = 1, \\ \Sigma x_i x_j &= su + v + p = su + v + 2 = \lambda, \\ \Sigma x_i x_j x_k &= pu + sv = 2u + sv = -6, \\ x_1 x_2 x_3 x_4 &= pv = 2v = -4. \end{aligned}$$

The first, third, and fourth of these equations enable us to calculate u and v, and the second determines λ. We obtain

$$v = -2, \quad s = 2, \quad u = -1, \quad \lambda = -2.$$

The numbers x_1 and x_2 are therefore the roots of the equation

$$x^2 - sx + p = x^2 - 2x + 2 = 0.$$

Similarly, x_3 and x_4 are the roots of the equation

$$x^2 - ux + v = x^2 + x - 2 = 0,$$

$$x_1 = 1 + i, \quad x_2 = 1 - i, \quad x_3 = 1, \quad x_4 = -2.$$

REMARK: We can equally well write that the two equations $A(x) = 0$ and $A(2/x) = 0$ have two common roots x_1 and x_2. To do this ,we express the fact that the greatest common divisor of the polynomials constituting the left-hand members is a polynomial of degree 2 and that the zeros of this polynomial are x_1 and x_2.

223

The polynomial part is equal to 1; the simple poles are j and j^2, and the double pole is -1. The decomposition is of the form

$$R(x) = 1 + \frac{A}{(x+1)^2} + \frac{B}{x+1} + \frac{C}{x-j} + \frac{\bar{C}}{x-j^2}.$$

(The coefficients C and $\bar{C}$ relative to the poles j and j^2 are conjugates since the coefficients of R are real.)

To evaluate A and B, we make the change of variable $x+1 = t$ and multiply by t^2. Then the left member of the above equation becomes

$$t^2 R(-1+t) = \frac{(t-1)^4 + 3(t-1)^2 + 1}{(t-1)^2 + (t-1) + 1} = \frac{5 - 10t + \cdots}{1 - t + \cdots},$$

$$5 - 10t = (1-t)(5-5t) + \cdots.$$

Then,

$$5 - 10t + \cdots = (1-t)(5-5t) + \cdots$$

and the part of the decomposition corresponding to the double pole -1 is

$$\frac{5}{(x+1)^2} - \frac{5}{x+1}.$$

The coefficient C relative to the simple pole $x-j$ is

$$C = [(x-j)R(x)]_{x=j} = \left[\frac{x^4 + 3x^2 + 1}{(x+1)^2(x-j^2)}\right]_{x=j} = \frac{j^4 + 3j^2 + 1}{(1+j)^2(j-j^2)}.$$

The relations $j^3 = 1$ and $1+j+j^2 = 0$ enable us to simplify the expression before replacing j with its numerical expression:

$$j^4 + 3j^2 + 1 = j + 3j^2 + 1 = 2j^2,$$

$$(1+j)^2 = (-j^2)^2 = j^4 = j,$$

$$C = \frac{2j^2}{j(j-j^2)} = \frac{2j}{j-j^2} = \frac{-1+i\sqrt{3}}{i\sqrt{3}} = 1 + i\frac{\sqrt{3}}{3},$$

$$R(x) = 1 + \frac{5}{(x+1)^2} - \frac{5}{x+1} + \frac{1+i\sqrt{3}/3}{x-j} + \frac{1-i\sqrt{3}/3}{x-j^2}.$$

To check the calculations, it is a good idea to assign to x a particular value, such as $x = 0$ in the present example. We can easily verify that

$$R(0) = 1 = 1 + 5 - 5 + \frac{1 + i\sqrt{3}/3}{-j} + \frac{1 - i\sqrt{3}/3}{-j^2}$$

$$= 1 - j^2\left(1 + \frac{i\sqrt{3}}{3}\right) - j\left(1 - i\frac{\sqrt{3}}{3}\right) = 1 - 2\mathrm{Re}\left[j^2\left(1 + \frac{i\sqrt{3}}{3}\right)\right].$$

REMARK: In certain problems (in particular, the search for a primitive of R), what we need is a decomposition into real elements. Then, all that we have to do is add the last two terms, which are conjugates if x is a real number.

One may also proceed directly by seeking a decomposition of the form

$$R(x) = 1 + \frac{5}{(x+1)^2} - \frac{5}{x+1} + \frac{\lambda x + \mu}{x^2 + x + 1}$$

and determining λ and μ by assigning to x particular numerical values:

$$x = 0 \Rightarrow 1 = 1 + \mu \Rightarrow \mu = 0,$$

$$x = 1 \Rightarrow \frac{5}{12} = 1 + \frac{5}{4} - \frac{5}{2} + \frac{\lambda}{3} \Rightarrow \lambda = 2.$$

224

First question:

If u is a zero of A of order α, it is a zero of A' of order $(\alpha - 1)$ and hence a pole of A'/A of order 1. The coefficient U of $1/(x-u)$ in the decomposition of A'/A is

$$U = \left[\frac{(x-u)A'(x)}{A(x)}\right]_{x=u} = \left\{\frac{(x-u)\left[\dfrac{(x-u)^{\alpha-1}}{(\alpha-1)!}A^{(\alpha)}(u) + \cdots\right]}{\dfrac{(x-u)^{\alpha}}{\alpha!}A^{(\alpha)}(u) + \cdots}\right\}_{x=u}$$

when we expand A' and A by Taylor's formula.

We can now divide the numerator and denominator by $(x-u)^{\alpha}$ and then assign to x the value u:

$$U = \frac{\dfrac{1}{(\alpha-1)!}A^{(\alpha)}(u)}{\dfrac{1}{\alpha!}A^{(\alpha)}(u)} = \alpha.$$

Thus, if u is a zero of A of order α, the element of the decomposition relative to this pole is $\alpha/(x-u)$ and can be treated as the sum of α elements $1/(x-u)$. If we assign to x the value a in the decomposition of A'/A, we obtain

$$\frac{A'(a)}{A(a)} = \sum_{i=1}^{n} \frac{1}{a - x_i} \quad \Rightarrow \quad \sum_{i=1}^{n} \frac{1}{x_i - a} = -\frac{A'(a)}{A(a)}.$$

Similarly, let us write the decomposition of the derivative of A'/A:

$$\frac{A''(x)A(x) - A'^2(x)}{A^2(x)} = \sum_{i=1}^{n} \frac{-1}{(x - x_i)^2}.$$

When we assign to x the value a, we have

$$\sum_{1}^{n} \frac{1}{(x_i - a)^2} = \frac{A'^2(a) - A(a)A''(a)}{A^2(a)}.$$

Let us decompose

$$\frac{2x^2 + 1}{(x^2 - 1)^2}$$

into partial fractions:

$$\frac{2x^2 + 1}{(x^2 - 1)^2} = \frac{3}{4}\frac{1}{(x - 1)^2} + \frac{1}{4}\frac{1}{x - 1} + \frac{3}{4}\frac{1}{(x + 1)^2} - \frac{1}{4}\frac{1}{x + 1}.$$

When we assign to x the values x_i and add, we obtain

$$S = \sum_{i=1}^{4} \frac{2x_i^2 + 1}{(x_i^2 - 1)^2} = \frac{3}{4}\sum_{i} \frac{1}{(x_i - 1)^2} + \frac{1}{4}\sum_{i} \frac{1}{x_i - 1} + \frac{3}{4}\sum_{i} \frac{1}{(x_i + 1)^2} - \frac{1}{4}\sum_{i} \frac{1}{x_i + 1}$$

and the calculation of S is reduced to finding the expressions evaluated in (1) (we take $a = 1$ and $a = -1$):

$$\frac{A'(x)}{A(x)} = \frac{4x^3 - 1}{x^4 - x + 1} \quad \text{and} \quad \frac{A'^2(x) - A(x)A''(x)}{A^2(x)} = \frac{4x^6 + 4x^3 - 12x^2 + 1}{(x^4 - x + 1)^2},$$

$$\sum_{i} \frac{1}{(x_i - 1)^2} = -3, \qquad \sum_{i} \frac{1}{x_i - 1} = -3,$$

$$\sum_{i} \frac{1}{(x_i + 1)^2} = -\frac{11}{9}, \qquad \sum_{i} \frac{1}{x_i + 1} = \frac{5}{3},$$

$$S = -\frac{9}{4} - \frac{3}{4} - \frac{11}{12} - \frac{5}{12} = -\frac{13}{3}.$$

225

The poles are the roots of the equation

$$x^{2q} = -1 = e^{i\pi}.$$

These are the numbers

$$x_k = e^{i(2k+1)\pi/2q} \qquad (k = 0, 1,..., 2q-1).$$

They are all simple poles. Therefore, the decomposition of R can be written

$$R(x) = \sum_{k=0}^{2q-1} \frac{A_k}{x - x_k}.$$

The value of A_k can be obtained by taking the quotient obtained by dividing the numerator of R (namely, 1) by the derivative of the denominator evaluated at $x = x_k$:

$$A_k = \frac{1}{2qx_k^{2q-1}} = \frac{x_k}{2qx_k^{2q}} = -\frac{x_k}{2q}.$$

None of these poles is real. Hence, they come in conjugate pairs; specifically, $\bar{x}_k = x_{2q-k-1}$. We can verify that (since R has real coefficients) corresponding to the conjugate poles are conjugate coefficients A_k, and we can write the decomposition of R in the form

$$R(x) = \sum_{k=0}^{q-1} \left(\frac{A_k}{x - x_k} + \frac{\bar{A}_k}{x - \bar{x}_k} \right) = \sum_{k=0}^{q-1} \frac{A_k(x - \bar{x}_k) + \bar{A}_k(x - x_k)}{(x - x_k)(x - \bar{x}_k)}.$$

Let us evaluate explicitly each term in the sum:

$$A_k(x - \bar{x}_k) + \bar{A}_k(x - x_k) = -\frac{1}{2q}[(x_k + \bar{x}_k)x - 2x_k\bar{x}_k]$$

$$= -\frac{1}{q}[x \cos [(2k + 1)\pi/2q] - 1],$$

$$(x - x_k)(x - \bar{x}_k) = x^2 - x(x_k + \bar{x}_k) + x_k\bar{x}_k = x^2 - 2x \cos[(2k + 1)\pi/2q] + 1,$$

$$R(x) = -\frac{1}{q} \sum_{k=0}^{q-1} \frac{x \cos [(2k + 1)\pi/2q] - 1}{x^2 - 2x \cos [(2k + 1)\pi/2q] + 1}.$$

REMARKS: (1) The evaluation of a primitive of R is made without difficulty by starting with the last formula obtained.

(2) If we replace x with $-x$, we permute the terms with subscript k and h $= q-k-1$ in the decomposition (since the decomposition of R into simple elements is unique and hence remains unchanged like R itself when we replace x with $-x$).

226

First question:

We know that

$$\lambda_i = \frac{A(x_i)}{B'(x_i)}.$$

If we denote by n the degree of B, we know that

$$B(x) = b \prod_{j=i}^{n} (x - x_j),$$

$$B'(x_i) = b \prod_{j \neq i} (x_i - x_j).$$

The sequence $B'(x_k)$ is alternating. When we replace x_i with x_{i+1}, the factor $x_i - x_{i+1}$ is replaced with its negative and the other factors keep their sign (obviously, we are assuming that the sequence $\{x_i\}$ is an increasing one).

A necessary and sufficient condition for all the λ_i to have the same sign is that the sequence $A(x_i)$ be alternating. Every interval $]x_i, x_{i+1}[$ then contains at least one zero of A. (Examination of the sign of the first factors of A yields this result immediately.) A has at least $n-1$ real zeros.

The degree of A does not exceed $n-1$ since our assumption is that the polynomial part of A/B is zero. The number of zeros of A cannot exceed $n-1$. The zeros of A are all real. Each of the intervals $]x_i, x_{i+1}[$ contains exactly one of them. Therefore, these zeros separate the poles.

The converse is immediate: If the zeros of A are real and if they separate the poles x_i, the sequence $\{A(x_i)\}$ is alternating and the λ_i all have the same sign.

Second question:

Let us use the decompositions of P and Q into prime factors:

$$P(x) = \prod_{i=1}^{n_1} (x - x_i)^{\alpha_i} \times \prod_{j=1}^{n_2} (x - u_j)^{\beta_j},$$

$$Q(x) = \prod_{i=1}^{n_1} (x - x_i)^{\alpha'_i} \times \prod_{h=1}^{n_3} (x - v_h)^{\gamma_h},$$

where the n_1 numbers x_i are the common zeros of P and Q, where the n_2 numbers u_j are the other zeros of P, and where the n_3 numbers v_h are the other zeros of Q.

The study made in exercise 224 enables us to write

$$\frac{P'(x)}{P(x)} = \sum_i \frac{\alpha_i}{x - x_i} + \sum_j \frac{\beta_j}{x - u_j},$$

$$\frac{Q'(x)}{Q(x)} = \sum_i \frac{\alpha_i'}{x - x_i} + \sum_h \frac{\gamma_h}{x - v_h},$$

$$R(x) = \frac{P'(x)}{P(x)} + k\frac{Q'(x)}{Q(x)} = \sum_i \frac{\alpha_i + k\alpha_i'}{x - x_i} + \sum_j \frac{\beta_j}{x - u_j} + \sum_h \frac{k\gamma_h}{x - v_h}.$$

R is a rational fraction whose polynomial part vanishes and whose poles are all real and simple. Finally, the coefficients in the decomposition are all positive. Thus, we know (from the proof of (1)) that the zeros of the numerator are all real and that they separate the poles. There are $n_1 + n_2 + n_3$ poles and hence the numerator has

$$n_1 + n_2 + n_3 - 1$$

real and simple zeros. These are also zeros of $P'Q + kPQ'$.

But we also know other zeros of the polynomial $P'Q + kPQ'$:

$$x_i \text{ is a zero of order } \alpha_i + \alpha_i' - 1;$$
$$u_j \text{ is a zero of order } \beta_j - 1;$$
$$v_h \text{ is a zero of order } \gamma_h - 1.$$

The sum of the orders of these different zeros is

$$\text{for the } x_i: \quad \alpha = \sum_{i=1}^{n_1} (\alpha_i + \alpha_i' - 1) = \sum_{i=1}^{n_1} (\alpha_i + \alpha_i') - n_1;$$

$$\text{for the } u_j: \quad \beta = \sum_{j=1}^{n_2} (\beta_j - 1) = \sum_{j=1}^{n_2} \beta_j - n_2;$$

$$\text{for the } v_h: \quad \gamma = \sum_{h=1}^{n_3} (\gamma_h - 1) = \sum_{h=1}^{n_3} \gamma_h - n_3.$$

In summary, the sum of the orders of all these zeros of $P'Q + kPQ'$ is

$$\alpha + \beta + \gamma + (n_1 + n_2 + n_3 - 1) =$$

$$= \sum_{i=1}^{n_1} (\alpha_i + \alpha_i') + \sum_{i=1}^{n_2} \beta_j + \sum_{h=1}^{n_3} \gamma_h - 1$$

$$= \left(\sum_{i=1}^{n_1} \alpha_i + \sum_{i=1}^{n_2} \beta_j \right) + \left(\sum_{i=1}^{n_1} \alpha_i' + \sum_{h=1}^{n_3} \gamma_h \right) - 1$$

$$= \deg P + \deg Q - 1$$

$$= \deg (P'Q + kPQ').$$

The polynomial $P'Q + kPQ'$ has no zeros other than these, and all of these are real.

Part 3

LINEAR AND MULTILINEAR ALGEBRA

Exercises dealing with the definitions and results of Chapters
VII-XI in the book *Mathématiques Générales* by C. Pisot and
M. Zamansky.

Introduction

The importance of this part led me to break it up into four chapters. In the first chapter, the exercises deal with fundamental concepts (vector spaces, bases, linear mappings). It is essential that the student understand these concepts thoroughly before beginning the study of matrices, in which the technical and formal aspects of certain calculations can somewhat hide the reasoning behind them.

In the second chapter, one begins calculation with matrices and determinants and the solution of systems of linear equations.

The third chapter has to do with eigenvalue problems and the reduction of matrices, the fourth with Euclidean spaces and with bilinear and quadratic forms.

In a systematic way, I have presented exercises that assume only a knowledge of the elementary concepts of analysis. The few exceptions are noted when they occur.

Thus, the reader will be able to familiarize himself with the basic algebraic concepts before studying the second part of the general mathematics program. This should enable him to use successfully algebraic methods in the study of problems in analysis. Also, the latter part of this book contains exercises in which algebra plays an essential role.

A review of certain results

Here, it is not possible even to list the important theorems, and all I can do is refer the student to his course in general mathematics.

However, I wish to emphasize certain results, often quite simple though frequently used in problems and in some cases, I feel, not completely understood by students.

First of all, the important concept of a dual space is often poorly understood and deserves some special thought.

Bases

Certain problems become obvious if the space is described in terms of a suitably chosen basis. To facilitate this choice, let us recall the following:

(1) Any set of n independent vectors, where n is the dimension of the space, constitutes a basis.

(2) To any set of independent vectors that do not constitute a basis, it is possible to add other independent variables such that the combined set of all these vectors does constitute a basis (the theorem on an incomplete basis).

(3) In a Euclidean space, the Gram–Schmidt orthonormalization procedure enables us to replace a basis $\{e_i\}$ with an orthonormal basis $\{u_i\}$ such that the subspace E_i generated by $e_1, e_2,..., e_i$ is identical to the subspace generated by $u_1, u_2,..., u_i$. (One can apply the method indicated in PZ, Book II, Chapter X, part 2, § 1, to the subspace $F = E_i$ and to the subspace E_{i-1} of E_i, an orthonormal basis of which is $\{u_1,..., u_{i-1}\}$.) The matrix of the transformation from one basis to the other is then triangular.

Dimensions of vector spaces

To determine the dimension of a vector space, we seek a basis for that space or we show that it is isomorphic to a vector space already studied.

If two vector spaces over a single field have the same dimension, they are isomorphic.

A (finite-dimensional) vector space has no proper subspace of the same dimension as itself. Therefore, to show that two vector spaces E and F are identical, it is sometimes convenient just to show that F is contained in E and that its dimension is that of E.

Linear mappings

To define a particular linear mapping, it is sufficient to define the images of the vectors of a basis (that is, n independent vectors, where the space on which the mapping is defined is of dimension n); these images are arbitrary.

To show that a linear mapping satisfies certain linear conditions, it is sufficient to show that these conditions are satisfied for the images of the basis vectors.

The rank of a linear mapping f on E into F is a basic concept. It can be defined in three different ways:

(a) the dimension of $f(E)$;

(b) the dimension of E minus the dimension of $f^{-1}(0)$;

(c) if $f(x)$ is defined in terms of its coordinates, that is, the linear forms $f_i(x)$, then the rank of f is the dimension in the dual space E^* of the vector subspace generated by the forms f_i.

We shall make use of the existence of these three definitions. In the first place, we shall, of course, choose whichever one of them is the simplest for the particular problem in question. We shall also use them, however, to obtain relationships between the dimensions of different vector spaces, for example, $f(E)$ and $f^{-1}(0)$.

The solution of linear systems

Determinants always provide a solution to such a problem. However, this solution is often more theoretical than practical. In many cases, especially when the coefficients are numbers, the method of successive eliminations is the simplest one.

If a homogeneous system (H) corresponding to a system (S) of n equations in n unknowns has only the trivial solution (with all the unknowns equal to zero) then (S) has one and only one solution. If the system (H) is indeterminate, then the system (S) is either impossible or indeterminate (the theorem of the alternative).

Finding the inverse of a matrix

Instead of evaluating the determinant and the minors of a matrix A, it is often simpler to write the equations represented by $AX = Y$, where X and Y are column matrices, and to solve these equations by successive eliminations (especially, if A is a matrix, the elements in which are given numbers).

Matrices

To study the properties of a matrix A, it is often convenient to consider the mapping $\tilde{A}$ associated with A. This is the mapping whose matrix relative to the given basis is A. Thus, the coordinates of $\tilde{A}(e_i)$ are the elements of the ith column of A, where e_i is the ith basis vector.

In particular, to determine the transform of the matrix corresponding to a change of coordinates, it is often easier to find the vectors transformed by $\tilde{A}$ of the new basis vectors than to use the transition matrices. However, we must distinguish carefully between A and $\tilde{A}$ since A is the matrix of $\tilde{A}$ only for the given basis.

With regard to a given basis, it is often convenient to assign to every vector x the column matrix X composed of the coordinates of that vector (but here again we need to distinguish between x and X). For this we have the notations

$$Y = AX \Leftrightarrow y = \tilde{A}(x)$$

$X^T A Y$ for the bilinear form of the matrix A.

Eigenvectors

If the eigenvalues of a matrix are all distinct, there exists a basis consisting of the eigenvectors.

A symmetric matrix always has a basis consisting of eigenvectors.

Quadratic forms

Giving a *symmetric* bilinear form is equivalent to giving the quadratic form that it generates since each of these determines the other completely. Therefore, we can reason in terms of the bilinear form or the quadratic form, depending on which we find the more convenient in a particular case.

Exercises

CHAPTER 1

Vector spaces and subspaces; bases, mappings, and linear forms

301

Let x_1, x_2, x_3, x_4, and x be the following vectors in the space $\mathbf{R}^4$:

$$x_1 = (1, 1, 2, 1), \qquad x_2 = (1, -1, 0, 1), \qquad x_3 = (0, 0, -1, 1),$$
$$x_4 = (1, 2, 2, 0), \qquad x = (1, 1, 1, 1).$$

(1) Show that the four vectors x_1, x_2, x_3, and x_4 constitute a basis for $\mathbf{R}^4$.
(2) Find the coordinates of x relative to the basis x_1, x_2, x_3, x_4.

302

Let a, b, and c denote three elements in a vector space E over the field of complex numbers. Define

$$u = b + c, \qquad v = c + a, \qquad w = a + b.$$

(1) Show that the vector subspaces generated by a, b, and c on the one hand and u, v, and w on the other are identical.
(2) Show that the vectors u, v, and w are independent if and only if a, b, and c are independent.

303

We recall that the set of continuous mappings of $\mathbf{R}$ into $\mathbf{R}$ has a vector space structure on $\mathbf{R}$ if we define the sum of two mappings and the product of a mapping by a real number in the usual way.
   Are the functions f_n, defined by

$$f_n(t) = \sin^n t,$$

independent in this space?

304

Let A and B denote finite-dimensional vector subspaces of a vector space E over the field K.

(1) Show that the set (denoted by $A+B$) consisting of all the sums of an element in A and an element in B is a vector subspace of E.

(2) Show that $A+B$ is finite-dimensional and that

$$\dim (A+B)+\dim A\cap B = \dim A +\dim B.$$

(Consider first the simple case in which $A\cap B$ consists only of the zero element.)

305

Consider the set E of functions x defined in R by

$$x(t) = a_0+ \sum_{k=1}^{n} (a_k \cos kt+b_k \sin kt),$$

where n is a natural number and $a_0, a_1,..., a_n, b_1,..., b_n$ are arbitrary real numbers.

(1) Show that E is a vector subspace of the vector space over R of mappings of R into R.

(2) Let m denote an integer such that $0\leqslant m\leqslant n$ and let E_m denote the set of functions x defined by

$$x(t) = a_0+ \sum_{k=1}^{m} (a_k \cos kt+b_k \sin kt).$$

Show by induction on m that, if x is the zero function belonging to E_m, the coefficients a_k and b_k are all 0. (Use the function $x'' +m^2x$, where x'' denotes the second derivative of the function x.)

(3) Exhibit a basis of the vector subspace E.

(4) Denote by u the linear mapping of E into E that assigns to every element x in E the element $y=u(x)$ defined by

$$y(t) = x(t+\tfrac{1}{4}\pi).$$

Show that u is a linear mapping and that u^8 is the identity mapping, where u^8 denotes the composite of eight applications of the mapping u.

Find the kernel of u and the image set $u(E)$.

306

Let ω and θ denote two algebraic numbers, that is, roots of equations with rational coefficients:

$$\omega^n = \sum_{k=0}^{n-1} \alpha_k\omega^k, \qquad \theta^m = \sum_{k=0}^{m-1} \beta_k\theta^k.$$

(1) Show that the field R of real numbers is a vector space over the field Q of rational numbers, where the sum of two real numbers and the product of a real number and a rational number (which is a particular case of the product of two real numbers) are defined in the usual way in R. (We denote this vector space by R_θ.)

(2) Show that the set E of real numbers x that are equal to a polynomial in ω and θ with rational coefficients is a vector subspace of R_θ.

(3) Show that every number in E is equal to a polynomial in ω and θ with rational coefficients the degrees of which with respect to ω and θ are at most $n-1$ and $m-1$ respectively. From this deduce that E is a vector subspace of finite dimension.

(4) Show that the number $\omega + \theta$ is the root of an algebraic equation of degree not exceeding nm with rational coefficients.

307

Consider the set $\mathscr{S}$ of complex sequences $\{u_n\}$ that verify the recursion relation

$$u_n = a_1 u_{n-1} + \cdots + a_k u_{n-k} \qquad \text{if} \qquad n \geqslant k,$$

where $a_1, a_2, \ldots, a_k$ are given complex numbers.

(1) Show that $\mathscr{S}$ is a vector subspace of the vector space of complex sequences over the field C.

We recall (see PZ, Book II, Chapter III, part 3, § 2) the definition of the operations

$$\{u_n\} + \{v_n\} = \{u_n + v_n\}, \qquad \lambda\{u_n\} = \{\lambda u_n\}.$$

(2) Exhibit a basis for $\mathscr{S}$ and show that the dimension of $\mathscr{S}$ is k.

308

(1) Let a, b, and c denote arbitrary rational numbers. Show that the polynomials $u^3 - 2$ and $a + bu + cu^2$ are relatively prime.

(2) Show that the set E of real numbers of the form

$$a + b\sqrt[3]{2} + c\sqrt[3]{4} \qquad (a, b \text{ and } c \text{ rational})$$

is a ring and (for example, by using the result of (1)) that this ring is a field.

(3) Show that E is a three-dimensional vector space over the field of rational numbers if we define the product of a number in E by a rational number as the number representing the product of these two real numbers.

309

(1) Show that the set consisting of polynomials of degree not exceeding n with complex coefficients together with the zero polynomial is a vector space of dimension $n+1$ over the field of complex numbers. (We denote this field by E_{n+1}.)

(2) Let $\{A_p\}$ denote any set of $n+1$ polynomials belonging to E_{n+1}, where A_p is of degree p for $p = 0, 1,...,n$. Show that these polynomials A_p constitute a basis for E_{n+1}.

(3) Let B denote a particular polynomial of degree $k \leqslant n$. Show that the set of polynomials obtained by multiplying B by polynomials of degree not exceeding $n-k$ (so that the product belongs to E_{n+1}) constitutes a vector subspace F_B of dimension $n-k+1$ and that the polynomials of degree not exceeding $k-1$ constitute a vector subspace G_k that is complementary to F_B (cf. PZ, Book II, Chapter VII, part 2, § 3). Show that infinitely many subspaces of E_{n+1} exist that are complementary to the subspace G_k. Show with a counterexample that the union of two vector subspaces is not in general a vector subspace.

310

We still denote by E_{n+1} the vector space of polynomials of degree not exceeding n with complex coefficients that we studied in exercise 309. We define polynomials U_k for $0 \leqslant k \leqslant n$ by

$$U_0 = 1, \qquad U_k(x) = x(x-1)\cdots(x-k+1) \qquad \text{if} \qquad k>0.$$

(1) Show that the polynomials U_k constitute a basis for E_{n+1}.

(2) Show that there exists one and only one linear mapping φ of E_{n+1} into itself that satisfies the $n+1$ conditions

$$\varphi(x^k) = U_k, \qquad (k = 0, 1,...,n)$$

and that φ is bijective.

(3) We define a mapping δ of E_{n+1} into itself by

$$[\delta(P)](x) = P(x+1)-P(x).$$

Show that δ is linear. Find the polynomials $\delta(U_k)$. What is the kernel of δ and what is the vector subspace $\delta(E_{n+1})$?

(4) If R is a polynomial with complex coefficients, show that the equation

$$P(x+1)-P(x) = R(x)$$

is satisfied by one and only one polynomial P that vanishes at $x = 0$. Find P if R is the polynomial $x^3 - 5x^2 + x + 1$.

(5) Define directly the mapping

$$d = \varphi^{-1} \circ \delta \circ \varphi$$

of E_{n+1} into itself.

311

Let E denote the vector space of polynomials with complex coefficients and let A denote a given polynomial of degree α. Let E_{n+1} denote the subspace of E composed of polynomials of degree not exceeding n. Let f be the linear mapping of E into itself defined by

$$f(P) = AP' - A'P,$$

where A' and P' are the derivatives of the polynomials A and P.

(1) Show that, if the degree k of P is not equal to α, then the degree of $f(P)$ is exactly $\alpha + k - 1$. Show that $f(A)$ is the zero polynomial and that the set $f(E_{\alpha+1})$ is identical to the set $f(E_\alpha)$. Find the kernel of f.

(2) What is the rank of the restriction of f to E_{n+1}? (See PZ, Book II, Chapter VII, part 3, §3).

(3) The set of polynomials Q that belong to E_{p+1} and to $f(E)$ is a vector subspace F_{p+1}. Show that $F_{p+1} = f(E_{p-\alpha+2})$ and find the dimension of F_{p+1}. (There will be different cases to treat separately according to the value of p.) Determine the subspace F_3 when $A = x^2$.

(4) Assume that the zeros of the polynomial A are all distinct. Then, show that a necessary and sufficient condition for the fraction Q/A^2 to be the derivative of a rational function is that the polynomial Q belong to $f(E)$.

Show that this is not true if A has multiple zeros. (Suggestion: Consider fractions with denominator x^4.)

312

Let E denote a finite-dimensional vector space over the field of real numbers. Let u and v denote linear mappings of E into itself. Assume that the kernel $u^{-1}(0)$ of u contains the kernel $v^{-1}(0)$ of v.

(1) Show that it is possible to choose a basis $e_1,\dots,e_n$ for E the first k vectors of which constitute a basis for $v^{-1}(0)$ and the first $k+h$ vectors constitute a basis for $u^{-1}(0)$. Show that the vectors $v(e_i)$ of index exceeding k are independent and that they constitute a basis for $v(E)$.

(2) Show that there exist linear mappings w of E into E such that

$$u = w \circ v.$$

(3) Let E denote the space of elementary geometry that we identify with the space $\mathbf{R}^3$. By definition, the images of a vector by v and u are the projections

of this vector onto a plane P passing through O and onto a line D passing through O and contained in the plane P. Give the definition of the restriction of w to the plane P. What theorem of elementary geometry do we have here?

313

Let E and F denote two vector spaces of dimension m and n respectively over the same field K. Let e_i denote the vectors of a basis of E (where $i = 1, 2,...,m$).

(1) Show that the set of linear mappings f on E into F constitutes a vector space $\mathscr{L}$ if we define

$$(f+g)(x) = f(x) + g(x),$$
$$(\lambda f)(x) = \lambda f(x).$$

(2) Show that a mapping f is defined when we give the vectors $f(e_i)$.

Show that, for an arbitrary basis $\{\varepsilon_i\}$ for the space F, there exists a linear mapping g of E onto F such that $g(e_i) = \varepsilon_i$ for every i. Determine the dimension of $\mathscr{L}$.

(3) Suppose now that the space F is the space E itself. By definition, the product of two mappings of $\mathscr{L}$ is the composite of these two mappings. Show that $\mathscr{L}$ has a ring structure.

(4) Use the preceding results to show that, for every linear mapping f of E into E, there exists a polynomial A with coefficients in K such that

$$A(f) = \sum_{i=0}^{l} a_i f^i = 0,$$

where f^0 is the identity mapping and 0 represents the zero mapping that maps every vector into the zero vector.

314

Suppose that E is an n-dimensional vector space over the field K. We recall that the linear mappings of E into itself form a ring $\mathscr{L}$ (see exercise 313).

Let f denote a particular element of $\mathscr{L}$ possessing the following property: The images $f^p(x_0)$ of a particular vector x_0 in E constitute a basis for E if we take $1 \leqslant p \leqslant n$.

(1) Show that the mapping f is bijective.

(2) Show that there exist numbers a_p in K such that

$$(f^n + a_{n-1}f^{n-1} + \cdots + a_0 e)(x_0) = 0,$$

where e denotes the identity mapping. From this show that the mapping $f^n + a_{n-1}f^{n-1} + \cdots + a_0 e$ is the zero mapping. (Show that this mapping transforms the basis vectors into the zero vector.)

(3) Suppose that a mapping g commutes with f, that is, that $gf = fg$. Show that it is possible to find numbers b_i in K such that

$$g(x_0) = (b_{n-1}f^{n-1} + b_{n-2}f^{n-2} + \cdots + b_0 e)(x_0)$$

and use this result to show that

$$g = b_{n-1}f^{n-1} + b_{n-2}f^{n-2} + \cdots + b_0 e.$$

315

Let E_{n+1} denote the vector space of polynomials of degree not exceeding n with complex coefficients (the space studied in exercise 309).

(1) Show that the value $A(x_0)$ assumed by the polynomial A at x_0 is a linear form on E_{n+1}.

(2) If the numbers $x_1, x_2,...,x_{n+1}$ are distinct, the linear forms $A(x_1), ..., A(x_{n+1})$ are independent and constitute a basis $\mathscr{B}^*$ of the dual E_{n+1}^* of E_{n+1}.

(3) Show that the Lagrange polynomials relative to the values $x_1, x_2, ..., x_{n+1}$ constitute a basis $\mathscr{B}$ of E_{n+1} that is the dual of the basis $\mathscr{B}^*$ of (2). (Two bases e_i and φ_j of a space E and its dual E^* respectively are said to be dual to each other if $\varphi_j(e_i) = 0$ whenever $j \neq i$ and $\varphi_i(e_i) = 1$.)

(4) Find the basis of E_{n+1}^* that is dual to the basis $1, x, ..., x^n$ of E_{n+1}.

(5) Suppose that $x_0, x_1, ..., x_{n+1}$ are given vectors of the space E_{n+1}. Show that there exist constants $\lambda_1, ..., \lambda_{n+1}$ such that, for every polynomial A of degree not exceeding n,

$$A'(x_0) = \lambda_1 A(x_1) + \lambda_2 A(x_2) + \cdots + \lambda_{n+1} A(x_{n+1}),$$

where A' is the derivative of A. Find explicit expressions for the constants λ_i.

316

A linear mapping f of $\mathbf{R}^3$ into itself is defined by giving the coordinates (X, Y, Z) of the vector $f(u)$ as a function of the coordinates (x, y, z) of the vector u:

$$X = (m-2)x + 2y - z,$$
$$Y = 2x + my + 2z,$$
$$Z = 2mx + 2(m+1)y + (m+1)z.$$

Show that the rank of f is equal to 3 except for particular values of m and determine these particular values. Find the ranks for these values and define the subspace $f(\mathbf{R}^3)$.

317

Let E denote an n-dimensional vector space over a field K and let E^* denote its dual space. Let Φ denote a set of vectors belonging to E and let F denote the vector subspace of E generated by elements of Φ.

(1) Show that the linear forms that vanish for all elements of Φ constitute a vector subspace F^* of E^*.

(2) Show that the zeros that are common to the linear forms of F^* are the vectors of F.

(3) Show that $\dim F + \dim F^* = n$.

(4) Let E_{n+1} denote the vector space of polynomials of degree not exceeding n with complex coefficients. Let us take for F the subspace of polynomials that are multiples of $(x-1)^2 (x-2)^3 (x-3)$. Show that F^* is the set of linear forms l defined by

$$l(A) = \lambda_1 A(1) + \mu_1 A'(1) + \lambda_2 A(2) + \mu_2 A'(2) + v_2 A''(2) + \lambda_3 A(3).$$

318

FIRST PART

Let Γ denote an additive abelian group. Let E denote the set of all finite formal sums $\sum a_\gamma \gamma$, where a_γ is a complex number and $\gamma \in \Gamma$, and assume that the dimension of the vector space E is n. (Alternatively, E may be regarded as the set of all weak mappings $\Gamma \to C$, where, by a weak mapping, we mean a mapping that is zero for all but a finite number of elements in Γ.)

(1) On this same group Γ, let us define multiplication by a real number as the restriction to the Cartesian product $\Gamma \times R$ of the multiplication that is defined, by hypothesis, in the Cartesian product $\Gamma \times C$. In other words, if we denote a single element of Γ by x when it is considered as a vector in E and by x' when we restrict ourselves to multiplication by a real number, the operations are defined by

$$x' + y' = (x+y)',$$
$$rx' = (rx)' \quad \text{if} \quad r \text{ is real.}$$

Show that this defines on Γ a vector space E' over R. Show that the elements x' and $(ix)'$ are independent and that the dimension of E' is $2n$.

(To have a specific example, one may take for E the vector space over C of polynomials of degree not exceeding 2 with complex coefficients.)

(2) Let φ denote the mapping of E' into E' defined by

$$\varphi(x') = (ix)'.$$

Show that φ is an isomorphism of E' onto E' (that is, that φ is linear and bijective) and that $\varphi \circ \varphi = -e$, where e denotes the identity mapping.

SECOND PART

Let F denote a $2n$-dimensional vector space over $\mathbf{R}$.

(1) Show that there exist isomorphisms f of F onto F such that $f \circ f = -e$. (One may treat first the case in which F is the set of complex numbers, which gives the result for $n = 1$, and then treat the general case.)

(2) In the additive abelian group of elements of F, we define multiplication by a complex number $\lambda + i\mu$ by

$$(\lambda + i\mu)x = \lambda x + \mu f(x),$$

where f denotes an arbitrary (but fixed) isomorphism of the type referred to in (1). Show that we thus obtain a vector space E over the field $\mathbf{C}$ of complex numbers. Show that the space E' derived from E as in the first part of this exercise is isomorphic to F and that the dimension of E is n.

319

Consider the differential equation

$$(E) \qquad\qquad\qquad y'' + py' + qy = 0,$$

where p and q are real constants. A real function f defined in $\mathbf{R}$ is a solution of equation (E) if it is everywhere twice differentiable and if its derivatives satisfy equation (E). A function defined from $\mathbf{R}$ into $\mathbf{C}$ is a solution of (E) if its real part and its imaginary part are both solutions of (E).

(1) Show that the solutions of (E) constitute a vector space (S) over the field $\mathbf{C}$. We assert (without proof) that the dimension of this vector space is 2.

(2) Show that if f is a function of (S), its derivative f' is also a function of (S) and show that the mapping l of (S) into (S) that assigns to every function f its derivative f' is linear.

(3) Show that, if l is proportional to the identity mapping, the dimension of (S) is equal to 1. Find the values of p and q for which l is not an isomorphism of (S) onto (S).

(4) Show that

$$l \circ l + pl + qe = 0,$$

where e denotes the identity mapping. Show that there exist numbers r_1 and r_2 such that

$$(l - r_1 e) \circ (l - r_2 e) = 0.$$

Show, finally, that neither of the mappings $l - r_1 e$ and $l - r_2 e$ has a single-valued inverse.

320

Let us define a linear mapping f of R^4 into itself by giving the coordinates X, Y, Z, T of the vector $f(u)$ as a function of the coordinates x, y, z, t of the vector u:

$$
\begin{aligned}
X &= x+y+z-t, \\
Y &= -x+y-z-t, \\
Z &= x-y-z-t, \\
T &= -x-y+z+3t.
\end{aligned}
$$

Determine the rank of f and define the space $f(R^4)$. Show that $f(R^4)$ has no points in common with the domain D defined by

$$
X>0, \qquad Y>0, \qquad Z>0, \qquad T>0,
$$

i.e., that there do not exist values for x, y, z, and t such that these four inequalities are verified.

(2) Let us define a linear mapping f of R^n into R^p by giving the coordinates of the vector $f(u)$, the image of the vector u:

$$
X_i = f_i(u) = \sum_{j=1}^{n} a_{ij}x_j, \qquad\qquad (i = 1, 2,..., p).
$$

Assume that the rank of f is at least equal to $p-1$. Show that a necessary and sufficient condition for the system of inequalities

$$(S) \qquad\qquad\qquad f_i(u)>0 \qquad\qquad\qquad (i = 1, 2,...,p)$$

to have no solution is that there exist positive numbers λ_i, not all zero, such that

$$
\sum_i \lambda_i f_i = 0.
$$

CHAPTER 2

Matrices, determinants, and linear equations

321

Assume in the following that all the numbers mentioned belong to a given field K. Let $M(s)$, $N(t)$, and $P(u)$ be matrices defined by

$$M(s) = \begin{pmatrix} s & 0 \\ 0 & \dfrac{1}{s} \end{pmatrix}, \qquad N(t) = \begin{pmatrix} 1 & 0 \\ t & 1 \end{pmatrix}, \qquad P(u) = \begin{pmatrix} 1 & u \\ 0 & 1 \end{pmatrix},$$

where the number s is different from zero.

Show that a necessary and sufficient condition for a matrix

$$X = \begin{pmatrix} a & b \\ c & d \end{pmatrix}$$

to be the product $M(s)N(t)P(u)$ is that

$$a \neq 0, \qquad ad - bc = 1.$$

If this condition is satisfied, show that the numbers s, t, and u are uniquely determined.

322

Let a, b, c, and d be numbers belonging to a field K. Define

$$A = \begin{pmatrix} a & b \\ c & d \end{pmatrix}.$$

(1) Find the matrix A^2 and show that there exist numbers α and β, expressible as functions of a, b, c, and d such that

$$A^2 - \alpha A - \beta I = 0.$$

What are the cases in which the coefficients α and β are not unique?

(2) Show that the set $\mathscr{A}$ of matrices B that can be put in the form

$$B = uI + vA,$$

where u and v are arbitrary numbers in K and I is the unit matrix, is a commutative ring with unity.

(3) *For this question and the following, suppose that K is the field $\mathbf{R}$ of real numbers.* Show that if B has an inverse B^{-1}, this inverse belongs to $\mathscr{A}$. Under what condition will the ring $\mathscr{A}$ be a field?

(4) Suppose that $a = d = 0$, $b = 1$, $c = -1$. Show that $\mathscr{A}$ is isomorphic to the field of complex numbers.

323

This exercise assumes knowledge of the results of question (2) of the preceding exercise. All the matrices in this exercise are assumed to belong to the ring of 2×2 matrices with complex elements.

(1) Define

$$M = \begin{pmatrix} \alpha & \beta \\ \gamma & \delta \end{pmatrix}$$

and assume that the element β is nonzero. Show that a necessary and sufficient condition for the matrix

$$B = \begin{pmatrix} x & y \\ z & t \end{pmatrix}$$

to commute with M, that is, for BM to be equal to MB, is that there exist a number k such that

$$y = k\beta, \qquad z = k\gamma, \qquad x - t = k(\alpha - \delta).$$

Show that the set of matrices B that commute with M is a two-dimensional vector space over the field $\mathbf{C}$ of complex numbers and is one of the rings $\mathscr{A}$ studied in exercise 322 (taking K as the field $\mathbf{C}$ of complex numbers).

(2) Suppose that the element β in the matrix M is zero. Find an expression for the set of matrices B that commute with M and show that it is also one of the rings studied in exercise 322.

(3) Show that every commutative subring of the ring of 2×2 matrices is a subring of one of the rings $\mathscr{A}$ studied in exercise 322.

324

(1) Let $\mathscr{M}$ denote the ring of 2×2 matrices with complex elements. Let K denote the set of matrices M defined by

$$M = \begin{pmatrix} m & n \\ -\bar{n} & \bar{m} \end{pmatrix},$$

where m and n are arbitrary complex numbers. Show that K is a division ring.

(2) Show that K is not a subspace of the vector space over C of 2×2 matrices with complex elements.

(3) Show that it is possible to define on K a vector space structure over R. Find a basis E_1, E_2, E_3, E_4 for this space and show that multiplication in K can be defined by giving the products $E_i E_j$.

325

(1) Let E denote a set on which an associative operation with a neutral element e is defined. Show that the subset E_1 of elements with inverses is a group. (As an example, the set of 2×2 matrices with complex elements with inverses constitutes a group G.)

(2) Let A denote a matrix belonging to G and let α denote the mapping of C^2 into itself defined by

$$\begin{pmatrix} x' \\ y' \end{pmatrix} = A \begin{pmatrix} x \\ y \end{pmatrix} = \begin{pmatrix} a & b \\ c & d \end{pmatrix} \begin{pmatrix} x \\ y \end{pmatrix}.$$

Show that the ratio x'/y' is a homographic function h_A of the ratio x/y. Let H denote the set of homographic functions (not decomposed into partial fractions). If we define a product in H as the composition of two functions, show that H is a group and that the mapping φ defined by $\varphi(A) = h_A$ is a group homomorphism, that is (see exercise 114 of Part 1),

$$\varphi(A) \, \varphi(B) = \varphi(AB).$$

Find the kernel of φ, that is, the set $\varphi^{-1}(e)$, where e denotes the neutral element of H, and more generally the sets $\varphi^{-1}(h)$ relative to an arbitrary homographic function h.

326

Let $\mathscr{A}$ denote the ring of $n \times n$ matrices whose elements belong to a field K.

(1) Show that the set E of matrices that commute with a given matrix A is a subring of $\mathscr{A}$.

(2) Suppose that A and B commute with each other and that P and Q are two polynomials with coefficients in K. Show that the matrices $P(A)$ and $Q(B)$ commute.

(3) If A has an inverse and if A commutes with B, show that the matrices A^{-1} and B commute with each other.

327

Show that if the rank of an $m \times n$ matrix $A = (a_{ij})$ is 1, there exist numbers u_i and v_j, not all zero, such that

$$a_{ij} = u_i v_j,$$

where a_{ij} denotes the element in the ith row and jth column. (These elements are assumed to belong to a field K.)

328

Throughout this exercise, all the numbers referred to are assumed to belong to a single field K.

An $n \times n$ matrix $T = (t_{ij})$ is called a triangular matrix if $t_{ij} = 0$ whenever $i > j$, that is, whenever all the elements below the principal diagonal are zero:

$$T = \begin{vmatrix} t_{11} & t_{12} \cdots \cdots \cdots \cdots & t_{1n} \\ 0 & t_{22} \cdots \cdots \cdots \cdots & t_{2n} \\ \cdot & \cdots \cdots \cdots \cdots \cdots \\ \cdot \\ 0 \cdots \cdots \cdots 0 & t_{ii} & t_{in} \\ \cdot & \cdots \cdots \cdots \cdots \cdots \\ \cdot \\ 0 \cdots \cdots \cdots \cdots \cdots 0 & t_{nn} \end{vmatrix}.$$

(1) Show that the $n \times n$ triangular matrices constitute a ring $\mathcal{A}$. Is this ring commutative?

(2) Let θ denote the linear mapping (of K^n into K^n) corresponding to T, that is, the linear mapping the matrix of which relative to the canonical basis e_i (for $i = 1, 2,...,n$) of K^n is T.

Show that a necessary and sufficient condition for the matrix T to be triangular is that the vector subspaces E_p, generated by the vectors $e_1, e_2,...,e_p$, are stable in the mapping θ, that is, that $\theta(E_p) \subset E_p$. Show that this characterization enables us to establish without any calculation that the set of triangular matrices constitutes a ring.

(3) Let X denote an $n \times 1$ matrix and let x_i (for $i = 1,...,n$) denote its elements. Let T denote a triangular matrix. Discuss the matrix equation $TX = 0$. What is the rank of the mapping θ corresponding to T?

329

The elements of the matrices considered in this question are assumed to belong to a field K.

Let M denote an $n \times n$ matrix built up from four matrices M_{ij}, as shown in the following diagram:

$$
\begin{array}{cc}
 & \overset{p}{} \quad \overset{q}{} \\
\begin{array}{c} p \\ q \end{array} \left(\begin{array}{c|c} M_{11} & M_{12} \\ \hline M_{21} & M_{22} \end{array} \right)
\end{array}
$$

where the matrix M_{11} has p rows and p columns, M_{12} has p rows and q columns, M_{21} has q rows and p columns, and M_{22} has q rows and q columns, the sum of p and q being n. If N is another $n \times n$ matrix of the same form, show that the product MN has the same formal expression as it would have if the matrices M_{ij} and N_{ij} were numbers.

330

In this exercise, we assume familiarity with the results of 329. It should be recalled that the set of real matrices of the form

$$
\begin{pmatrix} u & v \\ -v & u \end{pmatrix}
$$

constitutes a field $\mathscr{C}$ that is isomorphic to the field of complex numbers (cf. exercise 322, (4)).

In the ring $\mathscr{N}$ of 4×4 matrices with real elements consider the set $\mathscr{M}$ of matrices M defined by

$$
M = \begin{pmatrix}
a & b & c & d \\
-b & a & -d & c \\
c' & d' & a' & b' \\
-d' & c' & -b' & a'
\end{pmatrix},
$$

where a, b, c, d, a', b', c', and d' are real numbers.

(1) Show that $\mathscr{M}$ is an eight-dimensional vector space over R and that it is a ring.

(2) Show that this ring is isomorphic to the ring $\mathscr{A}$ of 2×2 matrices with complex coefficients.

(3) Is it possible to define on $\mathscr{A}$ a vector space structure that is isomorphic to that of the vector space $\mathscr{M}$?

(4) Let X denote a 4×1 matrix. Under what conditions does the equation $MX = 0$ have a nonzero solution?

331

In the following, the coefficients and the unknowns are complex numbers.
   (1) Solve and discuss the system (S):

$$\begin{cases} \lambda x_i + a_i x_{n+1} = b_i & (i = 1, 2,..., n) \\ \sum_{i=1}^{n} a_i x_i + \lambda x_{n+1} = b_{n+1}. \end{cases}$$

Evaluate the determinant D_n of the coefficient matrix of this system and find the values of λ for which the system (S) has a unique solution.

332

Generalization of a Vandermonde system

Solve and discuss the system of equations

$$x_1 + x_2\lambda_i + x_3\lambda_i^2 + x_4\lambda_i^4 + x_5\lambda_i^6 = 0 \qquad (i = 1, 2, 3, 4)$$
$$x_1 + x_2\mu + x_3\mu^2 + x_4\mu^4 + x_5\mu^6 = 1.$$

One may use the polynomial A defined by

$$A(u) = x_1 + x_2 u + \cdots + x_5 u^6$$

and then follow a procedure analogous to that shown in PZ, Book II, Chapter IX, part 3, § 1.

333

We define the following four linear forms on R^4:

$$\begin{aligned} X &= & ay + bz + ct, \\ Y &= ax & + cz + bt, \\ Z &= bx + cy & + at, \\ T &= cx + by + az. \end{aligned}$$

   (1) Find the rank of the system of forms X, Y, Z, T (that is, the rank of the mapping of R^4 into R^4 that they define).
   (2) What is the condition under which these forms will be independent? (Evaluate the determinant of the system.)
   (3) Discuss the existence of solutions of the system

$$X = a+b+c, \qquad Y = a, \qquad Z = b, \qquad T = c.$$

334

Show that the determinant of an antisymmetric matrix of odd order is zero. (A square matrix $A = (a_{ij})$ is said to be *antisymmetric* if

$$a_{ij} + a_{ji} = 0$$

for all i and j.)

335

Determinants of matrices with cyclically permuted elements

Consider a determinant D of a matrix with complex elements, each row of which is obtained from the preceding row by a cyclic permutation of elements:

$$D = \begin{vmatrix} a_1 & a_2 \dots & a_n \\ a_n & a_1 \dots a_{n-1} \\ a_{n-1} & a_n \dots a_{n-2} \\ \dots \dots \dots \dots \\ a_2 & a_3 \dots & a_1 \end{vmatrix}.$$

(1) Show that the determinant D, treated as a polynomial in a single variable a_1, for example, is divisible by

$$a_1 + a_2 u + \cdots + a_n u^{n-1},$$

where u is any of the nth roots of unity. Express D as a product of n factors.

(2) Evaluate the determinant D if $a_k \equiv k$.

(3) Evaluate the determinant D if

$$a_k = \begin{cases} 1 & \text{for} & k \leqslant p \\ 0 & \text{for} & k > p, \end{cases}$$

where p is a given number not exceeding n. You may use the following result: If p is prime to n, the numbers u_h^p are all distinct and, except possibly for order, they are equal to the u_h, where the u_h, for $h = 0, 1, 2, \dots, n-1$, are the n nth roots of unity.

336

A Vandermonde determinant is a determinant of the form $D = |d_{ij}|$ defined by $d_{ij} = \lambda_i^{j-1}$ for $j \geqslant 1$. We recall that

$$D = \prod_{i > j} (\lambda_i - \lambda_j)$$

(see PZ, Book II, Chapter IX, part 3, § 2).

Let us denote by P_1, P_2,..., P_{n+1} a set of $n+1$ polynomials of the form

$$P_i(x) = \sum_{j=1}^{n+1} a_{ij} x^{j-1},$$

where the a_{ij} are complex numbers. Let x_1, x_2,..., x_n denote $n+1$ distinct complex numbers. Let A, X, and Y be matrices defined by

$$A = (a_{ij}), \qquad X = (x_j^{i-1}), \qquad Y = (P_i(x_j)),$$

where the (outer) parentheses denote the element in the ith row and jth column. Show that $Y = AX$.

If the polynomials P_i are of the form $P_i(x) = (x+i)^n$, find as simple an expression as possible for the value of the determinant

$$\Delta_n = \begin{vmatrix} 1^n & 2^n \ldots\ldots (n+1)^n \\ 2^n & 3^n \ldots\ldots (n+2)^n \\ \ldots\ldots\ldots\ldots\ldots\ldots \\ (n+1)^n & (n+2)^n \ldots (2n+1)^n \end{vmatrix}.$$

337

Let $A = (a_{ij})$ denote an $n \times n$ matrix with complex elements and let A_{ji} denote the coefficient of a_{ij} (the coefficient of a_{ij} in the expansion of the determinant of A in terms of the elements of the jth column). We recall that, if B is the matrix (A_{ij}), then

$$BA = AB = (\det A)I$$

(see PZ, Book II, Chapter IX, part 2, § 2).

Suppose that the rank of A is $n-1$, that of B is 1, and that the rank of A is at most $n-2$. Show that B is the zero matrix.

338

Solve and discuss the system (S):

$$\frac{x}{a+\lambda_i} + \frac{y}{b+\lambda_i} + \frac{z}{c+\lambda_i} + \frac{t}{d+\lambda_i} - 1 = 0 \qquad (i = 1, 2, 3, 4).$$

Interpret this system by using the function f defined by

$$f(u) = \frac{x}{a+u} + \frac{y}{b+u} + \frac{z}{c+u} + \frac{t}{d+u} - 1.$$

339

Evaluate the determinant

$$
D_n =
\begin{vmatrix}
a_1+b_1 & b_1 & \ldots & b_1 \\
b_2 & a_2+b_2 & \ldots & b_2 \\
\ldots\ldots\ldots\ldots\ldots\ldots\ldots\ldots\ldots\ldots\ldots \\
b_n & b_n & \ldots & a_n+b_n
\end{vmatrix}
$$

(for example, by induction on n).

CHAPTER 3

Eigenvalues, reduction of matrices

340

Let A denote an $n \times n$ matrix with elements belonging to a field K. Let P denote the characteristic polynomial of A and assume that $P(0)$ is nonzero.

Show that A has an inverse A^{-1} and that the characteristic polynomial R of A^{-1} is defined by

$$R(\lambda) = \frac{(-1)^n \lambda^n}{P(0)} P\left(\frac{1}{\lambda}\right).$$

One may, for example, examine the determinant of $A(A^{-1} - \lambda I)$. If K is a subfield of C, one may use a triangular matrix B that is similar to A.

341

Let $\mathscr{A}$ denote the ring of $n \times n$ matrices with elements belonging to a field K. Consider the set E of matrices $A = (a_{ij})$ such that the sum

$$\sum_i a_{ij}$$

of the terms in a single column is a number $s(A)$ that is independent of j.

(1) Show that a necessary and sufficient condition for A to belong to E is that

$$VA = \lambda V,$$

where V denotes the row matrix $(1, 1, ..., 1)$.

(2) Show that E is a subring of $\mathscr{A}$ and that, for any nonsingular matrix A in E, the inverse matrix A^{-1} belongs to E.

(3) Show that $s(A)$ is a characteristic value of the matrix A.

(4) Suppose that the elements of A are positive numbers. Show that the characteristic values of A do not exceed $s(A)$ in absolute value.

342

(1) Find the characteristic values and the eigenvectors of the matrix

$$A = \begin{pmatrix} 1 & 3 & 0 \\ 3 & -2 & -1 \\ 0 & -1 & 1 \end{pmatrix}.$$

(2) Find a nonsingular matrix T such that the matrix $T^{-1}AT$ will be a diagonal matrix. Find the matrix T^{-1} and check your calculations by taking the product $T^{-1}AT$.

343

Find the characteristic values and the eigenvectors of the matrix

$$A = \begin{pmatrix} 1 & 1 & 0 \\ -1 & 2 & 1 \\ 1 & 0 & 1 \end{pmatrix}.$$

Find a nonsingular matrix T and its inverse T^{-1} such that $T^{-1}AT$ will be a diagonal matrix D. Write out D.

344

In this exercise, the matrices in question are assumed to belong to the ring $\mathscr{A}$ of $n \times n$ matrices with complex elements.

As we know, if we assign to every matrix M the linear mapping $\tilde{M}$ of C^n into C^n the matrix of which relative to the canonical basis of C^n is A, we define an isomorphism φ of the ring $\mathscr{A}$ onto the ring of linear mappings of C^n into itself.

We denote by A a given matrix the characteristic polynomial f of which has all distinct zeros.

(1) Show that if V is an eigenvector of A, then V belongs to the kernel of $f(\tilde{A})$ and show that $f(A) = 0$.

(2) Show that a necessary and sufficient condition for A and B to commute, that is, for AB to be equal to BA, is that the eigenvectors of A be eigenvectors of B.

(3) Show that, if B commutes with A, there exists a polynomial g of degree less than n with complex coefficients such that $B = g(A)$.

(4) Show that a necessary and sufficient condition for B to be nonsingular is that the polynomials f and g be relatively prime.

345

Let A be a 2×2 matrix with complex elements and with a double characteristic value λ. Show that there exists a matrix B similar to A (that is, $B = T^{-1}AT$) and equal to one of the two matrices

$$\begin{pmatrix} \lambda & 0 \\ 0 & \lambda \end{pmatrix}, \quad \begin{pmatrix} \lambda & 1 \\ 0 & \lambda \end{pmatrix}.$$

Find B^n.

346

(1) Find the characteristic values and the eigenvectors of the matrix

$$A = \begin{pmatrix} 8 & -1 & -5 \\ -2 & 3 & 1 \\ 4 & -1 & -1 \end{pmatrix}.$$

(2) Find a basis such that the transformed matrix of A, which we denote by B, will be triangular and find the matrix B.

(3) Show that, among all the bases that satisfy the conditions of (2), there exists at least one such that the matrix B has one and only one nonzero element off the principal diagonal and that this term is equal to 1. Find the corresponding matrix B_0.

(4) Find B_0^n. Show, without carrying out the calculations, how one can find A^n.

347

Define an $n \times n$ matrix A by

$$\alpha_{p,p-1} = 1, \qquad \alpha_{p,n} = a_{p-1}, \qquad \alpha_{ij} = 0 \quad \text{if} \quad i-j \neq 1 \quad \text{and} \quad j \neq n,$$

where i denotes the row and j the column of the element α_{ij} and where the elements a_{p-1} in the last column are given complex numbers:

$$A = \begin{vmatrix} 0. & \ldots\ldots\ldots & 0 & a_0 \\ 1 & & & a_1 \\ & \diagdown & 0 & \vdots \\ & & \diagdown & \vdots \\ 0 & & \diagdown & \vdots \\ & & 1 & a_{n-1} \end{vmatrix}.$$

Let $\tilde{A}$ denote the linear mapping (of C^n into itself) corresponding to A and denote by f the polynomial defined by

$$f(t) = t^n - a_{n-1}t^{n-1} - a_{n-2}t^{n-2} - \cdots - a_0.$$

(1) Find the vectors $\tilde{A}(e_i)$ and the vectors $\tilde{A}^i(e_1)$ for $1 \leqslant i \leqslant n$. Show that $f(\tilde{A})$ maps all the vectors e_i into 0 and show that $f(A)$ is zero. (The vectors e_l are the vectors of the canonical basis of C^n.)

(2) Show that the set $\mathscr{M}$ of matrices M that can be defined by $M = p(A)$, where p belongs to the set of polynomials with complex coefficients, constitutes a ring. Show that the set $\mathscr{I}$ of polynomials p such that the matrix $p(A)$ is zero is an ideal in the ring of polynomials and that this ideal contains a polynomial of degree n.

(3) Show that the characteristic polynomial of A is $(-1)^n f$.

(4) Show that a necessary and sufficient condition for the matrix $p(A)$ to be nonsingular is that the polynomials p and f be relatively prime.

CHAPTER 4

Euclidean spaces, quadratic forms

348

In an n-dimensional vector space, the determinant of the coordinates of n vectors is referred to as the determinant of those vectors relative to a given basis.

(1) Show that in a change of basis, the determinant of n vectors is multiplied by a factor that is independent of the vectors themselves.

(2) Show that, in a Euclidean space, the absolute value of the determinant of n vectors relative to an orthonormal basis is independent of the basis in question. Show that, in choosing a particular orthonormal basis, the absolute value of the determinant of n vectors does not exceed the product of the magnitudes of these vectors.

349

In an n-dimensional Euclidean space, consider a set of p vectors $V_1,...,V_p$, where p does not exceed n. To every element $X = (x_1,...,x_p)$ of R^p, we assign the number

$$\omega(X) = \left\| \sum_{i=1}^{p} x_i V_i \right\|^2.$$

Show that $\omega(X)$ is positive definite if and only if the vectors V_i are linearly independent. Use the preceding result to show that a necessary and sufficient condition for the vectors V_i to be independent is that the determinant D of the matrix $A = V_i \cdot V_j$ be nonzero. (The element in the ij position of the matrix A is the scalar product $V_i \cdot V_j$.) Show that D is nonnegative.

350

Denote by g a symmetric bilinear form defined on $R^n \times R^n$ and such that $g(x,x) \geq 0$ for every value of the vector x, including the value 0 (this quadratic form is not necessarily strictly positive definite).

(1) Show that

$$[g(x, y)]^2 \leq g(x, x)g(y, y).$$

(2) A necessary and sufficient condition for g to be nondegenerate (that is, for the linear form on R^n that maps y into $g(x, y)$, where x is fixed, to be nonzero when x is not the zero vector) is that the quadratic form $g(x, x)$ be positive definite.

(3) If g is degenerate, show that the elements x at which $g(x, x)$ vanishes, constitute a vector subspace of R^n.

(4) If g is nondegenerate, show that the linear mappings α of R^n into R^n that leave $g(x, y)$ invariant (that is, such that $g(\alpha x, \alpha y) = g(x, y)$) constitute a group G, where the group operation is the composition of mappings.

351

The second part of this exercise assumes a knowledge of the theory of the definite integral.

FIRST PART

Consider a Euclidean space E and a vector subspace F of E (these spaces being of finite or infinite dimension).

(1) Let x be a vector in E. By examining the distance from x to the vector $x_0 \times \lambda y$, show that a necessary and sufficient condition for x_0 in F to be at a minimum distance from x is that $x - x_0$ be orthogonal to all the vectors in F.

Show that two vectors in F, both at minimum distance from x, cannot exist.

(2) If F is finite-dimensional, show that there will indeed exist a vector x_0 in F at minimum distance from x.

SECOND PART

(1) Show that it is possible to define a scalar product by

$$x \cdot y = \int_0^{2\pi} x(t) y(t) \, dt$$

in the vector space over R of real functions that are continuous on $[0, 2\pi]$. What is the norm corresponding to this scalar product?

We shall denote by E the (infinite-dimensional) Euclidean space thus obtained.

(2) In this space, show that the functions 1, $\cos kt$, $\sin ht$ (where k and h are arbitrary natural numbers) are mutually orthogonal.

(3) Let F be the vector space of trigonometric polynomials of order n (that is, linear combinations of 1, $\cos kt$, $\sin ht$, where $k \leq n$ and $h \leq n$). Find the (trigonometric) polynomial x_0 of minimum distance from a function x.

352

This exercise assumes a knowledge of integration theory.

Let E denote the vector space over R of polynomials of degree not exceeding n and with real coefficients. We define the scalar product of two polynomials P and Q by

$$P \cdot Q = \tfrac{1}{2} \int_{-1}^{1} P(t)Q(t)\, dt.$$

(1) Show that E is a Euclidean space of dimension $n+1$. Show that there exist elements U_p of degree p (exactly) constituting an orthonormal basis (for example, by using the Gram–Schmidt orthonormalization procedure). Show that all these bases can be derived from any one of them by replacing U_p with $\varepsilon_p U_p$, where $\varepsilon_p = \pm 1$.

(2) Evaluate U_0, U_1, and U_2.

(3) Show that the polynomials U_p are characterized up to sign by the three properties

$$\deg U_p = p, \qquad \|U_p\| = 1.$$

U_p is orthogonal to every polynomial of degree less than p.
Show that

$$U_p = u_p \frac{d^p}{dt^p}(t^2 - 1)^p,$$

where u_p is a constant and evaluate $|u_p|$, for example, by p applications of the formula for integration by parts (see PZ, Book III, Chapter IV, part 3, § 4).

(4) Show that the zeros of the polynomial U_p are real and distinct and that they belong to the interval $]-1, 1[$.

353

Let φ and ψ denote two symmetric bilinear forms defined on $R^n \times R^n$ and denote by ω and Ω the corresponding quadratic forms. Assume that the quadratic form ω is positive definite.

(1) Show that

$$\varphi(x, y) = \frac{\omega(x+y) - \omega(x-y)}{4}, \qquad \psi(x, y) = \frac{\Omega(x+y) - \Omega(x-y)}{4}.$$

(2) Show that there exists a basis for R^n with respect to which the quadratic forms ω and Ω are both reduced to the sums of squares. One may use the fact that ω is positive definite in order to reduce this problem to the problem of reducing a single quadratic form via the orthonormal bases.

(3) Use the reduced forms obtained in (2) to find the values λ for which the bilinear form $\psi - \lambda \varphi$ is degenerate (that is, the values for which there exist nonzero vectors y such that $\psi(x, y) - \lambda \varphi(x, y) = 0$ for every x). Show that these numbers are the roots of the equation

$$\det (B - \lambda A) = 0,$$

where A and B are the matrices of φ and ψ in an arbitrary basis for R^n.

Finally, show that, if the roots of this equation are all distinct, one can determine directly the basis mentioned in (2).

(4) Show that the result of (2) does not hold if either of the two forms ω and Ω fails to be positive definite. For example, in R^2 take

$$\omega(x) = x_1^2 - x_2^2, \qquad \Omega(x) = 2x_1 x_2,$$

where x_1 and x_2 are the coordinates of x.

354

Let $\tilde{A}$ denote a linear mapping of an n-dimensional Euclidean space E into itself. Let A be its matrix relative to a given basis in E. Show that a necessary and sufficient condition for the transforms by $\tilde{A}$ of n independent vectors $u_1, u_2, ..., u_n$ to be mutually orthogonal is that, relative to the basis $u_1, ..., u_n$, the quadratic form

$$x \to ||\tilde{A}(x)||^2$$

reduce to a sum of squares. From this, show that there exists at least one system of n unit vectors $u_1, u_2, ..., u_n$ that are mutually orthogonal and whose transforms by $\tilde{A}$ are mutually orthogonal.

355

Let φ denote a symmetric bilinear form on $C^n \times C^n$. We recall that giving the quadratic form ω associated with φ is sufficient to determine φ.

(1) Show that the set E of vectors y such that $\varphi(x, y) = 0$ for every x is a vector subspace of C^n.

(2) Show that the form ω is given by

$$\omega(x) = \sum_{i=1}^{p} l_i^2(x),$$

where the linear forms l_i are independent.

Give the corresponding expression for $\varphi(x, y)$ and show that the number p of forms l_i is equal to $n - \dim E$. From this, show that the number p does not depend on the decomposition chosen.

(3) Show by induction on n that the quadratic for m ω can always be decomposed into a sum of squares, as was assumed in (2).

356

Consider a quadratic form ω on R^n. Considerations analogous to those made in exercise 355 enable one to show the following results, which we shall accept:

(a) There exist independent forms f_i such that

$$\omega = \sum_{i=1}^{p} \varepsilon_i f_i^2(x), \qquad \text{where} \qquad \varepsilon_i = \mp 1;$$

(b) The number p of forms f_i is the same for all decompositions of ω of the preceding type.

Show that the number of coefficients ε_i that are equal to 1 is the same for all decompositions of this form (and hence that the number of coefficients ε_i that are equal to -1 is also the same). For this, one may write the equality of two sums of squares and study the dimension of the v ector space of the zeros of these two sums.

Solutions

CHAPTER 1

301

First question:

R^4 is a 4-dimensional vector space. For a set of vectors equal in number to the dimension of a vector space, in the present case 4, to constitute a basis for the space, it is sufficient that either of the following propositions be true:

(1) the four vectors are independent,

(2) the four vectors span the space.

We shall establish that the second property holds. To do this, let us show that the vectors e_1, e_2, e_3, and e_4 of the canonical basis of R^4 are linear combinations of x_1, x_2, x_3, and x_4. The definition of e_1, e_2, e_3, and e_4 and the rules for calculating with the elements of a vector space enable us to write the two systems of equations:

$$(1) \begin{cases} x_1 = e_1 + e_2 + 2e_3 + e_4 \\ x_2 = e_1 - e_2 \quad\quad + e_4 \\ x_3 = \quad\quad\quad - e_3 + e_4 \\ x_4 = e_1 + 2e_2 + 2e_3 \end{cases} \Rightarrow (2) \begin{cases} 2e_1 = x_1 + x_2 - 2(e_3 + e_4) \\ 2e_2 = x_1 - x_2 - 2e_3 \\ x_3 = \quad\quad\quad - e_3 + e_4 \\ x_4 = e_1 + 2e_2 + 2e_3. \end{cases}$$

If we replace e_1 and e_2 with their expressions given by the first two equations of (2), we conclude that

$$x_3 = -e_3 + e_4,$$

$$x_4 = -e_3 - e_4 + \frac{3x_1 - x_2}{2},$$

which shows that e_3 and e_4 are linear combinations of x_1, x_2, x_3, and x_4. By using the first two of the equations in the system (2), we can show that the same is true of e_1 and e_2.

Second question:

We easily see that

$$x = e_1 + e_2 + e_3 + e_4 = x_1 - e_3$$

$$= x_1 + \frac{x_3 + x_4}{2} - \frac{3x_1 - x_2}{4} = \frac{x_1 + x_2 + 2x_3 + 2x_4}{4}.$$

Therefore, the coordinates are $\frac{1}{4}$, $\frac{1}{4}$, $\frac{1}{2}$, and $\frac{1}{2}$.

REMARKS: (1) By calculating the coordinates of x in the canonical basis by using its expression relative to the basis x_1, x_2, x_3, x_4, we obtain a check on our calculations.

(2) One can establish the independence of x_1, x_2, x_3, and x_4 by showing that the vector equation

$$\lambda_1 x_1 + \lambda_2 x_2 + \lambda_3 x_3 + \lambda_4 x_4 = 0$$

has only the solution $\lambda_1 = \lambda_2 = \lambda_3 = \lambda_4 = 0$. However, this method does not facilitate the calculation of new coordinates of x, as does the method that we used.

302

First question:

The vectors u, v, and w, which are linear combinations of a, b, and c, belong to the vector subspace F generated by a, b, and c. Hence, the vector subspace G generated by u, v, and w is contained in F (since every linear combination of u, v, and w is a linear combination of a, b, and c). On the other hand,

$$u + v + w = 2(a + b + c),$$

$$\begin{cases} a = \dfrac{u+v+w}{2} - u = \dfrac{v+w-u}{2}, \\[3mm] b = \dfrac{u+v+w}{2} - v = \dfrac{u+w-v}{2}, \\[3mm] c = \dfrac{u+v+w}{2} - w = \dfrac{u+v-w}{2}. \end{cases}$$

The vectors a, b, and c are linear combinations of u, v, and w. The vector subspace F is contained in G:

$$F \subset G \quad \text{and} \quad G \subset F \;\Rightarrow\; F = G.$$

Second question:

The statements

a, b, and c are independent   and   dim $F = 3$

are equivalent. Similarly, the statements

u, v, and w are independent   and   dim $G = 3$

are equivalent. We have seen that F and G are identical. Therefore,
a, b, and c are independent $\Leftrightarrow$ dim $F =$ dim $G = 3 \Leftrightarrow u, v,$ and w are independent.

REMARK: The above result will remain valid if we assume that E is a vector space over R or Q. However, this result is not general. For example, it is not true if the field K of coefficients is such that

$$1 + 1 = 0$$

(for example, if K is the field of integers mod 2). In this case, we see that u, v, and w are dependent since

$$u + v + w = 0.$$

303

The functions f_n are independent if the coefficients in every linear combination of the f_n that is equal to the neutral element of the vector space are zero. It is important to note that a linear combination of elements of a vector space is by definition a linear combination of a finite number of these. Therefore, if we denote by N the maximum index of the f_n that occur in the linear combination in question, we can write this combination in the form

$$\sum_1^N \lambda_n f_n.$$

Consider a linear combination of the f_n that is equal to the neutral element v of the space, that is, the zero function:

$$\sum_1^N \lambda_n f_n = v \qquad \text{or} \qquad \sum_1^N \lambda_n \sin^n t = 0 \qquad \forall t.$$

For showing that this condition implies that all the λ_n are zero, there are several procedures. For example,

(1) The polynomial $\sum_1^N \lambda_n u^n$ vanishes for all numbers u between -1 and 1.

There are infinitely many zeros. Therefore, this polynomial is the zero polynomial, and all the λ_n are equal to 0.

(2) Suppose that not all the λ_n are zero. Denote by λ_h the nonzero coefficient of smallest index. Then, we can write the given condition in the form

$$\sin^h t(\lambda_h + \lambda_{h+1} \sin t + \cdots + \lambda_N \sin^{N-h} t) = 0.$$

We know that $|\sin t| \geqslant |\sin^K t|$ for every positive integer K. Therefore,

$$|\lambda_{h+1} \sin t + \cdots + \lambda_N \sin^{N-h} t| \leqslant |\sin t| \, (|\lambda_{h+1}| + \cdots + |\lambda_{N-h}|)$$
$$|\lambda_h + \cdots + \lambda_N \sin^{N-h} t| \geqslant |\lambda_h| - |\sin t| \, (|\lambda_{h+1}| + \cdots + |\lambda_{N-h}|).$$

Thus, if $|\sin t|$ is small enough, the factor

$$\lambda_h + \cdots + \lambda_N \sin^{N-h} t$$

is nonzero, the factor $\sin^h t$ is also nonzero, and the equation is not satisfied.

304

First question:

A subset of E is a vector subspace if every linear combination of two elements of that subset is also an element of that subset. Therefore, let us set

$$z = x+y \qquad (x \in A \qquad \text{and} \qquad y \in B),$$
$$z' = x'+y' \qquad (x' \in A \qquad \text{and} \qquad y' \in B).$$

If λ and λ' are two numbers in K, we see that

$$\lambda z + \lambda' z' = (\lambda x + \lambda' x') + (\lambda y + \lambda' y').$$

The sum $\lambda x + \lambda' x'$ is an element of the subspace A, and $\lambda y + \lambda' y'$ is an element of the subspace B. Therefore, $\lambda z + \lambda' z'$ is an element of $A + B$.

Second question:

If $A \cap B$ consists only of the zero element, let us take

$$z = x+y = x'+y' \qquad \text{where} \qquad x \in A, \, x' \in A, \, y \in B, \, y' \in B.$$

From this, we deduce that $x - x' = y' - y$, where $x - x' \in A$ and $y' - y \in B$. The vector $x - x' = y' - y$ belongs to $A \cap B$ and hence is the zero element. The decomposition $z = x+y$ is unique, and we say then that $A + B$ is the direct sum of A and B.

If $a_1, \ldots, a_n$ and $b_1, \ldots, b_m$ are bases of A and B respectively and if z denotes an element of $A + B$, we see that

$$z = x+y = \sum_1^n \alpha_i a_i + \sum_1^m \beta_i b_i.$$

The coefficients α_i and β_i are unique. Otherwise, we would have two decompositions $z = x+y = x'+y'$. Therefore, the vectors $a_1,..., a_n, b_1,..., b_m$ constitute a basis for $A+B$. (A set of vectors e_i constitutes a basis for a vector space if every element of the space can be written in exactly one way as a linear combination of the vectors e_i.)

If $A \cap B$ consists of vectors other than the zero vector, consider a subspace A' complementary to $A \cap B$ in A (that is, a subspace A' such that A is the direct sum of $A \cap B$ and A'):

$$x \in A \Rightarrow x = u+x' \qquad \text{where} \qquad u \in A \cap B \qquad \text{and} \qquad x' \in A';$$

$$y \in B \Rightarrow x+y = x'+(u+y) \qquad \text{where} \qquad x' \in A' \qquad \text{and} \qquad u+y \in B.$$

Therefore, the subspace $A+B$ is contained in $A'+B$. Since it obviously contains $A'+B$, these two subspaces are identical.

The sum $A'+B$ is direct:

$$y \in A' \cap B \Rightarrow y \in A \cap B \Rightarrow y \in A' \cap (A \cap B) \Rightarrow y = 0.$$

Thus,

$$\dim A+B = \dim A' + \dim B = (\dim A - \dim A \cap B) + \dim B.$$

(The dimension of A' is obtained by writing that A is the direct sum of A' and $A \cap B$.)

305

First question:

Let us verify that

$$x \in E \qquad \text{and} \qquad y \in E \Rightarrow \lambda x + \mu y \in E.$$

We denote the coefficients in the expression for x by a_k and b_k and we denote those in the expression for y by α_k and β_k. Then,

$$(\lambda x + \mu y)(t) = \lambda x(t) + \mu y(t)$$

$$= \lambda [a_0 + \sum_{k=1}^{n} (a_k \cos kt + b_k \sin kt)] +$$

$$+ \mu [\alpha_0 + \sum_{k=1}^{n} (\alpha_k \cos kt + \beta_k \sin kt)]$$

$$= (\lambda a_0 + \mu \alpha_0) + \sum_{k=1}^{n} \{(\lambda a_k + \mu \alpha_k) \cos kt +$$

$$+ (\lambda b_k + \mu \beta_k) \sin kt\}.$$

The function $\lambda x + \mu y$ is indeed a function in E.

Second question:

The result is obvious if $m = 0$ since $x(t) = a_0$.

Let us suppose that the property is established for E_{m-1}. We denote by a_k and b_k the coefficients of the zero function, which is an element of E_m and which we now denote by x. The function x is infinitely differentiable and its derivatives are all zero. Therefore, the function $x'' + m^2 x$ is also the zero function. Now,

$$(x'' + m^2 x)(t) = m^2 a_0 + \sum_{k=1}^{m-1} (m^2 - k^2)(a_k \cos kt + b_k \sin kt).$$

(The linear combination chosen caused the coefficients of $\cos mt$ and $\sin mt$ to vanish.)

The function $x'' + m^2 x$ is the zero function. It is an element of E_{m-1}. All its coefficients are zero. Therefore,

$$x(t) = a_m \cos mt + b_m \sin mt.$$

By assigning to t the values 0 and $\pi/2m$, we obtain $a_m = b_m = 0$.

Third question:

The result of section 2 shows that the functions that map t into 1, $\cos kt$, and $\sin kt$ are independent. (If a linear combination of these functions is the zero function, the coefficients in that linear combination are zero.) These functions are the generators of E by the definition of E. Therefore, they constitute a $(2n+1)$-dimensional basis for E.

Fourth question:

Let x_1 and x_2 denote two elements of E and let λ_1 and λ_2 denote two real numbers. For every value t,

$$[u(\lambda_1 x_1 + \lambda_2 x_2)](t) = (\lambda_1 x_1 + \lambda_2 x_2)(t + \tfrac{1}{4}\pi) = \lambda_1 x_1(t + \tfrac{1}{4}\pi) + \lambda_2 x_2(t + \tfrac{1}{4}\pi)$$

$$= \lambda_1 [u(x_1)](t) + \lambda_2 [u(x_2)](t) = [\lambda_1 u(x_1) + \lambda_2 u(x_2)](t),$$

that is,

$$u(\lambda_1 x_1 + \lambda_2 x_2) = \lambda_1 u(x_1) + \lambda_2 u(x_2).$$

Thus, the mapping u is a linear mapping.

We see easily by induction on the integer p that

$$[u^p(x)](t) = x(t + p\tfrac{1}{4}\pi),$$

from which we get

$$[u^8(x)](t) = x(t+2\pi) = x(t),$$

that is, the mapping u^8 is the identity mapping.

If x is an element of the kernel of u, we have

$$u(x) = 0 \Rightarrow u^7[u(x)] = u^8 x = x = 0.$$

The kernel is zero, and therefore the image $u(E)$ is the space E itself: For a linear mapping of a finite-dimensional space X into a space of the same dimension, in particular into itself, to be bijective, it is sufficient that it be injective or that it be surjective because the dimension of the kernel plus the dimension of the image is equal to the dimension of X.

We might also note that the mapping u has an inverse v defined by

$$[v(x)](t) = x(t - \tfrac{1}{4}\pi),$$

which implies that u is bijective.

306

First question:

We note that R is an abelian group with respect to addition and that the product of a real number x by a rational number λ (being a special case of the product of two real numbers) satisfies the following:

$$\lambda(x+y) = \lambda x + \lambda y, \qquad (\lambda+\mu)x = \lambda x + \mu x,$$
$$\lambda(\mu x) = (\lambda\mu)x, \qquad\qquad 1x = x.$$

Second question:

Suppose that x and y are two numbers in E:

$$x = A(\omega, \theta), \qquad y = B(\omega, \theta),$$

where A and B are two polynomials with rational coefficients. Then, if λ and μ are two rational numbers, we define the polynomial $\lambda A + \mu B$ by

$$\lambda x + \mu y = \lambda A(\omega, \theta) + \mu B(\omega, \theta) = (\lambda A + \mu B)(\omega, \theta).$$

Since $\lambda x + \mu y$ is expressed as a polynomial in ω and θ with rational coefficients, it is a number belonging to E.

Third question:

We can verify by induction on p that every power ω^p of ω, where $p \geq n$, is a linear combination with rational coefficients of $1, \omega, \ldots, \omega^{n-1}$. Similarly, if

$q \geqslant m$, every power θ^q is a linear combination of $1, \theta,..., \theta^{m-1}$. Therefore, every product $\omega^p\theta^q$ is a linear combination, with rational coefficients, of the products $\omega^h\theta^k$, where $0 \leqslant h \leqslant n-1$ and $0 \leqslant k \leqslant m-1$. This result can be extended immediately to polynomials in ω and θ with rational coefficients.

This result can be interpreted by saying that the elements $\omega^h\theta^k$ (where $0 \leqslant h \leqslant n-1$ and $0 \leqslant k \leqslant m-1$) are generators of the subspace E. This subspace is therefore of finite dimension, the dimension being at most nm, which is the number of generators.

Fourth question:

The powers $(\omega+\theta)^l$ of $\omega+\theta$ are elements of E. At most, nm of these are independent. Therefore, a linear relationship with rational coefficients, not all zero, exists among the $nm+1$ elements

$$(\omega+\theta)^0 = 1, \qquad \omega+\theta, \qquad (\omega+\theta)^2,..., (\omega+\theta)^{nm},$$

that is, $\omega+\theta$ is a root of an algebraic equation with rational coefficients of degree at most nm.

307

First question:

Suppose that $\{u_n\}$ and $\{v_n\}$ are two sequences in $\mathscr{S}$:

$$u_n = a_1 u_{n-1} + \cdots + a_k u_{n-k},$$
$$v_n = a_1 v_{n-1} + \cdots + a_k v_{n-k}.$$

Then,

$$u_n - v_n = a_1(u_{n-1} - v_{n-1}) + \cdots + a_k(u_{n-k} - v_{n-k}),$$
$$\lambda u_n = a_1(\lambda u_{n-1}) + \cdots + a_k(\lambda u_{n-k}),$$

where λ is any complex number. The two sequences

$$\{u_n - v_n\} \text{ and } \{\lambda u_n\}$$

belong to $\mathscr{S}$. But, by the definitions of the operations,

$$\{u_n - v_n\} = \{u_n\} - \{v_n\} \text{ and } \{\lambda u_n\} = \lambda\{u_n\}.$$

We have shown that, if the sequences $\{u_n\}$ and $\{v_n\}$ belong to $\mathscr{S}$, so do the sequences $\{u_n\} - \{v_n\}$ and $\lambda\{u_n\}$. Thus, $\mathscr{S}$ is a vector subspace of the vector space over C of sequences with complex coefficients (see section 2 of the introduction to part 1).

Second question:

If the numbers u_0, u_1,..., u_{k-1} are given, the recursion relationship determines all the terms of the sequence uniquely. Therefore, there exists a sequence whose terms of rank $n < k$ (i.e., whose first $k-1$ terms) have arbitrary given values, and there is only one: two sequences whose terms of rank $n < k$ are equal are identical.

We shall define the k sequences $\{e_n^{(h)}\}$ for $h = 0, 1,..., k-1$, by their first k terms:

$$e_n^{(h)} = 0 \quad \text{if} \quad n < k \quad \text{and} \quad n \neq h,$$

$$e_h^{(h)} = 1.$$

These k sequences are independent. To see this, let us form the sequence

$$\sum_{h=0}^{k-1} \lambda_h \{e_n^{(h)}\} = \left\{ \sum_{h=0}^{k-1} \lambda_h \, e_n^{(h)} \right\}.$$

If $h < k$, the hth term of this sequence is equal to λ_h. By definition, if this sequence is the zero sequence, all its terms are zero and the λ_h are zero. Thus we have the definition of independent elements in a vector space.

Now, let $\{u_n\}$ denote an arbitrary sequence in $\mathscr{S}$. Corresponding to it, let us define the sequence

$$\{v_n\} = \sum_{h=0}^{k-1} u_h \{e_n^{(h)}\} = \left\{ \sum_{h=0}^{k-1} u_h \, e_n^{(h)} \right\}.$$

Obviously, if $n < k$, the term v_n is equal to u_n. The two sequences $\{u_n\}$ and $\{v_n\}$ are therefore identical.

Every sequence in $\mathscr{S}$ is a linear combination of the independent sequences $\{e_n^{(h)}\}$. These sequences constitute a basis of $\mathscr{S}$, and the dimension of $\mathscr{S}$ is the number k of these sequences.

308

First question:

If $u^3 - 2$ is divisible by $a + bu + cu^2$, the quotient is a first-degree polynomial with rational coefficients. The zero of this polynomial is rational and is a zero of $u^3 - 2$. This is impossible since the only real zero of $u^3 - 2$ is $\sqrt[3]{2}$, which is irrational.

If the greatest common divisor of $u^3 - 2$ and $a + bu + cu^2$ is of first degree, the zero x_0 of this polynomial is a zero of $u^3 - 2$. But we know that the coefficients of the greatest common divisor of two polynomials belong to the field of the coefficients of these polynomials, which in the present case is the field of rational numbers. Therefore, the zero x_0 is rational and we have seen that $u^3 - 2$ cannot have a rational zero.

Therefore, the polynomials $u^3 - 2$ and $a + bu + cu^2$ cannot have a common divisor.

Second question:

The difference between two numbers in E is obviously a number in E. Because of the distributivity of a product with respect to addition, the product of two numbers in E is a linear combination, with rational coefficients, of $\sqrt[3]{2}$, $\sqrt[3]{4}$, and their products, namely,

$$\sqrt[3]{2}\sqrt[3]{2} = \sqrt[3]{4}, \qquad \sqrt[3]{2}\sqrt[3]{4} = 2, \qquad \sqrt[3]{4}\sqrt[3]{4} = 2\sqrt[3]{2}.$$

The product of two numbers in E is also a number in E. The two results that we have just established show that E is a subring of the ring of real numbers (cf. section 2 of the introduction to part 1).

The polynomials $u^3 - 2$ and $a + bu + cu^2$ are relatively prime. Therefore (by Bézout's theorem), there exist unique polynomials U and V that satisfy the conditions

$$\begin{cases} U(u)(u^3 - 2) + V(u)(a + bu + cu^2) = 1, \\ \deg U < 2, \qquad \deg V < 3. \end{cases}$$

The coefficients of these polynomials are rational when those of $u^3 - 2$ and $a + bu + cu^2$ are rational. If we assign to u the value $\sqrt[3]{2}$, we obtain

$$V(\sqrt[3]{2})(a + b\sqrt[3]{2} + c\sqrt[3]{4}) = 1.$$

The reciprocal of $a + b\sqrt[3]{2} + c\sqrt[3]{4}$ is

$$V(\sqrt[3]{2}) = a' + b'\sqrt[3]{2} + c'\sqrt[3]{4}.$$

(The polynomial V is of at most second degree, and its coefficients a', b', c', are rational.) The reciprocal of a nonzero element of E belongs to E. Thus, the ring E is a field.

Third question:

We already know that E is an abelian group with respect to addition. The properties of multiplication of real numbers are such that the product of a number in E by a rational number verifies the axioms regarding multiplication of vectors by scalars. Therefore, E is a vector space over Q.

The numbers 1, $\sqrt[3]{2}$, $\sqrt[3]{4}$, are generators of the space (every element of E is a linear combination, with rational coefficients, of these three particular elements).

To show that these elements constitute a basis for E, we need to show that they are independent. Suppose that there exist rational numbers p, q, and r, not all zero, such that

$$p+q\sqrt[3]{2}+r\sqrt[3]{4} = 0.$$

Then, the polynomials u^3-2 and $p+qu+ru^2$ would have a common zero at $\sqrt[3]{2}$ and would therefore not be relatively prime. This situation, however, is impossible as was shown in (1).

309

First question:

We know that polynomials with complex coefficients constitute a vector space E over C.

The difference between two polynomials of degree not exceeding n is a polynomial of degree not exceeding n. Similarly, the product of a polynomial of degree not exceeding n by a complex number is also a polynomial of degree not exceeding n. Therefore, E_{n+1} is a vector subspace of E.

The polynomials 1, x,..., x^n are obviously independent and they generate the subspace E_{n+1}. They constitute a basis for E_{n+1}. Therefore, the dimension of E_{n+1} is $n+1$.

Second question:

Let us show by induction on k that x^k is a linear combination of A_0, A_1,..., A_k. This is obvious if $k = 0$ since A_0 is a nonzero constant (remember that the zero polynomial is not a polynomial of degree 0).

Suppose that the result holds for $k \leqslant p-1$. By hypothesis,

$$A_p = a_p x^p + A'_p,$$

where

$$a_p \neq 0 \quad \text{and} \quad \deg A'_p \leqslant p-1.$$

The polynomial A'_p, which is a linear combination of the powers x^k, where $k \leqslant p-1$, is a linear combination of the A_k for $k \leqslant p-1$:

$$A'_p = \sum_{k=0}^{p-1} \lambda_k A_k \Rightarrow x^p = \frac{1}{a_p} A_p - \sum_{k=0}^{p-1} \frac{\lambda_k}{a_p} A_k,$$

so that x^p is a linear combination of the A_k (for $k \leqslant n$), which completes our proof.

The polynomials A_0, A_1,..., A_n are generators of E_{n+1}: Every polynomial P in E_{n+1} is a linear combination of 1, x,..., x^n, and hence of A_0, A_1,..., A_n. We know that the dimension $n+1$ of the space that they generate is equal to the maximum number of independent elements that can be taken from the set A_0,..., A_n. In other words, the polynomials A_p are independent and constitute a basis for E_{n+1}.

Third question:

F_B is a vector subspace of E_{n+1}. Since

$$P = BQ \quad \text{and} \quad P' = BQ' \quad \Rightarrow \quad P - P' = B(Q - Q'),$$

$$P = BQ \quad\quad\quad\quad\quad \Rightarrow \quad\quad \lambda P = B(\lambda Q).$$

The polynomials B, Bx,..., Bx^{n-k}, are obviously independent and every multiple BQ of B of degree not exceeding n is a linear combination of these polynomials since Q is a linear combination of 1, x,..., x_{n-k}. Therefore, they constitute a basis for F. The dimension of F is $n-k+1$.

The division algorithm enables us to write

$$P = BQ + R, \quad \text{with} \quad \deg R < \deg B,$$

where the polynomials Q and R are unique. In other words, every polynomial in E_{n+1} is the sum of a polynomial in F_B and a polynomial of degree not exceeding $k-1$, this decomposition being unique.

On the basis of what we have just shown, if B is an arbitrary polynomial of degree k, the set F_B of multiples of B is a vector subspace complementary to G_k. Since two subspaces F_B and $F_{B'}$ are not identical unless B and B' are proportional, we have thus determined an infinite collection of spaces complementary to G_k.

Let us take for F_1 the vector subspace of multiples of $x+1$ and let us take for F_2 the subspace of multiples of x. The polynomial $(x+1) - x$, that is, the constant polynomial 1, is the difference between two elements of $F_1 \cup F_2$. However, it does not belong either to F_1 or to F_2 and hence does not belong to $F_1 \cup F_2$. Therefore, the set $F_1 \cup F_2$ is not a vector space.

310

First question:

The polynomials U_k verify the hypotheses made on the polynomials A_p in (2) of exercise 309. Therefore, they constitute a basis for E_{n+1}.

Second question:

The polynomials x^k constitute a basis for E_{n+1}. If P is an arbitrary polynomial in E_{n+1}, there exists a unique system of coefficients a_k such that

$$P = \sum_{k=0}^{n} a_k x^k.$$

Therefore, for φ to be a linear mapping, it is necessary that

$$\varphi(P) = \varphi\left(\sum_{0}^{n} a_k x^k\right) = \sum_{0}^{n} a_k \varphi(x^k) = \sum_{0}^{n} a_k U_k,$$

which proves the uniqueness of the function φ.

Conversely, the mapping φ that assigns to P the polynomial $\sum_{0}^{n} a_k U_k$, is obviously linear:

$$\varphi(P+P') = \sum_{0}^{n} (a_k + a'_k) U_k = \sum_{0}^{n} a_k U_k + \sum_{0}^{n} a'_k U_k = \varphi(P) + \varphi(P'),$$

$$\varphi(\lambda P) = \sum_{0}^{n} (\lambda a_k) U_k = \lambda \sum_{0}^{n} a_k U_k = \lambda \varphi(P),$$

which proves the existence of the function φ.

(Note that the only property used in the proof is the fact that the x^k constitute a basis of E_{n+1}.)

Let us show that φ is bijective by using the fact that the U_k constitute a basis for E_{n+1}. If Q is a polynomial in E_{n+1}, it has a unique representation of the form

$$\sum_{0}^{n} a_k U_k$$

and hence is the image under φ of one and only one polynomial P, namely, the polynomial

$$\sum_{0}^{n} a_k x^k.$$

In other words, for every Q, the equation $\varphi(P) = Q$ has one and only one solution. Thus, the mapping φ is bijective.

Third question:

Obviously,

$$\delta(P+Q) = \delta(P) + \delta(Q), \qquad \delta(\lambda P) = \lambda \delta(P).$$

Therefore, δ is linear.

$$\delta(U_k) = (x+1)x\cdots(x-k+2)-x(x-1)\cdots(x-k+2)(x-k+1)$$
$$= x(x-1)\cdots(x-k+2)[x+1-(x-k+1)] = kU_{k-1}$$

for $k \neq 0$; $\delta(U_0) = 0$.
For an arbitrary polynomial P in E_{n+1},

$$P = \sum_0^n p_k U_k \Rightarrow \delta(P) = \sum_0^n p_k \delta(U_k) = \sum_1^n k p_k U_{k-1}.$$

The U_k are independent. Therefore, $\delta(P)$ is zero if and only if $k p_k = 0$, that is, if $p_k = 0$ for $k = 1, 2,...,n$. The kernel is the subspace of constant polynomials, that is, polynomials proportional to U_0.

The image $\delta(E_{n+1})$ is the subspace generated by the polynomials U_{k-1}, that is, the subspace E_n of polynomials of degree not exceeding $n-1$.

Finally, if F is the subspace generated by the U_k of index $k \geqslant 1$, that is, the subspace of zero polynomials for $x = 0$, the restriction $\bar\delta$ of δ to F is a bijective mapping of F onto $\delta(E_{n+1})$, that is, onto E_n:

$$P = \sum_1^n p_k U_k \Rightarrow \bar\delta(P) = \sum_1^n k p_k U_{k-1},$$

$$Q = \sum_0^{n-1} q_k U_k \Rightarrow \bar\delta^{-1}(Q) = \sum_0^{n-1} \frac{q_k}{k} U_{k+1}.$$

Fourth question:

If P is of degree m, the polynomial Q defined by $Q(x) = P(x+1)-P(x)$ is of degree $m-1$. Therefore, if the degree of R is $n-1$, the solutions P of the equation belong to the vector space E_{n+1}, and we can use the study made in (3) of the operator δ: The restriction $\bar\delta$ of δ to F is a bijective mapping of F onto E_n. Therefore, there exists one and only one polynomial P in F (the zero polynomial for $x = 0$) that is a solution of the equation.

To determine P explicitly, let us express R in E_4 in terms of the basis U_0, U_1, U_2, U_3, and let us use the expression for $\bar\delta^{-1}$ given in (3):

$$R = x^3 - 5x^2 + x + 1 = x(x-1)(x-2) - 2x^2 - x + 1$$

$$= x(x-1)(x-2) - 2x(x-1) - 3x + 1 = U_3 - 2U_2 - 3U_1 + U_0,$$

$$\bar\delta^{-1}(R) = \frac{U_4}{4} - \frac{2U_3}{3} - \frac{3U_2}{2} + U_1$$

$$= \frac{x}{12}(3x^3 - 26x^2 + 39x - 4).$$

Fifth question:

The mapping d, being a composite of three linear mappings, is a linear mapping. It will be sufficient for us to study the images under d of the elements of a basis for E_{n+1}, and we shall take the basis consisting of the x^k:

$$x^k \xrightarrow{\varphi} U_k \xrightarrow{\delta} kU_{k-1} \xrightarrow{\varphi^{-1}} kx^{k-1}.$$

The mapping d maps each polynomial x^k into its derivative kx^{k+1}. Since differentiation is a linear operation, the result applies to any polynomial in E_{n+1}. The image under d is the derivative of the polynomial in question.

311

First question:

Suppose that the terms of highest degree in A and P are ax^α and hx^k. Then, the term of highest degree in $f(P)$ is $ah(k-\alpha)x^{\alpha+k-1}$. This term will not be zero if $k-\alpha \neq 0$.

$$f(A) = AA' - A'A = 0.$$

If P is a polynomial in $E_{\alpha+1}$, it is equal to $\lambda A + P_1$, where λ is a number and P_1 is a polynomial in E_α (this is the division algorithm). Therefore,

$$f(P) = \lambda f(A) + f(P_1) = f(P_1).$$

The set $f(E_{\alpha+1})$ is contained in $f(E_\alpha)$. But $E_{\alpha+1}$ contains E_α and hence $f(E_{\alpha+1})$ contains $f(E_\alpha)$. These two sets are identical.

If the degree k of P is not equal to α, the polynomial $f(P)$ is not the zero polynomial (since the term of degree $\alpha+k-1$ is not zero). If $k = \alpha$, we know that $f(P) = f(P_1)$, and $f(P_1)$ is not the zero polynomial if P_1 is not zero (since the degree of P_1 is not equal to α). A necessary and sufficient condition for P to belong to the kernel of f is that P be proportional to the polynomial A.

Second question:

The rank r_{n+1} of the restriction $\bar{f}$ of f to E_{n+1} is the dimension of $\bar{f}(E_{n+1})$. We know also that

$$r_{n+1} = (n+1) - \dim \bar{f}^{-1}(0).$$

The kernel $\bar{f}^{-1}(0)$ consists only of the element zero if $n < \alpha$. If $n \geqslant \alpha$, it contains the polynomials that are proportional to A. In the latter case, it is of dimension 1.

Thus,

$$r_{n+1} = \begin{cases} n+1 & \text{if} & n \leqslant \alpha - 1 \\ n & \text{if} & n \geqslant \alpha. \end{cases}$$

Third question:

First, F_{p+1}, being the intersection of the vector subspaces E_{p+1} and $f(E)$, is a vector subspace of E_{p+1}. A polynomial Q belonging to F_{p+1}, is the image under f of at least one polynomial P, and we can always choose for P a polynomial of degree other than α because, if Q belongs to $f(E_{\alpha+1})$, it also belongs to $f(E_\alpha)$. Therefore,

$$\deg Q = \deg f(P) = \deg P + \alpha - 1,$$

$$\deg Q \leqslant p \iff \deg P \leqslant p - \alpha + 1.$$

Hence, F_{p+1} is the image $f(E_{p-\alpha+2})$ of the subspace $E_{p-\alpha+2}$ of polynomials of degree not exceeding $p - \alpha + 1$.

If $p - \alpha + 1 < 0$, that is, if $p < \alpha - 1$, there does not exist a polynomial P verifying the condition obtained, so that F_{p+1} consists only of the zero polynomial and is of dimension 0.

If $p - \alpha + 1 \geqslant 0$, the dimension of F_{p+1}, that is, the dimension of $f(E_{p-\alpha+2})$, is the rank $r_{p-\alpha+2}$ of the restriction $\bar{f}$ of f to $E_{p-\alpha+2}$, and this rank was determined in (2): If $0 \leqslant p - \alpha + 1 \leqslant \alpha - 1$, that is, if $\alpha - 1 \leqslant p \leqslant 2(\alpha - 1)$, then F_{p+1} is of dimension $p - \alpha + 2$; if $p - \alpha + 1 \geqslant \alpha$, that is, if $p \geqslant 2\alpha - 1$, then the subspace F_{p+1} is of dimension $p - \alpha + 1$.

If $A = x^2$, $\alpha = 2$. Therefore, F_3 is the image of E_2. The polynomials Q in F_3 are the polynomials $x^2 P' - 2xP$, where P is an arbitrary polynomial of degree 1 or 0. The polynomials Q are therefore the polynomials of degree not exceeding 2 and that vanish at $x = 0$.

Fourth question:

Suppose that the zeros of A are simple. Let us denote by N/D the fraction (reduced) whose derivative is equal to Q/A^2. The zeros of D of order α are poles of order at least $\alpha + 1$ of the derivative. Since the poles of Q/A^2 are of order 2, the zeros of D are all simple and the expression

$$\frac{N'D - ND'}{D^2}$$

for the derivative is irreducible. Therefore, the polynomial D is a divisor of A and the ratio Q/A^2 is the derivative of a fraction of the form P/A, which implies that $Q = f(P)$. Conversely, if $Q = f(P) = AP' - A'P$, the fraction Q/A^2 is the derivative of the fraction P/A. This condition is necessary and sufficient.

If Q is an arbitrary polynomial of degree not exceeding 2, the fractions Q/x^4 are the derivatives of rational functions:

$$\frac{ax^2+bx+c}{x^4} = \frac{\mathrm{d}}{\mathrm{d}x}\left(-\frac{a}{x} - \frac{b}{2x^2} - \frac{c}{3x^3} \right)$$

Now, we saw in (3) that the set F_3 of polynomials belonging to E_3 and to $f(E)$ (where the function f is defined by the polynomial x^2) is not identical to the space E_3.

If A has multiple roots, it is again obvious that, if Q belongs to $f(E)$, the fraction Q/A^2 is the derivative of a rational function. However, the converse of this statement does not hold.

312

First question:

Let us take a basis $e_1,...,\,e_k$ for $v^{-1}(0)$. These vectors are independent by definition and they are contained in $u^{-1}(0)$ since $u^{-1}(0)$ contains $v^{-1}(0)$. The theorem on the completion of a basis (see PZ, Book II, Chapter VII, part 2, §3) tells us that we can find h independent vectors which, together with $e_1,...,\,e_k$, constitute a basis for $u^{-1}(0)$. The $(k+h)$ vectors e_i obtained in this way are independent. Therefore, we can find $n-(k+h)$ vectors that, together with these, constitute a basis for the space E.

Study of the independence of the vectors $v(e_i)$

$$\sum_{i=k+1}^{n} \lambda_i v(e_i) = 0 \;\Rightarrow\; v\left(\sum_{k+1}^{n} \lambda_i e_i \right) = 0$$

since the function v is linear. The vector

$$\sum_{k+1}^{n} \lambda_i e_i$$

then belongs to the kernel of v and is a linear combination of the basis vectors of the kernel:

$$\sum_{i=k+1}^{n} \lambda_i e_i = \sum_{i=1}^{k} \mu_i e_i.$$

The vectors e_i are independent. The coefficients in this equation, in particular the λ_i, are zero. The vectors $v(e_i)$ are independent and, since they generate the space $v(E)$, they constitute a basis for $v(E)$.

Second question:

The mappings u and $w{\circ}v$ are linear. They will be identical if they map the basis vectors into the same vectors. Therefore, the equations of the problem are

$$u(e_i) = (w{\circ}v)\,(e_i) = w[v(e_i)] \qquad (i = 1, 2,..., n)$$
$$i \leqslant k \Rightarrow e_i \in v^{-1}(0) \subset u^{-1}(0)$$
$$u(e_i) = v(e_i) = w[v(e_i)] = 0.$$

The vectors $v(e_i)$ for $i>k$ constitute a basis for $v(E)$. Every vector in $v(E)$ can be uniquely represented in the form

$$\sum_{k+1}^{n} \lambda_i v(e_i).$$

The mapping w_1 defined on $v(E)$ into E by

$$w_1 \left[\sum_{k+1}^{n} \lambda_i v(e_i) \right] = \sum_{k+1}^{n} \lambda_i u(e_i)$$

is linear and it maps $v(e_i)$ into $u(e_i)$. (This is true for $i>k$ by the definition of w_1, and it is true for $i \leqslant k$ since these vectors are the zero vector.)

A linear mapping w of E into E is defined when we give its restrictions w_1 and w_2 to two complementary subspaces. The restriction w_2 of w to a subspace complementary to $v(E)$ can be chosen arbitrarily since, for any w_2, the mapping w satisfies the conditions

$$w[v(e_i)] = w_1[v(e_i)] = u(e_i).$$

Third question:

The kernel $v^{-1}(0)$ is the line Δ perpendicular to P at O, and the kernel $u^{-1}(0)$ is the plane Π perpendicular to D and O, which does indeed contain the line Δ. To define w, let us take a basis for $\mathbf{R}^3$ consisting of

(1) a unit vector e_1 on Δ,
(2) a unit vector e_2 on the intersection of the planes P and Π, and
(3) a unit vector e_3 on D.

$$u(e_1) = v(e_1) = 0,$$
$$u(e_2) = 0 \quad \text{and} \quad v(e_2) = e_2 \quad \Rightarrow \quad w(e_2) = 0,$$
$$u(e_3) = e_3 \quad \text{and} \quad v(e_3) = e_3 \quad \Rightarrow \quad w(e_3) = e_3.$$

In the plane P, the mapping w is the projection onto the straight line D. (In the complement of P consisting of vectors proportional to e_1, w is arbitrary.) Thus we have the result: The projection of a vector OM onto a straight line D passing through O is also the projection onto D of the projection of the vector OM onto a plane P containing D (the theorem on three perpendicular lines).

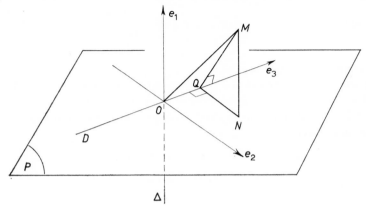

313

First question:

We know that the set $\mathscr{F}$ of functions mapping a set E into a vector space F over the field K is a vector space over K (see PZ, Book II, Chapter III, part 3, § 1). Let us show that $\mathscr{L}$ is a vector subspace of $\mathscr{F}$ by showing that $f-g$ and λf will be functions belonging to $\mathscr{L}$ if f and g are functions belonging to $\mathscr{L}$:

$$(f-g)(x+x') = f(x+x')-g(x+x')$$
$$\text{(from the definition of } f-g),$$
$$= f(x)+f(x')-[g(x)+g(x')]$$
$$\text{(since } f \text{ and } g \text{ are linear),}$$
$$= [f(x)-g(x)]+[f(x')-g(x')]$$
$$\text{(from the properties of a sum in } F),$$
$$= (f-g)(x)+(f-g)(x')$$
$$\text{(from the definition of } f-g),$$
$$(f-g)(\lambda x) = f(\lambda x)-g(\lambda x) = \lambda f(x)-\lambda g(x),$$
$$= \lambda[f(x)-g(x)] = \lambda[(f-g)(x)].$$

Thus, $f-g$ is linear and hence belongs to $\mathscr{L}$.

In the same way, one can verify that the function λf is linear. This proves that $\mathscr{L}$ is a vector subspace of $\mathscr{F}$.

Second question:

Giving the vectors $f(e_i)$ determines f by

$$f\left(\sum_{i=1}^{m} \lambda_i e_i\right) = \sum_{i=1}^{m} \lambda_i f(e_i).$$

The vectors $f(e_i)$ can be chosen arbitrarily: If the ε_i (for $i = 1,..., m$) are vectors belonging to F, the mapping g on E into F defined by

$$g\left(\sum_{i=1}^{m} \lambda_i e_i\right) = \sum_{i=1}^{m} \lambda_i \varepsilon_i$$

is linear and, by definition, $g(e_i) = \varepsilon_i$.

The sets $\{e_i\}$ of m vectors in F can be treated as elements of the vector space F^m, which is the Cartesian product of m spaces each equal to F. The mapping φ on $\mathscr{L}$ into F^m that assigns to every element f in $\mathscr{L}$ the set of m vectors $f(e_i)$ is linear and bijective:

$$\varphi(f+g) = \{(f+g)(e_i)\} = \{f(e_i)\} + \{g(e_i)\},$$
$$\varphi(\lambda f) = \{(\lambda f)(e_i)\} = \lambda\{f(e_i)\},$$
$$\varphi(f) = 0 \Rightarrow f(e_i) = 0 \qquad \forall i \Rightarrow f = 0.$$

The two vector spaces $\mathscr{L}$ and F^m are isomorphic and therefore of the same dimension, namely, mn. (The space F^m is isomorphic to the space K^{mn}.)

Third question:

$\mathscr{L}$ is an additive abelian group. The composition of mappings is a closed operation in $\mathscr{L}$ (that is, the composite of two linear mappings is a linear mapping). This operation is associative because the composition of functions is in general associative. Therefore, we need only show that the composition of linear mappings is distributive with respect to addition. (We must prove distributivity on both sides since composition is not commutative.)

The definition of the operations on the functions and the linearity of the function f enable us to write

$$[f(g+g')](x) = f[(g+g')(x)] = f[g(x)+g'(x)]$$
$$= f[g(x)]+f[g'(x)] = (fg)(x)+(fg')(x) = (fg+fg')(x),$$
$$[(f+f')g](x) = (f+f')[g(x)] = f[g(x)]+f'[g(x)]$$
$$= (fg)(x)+(f'g)(x) = (fg+f'g)(x).$$

Since these equations hold for every vector in the space,

$$f(g+g') = fg+fg', \qquad (f+f')g = fg+f'g.$$

Fourth question:

The vector space $\mathscr{L}$ of linear mappings of E into E is of dimension m^2. Therefore, m^2+1 elements of $\mathscr{L}$ are necessarily dependent.

Let us apply this result to the mappings f^i such that $0 \leqslant i \leqslant m^2$. We see that there exist coefficients a_i in the field K, not all zero, such that

$$\sum_{i=0}^{m^2} a_i f^i = 0.$$

REMARKS: (1) The polynomial A of the statement of the problem can always be chosen of degree not exceeding m^2.

(2) The set of polynomials A such that $A(f) = 0$ is an ideal $\mathscr{I}$ in the ring of polynomials with coefficients in K. (If $\mathscr{I}$ contains A_1 and A_2, it contains $A_1 - A_2$ and BA_1 for any polynomial B.)

314

First question:

The image $f(E)$ contains the vector $f^p(x_0)$, each of which is the image under f of the preceding one. Since these vectors constitute a basis for E, $f(E)$ is the space E itself. Therefore, the rank r of f is, by definition, n. But we know that

$$r = \dim E - \dim f^{-1}(0).$$

The dimension of $f^{-1}(0)$ is 0. Thus, f is a bijection of E onto E.

Second question:

The vectors $x_0, f(x_0), ..., f^{n-1}(x_0)$ are independent:

$$\sum_{i=0}^{n-1} \lambda_i f^i(x_0) = 0 \Rightarrow f\left(\sum_0^{n-1} \lambda_i f^i(x_0)\right) = \sum_0^{n-1} \lambda_i f^{i+1}(x_0) = 0.$$

The vectors $f^{i+1}(x_0)$ are independent by hypothesis. Therefore, the λ_i are all zero.

The vector $f^n(x_0)$ can be written as a linear combination of the basis vectors $x_0, f(x_0), ..., f^{n-1}(x_0)$. Therefore, there exist numbers a_p such that

$$f^n(x_0) + a_{n-1} f^{n-1}(x_0) + \cdots + a_0 x_0 = 0.$$

By applying f^p to these vectors, we obtain

$$f^p[f^n(x_0)] + a_{n-1} f^p[f^{n-1}(x_0)] + \cdots + a_0 f^p(x_0) = 0,$$

which, noting that $f^p f^q = f^q f^p$, we can rewrite

$$f^n[f^p(x_0)] + a_{n-1} f^{n-1}[f^p(x_0)] + \cdots + a_0 f^p(x_0) = 0.$$

The mapping

$$f^n + a_{n-1}f^{n-1} + \cdots + a_0 e$$

maps all the basis vectors $f^p(x_0)$ into zero and hence it is the zero mapping.

Third question:

The coefficients b_i can be obtained by writing the vector $g(x_0)$ as a linear combination of the basis vectors $x_0, f(x_0), \ldots, f^{n-1}(x_0)$:

$$g(x_0) = b_{n-1}f^{n-1}(x_0) + b_{n-2}f^{n-2}(x_0) + \cdots + b_0 x_0.$$

Let us denote by h the mapping

$$b_{n-1}f^{n-1} + \cdots + b_0 e.$$

This mapping commutes with f since it is the sum of mappings that commute with f. Therefore, the mapping $g-h$ commutes with f.

Since $(g-h)(x_0) = 0$, we conclude that

$$[f(g-h)](x_0) = [(g-h)f](x_0) = (g-h)[f(x_0)] = 0.$$

By induction on the integer p, we get

$$(g-h)[f^p(x_0)] = 0.$$

The mapping $g-h$, which maps the basis vectors $f^p(x_0)$ into 0, is the zero mapping.

315

First question:

Let us define a mapping l on E_{n+1} into the field $\mathscr{C}$ of complex numbers that maps every polynomial A in E_{n+1} into $A(x_0)$; that is, $l(A) = A(x_0)$. Then,

$$l(A+B) = (A+B)(x_0) = A(x_0) + B(x_0) = l(A) + l(B),$$

$$l(\lambda A) = (\lambda A)(x_0) \quad = \quad \lambda A(x_0) = \lambda l(A).$$

The mapping l is linear; it is a linear form on E_{n+1}.

Second question:

Let us suppose that the linear forms l_i, defined by $l_i(A) = A(x_i)$, are related by

$$\sum_i \lambda_i l_i = 0.$$

From the definition of the zero form, we have, for every polynomial A in E_{n+1},

$$\left(\sum_i \lambda_i l_i\right)(A) = \sum_i \lambda_i l_i(A) = \sum_i \lambda_i A(x_i) = 0.$$

Let us take for A the polynomial

$$A_j = \prod_{i \neq j}(x - x_i).$$

This polynomial is of degree n, and it vanishes for all the values x_i with the exception of x_j:

$$\sum_i \lambda_i A_j(x_i) = \lambda_j A_j(x_j) = 0 \Rightarrow \lambda_j = 0.$$

Since j represents any of the indices $1, 2, ..., n+1$, we have shown that $\lambda_j = 0$ for every j.

The forms l_i are independent. There are $n+1$ of them and the dimension of E_{n+1} and hence of E_{n+1}^* is $n+1$. They constitute a basis $\mathscr{B}^*$ of E_{n+1}^*.

Third question:

The study made in (2) shows that the polynomials

$$L_j = \frac{A_j}{A_j(x_j)}$$

constitute a basis dual to the basis $\mathscr{B}^*$:

$$l_i(L_j) = \frac{l_i(A_j)}{A_j(x_j)}.$$

Now, $l_i(A_j) = 0$ if $i \neq j$, and $l_j(A_j) = A_j(x_j)$. Thus, $l_i(L_j) = 0$ or 1 according as $i \neq j$ or $i = j$. We can verify immediately that the polynomials L_j are the Lagrange polynomials since

$$L_j(x) = \frac{\prod_{i \neq j}(x - x_i)}{\prod_{i \neq j}(x_j - x_i)}.$$

Fourth question:

The definition of the basis φ_j dual to the basis e_i shows that $\varphi_i(e_i)$ is the jth coordinate of e_i in the basis $e_1, e_2, ..., e_n$. The mapping that assigns to a vector x its jth coordinate in a given basis is a linear form that takes the same value as φ_j for all the basis vectors and is hence identical to φ_j. The value $\varphi_j(x)$ of the form φ_j for the vector x is therefore the jth coordinate of x in the basis $e_1, e_2, ..., e_n$.

Taylor's formula for polynomials,

$$A(x) = \sum_0^n \frac{A^{(k)}(0)}{k!} x^k,$$

shows that the coordinates of the polynomial A relative to the basis $1, x, \ldots, x^n$, are the linear forms φ_k defined by

$$\varphi_k(A) = \frac{A^{(k)}(0)}{k!}.$$

Thus, these forms φ_k are the elements of the basis that is dual to $1, x, \ldots, x^n$.

Fifth question:

The mapping ψ of E_{n+1} into $\mathscr{C}$, defined by $\psi(A) = A'(x_0)$, is a linear form on E_{n+1}:

$$\psi(A+B) = (A+B)'(x_0) = A'(x_0) + B'(x_0) = \psi(A) + \psi(B),$$
$$\psi(\lambda A) = (\lambda A)'(x_0) = \lambda A'(x_0) = \lambda \psi(A).$$

Thus, the form ψ is a linear combination of the elements of a basis of E^*_{n+1} and, in particular, of the basis $\mathscr{B}^*$ defined in (2). There exist complex numbers λ_i such that

$$\psi = \sum_1^{n+1} \lambda_i l_i, \qquad A'(x_0) = \psi(A) = \sum_1^{n+1} \lambda_i l_i(A) = \sum_1^{n+1} \lambda_i A(x_i).$$

We can obtain the values of the constants λ_i by taking for A successively the polynomials L_j of the basis $\mathscr{B}$ dual to the basis $\mathscr{B}^*$:

$$L'_j(x_0) = \sum_{i=1}^{n+1} \lambda_i l_i(L_j) = \lambda_j,$$

$$\lambda_j = L'_j(x_0) = L_j(x_0)\left(\sum_{i \neq j} \frac{1}{x_0 - x_i} \right) = \frac{\prod\limits_{i \neq j}(x_0 - x_i)}{\prod\limits_{i \neq j}(x_j - x_i)} \left(\sum_{i \neq j} \frac{1}{x_0 - x_i} \right).$$

316

The rank of the mapping f is the dimension of the vector subspace of the dual E of $\mathbf{R}^3$ generated by the linear forms X, Y, and Z (see PZ, Book II, Chapter VII, part 4, § 2). For this subspace to be of dimension 3, it is necessary and sufficient that the forms X, Y, and Z be independent.

Let us take for the basis of E the forms X, x, and y. This is a basis since the form z can be expressed as a linear combination of X, x, and y:

$$z = (m-2)x + 2y - X.$$

The expressions for X, Y, and Z in the new basis are

$$X = X,$$
$$Y = 2(m-1)x + (m+4)y - 2X,$$
$$Z = (m-1)(m+2)x + 4(m+1)y - (m+1)X.$$

These forms will be linearly dependent if and only if the forms $Y+2X$ and $Z+(m+1)X$ are linearly dependent, that is, proportional to each other.

For the rank of f to be less than 3, it is necessary and sufficient that

$$8(m-1)(m+1) - (m-1)(m+2)(m+4) = 0,$$

or

$$(m-1)(-m^2+2m) = 0.$$

The rank of f is equal to 3 unless m is 0, 1, or 2, in which case the rank is 2 since X and Y are always independent forms (Y cannot be proportional to X).

REMARK: One can also arrive at this result by noting that, if X, Y, and Z are independent, they constitute a basis for E. Therefore, it is possible to express x, y, and z as linear combinations of X, Y, and Z. It is then necessary to see whether the expressions for X, Y, and Z can be expressed in terms of the original basis forms x, y, and z.

Study of $f(\mathbf{R}^3)$ for particular values of m

The dimension of $f(\mathbf{R}^3)$ is 2. Therefore, there is a unique linear relationship among the coordinates of a vector of $f(\mathbf{R}^3)$.

(1) $m = 0$.

$$\begin{cases} Y+2X = -2x+4y \\ Z+X = -2x+4y \end{cases} \Rightarrow Y+X-Z = 0.$$

(2) $m = 1$.

$$\begin{cases} Y+2X = 5y \\ Z+2X = 8y \end{cases} \Rightarrow 6X+8Y-5Z = 0.$$

(3) $m = 2$.

$$\begin{cases} Y+2X = 2x+6y \\ Z+3X = 4x+12y \end{cases} \Rightarrow 2Y+X-Z = 0.$$

317

First question:

If f^* and g^* are two forms in F^* and λ is an element of K, we know that, for every element x in Φ,

$$f^*(x) = g^*(x) = 0 \Rightarrow (f^* - g^*)(x) = 0,$$
$$f^*(x) = 0 \Rightarrow \quad (\lambda f^*)(x) = 0.$$

F^* is a vector subspace of E^*.

Second question:

A form f^* of F^* vanishes for every vector of Φ and hence for every vector of F since every vector of F is a linear combination of vectors of Φ.

If a vector e_1 is not a vector of F, it, together with a basis $e_2,..., e_{p+1}$ for F, constitutes a system of independent vectors that can be completed to form a basis $e_1, e_2,..., e_n$ for the space E. Every vector x in E then has a unique decomposition of the form

$$x = \sum_{i=1}^{n} x_i e_i.$$

The linear form e^* defined by $e^*(x) = x_1$ is the zero form for the vectors $e_2,..., e_{p+1}$ and hence for every vector in F. It is a form in F^*. For e_1, on the other hand, it takes the value 1.

If e_1 is not a vector in F, there exists at least one form in F^* that does not vanish at e_1. The zeros that are common to the forms in F^* are therefore the vectors in F.

Third question:

Let us take a basis $f_1^*, f_2^*,..., f_q^*$ for F^* (where q is the dimension of F^*) and let us define a linear mapping l of E into K^q by setting

$$\xi_i = f_i^*(x), \qquad i = 1, 2,..., q.$$

The rank of the mapping l is the dimension of the vector subspace generated by the forms f_i^*, namely, q. But the rank of l is also equal to $\dim E - \dim l^{-1}(0)$. The kernel $l^{-1}(0)$ is the set of zeros that are common to the forms in F^*. It is F whose dimension was denoted by p. By expressing the rank of l in two different ways, we thus obtain

$$q = n - p \qquad \text{or} \qquad \dim F^* + \dim F = n.$$

Fourth question:

The polynomials in E_n that a given polynomial B divides constitute a vector subspace F of E_n. A complementary subspace F' consists of the set of polynomials of degree less than the degree of B (cf. exercise 309), and the dimension of F' is the degree of B. Here, the dimension of F' is 6 and that of F is $n+1-6 = n-5$. Thus, the dimension of F^* is 6.

The linear forms l of the type indicated are evidently the zero forms for the polynomials of F and hence they belong to F^*. Since the forms defined by $A(1),...,A(3)$ are independent, the dimension of the vector space L^* of the forms l is 6. The vector space L^* is contained in F^* and has the same dimension as does F^*. Therefore, $L^* = F^*$.

318

FIRST PART

First question:

The axioms of vector spaces are verified: Γ is an additive abelian group; multiplication by a real number has the required properties since these properties are satisfied in $\Gamma \times C$ and hence in the subset $\Gamma \times R$.

Let x' denote an element of E' other than the neutral element of the group Γ and let λ and μ be real numbers. Then,

$$\lambda x' + \mu(ix)' = (\lambda x') + (\mu ix)' = (\lambda x + \mu ix)' = [(\lambda + i\mu)x]',$$
$$\lambda x' + \mu(ix)' = 0 \Rightarrow [(\lambda + i\mu)x]' = 0 \Rightarrow \lambda + i\mu = 0 \Rightarrow \lambda = \mu = 0.$$

Thus, the elements x' and $(ix)'$ are independent.

Let $e_1, e_2,...,e_n$ be the basis vectors of E and let $e'_1, e'_2,..., e'_n$ be the corresponding vectors of E'. Reasoning analogous to that above shows that

$$x' = \sum_{p=1}^{n} \lambda_p e'_p + \sum_{p=1}^{n} \mu_p (ie_p)' \Leftrightarrow x = \sum_p (\lambda_p + i\mu_p)e_p.$$

Thus, a vector x' in E' has a unique decomposition as a linear combination of the vectors e'_p and $(ie_p)'$. These vectors constitute a basis for E', so that the dimension of E' is $2n$.

Second question:

The mapping φ is bijective. Indeed, the equation $y' = \varphi(x')$ has one and only one solution x' for any value of y':

$$y' = \varphi(x') \Leftrightarrow y' = (ix)' \Leftrightarrow y = ix.$$

The mapping φ is linear:

$$\varphi(x_1' + x_2') = [i(x_1 + x_2)]' = [ix_1 + ix_2]' = (ix_1)' + (ix_2)' = \varphi(x_1') + \varphi(x_2').$$

Finally, we note that the definition of $y' = \varphi(x')$ is $y' = (ix)'$.

$$(\varphi \circ \varphi)(x') = \varphi[\varphi(x')] = \varphi(y') = (iy)' = (-x)' = -x'.$$

SECOND PART

First question:

For any complex number x, we can define f by $f(x) = ix$. Then, f is an isomorphism of C onto itself, and $(f \circ f)(x) = i(ix) = -x$.

This isomorphism can be defined by setting $f(1) = i$ and $f(i) = -1$ for the elements of a basis of C. In this form, one can give the result for the space R^2 (or any 2-dimensional space). If e_1 and e_2 are the vectors of a basis for R^2, f will be the linear mapping defined by

$$f(e_1) = e_2, \qquad f(e_2) = -e_1.$$

Finally, a $2n$-dimensional space F that is isomorphic to R^{2n} can be treated as the product of n spaces R^2. If $e_1, e_2, ..., e_{2n}$ are the elements of a basis for E, we define the linear mapping f by

$$f(e_1) = e_2 \qquad \qquad f(e_{2p-1}) = e_{2p}$$
$$f(e_2) = -e_1 \qquad \qquad f(e_{2p}) = -e_{2p-1}$$

The image $f(F)$ contains all the basis vectors. It is identical to F. The rank of f is $2n$, and the kernel consists only of the element 0. (We recall that the rank of $f = \dim F - \dim f^{-1}(0)$.) The linear mapping f is bijective and is an isomorphism of F onto F.

Obviously, $(f \circ f)(e_i) = -e_i$. Thus, $(f \circ f)$ is the negative of the identity mapping.

Second question:

We know that E is an abelian group. Therefore, we need only show that the required properties regarding multiplication by scalars are satisfied:

(a) Distributivity with respect to the addition of vectors:

$$(\lambda + i\mu)(x + x') = \lambda(x + x') + \mu f(x + x') = \lambda x + \lambda x' + \mu f(x) + \mu f(x')$$
$$= (\lambda + i\mu)x + (\lambda' + i\mu')x'.$$

(b) Distributivity with respect to addition of complex numbers:

$$[(\lambda + i\mu) + (\lambda' + i\mu')]x = [\lambda + \lambda' + i(\mu + \mu')]x = (\lambda + \lambda')x + (\mu + \mu')f(x)$$
$$= \lambda x + \mu f(x) + \lambda' x + \mu' f(x) = (\lambda + i\mu)x + (\lambda' + i\mu')x.$$

(c) Associativity:

$$[(\lambda' + i\mu')(\lambda + i\mu)]x = [\lambda\lambda' - \mu\mu' + i(\mu'\lambda + \mu\lambda')]x$$
$$= (\lambda\lambda' - \mu\mu')x + (\lambda\mu' + \mu\lambda')f(x),$$

$$(\lambda' + i\mu')[(\lambda + i\mu)x] = (\lambda' + i\mu')[\lambda x + \mu f(x)]$$
$$= (\lambda' + i\mu')[\lambda x] + (\lambda' + i\mu')[\mu f(x)]$$
$$= \lambda'(\lambda x) + \mu' f(\lambda x) + \lambda'[\mu f(x)] + \mu' f[\mu f(x)]$$
$$= (\lambda\lambda' - \mu'\mu)x + (\mu'\lambda + \lambda'\mu)f(x).$$

Here, we use the two properties that f is linear and $f \circ f = e$. Then, in particular,

$$f[\mu f(x)] = \mu f[f(x)] = \mu(-x) = -\mu x.$$

(d) $1x = x$ (this property is true in F).

To obtain E', we need only restrict the multiplication by a scalar to real numbers and to set

$$rx' = (rx)'.$$

If we identify the vector x' with the vector x in F, we see that rx' is identified with rx and hence that E' can be identified with F. We know that if E is of dimension N, then E', and hence F, is of dimension $2N$, which proves that the dimension of E is n.

319

First question:

(S) is a subset of the vector space of functions on $\mathbf{R}$ into $\mathbf{C}$. To show that this is a vector subspace, we need to show that for any f and g belonging to (S), the vectors $f - g$ and λf belong to (S).

If f and g are twice differentiable, so are the vectors $f - g$ and λf; also,

$$(f - g)'' + p(f - g)' + q(f - g) = f'' + pf' + qf - (g'' + pg' + qg) = 0,$$
$$(\lambda f)'' + p(\lambda f)' + q(\lambda f) = \lambda f'' + \lambda p f' + \lambda q f = 0.$$

Second question:

By hypothesis, f is twice differentiable. The second derivative f'', being equal to $-pf' - qf$, is also differentiable. Then, when we write that the derivative of the function $f'' + pf' + qf$ is zero, we obtain

$$f''' + pf'' + qf' = (f')'' + p(f')' + qf' = 0.$$

The function f' is also a solution of (E).

The mapping l is linear since

$$l(f+g) = (f+g)' = f'+g' = l(f)+l(g),$$
$$l(\lambda f) = (\lambda f)' = \lambda f' = \lambda l(f).$$

Third question:

Suppose that the mapping l is equal to ke (where e is the identity mapping). Consider two arbitrary functions f and g belonging to (S). By hypothesis,

$$l(f) = f' = kf, \qquad l(g) = g' = kg,$$
$$f'g-fg' = 0 \Rightarrow \left(\frac{f}{g}\right)' = 0 \Rightarrow f = \lambda g.$$

The functions f and g are proportional and hence the degree of (S) is at most 1.

The mapping l will be an isomorphism of (S) onto (S) if it is bijective, that is, if its kernel consists only of 0. The kernel of l consists only of constant functions. For l to be an isomorphism, it is necessary and sufficient that a constant function not be a solution of (E), that is, that q be non-zero.

Fourth question:

By definition, $l(f) = f'$ and $(l \circ l)(f) = l(f') = f''$.
Therefore, for every function f in (S),

$$(l \circ l+pl+qe)(f) = f''+pf'+qf = 0.$$

The mapping $l \circ l+pl+qe$ is the zero mapping.
If r_1 and r_2 are the zeros of the polynomial u^2+pu+q, then

$$(l-r_1 e) \circ (l-r_2 e) = l \circ l-(r_1 e) \circ l-l \circ (r_2 e)+r_1 r_2 e.$$

(This is true because of the distributivity of the composition of linear mappings with respect to addition [cf. exercise 307, 3rd question].) Since $el = le = l$,

$$(l-r_1 e) \circ (l-r_2 e) = l \circ l-(r_1+r_2)l+r_1 r_2 e = l \circ l+pl+qe.$$

(More generally, the classical algebraic identities are valid in a commutative ring. Now, if the ring of mappings of (S) into (S) is not commutative, the subring used here of mappings that can be defined by a polynomial in l is commutative.)

If one of the mappings has an inverse mapping, the other is the zero mapping. However, this is impossible since, as we saw in (3), l is not proportional to e. In fact, for any f in (S),

$$[(l-r_1 e) \circ (l-r_2 e)](f) = (l-r_1 e)[(l-r_2 e)(f)] = 0.$$

Therefore, $(l-r_2 e)(f)$ is an element of the kernel of $l-r_1 e$.

If $l-r_1e$ has an inverse, its kernel consists only of zero. Therefore, $(l-r_2e)(f) = 0$ for every f in (S).

If $l-r_2e$ has an inverse, every element g in (S) is the image of a function f. Therefore, the kernel of $l-r_1e$ contains all the elements of (S). The mapping $l-r_1e$ is the zero mapping.

320

First question:

We can easily verify that $X+Y+Z+T = 0$. Therefore, the rank of f is at most 3.
However, if we take for our new coordinates in R^4

$$x, \quad y, \quad z, \quad s = x+y+z+t,$$

the expressions for X, Y, and Z become

$$X = s-2t, \qquad Y = -s+2y, \qquad Z = -s+2x.$$

These three forms are independent. The dimension of the vector space of the dual generated by X, Y, Z, and T, that is, the rank of f, is at least 3. Therefore, the rank of f is exactly 3.

$f(R^4)$ is the subspace defined by

$$X+Y+Z+T = 0.$$

This space has no point in common with D since the sum of four positive numbers cannot be 0.

Second question:

If the rank of f is p, then $f(R^p)$ is the space R^p itself. It contains, in particular, the points of the domain Δ defined by $X_i>0$ for $i = 1, 2,...,p$. The system (S) has solutions.

If the rank of f is $p-1$, there exist $p-1$ independent forms f_i, for example, the first $p-1$ forms, and the pth form is a linear combination of the others. There exist coefficients λ_i such that

$$f_p + \sum_{i=1}^{p-1} \lambda_i f_i = 0$$

and hence

$$f_p(u) + \sum_{i=1}^{p-1} \lambda_i f_i(u) = 0.$$

(a) If all the λ_i are nonnegative, the system (S) cannot have a solution because the terms $\lambda_i f_i(u)$ are nonnegative and $f_p(u)$ is positive, so that the sum cannot be zero.

(b) If one of the λ_i is negative (we assume that $\lambda_{p-1} < 0$), the system (S) has solutions. The $p-1$ independent forms f_i (for $i = 1, 2,...,p-1$) can assume arbitrary numerical values. (They define a surjective mapping of R^n into R^{p-1}.) Therefore, let us write the relation among the values of the forms f_i:

$$f_p(u) = |\lambda_{p-1}| f_{p-1}(u) - \sum_{i=1}^{p-2} \lambda_i f_i(u).$$

Let us assign to the $f_i(u)$ the value 1 if $i \leqslant p-2$. Then, we can choose $f_{p-1}(u)$ sufficiently large for the right, and hence the left, side of this equation to be positive.

A necessary and sufficient condition for the system (S) to have no solution is that the rank of f be $p-1$ and that condition (a) be satisfied. If we set $\lambda_p = 1$, there exist nonnegative coefficients λ_i (for $1 \leqslant i \leqslant p$), not all zero (we take $\lambda_p = 1$), such that

$$\sum_{i=1}^{p} \lambda_i f_i = 0.$$

CHAPTER 2

321

$$M(s)N(t) = \begin{pmatrix} s & 0 \\ \dfrac{t}{s} & \dfrac{1}{s} \end{pmatrix},$$

$$M(s)N(t)P(u) = \begin{pmatrix} s & su \\ \dfrac{t}{s} & \dfrac{ut+1}{s} \end{pmatrix}.$$

The conditions given are obviously necessary:

$$\begin{cases} s \neq 0 \text{ by hypothesis,} \\ s\dfrac{ut+1}{s} - \dfrac{t}{s}su = 1. \end{cases}$$

The last condition is an immediate consequence of the result concerning the determinant of a product of matrices:

$$\det\ [M(s)\ N(t)\ P(u)] = [\det M(s)]\ [\det N(t)]\ [\det P(u)],$$

and each of the determinants on the right is equal to 1.

To show that the given conditions are sufficient, we need to show that it is possible to determine s, t, and u if we know a, b, c, and d; that is, we need to solve the system

$$a = s, \qquad b = su, \qquad c = \frac{t}{s}, \qquad d = \frac{ut+1}{s}.$$

The first three equations yield

$$s = a \neq 0, \qquad u = b/a, \qquad t = ac,$$

and the fourth equation is satisfied since

$$\frac{ut+1}{s} = \frac{bc+1}{a} = d.$$

The solution obtained is unique. This decomposition can be accomplished in one and only one way.

322

First question:

$$A^2 = \begin{pmatrix} a^2 + bc & b(a+d) \\ c(a+d) & bc + d^2 \end{pmatrix} = (a+d)A + (bc - ad)I.$$

Thus, we obtain a solution $\alpha = a + d$, $\beta = bc - ad$. If there exists a second solution (α', β'), different from (α, β), we have

$$A^2 = \alpha A + \beta I = \alpha' A + \beta' I \Rightarrow (\alpha' - \alpha)A = (\beta - \beta')I.$$

The matrix A is proportional to I. It is necessary and sufficient that

$$a = d \quad \text{and} \quad b = c = 0.$$

Second question:

To show that $\mathscr{A}$ is a ring, that is, a subring of the ring of 2×2 square matrices, we shall follow the suggestions indicated in the introduction to Part 1 and show that

$$B \in \mathscr{A} \quad \text{and} \quad B' \in \mathscr{A} \Rightarrow B' - B \in \mathscr{A} \quad \text{and} \quad BB' \in \mathscr{A}.$$

Let us set $B = uI + vA$, $\quad B' = u'I + v'A$.
Then,

$$B' - B = (u' - u)I + (v' - v)A \Rightarrow B' - B \in \mathscr{A}$$
$$B'B = uu'I + (uv' + u'v)A + vv'A^2,$$
$$= uu'I + (uv' + u'v)A + vv'(\alpha A + \beta I)$$
$$= (uu' + \beta vv')I + (uv' + u'v + \alpha vv')A.$$

The matrix $B'B$ belongs to $\mathscr{A}$, and we can verify that $B'B = BB'$. Thus, the ring $\mathscr{A}$ is commutative and, finally, it contains the unit matrix I (which we obtain for $u = 1$, $v = 0$).

Third question:

Let us first take care of the case in which the matrix A is proportional to I. If $A = kI$, the ring $\mathscr{A}$ contains the matrices $(u + vk)I$, that is, all matrices proportional to I. Then, $\mathscr{A}$ is a field isomorphic to the field of real numbers.

Now assume that A is not proportional to I. The matrix B has an inverse if its determinant is nonzero, that is, if

$$(u+va)(u+vd)-v^2bc = u^2+\alpha uv - \beta v^2 \neq 0.$$

Let us now show that there exists in $\mathscr{A}$ a matrix $X=xI+yA$ that is the inverse of B. When we have done this, since the inverse of B is unique, we shall have established that it belongs to $\mathscr{A}$:

$$BX = XB = (ux+vy\beta)I + (uy+vx+vy\alpha)A.$$

In the case in question, A and I are not proportional. Therefore, it is necessary and sufficient that

$$\begin{cases} ux+v\beta y = 1, \\ vx+(u+v\alpha)y = 0, \end{cases}$$

and this system has a solution if and only if

$$u(u+v\alpha)-v^2\beta = u^2+\alpha uv - \beta v^2 \neq 0,$$

that is, if B has an inverse.

The ring $\mathscr{A}$ is a field if every element in it other than the zero matrix has an inverse. We have the zero matrix only when $u = 0$ and $v = 0$. (Recall that A is not proportional to I.) Therefore, for $\mathscr{A}$ to be a field, it is necessary and sufficient that $u^2+\alpha uv - \beta v^2$ not vanish when $u=v=0$. If the polynomial $z^2+\alpha z - \beta$ has a real zero z_1, the system $u=z_1$, $v=1$ makes $u^2+\alpha uv - \beta v^2$ vanish. If the polynomial $z^2+\alpha z - \beta$ has no real zero, the expression $u^2+\alpha uv - \beta v^2$ vanishes only when $u=v=0$. Therefore, a necessary and sufficient condition for $\mathscr{A}$ to be a field is that

$$\alpha^2+4\beta<0 \qquad \text{or} \qquad (a-d)^2+4bc<0.$$

Fourth question:

The ring $\mathscr{A}$ is a field since

$$(a-d)^2+4bc = -4.$$

The numbers α and β are equal to 0 and -1 respectively. If the matrices B are defined by the pair of real numbers (u, v) such that $B=uI+vA$, addition and multiplication are defined, in accordance with (2), by

$$(u, v)+(u', v') = (u+u', v+v'),$$
$$(u, v)(u', v') = (uu'-vv', uv'+u'v).$$

These formulas are identical to those defining the sum and product of complex numbers $u+iv$ and $u'+iv'$. The mapping on $\mathscr{A}$ into C that assigns to B the number $u+iv$ is a bijection that conserves both these operations. Thus, it is an isomorphism between the fields $\mathscr{A}$ and C.

323

First question:

Let us write the equality of the homologous terms of BM and MB:

(1)
$$\begin{cases} \alpha x + \gamma y = \alpha x + \beta z \\ \beta x + \delta y = \alpha y + \beta t \\ \alpha z + \gamma t = \gamma x + \delta z \\ \beta z + \delta t = \gamma y + \delta t \end{cases} \Leftrightarrow \begin{cases} \gamma y = \beta z, \\ (\alpha - \delta)y = (x-t)\beta, \\ (\alpha - \delta)z = (x-t)\gamma. \end{cases}$$

By hypothesis, β is nonzero. Therefore, the equation $y = k\beta$ defines k. Dividing by β if necessary, we obtain

$$z = k\gamma, \qquad x - t = k(\alpha - \delta).$$

The terms x, y, z, and t can be expressed in terms of the two parameters k and t. The matrices B are then written

$$B = \begin{pmatrix} t + k(\alpha - \delta) & k\beta \\ k\gamma & t \end{pmatrix} = t\begin{pmatrix} 1 & 0 \\ 0 & 1 \end{pmatrix} + k\begin{pmatrix} \alpha - \delta & \beta \\ \gamma & 0 \end{pmatrix}.$$

The two matrices

$$I = \begin{pmatrix} 1 & 0 \\ 0 & 1 \end{pmatrix} \quad \text{and} \quad A = \begin{pmatrix} \alpha - \delta & \beta \\ \gamma & 0 \end{pmatrix}$$

are independent (the term β being nonzero by hypothesis). The set of matrices B constituting the vector subspace generated by the matrices I and A is a two-dimensional space.

The set of matrices B is also one of the rings $\mathscr{A}$ studied in exercise 322. The parameters t and k play the role that the parameters u and v played in that exercise.

Second question:

If β is zero and γ is nonzero, we can proceed in a manner identical to the procedure followed in (1) provided we define the number k by $z = k\gamma$, and hence the results are the same.

If $\beta = \gamma = 0$, equations (1) reduce to

$$(\alpha - \delta)y = (\alpha - \delta)z = 0.$$

Two cases are possible:

If $\alpha = \delta$, the matrix M is proportional to the unit matrix I and it commutes with all the matrices in question. The matrix B is arbitrary.

If $\alpha \neq \delta$, the matrix M is a diagonal matrix that is not proportional to the unit matrix. The matrix B commutes with M if $y=z=0$, that is, if B is a diagonal matrix. The set $\mathscr{D}$ of diagonal matrices is also one of the rings $\mathscr{A}$ of exercise 322. We need only take

$$A = \begin{pmatrix} 1 & 0 \\ 0 & 0 \end{pmatrix}$$

for $\mathscr{A}$ to be the ring of diagonal matrices.

Third question:

Let us denote by $\mathscr{M}$ a commutative subring of the ring of 2×2 matrices.

If $\mathscr{M}$ contains only matrices proportional to I, then $\mathscr{M}$ is a subring of the ring of diagonal matrices.

If $\mathscr{M}$ contains a matrix M that is not proportional to I, the matrices in $\mathscr{M}$ commute with M, and $\mathscr{M}$ is a subset of the set $\mathscr{A}$ of matrices that commute with M. This set $\mathscr{A}$ is one of the rings of exercise 322. $\mathscr{M}$ is a subring of one of these rings $\mathscr{A}$.

In sum, the only commutative subrings of the ring of 2×2 matrices are the rings $\mathscr{A}$ consisting of matrices of the form $uI+vA$, where A is a given arbitrary matrix, and the subrings of these rings $\mathscr{A}$.

324

First question:

Let us show that K is a subring of $\mathscr{M}$. Following the outline indicated in the introduction to Part 1, we show that

$$M \in K \quad \text{and} \quad M' \in K \Rightarrow M - M' \in K \quad \text{and} \quad MM' \in K$$

$$M - M' = \begin{pmatrix} m & n \\ -\bar{n} & \bar{m} \end{pmatrix} - \begin{pmatrix} m' & n' \\ -\bar{n}' & \bar{m}' \end{pmatrix} = \begin{pmatrix} m - m' & n - n' \\ -(\overline{n-n'}) & (\overline{m-m'}) \end{pmatrix}$$

$$MM' = \begin{pmatrix} mm' - n\bar{n}' & mn' + n\bar{m}' \\ -(\bar{n}m' + \bar{m}\bar{n}') & -\bar{n}n' + \bar{m}\bar{m}' \end{pmatrix} = \begin{pmatrix} mm' - n\bar{n}' & mn' + n\bar{m}' \\ -(\overline{mn' + n\bar{m}'}) & (\overline{mm' - n\bar{n}'}) \end{pmatrix}.$$

The matrices $M - M'$ and MM' do belong to the set K, so that K does possess a ring structure.

This ring has a unity. To show this, we take $m=1$ and $n=0$. Then, we can see that the unit matrix belongs to K.

It is not commutative. To see this, note that

$$\begin{pmatrix} 1+i & 0 \\ 0 & 1-i \end{pmatrix}\begin{pmatrix} 0 & 1 \\ -1 & 0 \end{pmatrix} = \begin{pmatrix} 0 & 1+i \\ -1+i & 0 \end{pmatrix}$$

$$\begin{pmatrix} 0 & 1 \\ -1 & 0 \end{pmatrix}\begin{pmatrix} 1+i & 0 \\ 0 & 1-i \end{pmatrix} = \begin{pmatrix} 0 & 1-i \\ -1-i & 0 \end{pmatrix}.$$

(*It is not enough to say that the ring $\mathcal{M}$ is noncommutative, because this ring does contain commutative subrings, as was shown in exercise 323.*)

Finally, to show that K is a division ring, we need to show that every element M in K other than the zero element has an inverse and that this inverse M^{-1} belongs to K. Such an element M does have an inverse since $m\bar{m}+n\bar{n}=|m|^2+|n|^2$ can be zero only if $m=n=0$:

$$M^{-1} = \frac{1}{m\bar{m}+n\bar{n}}\begin{pmatrix} \bar{m} & -n \\ \bar{n} & m \end{pmatrix} = \begin{pmatrix} \mu & \nu \\ -\bar{\nu} & \bar{\mu} \end{pmatrix},$$

where

$$\mu = \frac{\bar{m}}{m\bar{m}+n\bar{n}}, \qquad \nu = \frac{-n}{m\bar{m}+n\bar{n}}.$$

Second question:

We have seen that K is an additive abelian group. However, the product (defined in $\mathcal{M}$) of an element M in K and a complex number z is not an element of K if z is not a real number:

$$zM = z\begin{pmatrix} m & n \\ -\bar{n} & \bar{m} \end{pmatrix} = \begin{pmatrix} zm & zn \\ -z\bar{n} & z\bar{m} \end{pmatrix},$$

$$\left.\begin{aligned} \overline{(zm)} &= \bar{z}\bar{m} \neq z\bar{m} \\ \overline{(zn)} &= \bar{z}\bar{n} \neq z\bar{n} \end{aligned}\right\} \text{except when } z \text{ is real.}$$

Third question:

K is an additive abelian group. Multiplication by a real number z, as defined in (2), maps every element in K into an element in K. This multiplication satisfies the vector-space axioms since it is the restriction to $K \times R$ of the multiplication by a scalar in the vector space $\mathcal{M}$ defined on $\mathcal{M} \times C$.

If we set $m=u+iv$ and $n=r+is$ (where u, v, r, and s are real), we have

$$M = \begin{pmatrix} m & n \\ -\bar{n} & \bar{m} \end{pmatrix} = u\begin{pmatrix} 1 & 0 \\ 0 & 1 \end{pmatrix} + v\begin{pmatrix} i & 0 \\ 0 & -i \end{pmatrix} + r\begin{pmatrix} 0 & 1 \\ -1 & 0 \end{pmatrix} + s\begin{pmatrix} 0 & i \\ i & 0 \end{pmatrix}.$$

The four elements on the right generate the space. We can verify immediately that they are independent. Therefore, they constitute a basis, which we denote by

$$E_1 = \begin{pmatrix} 1 & 0 \\ 0 & 1 \end{pmatrix}, \quad E_2 = \begin{pmatrix} i & 0 \\ 0 & -i \end{pmatrix}, \quad E_3 = \begin{pmatrix} 0 & 1 \\ -1 & 0 \end{pmatrix}, \quad E_4 = \begin{pmatrix} 0 & i \\ i & 0 \end{pmatrix}.$$

The elements in K can now be defined in terms of their four coordinates x_1, x_2, x_3, x_4 relative to the basis E_1, E_2, E_3, E_4. The distributivity of multiplication with respect to addition enables us now to write

$$\left(\sum_1^4 x_i E_i \right) \left(\sum_1^4 y_i E_i \right) = \sum_{i,j} x_i y_j (E_i E_j).$$

Therefore, we need only know the products of all pairs of the E_i:

$$E_1^2 = -E_2^2 = -E_3^2 = -E_4^2 = E_1 \qquad E_1 E_i = E_i E_1 = E_i$$

$$E_2 E_3 = -E_3 E_2 = E_4 \qquad E_3 E_1 = -E_1 E_3 = E_2 \qquad E_1 E_2 = -E_2 E_1 = E_3.$$

(The unit element E_1 is identified with the unit element in R and is denoted by 1.) The structure of the division ring K is then defined directly without the use of matrices.

325

First question:

The neutral element has an inverse.

If an element a has an inverse, this inverse a^{-1} has a as its inverse and hence belongs to E_1.

The product of two elements in E_1 is an element of E_1:

$$(ab)(b^{-1}a^{-1}) = a(bb^{-1})a^{-1} = e$$
$$(b^{-1}a^{-1})(ab) = b^{-1}(a^{-1}a)b = e.$$

Finally, the operation is associative in E_1 since it is assumed to be associative in E, of which E_1 is a subset.

Matrix multiplication is associative. The matrix

$$I = \begin{pmatrix} 1 & 0 \\ 0 & 1 \end{pmatrix}$$

is the neutral element. The set of nonsingular matrices is a group in accordance with the preceding result.

Second question:

If the ratios x'/y' and x/y are defined, that is, if

$$y \neq 0, \qquad cx+dy \neq 0 \qquad \text{or} \qquad x/y \neq -d/c,$$

$$\frac{x'}{y'} = \frac{ax+by}{cx+dy} = \frac{a(x/y)+b}{c(x/y)+d} = h_A\left(\frac{x}{y}\right).$$

By hypothesis, A has an inverse and hence $ad-bc \neq 0$.

The product of the matrices BA corresponds to the composition of the corresponding mappings α and β:

$$(x, y) \xrightarrow{\ \alpha\ } (x', y') \xrightarrow{\ \beta\ } (x'', y'').$$

Taking the ratios, we find

$$\frac{x}{y} \xrightarrow{\ h_A\ } \frac{x'}{y'} \xrightarrow{\ h_B\ } \frac{x''}{y''}.$$

By definition,

$$\frac{x''}{y''} = h_{BA}\left(\frac{x}{y}\right) = (h_B \circ h_A)\left(\frac{x}{y}\right).$$

Therefore,

$$h_{BA} = h_B \circ h_A.$$

The mapping φ conserves the operation and it is surjective. Also, the operation is associative since

$$(h_C h_B)h_A = h_{(CB)A} = h_{C(BA)} = h_C(h_B h_A);$$

the function h_I is a neutral element; finally, every function h has an inverse in H:

$$h_{A^{-1}} h_A = h_A h_{A^{-1}} = h_I.$$

Therefore, H is a group. (The results are shown for the functions h_A that are the images of an element in G and hence for every function in H since φ is surjective.)

The mapping φ is a homomorphism of G into H since

$$\varphi(A)\, \varphi(B) = \varphi(AB).$$

This is the relation we shall use systematically.

e is the identity mapping:

$$e\left(\frac{x}{y}\right) = \frac{x}{y}.$$

The matrices A of $\varphi^{-1}(e)$ are therefore defined by

$$a=d, \qquad b=c=0.$$

These are the matrices that are proportional to I.

Similarly, $\varphi^{-1}(h)$ consists of the matrices

$$\begin{pmatrix} a & b \\ c & d \end{pmatrix}$$

that are proportional to one of the matrices A_0 in the set.

We may say that the undecomposed homographic functions constitute a group that is isomorphic to the group Γ of nonsingular matrices defined up to a proportionality coefficient. (To be precise, it is necessary to show that we here define an equivalence relation that is compatible with the operation in accordance with the definition of exercise 111 of Part 1 and that the quotient set Γ thus has a group structure.)

326

First question:

Suppose that B and C are two matrices that commute with A. The distributivity of the product with respect to addition implies that

$$A(B-C) = AB-AC = BA-CA = (B-C)A.$$

The associativity of the product similarly implies that

$$A(BC) = (AB)C = (BA)C = B(AC) = B(CA) = (BC)A.$$

If B and C belong to E then $B-C$ and BC belong to E. Thus E is a subring of $\mathscr{A}$.

Second question:

By hypothesis, the ring E of matrices that commute with A contains B. Therefore, it also contains $Q(B)$ and A commutes with $Q(B)$. The ring of matrices that commute with $Q(B)$ contains A and hence $P(A)$. Thus, $P(A)$ and $Q(B)$ commute.

Third question:

If we multiply the product $A^{-1}B$ on the left by A, we obtain

$$A(A^{-1}B) = (AA^{-1})B = B = B(AA^{-1}) = (BA)A^{-1} = (AB)A^{-1} = A(BA^{-1}),$$

so that

$$A(A^{-1}B) = A(BA^{-1}).$$

Since A is regular (or as we can see from multiplying on the left by A^{-1}),

$$A^{-1}B = BA^{-1}.$$

327

If the rank of A is 1, the column vectors A_j generate, by definition, a vector subspace E of K^m of dimension 1. (The column vector A_j is the vector whose coordinates relative to the canonical basis of K^m are the elements a_{ij} of the jth column.)

X is a nonzero vector in E, the other vectors in E and, in particular, the A_j are proportional to X. Therefore, to every vector A_j there corresponds a number v_j such that

$$A_j = v_j X.$$

At least one of the vectors A_j is nonzero since the rank of the matrix A is 1: At least one of the v_j is nonzero. Let us denote by u_i the coordinates of X. The vector equation $A_j = v_j X$ is equivalent to the equations

$$a_{ij} = v_j u_i.$$

(We note that at least one of the u_i is nonzero since the vector X is nonzero.)

328

First question:

Suppose that

$$T = (t_{ij}) \qquad \text{and} \qquad T' = (t'_{ij})$$

are two triangular matrices. The elements of the difference $T - T'$ and of the product TT' are respectively

$$q_{i,j} = t_{i,j} - t'_{i,j}, \qquad p_{i,j} = \sum_h t_{i,h} t'_{h,j}.$$

Let us study those elements q_{ij} and p_{ij} for which $i > j$:

(a) $$t_{i,j} = t'_{i,j} = 0 \Rightarrow q_{i,j} = 0,$$

(b) $$h < i \Rightarrow t_{i,h} = 0,$$
$$h \geqslant i > j \Rightarrow h > j \Rightarrow t'_{h,j} = 0.$$

The product $t_{i,h} t'_{h,j}$ is zero for every h. Therefore, p_{ij} is zero.

We have shown that for triangular matrices T and T', the difference $T - T'$ and the product TT' are triangular matrices. Thus, the set of triangular matrices is a subring $\mathscr{A}$ of the ring of $n \times n$ matrices.

The ring $\mathscr{A}$ is not commutative. This can be seen for $n=2$, for example, from the 2×2 matrices,

$$\begin{pmatrix} 0 & 1 \\ 0 & 0 \end{pmatrix}\begin{pmatrix} 1 & 0 \\ 0 & -1 \end{pmatrix} = \begin{pmatrix} 0 & -1 \\ 0 & 0 \end{pmatrix} \quad \text{and} \quad \begin{pmatrix} 1 & 0 \\ 0 & -1 \end{pmatrix}\begin{pmatrix} 0 & 1 \\ 0 & 0 \end{pmatrix} = \begin{pmatrix} 0 & 1 \\ 0 & 0 \end{pmatrix}.$$

This example can also be considered valid for $n>2$. We need only fill out the matrices with zeros.

Second question:

The elements of the pth column of T are the coordinates of the vector $\theta(e_p)$. For T to be a triangular matrix, it is necessary and sufficient either

(1) that the coordinates t_{qp} of $\theta(e_p)$ of index $q>p$ be zero or

(2) that $\theta(e_p)$ be a linear combination of the e_i of index $i\leqslant p$, that is, that it belong to E_p.

Let us suppose that T is triangular:

$$i\leqslant p \implies \theta(e_i)\in E_i \subset E_p \implies \theta(e_i)\in E_p.$$

Since the e_i constitute a basis for E_p, $\theta(E_p)$ is generated by the $\theta(e_i)$, so that $\theta(E_p)\subset E_p$.

Conversely, if $\theta(E_p)\subset E_p$, the image of every vector in E_p and, in particular, in e_p belongs to E_p. Thus, the matrix T is triangular.

One can show without calculation that the operators θ constitute a subring of the ring of linear mappings on K^n into K^n. This implies the same property for the matrices T.

The space $(\theta-\theta')(E_p)$ is contained in the subspace generated by $\theta(E_p)$ and $\theta'(E_p)$ and hence in E_p.

Similarly, since the space $\theta'(E_p)$ is contained in E_p, we know that

$$(\theta\theta')(E_p) = \theta[\theta'(E_p)]\subset\theta(E_p)\subset E_p.$$

The mappings $\theta-\theta'$ and $\theta\theta'$ have the characteristic property, and their matrices $T-T'$ and TT' are triangular.

Third question:

Let us replace the matrix equation with the system of scalar equations, which is equivalent to it,

$$\begin{aligned}
t_{11}x_1 + \cdots\cdots + t_{1n}x_n &= 0, \\
t_{22}x_2 + \cdots\cdots + t_{2n}x_n &= 0, \\
t_{ii}x_i + \cdots + t_{in}x_n &= 0, \\
t_{nn}x_n \cdots \qquad\ &= 0.
\end{aligned}$$

If no term t_{ii} is zero, we can show by induction on $n-i$ that all the x_i are zero.

This is evident if $i=n$. Suppose that $x_n=x_{n-1}=\cdots=x_{i+1}=0$. Then, $t_{ii}x_i$, which is a linear combination of the x_j with index $j>i$, is zero. Since t_{ii} is nonzero, x_i is zero.

If $t_{ii}=0$ for the indices i_1, i_2, ..., i_p, an induction analogous to the preceding one shows that the unknowns x_{i_1}, x_{i_2},..., x_{i_p} can be taken arbitrarily. The others will then be determined. The solution matrices X thus form a p-dimensional vector space.

The solutions X of the equation $TX=0$ define the vectors in K^n that belong to the kernel of θ. Thus, θ has an inverse if no t_{ii} is zero. More generally,

$$\text{rank of } \theta = \dim K^n - \dim \theta^{-1}(0) = n-p,$$

where p is the number of terms t_{ii} that are equal to 0.

329

Let us decompose a column matrix X into the matrix X_1 of the first p rows and the matrix X_2 of the last q rows:

$$X = \begin{pmatrix} X_1 \\ X_2 \end{pmatrix}.$$

The matrix equation $Y=MX$ can then be replaced with the system

(1)
$$\begin{cases} Y_1 = M_{1,1}X_1 + M_{1,2}X_2, \\ Y_2 = M_{2,1}X_1 + M_{2,2}X_2. \end{cases}$$

which is obtained by separating, in each scalar equation, the terms containing the first p variables x_i from the other terms.

Similarly, if $X=NU$, the decomposition indicated yields

(2)
$$\begin{cases} X_1 = N_{1,1}U_1 + N_{1,2}U_2, \\ X_2 = N_{2,1}U_1 + N_{2,2}U_2. \end{cases}$$

Thus, we can replace X_1 and X_2 in the system (1) with these expressions. Then, the rules for matrix operations (associativity of the product when the product is defined and the distributivity of the product with respect to addition) enable us to write

(3)
$$\begin{cases} Y_1 = (M_{1,1}N_{1,1} + M_{1,2}N_{2,1})U_1 + (M_{1,1}N_{1,2} + M_{1,2}N_{2,2})U_2, \\ Y_2 = (M_{2,1}N_{1,1} + M_{2,2}N_{2,1})U_1 + (M_{2,1}N_{1,2} + M_{2,2}N_{2,2})U_2. \end{cases}$$

This equation is equivalent to the equation $Y=MNU$. The matrix MN is also defined by its decomposition, and

$$MN = \begin{pmatrix} M_{1,1}N_{1,1} + M_{1,2}N_{2,1} & M_{1,1}N_{1,2} + M_{1,2}N_{2,2} \\ M_{2,1}N_{1,1} + M_{2,2}N_{2,1} & M_{2,1}N_{1,2} + M_{2,2}N_{2,2} \end{pmatrix}.$$

REMARKS: (1) One does not actually need to calculate Y_1 and Y_2 as a function of U_1 and U_2. It is necessary only to note that the results of the calculation will remain valid if $p=q=1$. The formulas obtained must therefore still yield the formulas that we already know for the product of two 2×2 matrices.

(2) However, we should note the following difference: The products such as $M_{11}N_{11}$ or $M_{12}N_{21}$ ... depend on the order of the factors when the M_{ij} and N_{ij} are matrices.

(3) Finally, it is possible to consider more complicated decompositions into groups of $p_1, p_2, ..., p_h$ rows and columns, where the sum $p_1+p_2+ \cdots +p_h$ is equal to n. The product of two matrices can be expressed as if the matrices of the decomposition were numbers.

330

First question:

Study of the vector space structure of $\mathcal{M}$.

The difference between two matrices M_1 and M_2 defined by $(a_1, b_1,..., d_1')$ and $(a_2, b_2, ..., d_2')$ respectively is the matrix M defined by $(a_1 - a_2, b_1 - b_2, ... d_1' - d_2')$. If $\mathcal{M}$ contains M_1 and M_2, it also contains the difference $M_1 - M_2$. Thus, $\mathcal{M}$ is an additive subgroup of $\mathcal{N}$.

Similarly, the matrix rM_1 is the matrix M defined by $(ra_1, rb_1, ..., rd_1')$. It belongs to $\mathcal{M}$, which is a subspace of the vector space of 4×4 matrices.

Let us denote by $E_a, E_b, ..., E_{d'}$, the matrices M obtained by taking seven of the parameters $a, b, ..., d'$ equal to zero but the eighth equal to 1 (the *index* is the parameter to which the value 1 is assigned). Obviously, a matrix M has a unique decomposition of the form

$$M = aE_a + bE_b + \cdots + d'E_{d'}.$$

Thus, the matrices $E_a, E_b, ..., E_{d'}$ constitute a basis for $\mathcal{M}$, so that the dimension of $\mathcal{M}$ is 8.

To show that $\mathcal{M}$ is a subring of $\mathcal{N}$, we need to show that, for two matrices M and N in $\mathcal{M}$, the product MN also belongs to $\mathcal{M}$. To do this, we use the results of exercise 323, taking $p=q=2$:

$$M = \begin{pmatrix} M_{11} & M_{12} \\ M_{21} & M_{22} \end{pmatrix}, \qquad N = \begin{pmatrix} N_{11} & N_{12} \\ N_{21} & N_{22} \end{pmatrix}.$$

We note that a matrix M belongs to the set $\mathcal{M}$ if and only if the four matrices M_{ij} belong to the field $\mathscr{C}$ of matrices of the form $\begin{pmatrix} u & v \\ -v & u \end{pmatrix}$;

$$MN = \begin{pmatrix} M_{11}N_{11}+M_{12}N_{21} & M_{11}N_{12}+M_{12}N_{22} \\ M_{21}N_{11}+M_{22}N_{21} & M_{21}N_{12}+M_{22}N_{22} \end{pmatrix}.$$

Since $\mathscr{C}$ is a ring, the product of two matrices in $\mathscr{C}$ is a matrix in $\mathscr{C}$, and the same is true for the sum. Each of the submatrices of MN is thus a matrix in $\mathscr{C}$, and MN belongs to the set $\mathscr{M}$.

Second question:

We know that the field $\mathscr{C}$ of matrices of the form

$$\begin{pmatrix} u & v \\ -v & u \end{pmatrix}$$

is isomorphic to the field of complex numbers, that is, there exists a bijective mapping φ of $\mathscr{C}$ onto the field of complex numbers that conserves the operations of addition and multiplication.

To the matrix $M = (M_{ij})$ let us now assign the matrix

$$m = (m_{ij}) = (\varphi[M_{ij}]),$$

which we shall denote by $\psi(M)$. This mapping ψ is bijective since φ is bijective.

For two matrices M and N in $\mathscr{M}$, their sum and their product can be defined in terms of the matrices M_{ij} and N_{ij}. The formulas are the same as if M_{ij} and N_{ij} were numbers: If we take the sum and the product of m and n, we obtain the same result up to replacement of M_{ij} and N_{ij} with m_{ij} and n_{ij}; that is ,

$$\psi(M+N) = m+n = \psi(M)+\psi(N),$$
$$\psi(MN) = \quad mn = \psi(M)\psi(N).$$

Thus, ψ is an isomorphism of the rings $\mathscr{A}$ and $\mathscr{M}$.

Third question:

The mapping ψ is an isomorphism of the rings $\mathscr{A}$ and $\mathscr{M}$ and hence, in particular, of the additive groups. For ψ to be an isomorphism of the vector spaces, we need to define on $\mathscr{A}$ multiplication by a real number and to verify that $\psi(rM) = r\psi(M)$. This will be the case if

$$r\begin{pmatrix} m_{11} & m_{12} \\ m_{21} & m_{22} \end{pmatrix} = \begin{pmatrix} rm_{11} & rm_{12} \\ rm_{21} & rm_{22} \end{pmatrix}.$$

We thus define a vector space structure on $\mathscr{A}$. However, *this is not the usual structure defined on it*. The set $\mathscr{A}$ of 2×2 matrices with complex coefficients has a vector space structure E over C (cf. PZ, Book II, Chapter VIII, part 2, § 1). Here, however, we consider the vector space E' obtained from E by restricting the multiplication by a scalar to the case of a real scalar (in this connection, see exercise 318). Also, the dimension of E is 4 and the dimension of E', which is isomorphic to $\mathscr{M}$ is 8 (just as that of $\mathscr{M}$ is).

Fourth question:

By decomposing the matrices into groups of two lines or columns, we replace the matrix equation $MX=0$ with the system

$$M_{11}X_1 + M_{12}X_2 = 0,$$
$$M_{21}X_1 + M_{22}X_2 = 0.$$

We shall use the fact that the matrices M_{ij} belong to a field, namely, the field $\mathscr{C}$.

If the four matrices M_{ij} are all the zero matrix, that is, if M is the zero equation, X is obviously arbitrary.

Therefore, let us suppose that one of these matrices is nonzero, for example, the matrix M_{11}. Then,

$$\begin{cases} X_1 = -M_{11}^{-1}M_{12}X_2, \\ (-M_{21}M_{11}^{-1}M_{12}+M_{22})X_2 = 0. \end{cases}$$

In any field, in the present case the field $\mathscr{C}$, a nonzero element has an inverse and hence is regular (cf. problem 106). If the coefficient of X_2 in the second equation is not the zero matrix, X_2, and hence X_1, must be the zero vector. Thus, the only solution is the zero vector. If the coefficient of X_2 is the zero matrix, X_2 is arbitrary and the solutions X of the system constitute a two-dimensional vector subspace of R^4.

For the equation $MX=0$ to have a nonzero solution, it is necessary and sufficient that the coefficient of X_2 be the zero matrix. By using the commutativity of multiplication in $\mathscr{C}$ and multiplying by M_{11}, we obtain the condition

$$-M_{21}M_{12}+M_{11}M_{22}=0$$

or, setting the elements of this matrix equal to zero, we obtain the two conditions

$$\begin{cases} aa'+dd' = bb'+cc', \\ ab'+ba' = cd'+dc'. \end{cases}$$

REMARK: The study made in (4) can be presented in a different manner. Let us denote the system of four scalar equations, which is equivalent to the matrix equation $MX=0$, by

$$y_1 = 0, \qquad y_2 = 0, \qquad y_3 = 0, \qquad y_4 = 0.$$

These four equations can be replaced by the equivalent system in which the unknowns are the complex numbers $x+iy$ and $z+it$:

$$y_1+iy_2 = 0, \qquad y_3+iy_4 = 0.$$

Examination of this system is identical to that made above by virtue of the isomorphism between the field of matrices of the form

$$\begin{pmatrix} u & v \\ -v & u \end{pmatrix}$$

and the field of complex numbers.

331

First question:

The method employed is that of elimination. Let us suppose first that $\lambda \neq 0$.

$$x_i = \frac{b_i - a_i x_{n+1}}{\lambda}, \qquad (i = 1, 2, ..., n),$$

When we replace the x_i with the expression given for them in the last expression, we have

$$x_{n+1}\left(\lambda^2 - \sum_{i=1}^n a_i^2\right) = \lambda b_{n+1} - \sum_{i=1}^n a_i b_i.$$

If

$$\lambda^2 - \sum_1^n a_i^2 \neq 0,$$

the system (S) has a unique solution. The number x_{n+1} is given by the last equation, and the other x_i are given in terms of x_{n+1}.
    If

$$\lambda^2 - \sum_1^n a_i^2 = 0 \qquad \text{and} \qquad \lambda b_{n+1} - \sum_{i=1}^n a_i b_i \neq 0,$$

the system (S) has no solution.
    If

$$\lambda^2 - \sum_1^n a_i^2 = 0 \qquad \text{and} \qquad \lambda b_{n+1} - \sum_1^n a_i b_i = 0,$$

the last equation is satisfied whatever x_{n+1} may be. The choice of x_{n+1} determines those x_i of index $i \leq n$. Thus, the solutions depend on an arbitrary solution.
    Let us suppose now that $\lambda = 0$. The first n equations are compatible if and only if

$$\frac{b_1}{a_1} = \frac{b_2}{a_2} = \cdots = \frac{b_n}{a_n}.$$

(if one of the a_i is zero, the corresponding b_i must also be zero). The number x_{n+1} is equal to the common value of these ratios, and the last equation yields a relation between the x_i of index $i \leqslant n$. The solutions depend on $n-1$ arbitrary solutions.

Second question:

The determinant D_n is given by

$$D_n = \begin{vmatrix} \lambda & & & & a_1 \\ & \lambda & 0 & & a_2 \\ & & \ddots & & \vdots \\ 0 & & & & a_n \\ a_1 & a_2 & \cdots & a_n & \lambda \end{vmatrix}$$

and we can calculate without difficulty $D_1 = \lambda^2 - a_1^2$ and $D_2 = \lambda(\lambda^2 - a_1^2 - a_2^2)$. Let us determine D_n by induction and let us assume that

$$D_{n-1} = \lambda^{n-2}(\lambda^2 - a_1^2 - a_2^2 - \cdots - a_{n-1}^2),$$

a result established for the first two terms.

Let us expand D_n in terms of elements of the first row:

$$D_n = \lambda \Delta_1 + (-1)^{n+2} a_1 \Delta_{n+1},$$

where Δ_1 and Δ_{n+1} are the minors of λ and a_1 in D_n; that is,

$$\Delta_1 = \begin{vmatrix} \lambda & & & a_2 \\ & \ddots & 0 & \vdots \\ 0 & & & a_n \\ a_2 & \cdots & a_n & \lambda \end{vmatrix} \qquad \Delta_{n+1} = \begin{vmatrix} 0 & \lambda & 0 & 0 \\ & & \ddots & 0 \\ 0 & & & \lambda \\ a_1 & a_2 & \cdots & a_n \end{vmatrix}.$$

Δ_1 is a determinant of the same form as D_{n-1}, the parameters being $a_2, \ldots, a_n$. Therefore,

$$\Delta_1 = \lambda^{n-2}(\lambda^2 - a_2^2 - a_3^2 - \cdots - a_n^2).$$

In each of the first $n-1$ rows of Δ_{n+1}, there is a single nonzero term. In the expansion of this determinant, there is one and only one nonzero term. This is the one defined by the permutations $\left(\begin{smallmatrix} 1,2,\ldots,n \\ 1,2,\ldots,n \end{smallmatrix}\right)$ of the rows and $\left(\begin{smallmatrix} 1,2,\ldots,n-1,n \\ 2,3,\ldots,n,1 \end{smallmatrix}\right)$ of the columns. These permutations involve 0 and $n-1$ transpositions respectively:

$$\Delta_{n+1} = (-1)^{n-1}\lambda^{n-1}a_1$$
$$D_n = \lambda^{n-1}(\lambda^2 - a_2^2 - a_3^2 - \cdots - a_n^2) + (-1)^{2n-3}\lambda^{n-1}a_1^2$$
$$= \lambda^{n-1}(\lambda^2 - a_1^2 - a_2^2 - \cdots - a_n^2).$$

Thus, the formula that is valid for D_{n-1} remains valid for D_n.

A necessary and sufficient condition for the system (S) to have a unique solution is that D_n be nonzero, and for this it is necessary and sufficient that

$$\lambda \neq 0 \qquad \text{and} \qquad \lambda^2 - a_1^2 - a_2^2 - \cdots - a_n^2 \neq 0.$$

332

The system mentioned may be written

$$A(\lambda_1) = A(\lambda_2) = A(\lambda_3) = A(\lambda_4) = 0 \qquad A(\mu) = 1.$$

The numbers λ_1, λ_2, λ_3, and λ_4 are the zeros of A. We shall assume them to be distinct. The polynomial A is of degree 6 and it has two other zeros, which we denote by λ' and λ''.

To determine λ' and λ'', let us use the relationships between the zeros of a polynomial and its coefficients:

coefficient of u^5, $\qquad 0 = \lambda_1 + \lambda_2 + \lambda_3 + \lambda_4 + \lambda' + \lambda''$

coefficient of u^3, $\qquad 0 = \sum_{i,j,k} \lambda_i \lambda_j \lambda_k + (\lambda' + \lambda'') \sum_{i,j} \lambda_i \lambda_j + \lambda' \lambda'' \sum_i \lambda_i.$

We shall denote by s_1, s_2, s_3, and s_4 the sums of the products of all combinations of one, two, three, and four of the λ_i respectively and we shall assume that s_1 is nonzero:

$$\lambda' + \lambda'' = -s_1, \qquad \lambda' \lambda'' = \frac{s_1 s_2 - s_3}{s_1},$$

from which we get

$$A = a(\lambda - \lambda_1)(\lambda - \lambda_2)(\lambda - \lambda_3)(\lambda - \lambda_4)\left(\lambda^2 + s_1\lambda + \frac{s_1 s_2 - s_3}{s_1}\right).$$

Finally, we obtain the coefficient a by setting $A(\mu) = 1$ and assuming that none of the factors in the denominator is 0:

$$a = \frac{1}{(\mu - \lambda_1)(\mu - \lambda_2)(\mu - \lambda_3)(\mu - \lambda_4)\left(\mu^2 + s_1\mu + \dfrac{s_1 s_2 - s_3}{s_1}\right)}.$$

The values of the unknowns are the coefficients of the polynomial A, which we write

$$A = a(\lambda^4 - s_1\lambda^3 + s_2\lambda^2 - s_3\lambda + s_4)\left(\lambda^2 + s_1\lambda + \frac{s_1 s_2 - s_3}{s_1}\right).$$

Then, the values of the unknowns are given by

$$x_1 = a\frac{s_4(s_1 s_2 - s_3)}{s_1}, \qquad x_2 = a\frac{s_1^2 s_4 + s_3^2 - s_1 s_2 s_3}{s_1},$$

$$x_3 = a\frac{s_1 s_4 - s_1^2 s_3 + s_1 s_2^2 - s_2 s_3}{s_1}, \qquad x_4 = a\frac{2 s_1 s_2 - s_1^3 - s_3}{s_1}, \qquad x_5 = a.$$

Discussion: The above study assumes that the λ_i are distinct, that $s_1 = \sum_i \lambda_i$ is nonzero, and that μ does not cause any of the factors of A to vanish. Let us now consider the contrary cases.

If two of the λ_i are equal, there are only three zeros of A and we can take one of these arbitrarily. The solutions depend on an arbitrary solution (which is obvious since two of the equations are identical).

If $s_1 = 0$, the equation given by the coefficient of u^3 becomes $s_3 = 0$. The system is impossible if $s_3 \neq 0$. If $s_3 = 0$, the product $\lambda'\lambda''$ is indeterminate, and the solutions of the system depend on an arbitrary solution.

Finally, if $\mu = \lambda_i$ or if $\mu^2 + s_1\mu + (s_1 s_2 - s_3)/s_1 = 0$, the system is obviously impossible.

333

First question:

The rank of a system of linear forms is equal to the dimension of the vector space of the dual that these forms generate. We shall determine this number by a method of substitution, that is, by replacing the basis x, y, z, t with other bases of the dual.

Let us suppose first that $a \neq 0$. Then, we can solve the first two equations for x and y:

$$x = \frac{Y - cz - bt}{a}, \qquad y = \frac{X - bz - ct}{a}.$$

The forms X, Y, z, t constitute a basis for the dual space, and the other two forms are given by

$$Z = \frac{1}{a}[cX + bY - 2bcz + (a^2 - b^2 - c^2)t],$$

$$T = \frac{1}{a}[bX + cY + (a^2 - b^2 - c^2)z - 2bct].$$

The four forms X, Y, Z, and T will be independent if

$$D = 4b^2c^2 - (a^2 - b^2 - c^2)^2 \neq 0$$

$$D = (a+b+c)(a+b-c)(a+c-b)(b+c-a).$$

If $D=0$ but $bc \neq 0$, the three forms X, Y, and Z are independent. This is a system of rank 3.

If $D=0$ and $bc=0$, then $a^2 - (b^2 + c^2) = 0$ and Z and T are linear combinations of X and Y. This is a system of rank 2. (In this case, $b=0$ and $a = \pm c$ or $c=0$ and $a = \pm b$.)

Suppose now that $a=0$. Then, if $b^2 - c^2 \neq 0$, the forms X and Y are independent of x and y and the forms Z and T are independent of z and t. The rank of this system is 4.

If $b^2 - c^2 = 0$ but b and c are nonzero, the rank of the system is 2 since

$$cX - bY = 0 \quad \text{and} \quad cZ - bT = 0.$$

Finally, if $b=c=0$, all the forms are zero forms.

In sum, the rank of the system is 4 if $D \neq 0$; it is 3 if one of the factors of D is zero; it is 2 if two of the factors of D are zero.

REMARK: If we perform the same permutation on x, y, z, and t on the one hand and X, Y, Z, and T on the other, we obtain a system of the same form as the given system, with the parameters a, b, and c permuted. Therefore, we can make our analysis under the assumption that the parameter $a \neq 0$ and deduce from it the discussion for the case of $a=0$ without further calculations.

Second question:

To evaluate the determinant Δ of the system of forms, let us add to the elements in the first row the elements in the remaining rows multiplied by u, v, and w respectively:

$$\Delta = \begin{vmatrix} ua+vb+wc & a+vc+wb & b+uc+wa & c+ub+va \\ a & 0 & c & b \\ b & c & 0 & a \\ c & b & a & 0 \end{vmatrix}.$$

Each element in the first row is now equal to

$$\begin{aligned} a+b+c & \quad \text{if} & u = v = w = 1 \\ \pm(a-b-c) & \quad \text{if} & u = -v = -w = 1 \\ \pm(a+c-b) & \quad \text{if} & -u = v = -w = 1 \\ \pm(a+b-c) & \quad \text{if} & -u = -v = w = 1. \end{aligned}$$

The determinant Δ can be treated as a polynomial in a single variable a that ranges over R. This polynomial is divisible by each of the factors $a+b+c$, $a-b-c$, $a+c-b$, and $a+b-c$ and by their product if these factors are different (that is, if $bc(b+c)(b-c)\neq 0$). Since Δ is a fourth-degree polynomial, it is proportional to the product of the four factors

$$\Delta = k(a+b+c)(a-b-c)(a+c-b)(a+b-c)$$

and the coefficient k can be determined by setting the terms in a^4 equal to each other. This gives us $k=1$.

This identity is rigorously proven only if $bc(b+c)(b-c)\neq 0$. However, the result is valid for all values of a, b, and c. That is, the procedure used in the proof is not applicable except under the condition stated, but the algebraic identity remains valid anyway.

We again get the result that the forms are independent if

$$\Delta = -D = (a+b+c)(a-b-c)(a+c-b)(a+b-c) \neq 0.$$

Third question:

CASE 1: $D\neq 0$. The rank of the system is 4, and the system has a unique solution.

CASE 2: $D=0$ and $abc\neq 0$. The rank of the system is now 3. The four forms are dependent: If u, v, w are each equal to ± 1, the factors of D can be written $ua+vb+wc$, and we can verify that

$$ua+vb+wc = 0 \Rightarrow X+uY+vZ+wT = 0.$$

For the system to have a solution, it is necessary and sufficient that the right-hand members verify the same relation:

$$a+b+c+ua+vb+wc = a+b+c = 0.$$

SUBCASE 1. $u=v=w=1$. The a, b, and c verify the condition $a+b+c=0$. The relation between the right-hand members is satisfied, and the system has solutions that depend on an arbitrary solution.

SUBCASE 2.   Two of the numbers, u and v for example, are equal to 1 and the third is equal to -1. But the two conditions $a+b+c=0$ and $a+b-c=0$ then imply that $c=0$ in contradiction with the hypothesis that $abc\neq 0$. Thus, the system has no solution.

CASE 3:   One of the parameters is equal to 0. Let us suppose that this parameter is a. The rank of the system is 4 if $b^2-c^2\neq 0$, the rank is 2 if $b=uc$ or if $u=+1$. The forms X, Y, Z, and T then verify the two relations

$$X-uY = 0, \qquad Z-uT = 0.$$

For the system to have solutions, it is necessary and sufficient that

$$b+c = 0, \qquad b-uc = 0.$$

Thus, the system has solutions that depend on two arbitrary solutions if $b+c = 0$, but it has no solution if $b-c=0$.

334

The transpose A^T of A, defined by $A^T=(a_{ji})$ is in this case equal to $-A$, and we know that

$$\det A^T = \det A = \det(-A) = (-1)^n \det A.$$

If n is odd,

$$\det A = -\det A = 0.$$

335

First question:

The value of the determinant D is unchanged when we add to the elements of the first column those of the other columns multiplied respectively by $u, ..., u^{n-1}$. The elements in the first column then become

$$a_1 + a_2 u + \cdots + a_n u^{n-1},$$
$$a_n + a_1 u + \cdots + a_{n-1} u^{n-1},$$
$$\cdots\cdots\cdots\cdots\cdots\cdots$$
$$a_2 + a_3 u + \cdots + a_1 u^{n-1}.$$

They are all proportional if $u^n=1$ (each of them can be derived from the preceding one by multiplication by u), and D regarded as a polynomial in a_1 is divisible by the n factors obtained when we take successively for u the nth roots of unity: $u_h = e^{2h\pi i/n}$. Except for particular values of $a_2, ..., a_n$, the polynomial D is divisible by the product of these factors:

$$D = d \prod_{h=0}^{n-1} (a_1 + a_2 u_h + \cdots + a_n u_h^{n-1}),$$

where d is a constant since all the prime factors of D, which is of degree n, have been obtained. Therefore, d is equal to the coefficient of a_1^n in D, namely, 1.

Finally, this algebraic identity in $a_1, a_2, ..., a_n$, which we have proven except for particular values of $a_2, ..., a_n$, is valid regardless of what $a_1, a_2, ..., a_n$ are.

Second question:

In this case,

$$D = \prod_{h=0}^{n-1} (1+2u_h+\cdots+nu_h^{n-1}),$$

$$1+2u+\cdots+nu^{n-1} = \frac{d}{du}(1+u+\cdots+u^n) = \frac{d}{du}\left(\frac{u^{n+1}-1}{u-1}\right)$$

$$= \frac{nu^{n+1}-(n+1)u^n+1}{(u-1)^2}.$$

If u is a root of unity u_h other than 1,

$$\frac{nu_h^{n+1}-(n+1)u_h^n+1}{(u_h-1)^2} = \frac{nu_h-(n+1)+1}{(u_h-1)^2} = \frac{n}{u_h-1}$$

$$D = (1+2+\cdots+n) \prod_{h=1}^{n-1} \frac{n}{u_h-1}.$$

The u_h are the zeros of the polynomial u^n-1, and the $v_h=u_h-1$ are the zeros of the polynomial

$$(v+1)^n-1 = v[v^{n-1}+\cdots+n].$$

Therefore, the product of all the nonzero v_h is $(-1)^{n-1}n$.

$$D = \frac{n(n+1)}{2} \times n^{n-1} \frac{1}{(-1)^{n-1}n} = (-1)^{n-1} \frac{n^{n-1}(n+1)}{2}.$$

Third question:

In this case,

$$D = \prod_{h=0}^{n-1} (1+u_h+\cdots+u_h^{p-1})$$

if $\qquad u_h \neq 1 \qquad 1+u_h+\cdots+u_h^{p-1} = \frac{u_h^p-1}{u_h-1}$

$$D = p \prod_{h=1}^{n-1} \frac{u_h^p-1}{u_h-1}.$$

If p is not prime to n and if q is the greatest common divisor of p and n, then $p=qp_1$, $n=qn_1$, and

$$(e^{2n_1\pi i/n})^p = e^{2pn_1\pi i/n} = e^{2ip_1\pi} = 1.$$

The factor $u_{n_1}^p - 1$ is zero and

$$D = 0.$$

If p is prime to n, the numbers u_h^p (for $h=0, 1, ..., n-1$) are all distinct and

$$u_h^p = u_l^p \Leftrightarrow \frac{2hp\pi}{n} = \frac{2lp\pi}{n} \text{ mod } 2\pi$$

Here, n must divide the product $(h-l)p$. It is prime to p and hence must divide $h-l$, which is impossible since $h-l<n$. Since $u_0^p=1^p=1$, the numbers u_h^p (for $h=1, ..., n-1$) are all distinct and different from 1. They are the roots of unity (other than 1 itself). Thus, they are the u_h, possibly rearranged.

$$\prod_{h=1}^{n-1} (u_h^p-1) = \prod_{h=1}^{n-1} (u_h-1)$$

$$D = p.$$

336

The term y_{ij} in the product AX has the value

$$y_{ij} = \sum_h a_{ih}x_j^{h-1} = P_i(x_j).$$

The term δ_{ij} in Δ_n is equal to $(i+j-1)^n$ (for i and $j=1, 2, ..., n+1$). Therefore, we can set $\delta_{ij}=P_i(x_j)$ if $x_j=j-1$. In the notations of the preceding question, we have

$$\Delta_n = \det Y = \det A \det X.$$

The determinant of X is the Vandermonde determinant of the values $\lambda_j=j-1$. Therefore,

$$\det X = \prod_{i>j} (i-j) = (n!)[(n-1)!]\cdots(2!).$$

(Here, we take successively the terms for which $i=n$, $i=n-1$, ...)

The determinant of the matrix A of the coefficients in the polynomials $(x+i)^n$ is

$$
\begin{vmatrix}
1 & \binom{n}{1} \cdots\cdots\cdots\cdots\cdots \binom{n}{n-1} & 1 \\
2^n & \binom{n}{1}2^{n-1} \cdots\cdots 2\binom{n}{n-1} & 1 \\
\vdots & & \\
(n+1)^n & \binom{n}{1}(n+1)^{n-1}\cdots(n+1)\binom{n}{n-1} & 1
\end{vmatrix}
$$

$$\underbrace{}_{\det A}$$

$$
= \binom{n}{1}\binom{n}{2}\binom{n}{n-1}
\underbrace{
\begin{vmatrix}
1 & 1\cdots\cdots & 1 & 1 \\
2^n & 2^{n-1}\cdots\cdots & 2 & 1 \\
\vdots & & & \\
(n+1)^n & (n+1)^{n-1}\cdots n+ & 11
\end{vmatrix}
}_{\alpha}
$$

The terms of the determinant α are the terms of a Vandermonde determinant. But to have a true Vandermonde determinant, we must take care of the order of the columns; that is, we must make the permutation $\binom{1,\ 2,\ \cdots,\ n+1}{n+1,\ n,\ \cdots,\ 1}$ of the columns The number of transpositions in this permutation is

$$
n+(n-1)+(n-2)+\cdots+1 = \frac{n(n+1)}{2}
$$

$$
\alpha = (-1)^{n(n+1)/2}
\begin{vmatrix}
1 & 1\cdots\cdots & 1 \\
1 & 2\cdots\cdots & 2^n \\
\vdots & & \\
1 & n+1\cdots(n+1)^n
\end{vmatrix}
= (-1)^{n(n+1)/2}\prod_{i>j}(i-j).
$$

The differences $i-j$ are the same as for the determinant of X. Therefore,

$$
\alpha = (-1)^{n(n+1)/2}(n!)[(n-1)!]\cdots(2!).
$$

The preceding results enable us to write

$$
\Delta_n = (-1)^{n(n+1)/2}\{(n!)[(n-1)!]\cdots(2!)\}^2\binom{n}{1}\cdots\binom{n}{n-1}
$$

$$
= (-1)^{n(n+1)/2}\{n![(n-1)!]\cdots(2!)\}^2\frac{n!}{(n-1)!1!}\frac{n!}{(n-2)!2!}\cdots\frac{n!}{1!(n-1)!}.
$$

Simplifying this, we get

$$
\Delta_n = (-1)^{n(n+1)/2}(n!)^{n+1}
$$

337

$$\text{rank } A = n-1 \Rightarrow \det A = 0 \Rightarrow AB = 0.$$

If a and b are the linear mappings assigned to A and B, the image $(ab)(x)$ of every vector x is 0. Therefore, the image $b(x)$ belongs to the kernel of a for every x, and the vector space $b(C^n)$ is contained in the kernel of a.

We know that the mappings a and b have the same rank as A and B. Since the rank of a is $n-1$, the dimension of its kernel is 1. The dimension of $b(C^n)$, that is, the rank of b, is at most 1. In fact, it is exactly 1 since at least one of the elements A_{ij} is nonzero (the rank of A is $n-1$).

If the rank of A does not exceed $n-2$, all the coefficients A_{ij} are zero and B is the zero matrix.

338

The equations of the system (S) express the fact that the numbers λ_i are the zeros of the rational function f, that is, of the zeros of the polynomial constituting the numerator.

Let us suppose that the numbers a, b, c, and d are distinct and that the λ_i are distinct.

$$f(u) = \frac{F(u)}{(a+u)(b+u)(c+u)(d+u)}.$$

F is a fourth-degree polynomial whose fourth-degree term is $-u^4$.

$$F(u) = -(u-\lambda_1)(u-\lambda_2)(u-\lambda_3)(u-\lambda_4).$$

Here, x, y, z, and t are the coefficients in the decomposition of the fraction f into simple elements. We find

$$x = \frac{F(-a)}{(b-a)(c-a)(d-a)} = -\frac{(a+\lambda_1)(a+\lambda_2)(a+\lambda_3)(a+\lambda_4)}{(b-a)(c-a)(d-a)}$$

and analogous formulas for y, z, and t.

Discussion: If the numbers a, b, c, and d are distinct and if $\lambda_1 = \lambda_2$, we can determine only three zeros of F. The fourth is arbitrary. The system (S) has solutions that depend on an arbitrary solution. (This result is obvious since the four equations are reduced to three.)

If $a = b$, the polynomial constituting the numerator of f is of only third degree and cannot have four zeros. The system is impossible if the λ_i are distinct.

In general, if all of the four numbers a, b, c, and d are equal, the fraction to which f is reduced when the possible cancellations are made has a numerator and denominator of degree $4-p$. If the number q of distinct values λ_i exceeds $4-p$, the problem is impossible since the numerator cannot have more than $4-p$ zeros. If $q \leqslant 4-p$, the system has solutions depending on $4-p-q$ arbitrary solutions. (These results could also be easily found by a direct examination of the system (S).)

339

$$D_2 = a_1 a_2 + a_1 b_2 + a_2 b_1$$

$$D_3 = a_1 a_2 a_3 + a_1 a_2 b_3 + a_2 a_3 b_1 + a_3 a_1 b_2.$$

Thus, we shall assume that D_{n-1} can be expressed in the form

$$D_{n-1} = \prod_{i=1}^{n-1} a_i + \sum_{i=1}^{n-1} \left(b_i \prod_{\substack{j \neq i \\ j \leqslant n-1}} a_j \right)$$

(this result being established for $n-1=2$ and $n-1=3$).

To evaluate D_n, we shall use the following property: *A determinant is a linear function of the elements of any one of its columns* (in the present case, the nth column).

$$D_n = \begin{vmatrix} a_1+b_1 & b_1 & \cdots b_1 \\ b_2 & a_2+b_2 & \cdots b_2 \\ \cdots & \cdots & \cdots \\ b_n & b_n & b_n \end{vmatrix} + \begin{vmatrix} a_1+b_1 & b_1 & \cdots b_1 & 0 \\ b_2 & a_2+b_2 & \cdots b_2 & 0 \\ \cdots & \cdots & \cdots & \cdots \\ b_n & b_n & \cdots b_n & a_n \end{vmatrix} = D'_n + D''_n.$$

To evaluate the first determinant D'_n, let us subtract the elements in the last column from those in the other columns:

$$D'_n = \begin{vmatrix} a_1 & 0 & 0 \cdots b_1 \\ 0 & a_2 & 0 \cdots b_2 \\ \vdots & \vdots & \vdots \\ 0 & 0 & \cdots b_n \end{vmatrix} = a_1 a_2 \cdots a_{n-1} b_n = b_n \prod_{i=1}^{n-1} a_i.$$

By expanding the second determinant D''_n in terms of the elements of the last column, we see that

$$D''_n = a_n D_{n-1}.$$

Thus,

$$D_n = b_n \prod_1^{n-1} a_i + a_n \left[\prod_1^{n-1} a_i + \sum_{i=1}^{n-1} (b_i \prod_{\substack{j \neq i \\ j \leq n-1}} a_j) \right]$$

$$= \prod_1^n a_i + \sum_{i=1}^n (b_i \prod_{\substack{j \neq i \\ j \leq n}} a_j).$$

D_n has the same expression as D_{n-1} (except for replacement of $n-1$ with n). Thus, we have proven the formula by induction on n.

The determinant of A is equal to $P(0)$ and, by hypothesis, is nonzero. Therefore, the matrix A is nonsingular.

First method:

$$A(A^{-1} - \lambda I) = I - \lambda A = -\lambda(A - \frac{1}{\lambda} I).$$

Since the determinant of the product of two matrices is the product of the determinant of these matrices, we have

$$\det[A(A^{-1} - \lambda I)] = (\det A)[\det(A^{-1} - \lambda I)] = P(0)R(\lambda).$$

If we multiply a matrix by μ, we must multiply its determinant by μ^n:

$$\det[(-\lambda)(A - \frac{1}{\lambda} I)] = (-\lambda)^n \det(A - \frac{1}{\lambda} I) = (-1)^n \lambda^n P\left(\frac{1}{\lambda}\right).$$

Therefore,

$$P(0)R(\lambda) = (-1)^n \lambda^n P\left(\frac{1}{\lambda}\right).$$

Second method:

We know that if K is a subfield of C, there exists a nonsingular matrix T such that the matrix $B = T^{-1}AT$ will be a triangular matrix. The terms b_{ii} of the principal diagonal are the characteristic values λ_i, each appearing a number of times equal to its multiplicity. By hypothesis, none of these is zero (we assume that $P(0) \neq 0$).

We see immediately that $B^{-1} = T^{-1}A^{-1}T$. The characteristic polynomials of B^{-1} and A^{-1} are therefore identical. A simple calculation shows that the elements along the principal diagonal of B^{-1} are equal to

$$\frac{1}{b_{ii}} = \frac{1}{\lambda_i}.$$

From this, we conclude that, for the characteristic polynomial R of B,

$$R(\lambda) = (-1)^n \prod_{i=1}^{n} \left(\lambda - \frac{1}{\lambda_i}\right) = (-1)^n \prod_{i=1}^{n} \frac{\lambda}{\lambda_i}\left(\lambda_i - \frac{1}{\lambda}\right),$$

but

$$\prod_{i=1}^{n} \left(\lambda_i - \frac{1}{\lambda}\right) = (-1)^n \prod_{i=1}^{n} \left(\frac{1}{\lambda} - \lambda_i\right) = P\left(\frac{1}{\lambda}\right)$$

$$\prod_{i=1}^{n} \lambda_i = \text{constant term of } P = P(0)$$

$$R(\lambda) = \frac{(-1)^n \lambda^n P(1/\lambda)}{P(0)}.$$

341

First question:

By the definition of the product of two matrices, we have

$$VA = \left(\sum_i a_{i1}, \sum_i a_{i2}, ..., \sum_i a_{in}\right).$$

If A belongs to E, we have $VA = s(A)V$. Conversely, if $VA = \lambda V$, the sum $\sum_i a_{ij}$ of the elements in a column is equal to λ for every j. A belongs to E and $s(A) = \lambda$.

Second question:

Let us show, by using the characteristic property of (1), that, if A and B belong to E, then $A - B$ and AB also belong to E:

$$V(A - B) = VA - VB = s(A)V - s(B)V = [s(A) - s(B)]V$$
$$V(AB) = (VA)B = s(A)VB = s(A)s(B)V.$$

Thus, we have shown that E is a subring of $\mathscr{A}$ and that the mapping s of E into K conserves the operations. This mapping is a ring homomorphism.

Let us multiply the two members of the equation $AA^{-1} = A$ on the left by V:

$$V(AA^{-1}) = (VA)A^{-1} = s(A)VA^{-1} = V.$$

A^{-1} belongs to E, and

$$s(A^{-1}) = \frac{1}{s(A)}.$$

Third question:

The eigenvector equations are

$$X_i = \sum_{j=1}^{n} a_{ij}x_j - \lambda x_i = 0, \qquad (i = 1, 2, ..., n).$$

If we add these equations, we obtain

$$\sum_{i=1}^{n} X_i = \sum_{i=1}^{n}\left(\sum_{j=1}^{n} a_{ij}x_j\right) - \lambda \sum_{i=1}^{n} x_i = \sum_{j=1}^{n} x_j\left(\sum_{i=1}^{n} a_{ij}\right) - \lambda\left(\sum_{i=1}^{n} x_i\right)$$

$$= s(A)\left(\sum_{j=1}^{n} x_j\right) - \lambda\left(\sum_{i=1}^{n} x_i\right) = (s(A) - \lambda)\left(\sum_{i=1}^{n} x_i\right).$$

Here, we used the commutativity and associativity of addition in K.

If $\lambda = s(A)$, there is at least one linear relationship between the left-hand members of the equations

$$\sum_i X_i = 0.$$

The rank of the system is at most $n-1$, and the system admits nonzero solutions. A characteristic value of it is $s(A)$.

Fourth question:

The eigenvector equations of (3) imply that

$$|\lambda|\,|x_i| = \left|\sum_{j=1}^{n} a_{ij}x_j\right| \leqslant \sum_{j=1}^{n} |a_{ij}|\,|x_j| = \sum_{j=1}^{n} a_{ij}|x_j|.$$

If we add these inequalities, we obtain

$$|\lambda| \sum_{i=1}^{n} |x_i| \leqslant s(A)\left(\sum_{j=1}^{n} |x_j|\right)$$

$$|\lambda| \leqslant s(A).$$

342

First question:

The eigenvector equations are

$$\begin{cases} (1-\lambda)x + & 3y & = 0 \\ & 3x - (2+\lambda)y & -z = 0 \\ & -y + (1-\lambda)z & = 0. \end{cases}$$

CASE 1: $\lambda=1$. Here, the system reduces to $y=0$, $z=3x$. The number 1 is a characteristic value and $(1, 0, 3)$ is an eigenvector.

CASE 2: $\lambda \neq 1$. The system is equivalent to

$$
\begin{cases}
y = (1-\lambda)z \\
x = -\dfrac{3y}{1-\lambda} = -3z \\
z[-9-(2+\lambda)(1-\lambda)-1] = 0.
\end{cases}
$$

The coefficient of z must be zero. Then,

$$\lambda^2+\lambda-12 = 0 \Rightarrow \lambda = 3 \text{ or } \lambda = -4.$$

We obtain eigenvectors by assigning to one of the variables, z for example, an arbitrary value, here, -1.

Therefore, the characteristic values are 1, 3, and -4, and the coordinates of the corresponding eigenvectors are

$$
V_1 \begin{cases} 1 \\ 0 \\ 3 \end{cases}
\qquad
V_2 \begin{cases} 3 \\ 2 \\ -1 \end{cases}
\qquad
V_3 \begin{cases} 3 \\ -5 \\ -1. \end{cases}
$$

(Each of these vectors can obviously be replaced by a vector proportional to it.)

We can easily verify that these three vectors are independent and that they constitute a basis for $\mathbf{R}^3$. We know this *a priori* since the characteristic values are all distinct.

Second question:

The mapping of $\mathbf{R}^3$ into $\mathbf{R}^3$ corresponding to A, that is, the mapping of which the matrix with the canonical basis of $\mathbf{R}^3$ is A, can be defined by representing the vectors with column matrices X and Y made up of their coordinates in the canonical basis of $\mathbf{R}^3$ by using the relation $Y=AX$.

A change of coordinates will be defined by a nonsingular matrix T if we set

$$X = TX', \qquad Y = TY'.$$

Then,

$$Y' = T^{-1}ATX'.$$

For the matrix $T^{-1}AT$ to be diagonal, it is necessary and sufficient that the new basis vectors be eigenvectors. Let us denote such eigenvectors by V_1, V_2, and V_3:

$$X'_1 = \begin{pmatrix} 1 \\ 0 \\ 0 \end{pmatrix} \Rightarrow X_1 = TX'_1 = \begin{pmatrix} 1 \\ 0 \\ 3 \end{pmatrix}.$$

But the matrix TX'_1 is the first column of the matrix T. Therefore, the terms in this column are the coordinates of V_1.

Similarly, the terms in the second column are the coordinates of V_2, and those in the third column are the coordinates of V_3:

$$T = (V_1|V_2|V_3) = \begin{pmatrix} 1 & 3 & 3 \\ 0 & 2 & -5 \\ 3 & -1 & -1 \end{pmatrix}.$$

To determine T^{-1}, we can calculate the minors and the determinant of T or we can solve the system

$$\begin{cases} x = x' + 3y' + 3z' \\ y = 2y' - 5z' \\ z = 3x' - y' - z'. \end{cases}$$

Either way, we obtain

$$T^{-1} = \begin{pmatrix} \frac{1}{10} & 0 & \frac{3}{10} \\ \frac{3}{14} & \frac{1}{7} & -\frac{1}{14} \\ \frac{3}{35} & -\frac{1}{7} & -\frac{1}{35} \end{pmatrix} = \frac{1}{70} \begin{pmatrix} 7 & 0 & 21 \\ 15 & 10 & -5 \\ 6 & -10 & -2 \end{pmatrix}.$$

REMARK: Obviously, T is not unique. The V_1, V_2, and V_3 that we have been using are determined only up to a constant factor, and, what is more, their order is arbitrary.

Therefore, the results given here are by no means the only ones possible.

However, in any case, we need to verify that the columns of the matrix T that we have obtained are indeed the coordinates of eigenvectors of A, that is, that they are proportional to the vectors V_1, V_2, and V_3 given in this solution.

CHECK:

$$AT = \begin{pmatrix} 1 & 9 & -12 \\ 0 & 6 & 20 \\ 3 & -3 & 4 \end{pmatrix}$$

$$T^{-1}AT = \frac{1}{70}\begin{pmatrix} 70 & 0 & 0 \\ 0 & 210 & 0 \\ 0 & 0 & -280 \end{pmatrix} = \begin{pmatrix} 1 & 0 & 0 \\ 0 & 3 & 0 \\ 0 & 0 & 4 \end{pmatrix}.$$

Thus, we have indeed obtained a diagonal matrix every element of which is the characteristic value relative to the eigenvector whose coordinates appear in the corresponding column of T.

343

The problem is the same as the one treated in exercise 342. Therefore, we shall give the results without comment.

(1) *Determination of the eigenvectors.*

$$\begin{cases} (1-\lambda)x + y = 0 \\ -x+(2-\lambda)y+z = 0 \\ x+(1-\lambda) \quad z = 0 \end{cases} \Leftrightarrow \begin{cases} x = (\lambda-1)z \\ y = (\lambda-1)x = (\lambda-1)^2z \\ z(2-\lambda)[1+(\lambda-1)^2] = 0. \end{cases}$$

Characteristic values: $\lambda_1=2$, $\lambda_2=1+i$, $\lambda_3=1-i$.
The corresponding eigenvectors:

$$V_1\begin{cases} 1 \\ 1 \\ 1 \end{cases} \qquad V_2\begin{cases} i \\ -1 \\ 1 \end{cases} \qquad V_3\begin{cases} -i \\ -1 \\ 1. \end{cases}$$

(2) *The matrices T and T^{-1}.* For T, we take the matrix whose column vectors are V_1, V_2, V_3:

$$T = \begin{pmatrix} 1 & i & -i \\ 1 & -1 & -1 \\ 1 & 1 & 1 \end{pmatrix} \Rightarrow T^{-1} = \frac{1}{4}\begin{pmatrix} 0 & 2 & 2 \\ -2i & -1+i & 1+i \\ 2i & -(1+i) & 1-i \end{pmatrix}.$$

(3) *The reduced matrix $T^{-1}AT$.*

This is a diagonal matrix and every element in it is the characteristic value relative to the eigenvector whose coordinates constitute the corresponding column of T:

$$T^{-1}AT = \begin{pmatrix} 2 & 0 & 0 \\ 0 & 1+i & 0 \\ 0 & 0 & 1-i \end{pmatrix}.$$

As in exercise 342, we note that V_1, V_2, and V_3 can be replaced with vectors proportional to them and that the columns of T can be rearranged.

344

First question:

For an eigenvector V, we have $\tilde{A}(V) = \lambda V$,

$$\tilde{A}^2(V) = \tilde{A}[\tilde{A}(V)] = \tilde{A}(\lambda V) = \lambda \tilde{A}(V) = \lambda^2 V,$$

and, by induction on p, $\tilde{A}^p(V) = \lambda^p V$.

By taking a linear combination of these equations, we see that, for any polynomial p, and, in particular, for f,

$$[p(\tilde{A})](V) = p(\lambda)V$$

$$f(\lambda) = 0 \Rightarrow [f(\tilde{A})](V) = 0.$$

The eigenvectors V_i of A constitute a basis for the space since the characteristic values are all distinct. The kernel of $f(\tilde{A})$ that contains all the V_i contains the vector space that they generate, that is, the entire space, so that the mapping $f(\tilde{A})$ is the zero mapping.

If p is a polynomial, the mapping corresponding to $p(A)$ is $p(\tilde{A})$ since φ is a ring isomorphism. In particular, the matrix $f(A)$ of the mapping $f(\tilde{A})$ is the zero matrix.

Second question:

For AB to be equal to BA (or $\tilde{A}\tilde{B}$ to be equal to $\tilde{B}\tilde{A}$), it is necessary and sufficient that, for every eigenvector V_i of A,

$$(\tilde{A}\tilde{B})(V_i) = (\tilde{B}\tilde{A})(V_i).$$

This is true because the V_i constitute a basis for C^n. The two mappings $\tilde{A}\tilde{B}$ and $\tilde{B}\tilde{A}$, which transform the basis vectors in the same way, are identical.

Thus, let us write down these conditions and note that by the definition of the vectors V_i

$$(\tilde{B}\tilde{A})(V_i) = \tilde{B}[\tilde{A}(V_i)] = \tilde{B}(\lambda_i V_i) = \lambda_i \tilde{B}(V_i)$$

$$(\tilde{A}\tilde{B})(V_i) = \tilde{A}[\tilde{B}(V_i)],$$

which yields

$$\tilde{A}[\tilde{B}(V_i)] = \lambda_i \tilde{B}(V_i), \qquad (i = 1, 2, ..., n).$$

If $\tilde{B}(V_i)$ is not the zero vector, it is an eigenvector of $\tilde{A}$, and the corresponding characteristic value is λ_i. Since the characteristic values are distinct, to each of these corresponds a one-dimensional variety of eigenvectors, and $\tilde{B}(V_i)$ is proportional to V_i. This result remains valid if $\tilde{B}(V_i)$ is the zero vector. In this case, the proportionality constant is zero.

Thus, in either case, we obtain

$$\tilde{B}(V_i) = \mu_i V_i.$$

Obviously, this condition is sufficient.

Third question:

For B to be equal to $g(A)$ (or $\tilde{B}$ to be equal to $g(\tilde{A})$), it is necessary and sufficient that, for every basis vector V_i,

$$\tilde{B}(V_i) = \mu_i V_i = [g(\tilde{A})](V_i) = g(\lambda_i)V_i,$$

so that

$$\mu_i = g(\lambda_i), \qquad (i = 1, 2, ..., n).$$

We know there exists one and only one polynomial g of degree not exceeding $n-1$ that assumes n given values μ_i for given values λ_i. This polynomial is Lagrange's interpolational polynomial (cf. PZ, Book II, Chapter IV, part 5). Thus, the problem posed has one and only one solution.

The following facts should be noted:

(a) The mappings $\tilde{B}$ that commute with $\tilde{A}$ constitute an n-dimensional vector space E over C ($\tilde{B}$ is defined by $\mu_1, \mu_2, ..., \mu_n$).

(b) If g is a polynomial of degree not exceeding $n-1$, the mappings $g(\tilde{A})$ constitute a vector subspace E' of E.

(c) A mapping $g(\tilde{A})$ cannot be the zero mapping unless g is the zero polynomial. We need only examine the vectors $[g(\tilde{A})](V_i)$. Thus, E' is n-dimensional and identical to E.

Fourth question:

For B or $\tilde{B}$ to be nonsingular, it is necessary and sufficient that none of the characteristic values $g(\lambda_i)$ be zero. This result is obvious if we consider the matrix of $\tilde{B}$ relative to the basis V_i. This matrix is diagonal and its elements are the numbers $g(\lambda_i)$.

Therefore, for B to be nonsingular, it is necessary and sufficient that no zero of f be a zero of g, that is, it is necessary and sufficient that f and g be relatively prime.

345

To the characteristic value λ there corresponds at least one eigenvector V. If we take a new basis consisting of V and another vector W, we transform th e matrix A into

$$A_1 = \begin{pmatrix} \lambda & c \\ 0 & \lambda \end{pmatrix}.$$

(The terms in the first column of A_1 are the coordinates of the vector $\lambda V = \tilde{A}(V)$, where $\tilde{A}$ denotes the mapping corresponding to A.)

If $c \neq 0$, let us replace the vector W with the vector kW. The component of $\tilde{A}(kW)$ along V is multiplied by k and hence c is replaced by ck. If we take $k = 1/c$, we obtain

$$B = \begin{pmatrix} \lambda & 1 \\ 0 & \lambda \end{pmatrix}.$$

(a)
$$B = \begin{pmatrix} \lambda & 0 \\ 0 & \lambda \end{pmatrix} \Rightarrow B^n = \begin{pmatrix} \lambda^n & 0 \\ 0 & \lambda^n \end{pmatrix},$$

(b)
$$B = \begin{pmatrix} \lambda & 1 \\ 0 & \lambda \end{pmatrix} \Rightarrow B^n = \begin{pmatrix} \lambda^n & n\lambda^{n-1} \\ 0 & \lambda^n \end{pmatrix}.$$

Proof by induction on n: The result is clearly true if $n=1$. Then,

$$B^n = \begin{pmatrix} \lambda^{n-1} & (n-1)\lambda^{n-2} \\ 0 & \lambda^{n-1} \end{pmatrix} \begin{pmatrix} \lambda & 1 \\ 0 & \lambda \end{pmatrix} = \begin{pmatrix} \lambda^n & n\lambda^{n-1} \\ 0 & \lambda^n \end{pmatrix}.$$

346

First question:

The characteristic polynomial f of A is defined by

$$-f(\lambda) = -\det(A - \lambda I) = \lambda^3 - 10\lambda^2 + 32\lambda - 32 = (\lambda - 2)(\lambda - 4)^2.$$

To each of these two eigenvalues, there corresponds a one-dimensional variety of eigenvectors, which can be defined in terms of one of these vectors:

> To $\lambda = 2$ corresponds the eigenvector V_1 (1, 1, 1);
> To $\lambda = 4$ corresponds the eigenvector V_2 (1, −1, 1).

Second question:

One of the characteristic values of A is a double value. The eigenvectors do not constitute a basis for R^3. For basis vectors we take V_1, V_2, and some other vector, for example, the third vector e_3 of the canonical basis for R^3.

The first two columns of the transformed matrix B yield the coordinates in the new basis of the vectors $2V_1$ and $4V_1$, which are the images of V_1 and V_2 under the mapping $\tilde{A}$ corresponding to A. The terms along the diagonal are equal to 2 and 4, and the other terms are zero. To get the terms in the third column of B, we need to know the coordinates X, Y, and Z, in the new basis, of the vector $\tilde{A}(e_3)$, which are defined by

$$\tilde{A}(e_3) = -5e_1 + e_2 - e_3 = XV_1 + YV_2 + Ze_3.$$

By expressing all the vectors in terms of the basis e_1, e_2, e_3, we obtain

$$\begin{cases} -5 = X + Y \\ 1 = X - Y \\ -1 = X + Y + Z \end{cases} \Rightarrow \begin{cases} X = -2 \\ Y = -3 \\ Z = 4 \end{cases}$$

$$B = \begin{pmatrix} 2 & 0 & -2 \\ 0 & 4 & -3 \\ 0 & 0 & 4 \end{pmatrix}.$$

Third question:

Let us change the third basis vector and take as our new vector

$$\varepsilon_3 = aV_1 + bV_2 + ce_3,$$

the coefficient c not being zero (because otherwise V_1, V_2, and ε_3 would be dependent).

The matrix B', which is the image of B, is obtained as in (2). The elements α, β, and γ in the third column are defined by

$$\tilde{A}(\varepsilon_3) = 2(a-c)V_1 + (4b-3c)V_2 + 4ce_3 = \alpha V_1 + \beta V_2 + \gamma \varepsilon_3$$

$$\begin{cases} \alpha + \gamma a = 2(a-c) \\ \beta + \gamma b = 4b - 3c \\ \gamma c = 4c \end{cases} \Rightarrow \begin{cases} \alpha = -2(a+c) \\ \beta = -3c \\ \gamma = 4 \end{cases} \qquad \text{since } c \neq 0.$$

$$\begin{cases} \alpha = 0 & \text{if} \quad a + c = 0 \\ \beta = 1 & \text{if} \quad c = -\tfrac{1}{3} \end{cases} \quad \text{so that} \quad a = \tfrac{1}{3},$$

and b is arbitrary (we take it equal to zero).

The coordinates of the vectors of the new basis relative to the canonical basis e_1, e_2, e_3 of R^3 are

$$V_1\begin{Bmatrix}1\\1\\1\end{Bmatrix} \qquad V_2\begin{Bmatrix}1\\-1\\1\end{Bmatrix} \qquad \varepsilon_3\begin{Bmatrix}\frac{1}{3}\\\frac{1}{3}\\0,\end{Bmatrix}$$

and the matrix B_0 is

$$B_0 = \begin{pmatrix}2 & 0 & 0\\0 & 4 & 1\\0 & 0 & 4\end{pmatrix}.$$

Fourth question:

Calculation of B_0^n is identical to the calculation of B^n in exercise 345. We obtain easily

$$B_0^n = \begin{pmatrix}2^n & 0 & 0\\0 & 4^n & n4^{n-1}\\0 & 0 & 4^n\end{pmatrix}.$$

If T is the matrix whose column vectors are V_1, V_2, and ε_3, we know that T is nonsingular and that

$$B_0 = T^{-1}AT.$$

We can then show by induction on n that

$$B_0^n = T^{-1}A^nT.$$

(This result is actually known in advance since the function that assigns to a matrix M the matrix $T^{-1}MT$ is a ring isomorphism [cf. PZ, Book II, Chapter VIII, part 3, § 3].) Then,

$$A^n = TB_0^nT^{-1}.$$

We know T. We can derive T^{-1} by an elementary calculation. Then, A^n is the product of three known matrices.

347

First question:

The coordinates of the vector $\tilde{A}(e_i)$ are the elements of the ith column of A:

$$\tilde{A}(e_i) = e_{i+1} \quad \text{if} \quad i<n,$$
$$\tilde{A}(e_n) = a_0e_1 + a_1e_2 + \cdots + a_{n-1}e_n.$$

By induction on i, we can easily show that

$$\tilde{A}_i(e_1) = e_{i+1} \qquad \text{if} \qquad i < n,$$

$$\tilde{A}_n(e_1) = \tilde{A}(e_n).$$

Obviously,

$$[f(\tilde{A})][e_1] = \tilde{A}^n(e_1) - \sum_{i=0}^{n-1} a_i \tilde{A}^i(e_i) = \tilde{A}(e_n) - \sum_{i=0}^{n-1} a_i e_{i+1} = 0,$$

$$[f(\tilde{A})][e_1] = [f(\tilde{A})\tilde{A}^{i-1}][e_1] = [\tilde{A}^{i-1}f(\tilde{A})][e_1] = 0.$$

The mapping $f(\tilde{A})$ maps all the basis vectors into the zero vector. Thus, it is the zero mapping and its matrix $f(A)$ is the zero matrix.

Second question:

$\mathscr{M}$ is a subring of the ring of $n \times n$ matrices since it contains the difference and the product of any two of its elements:

$$p(A) - q(A) = (p-q)(A)$$

$$p(A)q(A) = pq(A).$$

Furthermore, the mapping p on the ring of polynomials into $\mathscr{M}$ conserves the operations. It is a ring homomorphism. The ring $\mathscr{M}$ is commutative.

Let us denote by p and q two polynomials in $\mathscr{I}$ and by r an arbitrary polynomial. Then,

$$(p-q)(A) = p(A) - q(A) = 0 \Rightarrow p - q \in \mathscr{I}$$

$$(rp)(A) = r(A)p(A) = 0 \Rightarrow rp \in \mathscr{I},$$

$\mathscr{I}$ is a stable additive subgroup under multiplication by an arbitrary element of $\mathscr{M}$. Therefore, it is an ideal.

The ideal $\mathscr{I}$ contains the nth-degree polynomial f in accordance with (1).

Third question:

The characteristic polynomial of A is the polynomial φ defined by

$$\varphi(\lambda) = \det(A - \lambda I) = \begin{vmatrix} -\lambda & & & & a_0 \\ 1 & & & 0 & \vdots \\ & & & & \vdots \\ 0 & & & & \vdots \\ & & & -\lambda & a_{n-2} \\ & & & 1 & a_{n-1} - \lambda \end{vmatrix}.$$

To evaluate this determinant, we can add to the first row the elements of the other rows multiplied respectively by $\lambda, \lambda^2, ..., \lambda^{n-1}$:

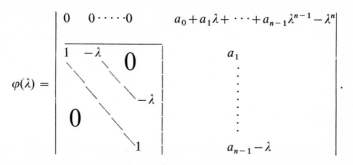

In the first row, only the last element is nonzero. Its minor, boxed off by the solid horizontal and dashed vertical lines, is the determinant of a triangular matrix. Thus, it is equal to the product of the terms along the diagonal, namely 1. Therefore,

$$\varphi(\lambda) = (-1)^{n+1}(a_0 + a_1\lambda + \cdots + a_{n-1}\lambda^{n-1} - \lambda^n) = (-1)^n f(\lambda).$$

Fourth question:

First, suppose that the polynomials p and f are relatively prime. Then, we know (on the basis of Bézout's theorem) that there exist polynomials q and g such that

$$pq + fg = 1.$$

The mapping p is a ring homomorphism; that is, it conserves the operations of addition and multiplication. Therefore, we can write

$$p(A)q(A) + f(A)g(A) = I,$$

where I denotes the unit matrix. But the matrix $f(A)$ is the zero matrix and the above equation reduces to

$$p(A)q(A) = I,$$

so that the matrix $p(A)$ is nonsingular.

If the polynomials p and f are not prime, they have at least one zero in common, which we denote by λ_0. To this characteristic value (a zero of f) there corresponds at least one eigenvector V_0. Therefore,

$$\tilde{A}(V_0) = \lambda_0 V_0 \Rightarrow [p(\tilde{A})](V_0) = p(\lambda_0)V_0 = 0.$$

The kernel of $p(\tilde{A})$ is not zero, so that $p(\tilde{A})$ does not have an inverse. Hence, the matrix $p(A)$ does not have an inverse. V_0 is an eigenvector and the characteristic value corresponding to it is zero.

CHAPTER 4

348

First question:

With a change of basis is associated a nonsingular matrix T such that the column matrices X and X' of the coordinates of a vector V in the two bases are connected by the relation $X = TX'$ (cf. PZ, Book II, Chapter VIII, part 3, § 3).

The matrix of the coordinates of the n vectors V_i can be written $(X_1, X_2, ..., X_n)$, this notation expressing the fact that the elements of the ith column are those of the matrix X_i. Then, the rules for calculating with matrices imply that the matrices $(X_1, X_2, ..., X_n)$ and $(X'_1, X'_2, ..., X'_n)$ of the coordinates of the vectors V_i relative to the bases in question satisfy the relation

$$(X_1, X_2, ..., X_n) = T(X'_1, X'_2, ..., X'_n)$$

and hence that

$$\det(X_1, X_2, ..., X_n) = [\det T][\det(X'_1, X'_2, ..., X'_n)].$$

The determinant of the n vectors V_i is multiplied by the factor $[\det T]^{-1}$ when we make a change of basis.

Second question:

In an orthonormal change of basis, the matrix T is an orthogonal matrix and $\det T = \pm 1$:

$$\det(X_1, X_2, ..., X_n) = \pm \det(X'_1, X'_2, ..., X'_n).$$

The absolute value of the determinant of n vectors is invariant.

If the vectors V_i are linearly dependent, the determinant of these vectors is zero and

$$0 \leqslant \prod_i l_i.$$

If the vectors V_i are independent, the Gram–Schmidt orthonormalization procedure (cf. PZ, Book II, Chapter X, part 2, § 1) enables us to define an orthonormal basis e_i such that, for every i, the vector spaces generated by e_1, e_2, ..., e_i and $V_1, V_2, ..., V_i$ are identical. The coordinates of V_i of index greater than i are then zero and the matrix $(X_1, ..., X_n)$ of the coordinates of the vectors V_i is triangular.

The determinant of the matrix $(X_1, ..., X_n)$ is then the product of the diagonal elements

$$|\det (X_1, X_2, ..., X_n)| = |x_{11}| \, |x_{22}| \cdots |x_{nn}|.$$

By the definition of the length of a vector,

$$l_i = \sqrt{\sum_j x_{ji}^2} \geqslant |x_{ii}|$$

$$|\det (X_1, X_2, ..., X_n)| = \prod_{i=1}^{n} |x_{ii}| \leqslant \prod_{i=1}^{n} l_i.$$

349

The quadratic form $\omega(X)$ is equal to 0 if and only if the vector

$$\sum_{1}^{p} x_i V_i$$

is the zero vector. Thus, a necessary and sufficient condition for $\omega(X)$ to vanish for some X is that the vectors V_i be linearly dependent. Hence, a necessary and sufficient condition for $\omega(X)$ *not* to vanish for any X is that the vectors V_i be linearly *independent*. Since $\omega(X) \geqslant 0$, the assertion of the problem follows.

By expanding the form ω, we obtain

$$\omega(X) = \left(\sum_{1}^{p} x_i V_i \right) \cdot \left(\sum_{1}^{p} x_i V_i \right) = \sum_{\substack{i \leqslant p \\ j \leqslant p}} x_i x_j (V_i \cdot V_j).$$

The matrix $A = V_i \cdot V_j$ is the matrix of the quadratic form ω. The coefficients of the reduced form are therefore the characteristic values of A. These are all nonnegative since the form ω is nonnegative.

For ω to be positive definite, it is necessary and sufficient that 0 not be a characteristic value. Then, all the characteristic values of A are positive. The constant term of the characteristic polynomial of A, namely $\det (A - \lambda I)$ is $\det A$, and this number is the product of the zeros of the polynomial since the coefficient of the term of highest degree λ^p is $(-1)^p$.

Therefore, for ω to be defined, that is, for the vectors V_i to be independent, it is necessary and sufficient that

$$\det A \neq 0.$$

Then, the characteristic values are positive and hence their product is positive:

$$\det A > 0.$$

REMARK: We might equally well take an orthonormal basis in the vector subspace generated by the V_i. If the coordinates of V_i in this basis are denoted by v_{ij}, we can easily show that det A is the square of the determinant of the vectors V_i:

$$\det A = \left(\begin{vmatrix} v_{11} & v_{12} \cdots v_{1p} \\ \vdots & \\ v_{p1} & v_{p2} \cdots v_{pp} \end{vmatrix} \right)^2,$$

and the required results are immediately deduced from this.

350

First question:

The proof is identical to the proof of the Schwarz inequality:

$$0 \leqslant g(x+\lambda y, x+\lambda y) = g(x, x) + 2\lambda g(x, y) + \lambda^2 g(y, y) \qquad \forall \lambda,$$

so that

$$[g(x, y)]^2 - g(x, x)g(y, y) \leqslant 0.$$

Second question:

If g is defined, it is nondegenerate. To see this, note that the linear form that maps y into $g(x, y)$ is the zero form:

$$g(x, y) = 0 \qquad \forall y \Rightarrow g(x, x) = 0$$

(when we take $y=x$), and this implies that $x=0$.

If g is nondegenerate, the form that maps x into $g(x, x)$ is defined. To see this, suppose that $g(x, x)=0$. The inequality in (1) implies that, for any y,

$$[g(x, y)]^2 \leqslant g(x, x)g(y, y) = 0 \Rightarrow g(x, y) = 0.$$

Then, by hypothesis, x is zero.

Third question:

We have an equivalence between $g(x, x)=0$ and $g(x, y)=0$ $\forall y$. Now, it is obvious that the elements that satisfy the second condition constitute a vector subspace N. To see this, suppose that x and x' are two elements in N and that λ is a real number:

$$g(x-x', y) = g(x, y) - g(x', y) = 0 \qquad \forall y$$

$$g(\lambda x, y) = \lambda g(x, y) = 0 \qquad \forall y.$$

Thus, $x-x'$ and λx are also elements of N.

Fourth question:

Let us show that the group axioms are satisfied:

(a) The composition of mappings is an associative operation.

(b) The neutral element (which is the identity mapping) is an element of G.

(c) The product of two mappings in G is a mapping in G:

$$g[\alpha\beta x, \alpha\beta y] = g[\alpha(\beta x), \alpha(\beta y)] = g[\beta x, \beta y] = g(x, y).$$

(d) Every mapping α has an inverse α^{-1}, which is an element in G. Let us show that the kernel of α consists only of zero. If x is any element in the kernel of α,

$$\alpha x = 0 \Rightarrow g(\alpha x, \alpha y) = g(x, y) = 0 \quad \forall y \Rightarrow x = 0.$$

Thus, the inverse α^{-1} exists and, since α is an element of G,

$$g(\alpha^{-1}x, \alpha^{-1}y) = g[\alpha(\alpha^{-1}x), \alpha(\alpha^{-1}y)] = g(x, y).$$

351

FIRST PART

First question:

The rules for calculation with scalar products yield

$$||x - (x_0 + \lambda y)||^2 = ||(x - x_0) + \lambda y||^2 = ||x - x_0||^2 + 2\lambda y \cdot (x - x_0) + \lambda^2 ||y||^2.$$

If y is a vector in F, this distance must be minimized for $\lambda = 0$. The first-degree term in the trinomial is zero. Therefore,

$$y \cdot (x - x_0) = 0 \Rightarrow y \text{ is orthogonal to } (x - x_0).$$

Suppose that $x - x_0$ is orthogonal to every vector in F. If x_1 is a vector in F, so is $x_0 - x_1$ and

$$||x - x_1||^2 = ||(x - x_0) + (x_0 - x_1)||^2 = ||x - x_0||^2 + ||x_0 - x_1||^2,$$

so that

$$||x - x_1|| > ||x - x_0||.$$

The last inequality shows also that, if $||x - x_0||$ is minimum and hence $x - x_0$ is orthogonal to every vector in F, there exists no other vector x_1 such that the distance $||x - x_1||$ is equal to $||x - x_0||$.

Second question:

Let us take an orthonormal basis $e_1, ..., e_n$ for F and let us set

$$x_0 = \sum_{i=1}^{n} \xi_i e_i.$$

For $x - x_0$ to be orthogonal to all vectors in F, it is necessary and sufficient that it be orthogonal to all vectors of a basis for F, for example, $e_1, ..., e_n$. Then, the equations of the problem are

$$(x - x_0) \cdot e_k = 0, \qquad (k = 1, 2, ..., n)$$

or

$$x \cdot e_k = x_0 \cdot e_k = \xi_k, \qquad (k = 1, 2, ..., n)$$

so that

$$x_0 = \sum_{k=1}^{n} (x \cdot e_k) e_k.$$

SECOND PART

First question:

The mapping on $E \times E$ into R defined by $(x, y) \rightarrow x \cdot y$ is clearly bilinear and symmetric (by virtue of the properties of an integral and the commutativity of the product fg). This is a scalar product if the quadratic form $x \cdot x$ is positive definite.

Let us suppose that

$$x \cdot x = \int_0^{2\pi} [x(t)]^2 dt = 0.$$

The function x^2 is continuous and nonnegative. If its integral is zero, it is the zero function (cf. PZ, Book III, Chapter IV, part 3, § 4). Thus, if $x \cdot x = 0$, x is the zero element of E, so that the quadratic form $x \cdot x$ is positive definite.

E is a Euclidean space, the norm in which is

$$\|x\| = \sqrt{x \cdot x} = \sqrt{\int_0^{2\pi} [x(t)]^2 dt}.$$

This space is infinite-dimensional since the elements 1, $\cos kt$, and $\sin ht$ are independent. The proof of this will be carried out in (2). We know that nonzero elements that are mutually orthogonal are independent.

Second question:

The verification is immediate:

$$\int_0^{2\pi} 1 \cdot \cos kt \, dt = \left[\frac{\sin kt}{k}\right]_0^{2\pi} = 0, \qquad \int_0^{2\pi} 1 \cdot \sin ht \, dt = \left[\frac{-\cos ht}{h}\right]_0^{2\pi} = 0$$

$$\int_0^{2\pi} \cos kt \cos ht \, dt = \int_0^{2\pi} \frac{\cos(k+h)t - \cos(k-h)t}{2} \, dt = 0 \qquad \text{if} \qquad k \neq h$$

$$\int_0^{2\pi} \sin kt \sin ht \, dt = \int_0^{2\pi} \frac{\cos(k-h)t - \cos(k+h)t}{2} \, dt = 0 \qquad \text{if} \qquad k \neq h$$

$$\int_0^{2\pi} \cos kt \sin ht \, dt = \int_0^{2\pi} \frac{\sin(h+k)t - \sin(h-k)t}{2} \, dt = 0.$$

Finally, let us evaluate the norms of these functions:

$$\int_0^{2\pi} 1 \, dt = 2\pi \qquad\qquad \Rightarrow \qquad \|1\| = \sqrt{2\pi}$$

$$\int_0^{2\pi} \cos^2 kt \, dt = \int_0^{2\pi} \frac{1 + \cos 2kt}{2} \, dt = \pi \Rightarrow \|\cos kt\| = \sqrt{\pi}$$

$$\int_0^{2\pi} \sin^2 ht \, dt = \int_0^{2\pi} \frac{1 - \cos 2ht}{2} \, dt = \pi \Rightarrow \|\sin ht\| = \sqrt{\pi}.$$

Third question:

We need only apply the result of (2) of the first part and remember that an orthonormal basis for F is constituted by the $2n+1$ functions e_k defined by

$$e_0(t) = \frac{1}{\sqrt{2\pi}}, \qquad e_{2k}(t) = \frac{\cos kt}{\sqrt{\pi}} (1 \leqslant k \leqslant n), \qquad e_{2h-1}(t) = \frac{\sin ht}{\sqrt{\pi}} (1 \leqslant h \leqslant n).$$

Let us set

$$x_0(t) = \frac{\xi_0}{\sqrt{2\pi}} + \sum_{k=1}^{n} \xi_k \frac{\cos kt}{\sqrt{\pi}} + \sum_{h=1}^{n} \eta_h \frac{\sin ht}{\sqrt{\pi}}.$$

The coefficients ξ_k and η_h are defined by

$$\xi_0 = x \cdot e_0 = \frac{1}{\sqrt{2\pi}} \int_0^{2\pi} x(u) du$$

$$\xi_k = x \cdot e_{2k} = \frac{1}{\sqrt{\pi}} \int_0^{2\pi} x(u) \cos ku \, du$$

$$\eta_h = x \cdot e_{2h-1} = \frac{1}{\sqrt{\pi}} \int_0^{2\pi} x(u) \sin hu \, du$$

$$x_0(t) = \frac{1}{2\pi} \int_0^{2\pi} x(u) du + \sum_{k=1}^{n} \frac{\cos kt}{\pi} \int_0^{2\pi} x(u) \cos ku \, du$$

$$+ \sum_{h=1}^{n} \frac{\sin kt}{\pi} \int_0^{2\pi} x(u) \sin hu \, du.$$

The trigonometric polynomial x_0 thus defined is that polynomial, out of all the nth-order polynomials, for which the distance $||x - x_0||$ is minimum. This distance is called the *mean square distance*. We can verify easily that

$$||x||^2 = ||x_0 + (x - x_0)||^2 = ||x_0||^2 + ||x - x_0||^2$$

$$||x - x_0||^2 = ||x||^2 - ||x_0||^2 = \int_0^{2\pi} x^2(t) dt - \xi_0^2 - \sum_{k=1}^{n} \xi_k^2 - \sum_{h=1}^{n} \eta_h^2.$$

352

First question:

The verification is identical to that made in exercise 351 (second part, (1)) to which we refer the reader.

We shall employ the Gram–Schmidt orthonormalization procedure (cf. PZ, Book II, Chapter X, part 2, § 2), and use the sequence of vector subspaces E_p of polynomials of degree less than p to define the sequence $\{U_p\}$.
E_1 admits 1 for a basis and, for a unitary basis, we take

$$\frac{1}{||1||} = 1 = U_0.$$

Let us assume that E_p has an orthonormal basis U_k (for $0 \leqslant k \leqslant p-1$), where the polynomial U_h is of degree k. We can determine the constants λ_i so that the polynomial

$$A = t^p + \sum_{k=0}^{p-1} \lambda_k U_k$$

will be orthogonal to the U_k. It is necessary and sufficient that

$$A \cdot U_k = (t^p) \cdot U_k + \lambda_k = 0, \qquad (0 \leqslant k \leqslant p-1).$$

The polynomial A thus obtained is not the zero polynomial since t^p does not belong to the space E_p generated by the U_k (for $0 \leqslant k \leqslant p-1$). The polynomial $A/\|A\|$ then constitutes, along with the U_k, a basis for E_{p+1}. Every polynomial in E_{p+1} is a linear combination of t^p and the U_k – hence of $A/\|A\|$ and the U_k of index $k<p$.

We can set $U_p = A/\|A\|$ and proceed inductively.

The same reasoning shows that U_p is unique except for sign.

U_0, a unitary basis of E_1, cannot be anything but ± 1.

Suppose that the result is established for $k \leqslant p-1$. Let B be a polynomial belonging to E_{p+1} that, together with the U_k of index $k \leqslant p-1$ (or their negatives, which changes nothing), constitutes an orthonormal basis for E_{p+1}.

Since B is an element of E_{p+1}, it can be expressed in terms of the basis U_k:

$$B = \sum_0^p b_k U_k$$

$$B \cdot U_k = 0 \Rightarrow \quad b_k = 0 \qquad \text{if} \qquad 0 \leqslant k \leqslant p-1,$$

$$\|B\| = \|U_p\| = 1 \Rightarrow |b_p| = 1 \qquad \text{or} \qquad b_p = \pm 1.$$

Second question:

Let us determine the U_i by the procedure that we used to prove their existence. We know that $U_0 = 1$. Therefore, let us take $A_1 = t + \lambda U_0 = t + \lambda$.

$$A_1 \cdot U_0 = (t) \cdot U_0 + \lambda = \tfrac{1}{2} \int_{-1}^{1} t \, dt + \lambda = \lambda = 0$$

$$\|A_1\|^2 = \tfrac{1}{2} \int_{-1}^{1} t^2 \, dt = \tfrac{1}{3},$$

$$U_1 = \frac{A_1}{\|A_1\|} = t\sqrt{3}.$$

Similarly,

$$A_2 = t^2 + \lambda_0 U_0 + \lambda_1 U_1$$

$$A_2 U_0 = (t^2) \cdot U_0 + \lambda_0 = \tfrac{1}{2} \int_{-1}^{1} t^2 \, dt + \lambda_0 = \tfrac{1}{3} + \lambda_0 = 0,$$

$$A_2 U_1 = (t^2) \cdot U_1 + \lambda_1 = \tfrac{1}{2} \int_{-1}^{1} t^3 \sqrt{3} \, dt + \lambda_1 = \lambda_1 = 0,$$

$$A_2 = t^2 - \tfrac{1}{3} \;\Rightarrow\; \|A_2\|^2 = \tfrac{1}{2} \int_{-1}^{1} (t^2 - \tfrac{1}{3})^2 \, dt = \frac{4}{45},$$

$$U_2 = \frac{A_2}{\|A_2\|} = \frac{3\sqrt{5}}{2} (t^2 - \tfrac{1}{3}).$$

Third question:

Proof by induction on p. For U_p to be orthogonal to every polynomial of degree less than p, it is necessary and sufficient that U_p be orthogonal to $U_0, U_1, \ldots, U_{p-1}$. We have seen (at the end of (2)) that there exists, up to sign, one and only one polynomial in E_{p+1} satisfying these conditions and having norm equal to 1.

The formula for integration by parts repeated p times is written

$$\int_{-1}^{1} x(t) y^{(p)}(t) \, dt = \left[\sum_{k=0}^{p-1} (-1)^k x^{(k)}(t) y^{(p-k-1)}(t) \right]_{-1}^{1} + (-1)^p \int_{-1}^{1} x^{(p)}(t) y(t) \, dt.$$

If we take for y the polynomial $(x^2 - 1)^p$, the first $p-1$ derivatives of y vanish at -1 and 1. All the terms in the brackets have one of these derivatives as a vector, and hence the bracketed expression is equal to 0. Therefore,

$$(R) \qquad \int_{-1}^{1} x(t) \left[\frac{d^p}{dt^p} (t^2 - 1)^p \right] dt = (-1)^p \int_{-1}^{1} x^{(p)}(t)(t^2 - 1)^p \, dt.$$

If x is a polynomial of degree less than p, the derivative $x^{(p)}$ is the zero function and hence,

$$x \cdot \left[\frac{d}{dt^p} (t^2 - 1)^p \right] = 0.$$

If we choose u_p in such a way that the polynomial

$$u_p \frac{d^p}{dt^p} (t^2 - 1)^p$$

will have a norm equal to 1, this polynomial will satisfy the characteristic properties of U_p. It is equal to $\pm U_p$.

$$U_p = u_p \frac{d^p}{dt^p}(t^2-1)^p \quad \text{if} \quad |u_p|\left\|\frac{d^p}{dt^p}(t^2-1)^p\right\| = 1.$$

In the relation (R), let us take for x the polynomial $d^p(t^2-1)^p/dt^p$:

$$x^{(p)} = \frac{d^{2p}}{dt^{2p}}(t^2-1)^p = 2p!,$$

so that

$$2\left\|\frac{d^p}{dt^p}(t^2-1)^p\right\|^2 = (-1)^p\int_{-1}^{1}(2p!)(t^2-1)^p\,dt = (-1)^p(2p!)I_p.$$

Integrating by parts, we obtain a relationship between I_k and I_{k-1}:

$$I_k = \int_{-1}^{1}(t^2-1)^k\,dt = [t(t^2-1)^k]_{-1}^{1} - \int_{-1}^{1}2kt^2(t^2-1)^{k-1}\,dt = -2k(I_k+I_{k-1})$$

$$(2k+1)I_k = -2kI_{k-1}.$$

Multiplying these equations and letting k vary from 1 to p, we obtain

$$I_p = (-1)^p\frac{2\cdot4\cdots2p}{1\cdot3\cdots2p+1}I_0 = 2(-1)^p\frac{(2\cdot4\cdots2p)^2}{(2p+1)!}$$

$$\left\|\frac{d^p}{dt^p}(t^2-1)^p\right\| = \frac{2\cdot4\cdots2p}{\sqrt{2p+1}} = \frac{2^p p!}{\sqrt{2p+1}} = \frac{1}{|u_p|}.$$

We can verify that, for $p=1$ and $p=2$, we do actually get U_1 and U_2.

Fourth question:

Let us show by induction on k that the kth derivative of $(t^2-1)^p$ has at least k distinct zeros in $]-1, 1[$ if $k\leqslant p$. The result is clearly true for $k=0$. Let us suppose it true for the $(k-1)$st derivative, which we denote by $C^{(k-1)}$. By hypothesis, this derivative has $k-1$ distinct zeros in $]-1, 1[$. The numbers -1 and 1 are also zeros of the derivative $C^{(k-1)}$. These $k+1$ zeros define k intervals, each containing at least one zero of the derivative of $C^{(k-1)}$, that is, of $C^{(k)}$. These k zeros are distinct since they lie in the interiors of the intervals and $C^{(k)}$ has at least k distinct zeros in $]-1, 1[$.

If we assign to k the value p, we see that U_p has at least p distinct zeros in $]-1, 1[$ and, since U_p is of degree p, all the zeros of U_p are real and distinct and they lie between -1 and 1.

353

First question:

We verify that, if $\omega(x) = \varphi(x, x)$,

$$\omega(x+y) = \varphi(x+y, x+y) = \varphi(x, x) + \varphi(x, y) + \varphi(y, x) + \varphi(y, y),$$
$$\omega(x-y) = \varphi(x-y, x-y) = \varphi(x, x) - \varphi(x, y) - \varphi(y, x) + \varphi(y, y),$$
$$\omega(x+y) - \omega(x-y) = 2[\varphi(x, y) + \varphi(y, x)] = 4\varphi(x, y).$$

Thus, giving the value of ω is sufficient to determine the symmetric bilinear form φ.

Second question:

The positive definite quadratic form ω enables us to define a Euclidean space structure E on R^n, where the scalar product of two vectors x and y is $\varphi(x, y)$. However, it should be noted that this is not the structure that is usually considered. In fact, *the canonical basis of R^n in this case is not an orthonormal basis.*

If we choose in E an orthonormal basis u_i and if we denote by ξ_i and η_i the coordinates of the vectors x and y relative to the basis u_i, we know that

$$\varphi(x, y) = \sum_{i=1}^{n} \xi_i \eta_i$$

$$\psi(x, y) = \sum_{i,j} a_{ij} \xi_i \eta_j,$$

where the matrix (a_{ij}) is also symmetric.

The theory of the reduction of quadratic forms can now be applied. We know that there exists a new basis v_i that is orthonormal with respect to the scalar product φ and such that ψ is reduced to a sum of squares. If we denote the coordinates of x and y in the basis v_i by ζ_i and χ_i, the new expressions for φ and ψ are

$$\varphi(x, y) = \sum_{i=1}^{n} \zeta_i \chi_i$$

$$\Omega(y) = \sum_{i=1}^{n} \lambda_i \zeta_i^2 \Leftrightarrow \psi(x, y) = \sum_{i=1}^{n} \lambda_i \zeta_i \chi_i.$$

Third question:

The expressions obtained imply that

$$\psi(x, y) - \lambda \varphi(x, y) = \sum_{i=1}^{n} (\lambda_i - \lambda) \zeta_i \chi_i.$$

Thus, the form $\psi - \lambda\varphi$ is degenerate if and only if λ is equal to one of the numbers λ_i. Furthermore,

$$\psi(x, v_i) - \lambda_i\varphi(x, v_i) = 0 \qquad \forall x,$$

and this property characterizes v_i if the numbers λ_i are all distinct.

If A and B are the matrices of φ and ψ relative to a basis for $\mathbf{R}^n$ (for example, relative to the canonical basis) and if X and Y are column matrices of the coordinates of x and y relative to that basis, we can write

$$\psi(x, y) - \lambda\varphi(x, y) = X^T(B - \lambda A)Y.$$

For the form $\psi - \lambda\varphi$ to be degenerate, it is necessary and sufficient that the dimension of the kernel of the mapping $\tilde{B} - \lambda\tilde{A}$ corresponding to the matrix $B - \lambda A$ be at least 1 and hence that

$$\det(B - \lambda A) = 0.$$

The numbers λ_i are the roots of this equation. If they are all distinct, the kernel of $B - \lambda_i A$ is a one-dimensional space and the vector v_i is one of the vectors in that space. Thus, it is completely determined up to a proportionality constant.

Fourth question:

The study made in (3) requires no particular hypothesis on ω. Therefore, it remains valid if the quadratic form ω is not positive definite. Suppose that, in a basis v_i, we have

$$\varphi(x, y) = \sum \mu_i \zeta_i \chi_i, \qquad \psi(x, y) = \sum \rho_i \zeta_i \chi_i.$$

The form $\psi - \lambda\varphi$ would be degenerate for the values $\lambda_i = \rho_i/\mu_i$ and at these values, $\det(B - \lambda A)$ vanishes.

If $\omega(x) = x_1^2 - x_2^2$ and $\Omega(x) = 2x_1 x_2$,

$$A = \begin{pmatrix} 1 & 0 \\ 0 & -1 \end{pmatrix}, \qquad B = \begin{pmatrix} 0 & 1 \\ 1 & 0 \end{pmatrix}$$

$$\det(B - \lambda A) = \det\begin{pmatrix} -\lambda & 1 \\ 1 & \lambda \end{pmatrix} = -(1 + \lambda^2).$$

There do not exist values of λ. The problem has no solution.

354

Proof of the necessity: If the vectors $\tilde{A}(u_i)$ are mutually orthogonal and if

$$x = \sum_{i=1}^{n} \xi_i u_i,$$

then

$$\|\tilde{A}(x)\|^2 = \left\| \sum_{i=1}^{n} \xi_i \tilde{A}(u_i) \right\|^2 = \sum_{i=1}^{n} \xi_i^2 \|\tilde{A}(u_i)\|^2.$$

Conversely, suppose that

$$x = \sum_{i=1}^{n} \xi_i u_i \;\Rightarrow\; \|\tilde{A}(x)\|^2 = \sum_{i=1}^{n} \lambda_i \xi_i^2.$$

The scalar product $\tilde{A}(x) \cdot \tilde{A}(y)$ defines a symmetric bilinear form, which generates the quadratic form $\|\tilde{A}(x)\|^2$. And this bilinear form is the symmetric bilinear form $\|\tilde{A}(x)\|^2$ since we know that this form is unique. Thus, if

$$x = \sum_{i=1}^{n} \xi_i u_i \quad \text{and} \quad y = \sum_{i=1}^{n} \eta_i u_i,$$

then

$$\tilde{A}(x) \cdot \tilde{A}(y) = \sum_{1}^{n} \lambda_i \xi_i \eta_i.$$

In particular,

$$\tilde{A}(u_i) \cdot \tilde{A}(u_j) = 0 \quad \text{if} \quad i \neq j.$$

The vectors u_i that we are seeking constitute an orthonormal basis for the space with respect to which the quadratic form $\|\tilde{A}(x)\|^2$ is reduced to a sum of squares. We know that at least one such basis always exists and that the vectors of this basis are the eigenvectors of the symmetric matrix of the quadratic form.

If X is a column matrix of the coordinates of x in the given basis, the quadratic form $\|\tilde{A}(x)\|^2$ can be defined by

$$\|\tilde{A}(x)\|^2 = \tilde{A}(x) \cdot \tilde{A}(x) = (AX)^T AX = X^T A^T AX.$$

The matrix $A^T A$ is symmetric:

$$(A^T A)^T = A^T (A^T)^T = A^T A.$$

Thus, the vectors u_i are the eigenvectors of the symmetric matrix $A^T A$.

355

First question:

Obviously, we need only verify that

$$y \in E \text{ and } y' \in E \;\Rightarrow\; \varphi(x, y-y') = \varphi(x, y) - \varphi(x, y') = 0,$$
$$y \in E \text{ and } \lambda \in C \;\Rightarrow\; \varphi(x, \lambda y) = \lambda \varphi(x, y) = 0.$$

Second question:

We know that

$$2\varphi(x, y) = \omega(x+y) - \omega(x) - \omega(y) = \sum_{i=1}^{p} \{[l_i(x+y)]^2 - l_i^2(x) - l_i^2(y)\}$$

$$= \sum_{i=1}^{p} \{[l_i(x) + l_i(y)]^2 - l_i^2(x) - l_i^2(y)\},$$

so that

$$\varphi(x, y) = \sum_{i=1}^{p} l_i(x)l_i(y).$$

The forms l_i are independent by hypothesis. For $\varphi(x, y)$ to be equal to zero for every x, it is necessary and sufficient that

$$l_i(y) = 0, \qquad (i = 1, 2, ..., p).$$

The vectors y that are solutions of this system constitute a vector space of dimension $n-p$. Now, this space is the space E:

$$n - p = \dim E \qquad \text{or} \qquad p = n - \dim E.$$

The number p of the forms l_i is determined directly from the properties of the bilinear form φ. Thus, it is the same for all decompositions of φ.

Third question:

As indicated in the statement of the problem, we shall prove the assertion by induction on n. The result is obvious for C^1:

$$w(z) = az^2 = (\alpha z)^2$$

if α is one of the two numbers such that $\alpha^2 = a$.
Let us assume the assertion true for C^{n-1}. Let us express the quadratic form ω in terms of the coordinates $x_1, x_2, ..., x_n$ of x and suppose first that ω contains a term in x_n^2

$$\omega(x) = a_{n,n}x_n^2 + 2\sum_{i=1}^{n-1} a_{i,n}x_i x_n + \sum_{\substack{i<n \\ j<n}} a_{i,j}x_i x_j$$

$$= a_{n,n}\left(x_n + \sum_{i=1}^{n-1} \frac{a_{i,n}}{a_{n,n}}x_i\right)^2 + \sum_{\substack{i<n \\ j<n}} a'_{ij}x_i x_j.$$

Only coordinates of index not exceeding $n-1$ appear in the quadratic form

$$\sum_{\substack{i<n \\ j<n}} a'_{ij}x_i x_j.$$

Therefore, we can consider this form as a form Ω on C^{n-1}, and apply to it the result established for C^{n-1}. There exist p independent linear forms l_i such that

$$\Omega = \sum_{i=1}^{p} l_i^2(x).$$

If α is one of the square roots of a_{nn}, the first term is the square of the form l defined by

$$l(x) = \alpha(x_n + \sum_{i=1}^{n-1} \frac{a_{i,n}}{a_{n,n}} x_i)$$

and this form l is independent of the forms l_i that appear in the decomposition of Ω since only l contains the variable x_n.

If ω does not contain any term in x_n^2 but contains a term in x_{n-1}, the same reasoning can be used to go from x_n to x_{n-1}.

Finally, if ω contains no squares, it contains at least one nonzero term $a_{ij}x_i x_j$, and the substitution

$$x_i' = \frac{x_i + x_j}{2}, \qquad x_j' = \frac{x_i - x_j}{2}$$

transforms this term into $a_{ij}(x_i'^2 - x_j'^2)$. The other terms in ω do not contain $x_i'^2$. Thus, after we make this substitution, we do have a square term in ω and the procedure used above can be applied.

356

Suppose that

$$(1) \qquad \omega(x) = \sum_{i=1}^{h} f_i^2(x) - \sum_{i=h+1}^{p} f_i^2(x) = \sum_{i=1}^{k} g_i^2(x) - \sum_{i=k+1}^{p} g_i^2(x),$$

where p is the (common) total number of the forms f_i and g_i (according to (b), this number is the same for both decompositions) and where h and k are the numbers of coefficients ε that are equal to 1 in the two decompositions respectively. Changing the order of the two decompositions if necessary, let us suppose that $h \leqslant k$.

Equation (1) implies that

$$(2) \qquad \sum_{i=1}^{h} f_i^2(x) + \sum_{i=k+1}^{p} g_i^2(x) = \sum_{i=h+1}^{p} f_i^2(x) + \sum_{i=1}^{k} g_i^2(x).$$

The first member of (2) contains $p - k + h = p - (k - h)$ linear forms. The zeros that are common to these two forms thus constitute a vector space E of dimension at least equal to

$$n-[p-(k-h)] = n-p+(k-h).$$

(Equality holds whether these forms are independent or not.)

For vectors in E, the left-hand member of (2) vanishes, and hence so does the right-hand member and, consequently, so do all the forms in the right-hand member since this member is the sum of squares. Therefore, the vectors in E make all the forms f_i vanish. Since these p forms are independent, the dimension of E is at most $n-p$.

Finally,

$$n-p+(k-h) \leqslant \dim E \leqslant n-p.$$

Therefore, $k-h \leqslant 0$, which, together with the hypothesis made, implies that $h=k$.

In the decomposition of a quadratic form ω defined on R^n as a sum of squares, the number of positive squares and the number of negative squares are invariant.

Part 4

REAL FUNCTIONS OF A REAL VARIABLE; CONTINUITY; DIFFERENTIATION, SEQUENCES OF NUMBERS AND FUNCTIONS

Exercises dealing with the material of Chapters I, II, and III of Book III and of Chapter I and the first portion of Chapter IV of Book IV of *Mathématiques Générales* by C. Pisot and M. Zamansky.

Introduction

The study of functions requires, in addition to the general theory, a certain number of techniques. For this reason, the exercises presented assume that the reader is familiar with the definition of, and the rules for manipulating the symbols o and O and with the asymptotic expansions of functions. To be able to vary the examples, I have used trigonometric functions and their inverses although the definition of these functions is not made until after the definition of the complex exponential function is made. However, their properties are well-known and will be proven later in the discussion.

This part has been divided into two chapters: The first chapter deals primarily with upper and lower bounds, limits, and the properties of continuous functions. It includes simple examples of uniformly convergent sequences of functions.

In the second chapter, we study the properties of derivatives and use these properties for the study of functions (maxima and minima, asymptotic expansions, limits, etc.).

Preliminary suggestions

Before turning to particular items, we must emphasize the need for great care regarding rigor. The reasoning should be complete, a careful examination should be made to assure that the hypotheses of the theorems used are verified, and no calculation should be made without justification. In this way, the definitions will be thoroughly understood and the theorems will be learned with precision. Subsequent study of analysis will thereby be made much easier.

Finally, let us recall certain results that will be used in this part.

Numerical sequences

Cauchy's criterion: *A necessary and sufficient condition for a sequence $\{x_n\}$ to be convergent is that the double sequence $\{x_p - x_q\}$ approach the limit 0.*

A nondecreasing sequence that is bounded above converges, and its limit is the least upper bound of the set of terms in the sequence.

Limits of functions

Three equivalent definitions can be given (PZ, Chapter III, 1st part, § 3). Which of these definitions should be used varies from problem to problem. In particular, the definition in terms of convergent denumerable sequences is often very convenient. (A necessary and sufficient condition for a function f to approach a limit l as its argument approaches x_0 is that, for every sequence $\{x_n\}$ that converges to x_0 without assuming the value x_0, the sequence $\{f(x_n)\}$ converges to l.)

The same remark holds for the definition of continuity of a function at a value x_0 since by definition, the function f must approach $f(x_0)$ as its argument approaches x_0.

Continuous functions

The properties of continuous functions *on a closed interval* $[a, b]$ are basic. Let us recall the more important of these:

The image of a closed bounded set is a closed bounded set. Thus, it contains its greatest lower and least upper bounds (the extreme-value theorem).

The image of an interval is an interval. In particular, if y_1 and y_2 are two values of the function, then every value between y_1 and y_2 is also a value of the function (the intermediate-value theorem).

Continuity implies uniform continuity.

Monotonic functions

Such a function has a right-hand limit and a left-hand limit at every value x. If f is a nondecreasing function, then

$$f(x+0) = \inf_{t>x} f(t), \qquad f(x-0) = \sup_{t<x} f(t).$$

Uniform convergence in an interval I

This concept is used frequently in the course and in these problems. Therefore, it calls for serious study. It can be defined in two equivalent ways:

$\forall\, \varepsilon,\, \exists\, N$ such that $n > N \Rightarrow |f_n(x)-f(x)| < \varepsilon,$ independent of $x \in I$;

$\forall\, \varepsilon,\, \exists\, N$ such that $\|f_n-f\| = \sup_{x \in I} |f_n(x)-f(x)| < \varepsilon.$

The second of these is in general more convenient. Since $\|f_n-f\|$ is a number, the study to be made has to do with a numerical sequence. On the other hand, with the first definition, to every value of x there correspond infinitely many possible numbers N, and we need to show that these numbers include some that satisfy the implication for all values of x in I.

We shall also use Cauchy's criterion frequently: A sequence $\{f_n\}$ converges uniformly if and only if the double sequence $\|f_p-f_q\|$ converges to 0.

Note that the norm $\|f\|$ depends on the interval I in question.

Differentiation of composite functions

Differential notation (the quantity $df=f'(x)dx$, where f is a differentiable function, is called the *differential* of f) makes calculation particularly simple:

$$d(g \circ f) = g'[f(x)]f'(x)dx = g'[f(x)]df.$$

Thus, the differential is written in the same way whether f is a variable or a function. For example,

$$d\left(\log \operatorname{tg}\frac{x}{2}\right) = \frac{d(\operatorname{tg} x/2)}{\operatorname{tg} x/2} = \frac{1}{\operatorname{tg} x/2}\frac{1}{\cos^2 x/2} d\left(\frac{x}{2}\right) = \frac{1}{\operatorname{tg} x/2}\frac{1}{\cos^2 x/2}\frac{dx}{2}$$

$$= \frac{dx}{\sin x}.$$

Taylor's formulas

Although they resemble each other formally, there are two formulas of Taylor, which are quite different since the areas of their applications have nothing in common:

The formula that generalizes the theorem of the mean, and that is valid if the function f in question is n times differentiable on the interval $[a, x]$, is

$$f(x) = f(a)+(x-a)f'(a)+ \cdots +\frac{(x-a)^n}{n!}f^{(n)}(c)$$

(see exercises 442 and 447).

The formula used to evaluate limits or to obtain asymptotic expansions:

$$f(x) = f(a)+(x-a)f'(a)+ \cdots +\frac{(x-a)^n}{n!}f^{(n)}(a)+o[(x-a)^n].$$

Here, the only information given is that the remainder is of the form $o[(x-a)^n]$, that is, that the quotient of is the remainder by $(x-a)^n$ approaches 0 as x approaches a (see exercises 440, 444, 445, 446, and 449).

Logarithmic and exponential functions

It is often convenient to express all logarithms in terms of the natural logarithm and all exponentials in terms of the exponential to the base e:

$$\log_a x = \frac{\ln x}{\ln a}; \qquad a^x = e^{x \ln a}.$$

In evaluating limits involving these functions, we should know that the following holds for $\alpha > 0$:

$$\text{when} \quad x \to +\infty, \quad \frac{\ln x}{x^\alpha} \to 0 \quad \text{and} \quad \frac{e^x}{x^\alpha} \to +\infty;$$

$$\text{when} \quad x \to 0, \quad x^\alpha \ln x \to 0.$$

Landau's symbols o and O

Let f be a function that approaches 0 as its argument approaches x_0 from the right.

(a) A function h is said to be an $o(f)$ and we write $h = o(f)$ if, for every $\varepsilon > 0$, there exists a number $x_1 > x_0$ such that

$$x_0 < x < x_1 \implies |h(x)| \leqslant \varepsilon |f(x)|.$$

(b) A function h is said to be an $O(f)$ and we write $h = O(f)$ if there exists a number $x_1 > x_0$ and a constant M such that

$$x_0 < x < x_1 \implies |h(x)| \leqslant M |f(x)|.$$

The definitions are analogous for approach to x_0 from the left or from both directions and for approach to $\pm \infty$.

The properties most often used are the following:

$$h = o(f) \implies h = O(f),$$
$$o(f) + o(f) = o(f), \qquad O(f) + O(f) = O(f),$$
$$o(f)O(g) = o(fg), \qquad O(f)O(g) = O(fg),$$
$$o(f)o(g) = o(fg),$$
$$o[O(f)] = O[o(f)] = o(f), \qquad O[O(f)] = O(f).$$

These notations are symbolic. We explain two of them:

The relation $o(f) + o(f) = o(f)$ means that, as x approaches x_0 from the left (for example),

$$h_1 = o(f) \quad \text{and} \quad h_2 = o(f) \implies h_1 + h_2 = o(f).$$

The relation $O[o(f)] = o(f)$ means that, as x approaches $+\infty$ (for example),

$$h = O(g) \quad \text{and} \quad g = o(f) \implies h = o(f).$$

Exercises

CHAPTER 1

Bounded sets, sequences, limits, continuous functions, uniform convergence

401

(1) Let a and b denote two real numbers. Suppose that, for every number $x > b$, the inequality $a \leq x$ is verified. Show that $a \leq b$.

(2) If a sequence $\{u_n\}$ has a limit l and if, for every number $x > b$, the inequality $u_n \leq x$ is satisfied for all n from some integer on, then the limit l does not exceed b.

402

Suppose that A and B are two nonempty sets of real numbers, that A is bounded above, and that B is contained in A. Show that B is bounded above and that $\sup B \leq \sup A$.

403

Suppose that A and B are two nonempty bounded sets of real numbers. Show that $A \cup B$ is bounded and that

$$\sup (A \cup B) = \sup (\sup A, \sup B), \qquad \inf (A \cup B) = \inf (\inf A, \inf B).$$

APPLICATIONS: (1) Show that a sequence $\{x_n\}$ of real numbers that possesses a finite limit l is bounded.

(2) A continuous mapping f on $0, +\infty$ into R that approaches a finite limit l as its argument approaches $+\infty$ is bounded.

404

Suppose that A and B are two nonempty bounded sets of real numbers. Show that $A \cap B$ is bounded and that

$$\sup (\inf A, \inf B) \leq \inf (A \cap B) \leq \sup (A \cap B) \leq \inf (\sup A, \sup B).$$

Give an example in which the inequalities are strict inequalities (the sets A and B may be taken as finite).

405

(1) Suppose that A and B are two nonempty sets of real numbers that are bounded above. Define a set C by

$$z \in C \Leftrightarrow \exists x \in A \quad \text{and} \quad \exists y \in B : z = x+y.$$

Show that C is bounded above and that

$$\sup C = \sup A + \sup B.$$

(2) Can we give an analogous result for the set $D = AB$ of products of an element in A and an element in B?

(3) Suppose that f and g are two bounded functions defined on a set X contained in R. Show that the mapping $f+g$ is bounded above and that

$$\sup (f+g) \leqslant \sup f + \sup g.$$

Give an example in which the inequality is a strict inequality.

406

Consider the function f defined on R by

$$f(x) = \begin{cases} 0 & \text{if } x=0, \\ 1/x & \text{if } x \text{ is a nonzero rational number,} \\ x & \text{if } x \text{ is irrational.} \end{cases}$$

(1) Show that the function f is a bijective mapping on R onto itself.
(2) Show that the function f is continuous at only two values.

407

Let f be a continuous mapping of the closed interval $[a, b]$ into R and let y be a number between $f(a)$ and $f(b)$.

(1) We define two sets A and B by

$$x \in A \Leftrightarrow a \leqslant x \leqslant b \quad \text{and} \quad f(x) < y,$$
$$x \in B \Leftrightarrow a \leqslant x \leqslant b \quad \text{and} \quad f(x) > y.$$

Show that the sets A and B are open. (A set E is said to be open if every element of E is the center of an open interval contained in E.)

(2) Show that the set $f^{-1}(y)$, that is, the set of numbers x in $[a, b]$ that satisfy the equation $f(x)=y$, is nonempty and that it has a smallest and a greatest element.

408

First Part

Simplify the expression

$$u_n = \sum_{k=1}^{n} \left(\frac{1}{k} - \frac{1}{k+1} \right)$$

and show that the sequence $\{u_n\}$ converges.

Use Cauchy's criterion to show that the sequence

$$v_n = \sum_{k=1}^{n} \left(\frac{1}{k} - \frac{1}{k+1} \right) \sin k\alpha$$

converges, where α is a given parameter.

Second Part

Let f denote the function defined in the interval $]0, 1]$ by

(a) $f(1)=1$;

(b) f is continuous at every point of $]0, 1]$;

(c) the restriction of f to the interval $[1/(k+1), 1/k]$ is, for every natural number k, a linear function defined by

$$f(x) = x \sin k\alpha + c_k,$$

where c_k denotes a constant (that is a real number depending only on k).

(1) Construct the graph of the restriction of f to the interval $[\frac{1}{5}, 1]$ if $\alpha=\frac{1}{4}\pi$.

(2) Show that the sequence $f(1/n)$ converges.

(3) Show that the function f has a limit at 0.

Third Part

Let g be a function defined in $]0, 1]$. Suppose that there exists a constant M such that, for every pair (x, x') of numbers in $]0, 1]$,

$$g(x') - g(x) \leqslant M \, |x' - x|.$$

(1) Show that the function f of the second part above fulfils the hypotheses made regarding the function g.

(2) Are the results of the questions 2 and 3 in the second part still valid for the function g?

409

(1) Let $\{u_n\}$ denote a nondecreasing sequence and let $\{v_n\}$ denote a nonincreasing sequence. Suppose that $u_n \leqslant v_n$ for every integer n. Show that the sequences $\{u_n\}$ and $\{v_n\}$ converge and that $\lim u_n \leqslant \lim v_n$. If the sequence $\{v_n - u_n\}$ converges to 0, show that the two sequences $\{u_n\}$ and $\{v_n\}$ have the same limit.

(2) Suppose that u_0 and v_0 are given and that $0 < u_0 < v_0$. Define two sequences $\{u_n\}$ and $\{v_n\}$ inductively by

$$u_{n+1} = \sqrt{u_n v_n}, \qquad v_{n+1} = \frac{u_n + v_n}{2}.$$

Use the results of (1) to show that the two sequences converge to the same limit.

410

Inductively defined sequences

Suppose that a function f is defined and continuous on an interval $[a, b]$. Consider the sequence $\{u_n\}$ defined by giving the value u_0 and the recursion relation

$$u_n = f(u_{n-1}).$$

In what follows we shall assume that all the terms of the sequence are numbers in the interval $[a, b]$ (since otherwise, the sequence would not be defined). *This condition will be the first condition to verify in each particular problem.* The verification will usually be by induction.

(1) Show that if the function f is nondecreasing, the sequence $\{u_n\}$ is mono-tonic and it approaches a limit u that is a root of the equation

$$u = f(u).$$

(2) Show that if the function f is nonincreasing, the sequence of even-numbered terms $\{u_{2n}\}$ and the sequence of odd-numbered terms $\{u_{2n+1}\}$ are monotonic convergent sequences.

(3) Apply the above results to an examination of the particular sequences

$\{u_n\}$, $\{v_n\}$, and $\{w_n\}$ defined by

$$u_0 = 0, \qquad u_n = \frac{u_{n-1}+1}{u_{n-1}+2},$$

$$v_0 = 0, \qquad v_n = \cos v_{n-1},$$

$$w_0 = \tfrac{1}{2}, \qquad w_n = (1-w_{n-1})^2.$$

411

This exercise also deals with an inductively defined sequence, but the hypotheses are different from those of exercise 410, and its interest is more theoretical than practical.

Denote by f a mapping of the closed interval $[a, b]$ into itself such that, for every u and v in $[a, b]$,

$$|f(u)-f(v)| < |u-v|.$$

(1) Show that f is a continuous mapping and that the equation

$$f(t) = t$$

has one and only one solution θ. (Study the behavior of the function F defined by $F(t)=f(t)-t$.)

(2) Define a sequence $\{x_n\}$ in terms of x_0 by

$$x_n = f(x_{n-1}).$$

(a) Show that the sequence $|x_n-\theta|$ is a decreasing sequence and that it has a limit l.

(b) Show that from the sequence $\{x_n\}$ we can choose a subsequence that converges to $\theta+\varepsilon l$ where ε is equal to $+1$ or -1.

(c) Show that

$$|f(\theta+\varepsilon l)-\theta| = l,$$

and derive from this the fact that $l=0$ and hence that the sequence $\{x_n\}$ converges to θ.

412

An example of a nonelementary function satisfying the hypotheses of exercise 411:

We recall that the limit of the sequence $\{u_n\}$ defined by

$$u_n = \sum_1^n \frac{1}{k(k+1)} = \sum_1^n \left(\frac{1}{k} - \frac{1}{k+1} \right)$$

is 1.

(1) Show by use of Cauchy's criterion that the sequence $\{S_n(t)\}$, defined by

$$S_n(t) = \sum_{k=1}^n \frac{\sin kt}{k^2(k+1)},$$

converges to a limit $S(t)$.

(2) Show that if $u \neq v$,

$$|S(v) - S(u)| < |v - u|.$$

Show from this that this function $S(t)$ defined in $[-\pi, \pi]$ satisfies the hypotheses of exercise 411.

413

(1) Let $\{x_n\}$ denote a real sequence with limit l. Show that the sequence $\{y_n\}$ defined by

$$y_n = \frac{x_1 + x_2 + \cdots + x_n}{n}$$

also converges to l.

(2) Show that if the sequence $\{x_{n+1} - x_n\}$ converges to l, then the sequence $\{x_n/n\}$ also converges to l.

(3) In this question, the properties of the logarithm and the exponential are assumed known.

If a sequence $\{x_n\}$ with positive terms converges to a limit l, show that the sequence $\sqrt[n]{x_1, ..., x_n}$ also converges to l.

Show that if $\{x_n\}$ is a sequence with positive terms and if the sequence $\{x_{n+1}/x_n\}$ converges to a limit l, then the sequence $\sqrt[n]{x_n}$ also converges to l.

(4)   Test the sequences $\{u_n\}$ and $\{v_n\}$ defined by

$$u_n = \frac{1}{n} + \frac{1}{2n} + \cdots + \frac{1}{n^2} = \sum_{p=1}^n \frac{1}{pn},$$

$$v_n = \sqrt[n]{n^3 + n^2 - 1},$$

for convergence and evaluate the limits if they exist.

414

Define a function f on $[0, 1]$ into R by

$$f(x) = \begin{cases} 0 & \text{if } x \text{ is irrational,} \\ 1/q & \text{if } x \text{ is a nonzero rational represented in its simplest form by } p/q, \\ 1 & \text{if } x=0. \end{cases}$$

Show that this function f is discontinuous at every rational value of x and continuous at every irrational value. (One can show that the values of x such that $f(x) \geqslant \varepsilon > 0$ are finite in number.) Show that if x_0 is rational, the function f has right- and left-hand limits at x_0.

415

Show directly (without using the general theorems) that the function f defined on the intervals $[0, 1]$ by $f(x) = \sqrt{x}$ is uniformly continuous.

416

Show that if a function f is defined and nondecreasing in the interval $[a, b]$ and assumes every value between $f(a)$ and $f(b)$ at least once, it is continuous at every value in the interval $[a, b]$.

Show by constructing a counterexample that the preceding result does not always hold for nonmonotonic functions.

417

(1) For every number θ in the interval $0 \leqslant \theta \leqslant \pi$, define $x = 2 \cos \theta$ and $y = 2 \cos n\theta$. Show by induction on n that this defines a polynomial function A_n on the interval $[-2, 2]$ such that $y = A_n(x)$. Show that the term of highest degree in A_n is x^n.

(2) Show that the zeros of the polynomials $A_n - 2$ and $A_n + 2$ are real and that they belong to the interval $[-2, 2]$.

(3) Show that, if f is a function defined and continuous on the interval $[-2, 2]$ and if $|f(x)| < 2$ for every x in $[-2, 2]$, then the function $f - A_n$ vanishes at least n times in that interval.

(4) Show that, if F is a polynomial and if the coefficient of the term of highest degree is 1, then there exists at least one number c in the interval $[-2, 2]$ such that $|F(c)| \geqslant 2$.

(5) Show that if G is a polynomial whose term of highest degree is x^n, then the function $|G|$ assumes in every interval $[a, b]$ at least one value exceeding $2[(b-a)/4]^n$. From this, show that the oscillation of G in every interval $[a, b]$ is at least equal to $4[(b-a)/4]^n$, the oscillation of G being defined as the number
$$\omega = \sup_{a \leqslant x \leqslant b} G(x) - \inf_{a \leqslant x \leqslant b} G(x).$$

418

This is an exercise on the symbols o and O and on equivalent functions. It should be worked without use of the theory of derivatives.

(1) Suppose that u is a real function of a numerical variable x defined in a neighborhood of 0 and that $u(x)$ tends to 0 as x tends to 0. By using the binomial theorem, show that
$$(1+u)^n = 1+nu+O(u^2).$$

Use this result to show that
$$\sqrt[n]{1+u} = 1+\frac{u}{n}+O(u^2).$$

(2) Show that the two functions f and g defined by
$$f(x) = \sqrt[2]{1+2x+3x^2} - \sqrt[3]{1+x+x^2},$$
$$g(x) = \sqrt[4]{1+x+x^2} - \sqrt[4]{1+x},$$

are equivalent as x tends to 0.

419

(1) Show that
$$\cot x = \frac{1}{x}+o(1)$$

as x approaches 0 (use the fact that $|\sin x| < |x| < |\tan x|$).

(2) Under what condition does the function f defined by
$$f(x) = \sum_{k=1}^{n} a_k \cot kx,$$

where the coefficients a_k are real numbers, have a finite limit as x approaches 0? Evaluate that limit.

420

(1) Show by induction on n that, if $x > 1$,

$$x^n \geqslant 1 + n(x-1).$$

Evaluate the limit of the sequence $\{x^n\}$, where x is a given positive number.
 (2) Show that the sequence of functions f_n defined by

$$f_n(x) = x^n \qquad \text{for} \qquad o \leqslant x \leqslant 1$$

converges uniformly in every interval $[0, h]$, where $0 < h < 1$.
 (3) Show that there does not exist a number N such that

$$n > N \Rightarrow f_n(x) < \varepsilon$$

for every x in $[0, 1[$. Does the sequence of functions $\{f_n\}$ converge uniformly in the interval $[0, 1]$? Could this result have been predicted without calculation?
 (4) Show that the sequence of functions $\{g_n\}$ defined by

$$g_n(x) = x^n(1-x) \qquad \text{for} \qquad 0 \leqslant x \leqslant 1$$

converges uniformly to the zero function in the interval $[0, 1]$.

421

A sequence of functions $\{u_n\}$ on the interval $[-\tfrac{1}{2}\pi, \tfrac{1}{2}\pi]$ is defined by

$$u_0(x) = x, \qquad u_n(x) = \sin [u_{n-1}(x)].$$

Show that this sequence converges uniformly to zero on $[-\tfrac{1}{2}\pi, \tfrac{1}{2}\pi]$.

422

A sequence of functions $\{f_n\}$ on R into R is defined by

$$f_n(x) = \frac{(x+1)^{2n+1} + (x-1)^{2n+1}}{(x+1)^{2n+1} - (x-1)^{2n+1}}.$$

(1) Show that the function f_n is continuous and that it is an odd function. Study the behavior of f_n. (Think of f_n as a composite of simpler functions.)
 (2) Show that there exist constants a and b such that $f_n(x) - (ax+b)$ approaches zero as x approaches $\pm \infty$ and construct the graph of f_n.
 (3) Show that, for every x, the sequence $\{f_n(x)\}$ converges to a limit $f(x)$ and show that this convergence is uniform in every interval $[a, b]$ with either both negative or both positive end-points. Is the converge uniform in the interval $[-a, a]$?

423

(1) Let f denote a real function that is defined and continuous on the interval $[a, b]$ and let u and v denote two numbers belonging to $[a, b]$. Find the expression for the first-degree polynomial l whose values at u and v are $f(u)$ and $f(v)$. Show that, for every value in $[u, v]$, there exists at least one number ξ in $[u, v]$ such that

$$l(x) = f(\xi).$$

Show, finally, that

$$\sup_{u \leqslant x \leqslant v} |l(x) - f(x)| \leqslant \omega,$$

where ω is the oscillation of f in $[u, v]$.

(2) Let Λ denote the set of continuous functions λ possessing the property that, to every function λ, there corresponds a finite partition

$$a = x_0 < x_1 < x_2 < \cdots < x_{n-1} < x_n = b$$

of the interval $[a, b]$ and the restriction of f to each of the intervals $[x_i, x_{i+1}]$ of this partition is a polynomial of degree not exceeding 1 (a linear or affine function).

Show that Λ is a vector space on $\mathbf{R}$.

Draw the graph of a function λ.

(3) Show that every function that is continuous in $[a, b]$ is the uniform limit of a sequence of functions belonging to Λ.

424

We define the function $[x]$ on the set $\mathbf{R}$ of real numbers x onto the set of *integers* by

$$[x] \leqslant x < 1 + [x].$$

Then, we define

$$D(x) = x - [x].$$

(1) Show that the function D is periodic. At what values is it continuous? Show that D is a regulated function on every finite interval. (A function is said to be *regulated* on an interval if it is the limit function to which some sequence of step functions converges uniformly.)

(2) Consider the sequence of functions u_n defined by

$$u_n(x) = \sum_{k=1}^{n} \frac{D(kx)}{2^k}.$$

Show that this sequence converges uniformly in the set R of real numbers. (One may use Cauchy's criterion.) Denote the limit function by u.

(3) Show that the function u is continuous at every irrational value of x.

(4) Show that the function u has left- and right-hand limits $u(x_0 - 0)$ and $u(x_0 + 0)$ at every rational value x_0. Show also that

$$u(x_0 - 0) = \lim u_n(x_0 - 0), \qquad u(x_0 + 0) = \lim u_n(x_0 + 0).$$

Finally, show that the function u is continuous from the right but discontinuous from the left at rational values of x.

425

Consider a real function f that is defined and continuous on the set R of reals such that

$$|f(x)| < |x| \qquad \text{for every } x \neq 0.$$

(1) Show that $f(0) = 0$.

(2) Let ε and M be given numbers such that $0 < \varepsilon < M$. Show that there exists a number $k < 1$ such that

$$|f(x)| \leqslant k|x| \qquad \text{for} \qquad \varepsilon \leqslant |x| \leqslant M.$$

Find a counterexample showing that such a bound may not exist for the entire interval $|x| \leqslant M$.

(3) Consider the sequence of functions f_n defined by

$$f_0(x) = f(x), \qquad f_1(x) = f[f_0(x)], \dots, \qquad f_{n+1}(x) = f[f_n(x)].$$

If ε and M are two numbers satisfying the double inequality $0 < \varepsilon < M$ and if k is as defined in (2), show that

$$|x| \leqslant M \Rightarrow |f_n(x)| \leqslant \sup(\varepsilon, k^{n+1}M).$$

(Use induction on n and consider separately the two cases $|f_n(x)| < \varepsilon$ and $|f_n(x)| \geqslant \varepsilon$.) Show from this that the sequence of the functions f_n converges uniformly to 0 on the interval $|x| \leqslant M$.

426

Consider the function f defined on $[0, 1/\pi]$ by

$$f(x) = \begin{cases} \sin(1/x) & \text{if } x \neq 0 \\ y_0 & \text{if } x = 0 \text{ (where } y_0 \text{ is a given real number).} \end{cases}$$

(1) Study the continuity of the function f.

(2) Show that there is no step-function φ such that

$$|f(x)-\varphi(x)| < \tfrac{1}{2}, \qquad \forall x \in \left[0, \frac{1}{\pi}\right].$$

From this conclude that f is not a regulated function on $[0, 1/\pi]$. Could this result have been predicted?

CHAPTER 2

Exponentials; derivatives; the logarithm and other common functions

427

Suppose that the function f is defined and continuous on $[a, b]$, where $a < b$, that it is differentiable in the interval $]a, b]$, and that the derivative f' approaches a limit l as x approaches a from the right. Show that f has a right-hand derivative at a that is equal to l.

EXAMPLE: $a = 0$, $b = \pi$, $f(x) = \sqrt{x} \sin x$.

428

Show that the function f defined on $[0, 1/\pi]$ by

$$f(x) = x^2 \cos \frac{1}{x} \quad \text{if} \quad x \neq 0; \quad f(0) = 0$$

is differentiable. Show from this that the derivative of a regulated function is not always a regulated function. (Use the result of exercise 426.)

429

(1) Study the behavior of the functions f and g defined by

$$f(x) = \frac{x^2 \sqrt{2}}{2\sqrt{x^4 - 2x^2 + 2}}, \qquad g(x) = \arcsin f(x).$$

(2) Show that g is differentiable everywhere except possibly at $-\sqrt{2}$ and $\sqrt{2}$. Evaluate g'. Show that g has a right-hand derivative and a left-hand derivative at the two exceptional values mentioned. Either do this by direct calculation or use the result of exercise 427.

Construct the graph of g.

430

Consider the function f of a real variable x defined by

$$f(x) = \frac{2x+1}{x^2+1} + 2 \arctan \frac{1-x}{1+x}$$

at all values at which the above expression is meaningful.

(1) At what values of x is this function f defined, continuous, and differentiable? Find an expression for f' at these values of x. Study the behavior of f and construct its graph.

(2) Show that the $(n-1)$st derivative of f is a rational fraction

$$f^{(n-1)}(x) = \frac{A_n(x)}{(x^2+1)^n}$$

and that the numerator A_n is an nth-degree polynomial. Show that the zeros of A_n are real and distinct, for example, by studying the behavior of the successive derivatives.

431

Suppose that a and α are two given numbers, where $a > 1$.

(1) Discuss the number of positive roots of the equation

$$a^x = x^\alpha.$$

(2) Show that the equation

$$a^{(a^x)} = x$$

has the same roots as the equation $a^x = x$. From this, find the number of solutions of the latter equation.

432

Study the variations and construct the graph of the function f defined for $x > 0$ by

$$f(x) = x^x.$$

433

Let m and n denote given real numbers. Consider the pairs of positive numbers (x, y) that satisfy the system

$$(S) \quad \begin{cases} x^{x+y} = y^n \\ y^{x+y} = x^m. \end{cases}$$

Exclude the trivial solution $(1, 1)$.

(1) Show that the system (S) has no solutions if m and n are of opposite sign.

(2) Show that, if m and n are both positive, the system (S) has a nontrivial solution except at particular values of m and n.

(3) If m and n are negative, show that the system has no solution other than the trivial solution except at particular values of m and n, where it has one or two solutions other than the trivial solution.

434

(1) Denote by f a convex function defined on $[a, b]$ and denote by p_i (where $i = 1, 2, ..., n$) positive numbers whose sum

$$\sum_{i=1}^{n} p_i$$

is equal to 1.

Show that

$$f\left(\sum_{i=1}^{n} p_i x_i\right) \leqslant \sum_{i=1}^{n} p_i f(x_i).$$

(2) Show that, if $\alpha > 1$, the function f defined for $x \geqslant 0$ by $f(x) = x^\alpha$ is convex and show that, if $x_1, x_2, ..., x_n$ are positive numbers, then

$$\left(\sum_{i=1}^{n} x_i\right)^\alpha \leqslant n^{\alpha-1}\left(\sum_{i=1}^{n} x_i^\alpha\right).$$

Can equality hold in this relation?

(3) Show that, if $\alpha > 1$,

$$\sum_{i=1}^{n} x_i^\alpha < \left(\sum_{i=1}^{n} x_i\right)^\alpha,$$

where $x_1, x_2, ..., x_n$ still denote positive numbers.

435

We denote by x, y, α, and β positive numbers such that $\alpha+\beta=1$.

(1) Show that

$$x^{\alpha}y^{\beta} \leqslant \alpha x+\beta y.$$

Under what conditions can equality hold?

(2) Apply the preceding result to the pairs of values

$$x_i = \frac{a_i^{1/\alpha}}{a}, \quad y_i = \frac{b_i^{1/\beta}}{b}, \quad (i = 1, 2, ..., n)$$

and add the inequalities obtained.

Show that, for a suitable choice of a and b,

$$\sum_{i=1}^{n} a_i b_i \leqslant \left(\sum_{i=1}^{n} a_i^{1/\alpha} \right)^{\alpha} \left(\sum_{i=1}^{n} b_i^{1/\beta} \right)^{\beta}.$$

436

(1) Let f denote a function that is defined, continuous, and differentiable on a closed interval $[x_1, x_2]$. Let us define two other functions φ and ψ in terms of it as follows:

φ is defined in $]x_1, x_2]$ by $\varphi(x) = \dfrac{f(x)-f(x_1)}{x-x_1}$;

ψ is defined in $[x_1, x_2[$ by $\psi(x) = \dfrac{f(x)-f(x_2)}{x-x_2}$.

Show that it is possible to assign values $\varphi(x_1)$ and $\psi(x_2)$ such that the functions φ and ψ are defined and continuous in $[x_1, x_2]$.

If m is a number lying in the open interval between $f'(x_1)$ and $f'(x_2)$, show that at least one of the equations

$$\varphi(x) = m, \quad \psi(x) = m$$

has a solution.

(2) Suppose that the function f is defined, continuous, and differentiable in a closed interval $[a, b]$. Show that if the derivative f' assumes the values α and β, it assumes every value between α and β.

(3) Suppose that the function f satisfies the hypotheses of (2) and in addition is convex. Show that the derivative f' is continuous.

437

If a function f is defined in an interval containing x_0 as an interior point and if the ratio

$$\frac{f(x_0+h)-f(x_0-h)}{2h}$$

approaches a limit as h approaches 0, this limit is called the *symmetric derivative* of the function f at x_0 and is denoted by $f_s'(x_0)$.

(1) Show that if the function f has right- and left-hand derivatives at x_0, it has a symmetric derivative at x_0.

(2) Show that the function f defined by

$$f(0) = 0 \quad \text{and} \quad f(x) = x \sin \frac{1}{x} \quad \text{if} \quad x \neq 0$$

possesses neither a right- not a left-hand derivative at 0 but that it does possess a symmetric derivative there.

(3) Show that if the function f is nondecreasing and has a symmetric derivative, this symmetric derivative is nonnegative.

(4) Show that if the functions f and g are continuous at a point x_0 and have symmetric derivatives at that point, then the sum $f+g$ and the product fg also have symmetric derivatives at x_0. Find expressions for these symmetric derivatives.

438

This exercise assumes knowledge of the definition of a symmetric derivative of a function (as given at the beginning of exercise 437) and also of its properties (as shown in (1) and (4) of that exercise).

(1) Let us denote by φ a function that is defined and continuous on a closed interval $[x_1, x_2]$ and which has a symmetric derivative at every point in the open interval $]x_1, x_2[$. Suppose in addition that

$$x_1 < x_2 \quad \text{and} \quad \varphi(x_1) > \varphi(x_2).$$

Let λ denote a number verifying the double inequality $\varphi(x_1) > \lambda > \varphi(x_2)$. Let E_λ denote the set of numbers x in $[x_1, x_2]$ such that $\varphi(x) > \lambda$.

Show that the least upper bound ξ of E_λ verifies the double inequality $x_1 < \xi < x_2$ and that every open interval containing ξ also contains numbers u and v such that $\varphi(u) > \lambda$ and $\varphi(v) \leqslant \lambda$. From this show that

$$\varphi(\xi) = \lambda \quad \text{and} \quad \varphi_s'(\xi) \leqslant 0.$$

(2) Let f denote a function that is defined and continuous on the closed interval $[a, b]$ and which has a positive symmetric derivative at every point in the open interval $]a, b[$. Show by contradiction that the function f is a strictly increasing function.

(3) Suppose that the function f, defined and continuous on $[a, b]$, has a symmetric derivative $f_s'(x)$ at every point x in $]a, b[$. Denote by x_1 and x_2 two numbers such that $a \leqslant x_1 < x_2 \leqslant b$. Show that, if there exists a number k such that $f_s'(x) < k$ for every x in $]a, b[$, then

$$f(x_2) - f(x_1) \leqslant k(x_2 - x_1).$$

More precisely, show that if the function representing the symmetric derivative is bounded in $]a, b[$ with greatest lower and least upper bounds m and M respectively, then

$$m(x_2 - x_1) \leqslant f(x_2) - f(x_1) \leqslant M(x_2 - x_1).$$

From this result, show, in particular, that, if $f_s'(x) \geqslant 0$, the function f is strictly increasing unless there is an interval at every point of which $f_s' = 0$.

439

(1) Show that the function f defined by

$$f(x) = \frac{\sin x}{x} \quad \text{if} \quad 0 < x \leqslant \frac{\pi}{2} \quad \text{and} \quad f(0) = 1$$

is continuous and differentiable in the interval $[0, \frac{1}{2}\pi]$ and evaluate its derivative.

(2) Show that if $0 \leqslant x \leqslant \frac{1}{2}\pi$, then

$$\frac{2x}{\pi} \leqslant \sin x \leqslant x.$$

(These inequalities are frequently used to majorize or minorize trigonometric functions.)

440

SIMPLE EXAMPLES OF THE USE OF ASYMPTOTIC EXPANSIONS IN INVESTIGATION OF EQUIVALENT FUNCTIONS OR LIMITS

(1) Find the limit of

$$\frac{\arctan x - \sin x}{\operatorname{tg} x - \arcsin x}$$

as x approaches 0.

(2) Find the limit l of the function f defined by

$$f(x) = \frac{1}{x} \ln \frac{e^x - 1}{x}$$

as x tends to 0 and then that of the expression

$$\frac{f(x) - l}{x}.$$

In the following exercise, the limit of the function f as x tends to $+\infty$ is to be found by a method that does not involve asymptotic expansions.

(3) Find the coefficients a and b such that the function g defined by

$$g(x) = \cos x - \frac{1 + ax^2}{1 + bx^2}$$

will be an infinitesimal of order as great as possible as x approaches 0. Then, determine the principal part of g.

(4) Find the limit as x tends to $+\infty$ of the function h defined by

$$h(x) = x\left[\frac{1}{e} - \left(\frac{x}{x+1}\right)^x\right].$$

441

(1) Show that if the functions f and g both approach $+\infty$ or if they both approach 0 as x tends to x_0 (or as x tends to $+\infty$) and if these two functions are equivalent, then the functions $\ln f$ and $\ln g$ are also equivalent.

(2) Show that the functions $\ln (e^x - 1)$ and $\ln e^x$ are equivalent as x tends to $+\infty$. Find the limit of the function f defined by

$$f(x) = \frac{1}{x} \ln \left(\frac{e^x - 1}{x}\right).$$

442

Show that, for every positive number x,

$$x - \frac{x^2}{2} < \ln (1+x) < x - \frac{x^2}{2} + \frac{x^3}{3},$$

$$1 + \frac{3}{2}x + \frac{3}{8}x^2 - \frac{x^3}{16} < (1+x)^{3/2} < 1 + \frac{3}{2}x + \frac{3}{8}x^2.$$

443

Find the limit of the sequence whose general term is

$$u_n = \prod_{p=1}^{n}\left(1+\frac{p}{n^2}\right) = \left(1+\frac{1}{n^2}\right)\left(1+\frac{2}{n^2}\right)\cdots\left(1+\frac{p}{n^2}\right)\cdots\left(1+\frac{n}{n^2}\right).$$

444

(1) Define functions g and h by

$$g(x) = \cos\sqrt{x} \qquad \text{if} \qquad x > 0,$$
$$h(x) = \cosh\sqrt{-x} \qquad \text{if} \qquad x < 0.$$

Show that the functions g and h are continuous and have derivatives of the first two orders. Evaluate these derivatives.

(2) Define a function f by

$$f(x) = g(x) \quad \text{if} \quad x > 0, \quad f(0) = 1, \quad f(x) = h(x) \quad \text{if} \quad x < 0.$$

Show that f is continuous and twice differentiable everywhere. Show that f and its first two derivatives are related by

$$4xf''(x)+2f'(x)+f(x) = 0.$$

445

Let f denote the function defined in the open intervals $]0, 1[$ and $]1, +\infty[$ by

$$f(x) = \frac{x \ln x}{x^2-1}.$$

(1) Show that $f(0)$ and $f(1)$ can be chosen in such a way that f will be continuous and differentiable for every $x \geqslant 0$.

(2) Study the behavior of the function f and construct its graph.

446

Define a function f by

$$f(x) = \frac{x}{1+e^{1/x}} \qquad \text{if} \qquad x \neq 0 \qquad \text{and} \qquad f(0) = 0.$$

(1) Show that the function f is continuous and has right- and left-hand derivatives at every real value of x.

(2) Study the behavior of f.

(3) Find the asymptotic expansion of f (including the x^{-3}-term) in powers of $1/x$ as x tends to $\pm\infty$. Construct the graph of f.

447

(1) Let f denote a function that is continuous and infinitely differentiable on an interval $[a, b]$ where $a<0<b$. Suppose that, for every integer n

$$f^{(n)}(0) = 0, \qquad \sup_{a\leqslant x\leqslant b}|f^{(n)}(x)| \leqslant n!\, k^n,$$

where k is a given number. Show that f vanishes in the interval $[-1/k, 1/k]$ and from this show that f vanishes in the interval $[a, b]$. (Use Maclaurin's formula.)

(2) *An example of a nonzero infinitely differentiable function all the derivatives of which vanish at zero.*

Define a function g by

$$g(x) = e^{-1/x^2} \qquad \text{if} \qquad x \neq 0 \qquad \text{and} \qquad g(0) = 0.$$

Show that g is infinitely differentiable except possibly at 0 and that $g^{(n)}(x)$ is of the form

$$G_n\left(\frac{1}{x}\right)e^{-1/x^2},$$

where G_n is a polynomial. From this show that the function $g^{(n)}$ approaches 0 as x approaches 0.

Finally, show that $g^{(n)}(0)$ exists and is equal to 0 for every n.

448

Solution of equations by the method of inductively defined sequences

In this exercise, one may use the results of exercise 410.

Suppose that a function f is continuous and differentiable on the closed interval $[a, b]$ and that the equation

$$x = f(x)$$

has a unique root x_0. Define a sequence $\{u_n\}$ in terms of u_0, where $a\leqslant u_0\leqslant b$, by the recursion relation

$$u_n = f(u_{n-1}).$$

(1) Show that if, for every x in $[a, b]$,

$$0 \leqslant f'(x) \leqslant 1,$$

then u_n belongs to $[a, b]$ for every n and that the sequence $\{u_n\}$ is monotonic and converges to x_0.

(2) Show that if, for every x in $[a, b]$,

$$-1 \leqslant f'(x) \leqslant 0,$$

and if u_1 belongs to $[a, b]$, then u_n belongs to $[a, b]$ for every n and the sequences whose general terms are u_{2n} and u_{2n+1} are monotonic and converge to the roots of the equation

$$x = (f \circ f)(x).$$

Finally, show that this equation has no root other than x_0 if $-1 < f'(x)$ or even (and this is somewhat more difficult) if $f'(x)$ assumes the value -1, if there is no interval throughout which $f'(x) = -1$.

(3) Show that the sequence $\{u_n\}$ does not converge to x_0 if the greatest lower bound of $|f'|$ in some neighborhood of x_0 is at least equal to 1.

(4) Suppose that

$$\sup_{a \leqslant x \leqslant b} |f'(x)| = k < 1.$$

Exhibit for $|u_n - x_0|$ an upper bound expressed as a function of $b - a$, k, and n.

(5) Find the solution of the equation

$$x = 32 \sin_{400} x + 80 = f(x),$$

within an accuracy of 10^{-3}, where the notation $\sin_{400} x$ indicates the sine of the arc of which x is the measure in grads (a grad being $\frac{1}{400}$ of a circle). (We might also write $\sin (\pi x / 200)$ to express the radian measure of the arc in question.)

449

Study of the preceding problem in certain particular cases

(The present study is of theoretical interest, but unlike the preceding one, it is not applicable to numerical computation.)

We assume known the results of exercise 448 and, for (4) and (5), those of exercise 413.

Let f denote a function with continuous derivatives of the first three orders in a neighborhood of the value x_0 and assume that $f(x_0) = x_0$.

Define a sequence $\{u_n\}$ inductively by $u_n = f(u_{n-1})$, where u_0 is given. In all that follows, assume that u_0 is sufficiently close to x_0 for the necessary hypotheses to be satisfied throughout the interval $[u_0, x_0]$.

(1) Suppose that

$$f'(x_0) = 1 \quad \text{and} \quad f''(x_0) > 0.$$

Show that the limit of the sequence $\{u_n\}$ is x_0 if $u_0 < x_0$ and that the study is impossible if $u_0 > x_0$. What can we say if $f'(x_0)=1$ but $f''(x_0)<0$?

(2) Show that, if

$$f'(x_0) = 1 \quad \text{and} \quad f''(x_0) = 0,$$

then the sequence $\{u_n\}$ converges to x_0 if $f'''(x_0)<0$ but does not converge to x_0 if $f'''(x_0)>0$.

(3) If $f'(x_0)=-1$, what do we obtain by applying to the function $f \circ f$ the preceding results?

(4) Suppose that f is twice differentiable at a point x_1. Find the limit of

$$\frac{1}{f(x)-f(x_1)} - \frac{1}{(x-x_1)f'(x_1)}$$

as x approaches x_1. From this, show that the three relations $f'(x_0)=1, f''(x_0)>0$, and $u_0 < x_0$ together imply

$$u_n - x_0 = O\left(\frac{1}{n}\right).$$

(5) Suppose that f is twice differentiable in a neighborhood of x_1 and three times differentiable at the point x_1. Suppose that $f''(x_1)=0$. What is the limit of

$$\frac{1}{[f(x)-f(x_1)]^2} - \frac{1}{[(x-x_1)f'(x_1)]^2}$$

as x approaches x_1? From this, show that the three relations $f'(x_0)=1, f''(x_0)=0$, and $f'''(x_0)<0$ imply

$$u_n - x_0 = O\left(\frac{1}{\sqrt{n}}\right).$$

(6) What can we say of $f'(x_0)=-1$?

450

(1) Suppose that a function φ is continuous and n times differentiable in $[a, b]$ and that it vanishes for at least $n+1$ distinct values in $[a, b]$. Show that its nth derivative vanishes at least once in $]a, b[$.

(2) Let f denote a function that is continuous and n times differentiable in $[a, b]$ and denote by A that polynomial of degree not exceeding $n-1$ that assumes at n distinct fixed values in $[a, b]$ (let us denote these values by

$x_1, x_2, \ldots, x_n)$ the same value as does f. (For the existence of this polynomial A, see PZ, Book II, Chapter IV, part 5.)

Show that, for every value x in $[a, b]$, there corresponds at least one number c in $]a, b[$ such that

$$f(x) - A(x) = \frac{(x - x_1) \cdots (x - x_n)}{n!} f^{(n)}(c).$$

This can be done by considering the function φ defined by

$$\varphi(t) = f(t) - A(t) - C(t - x_1) \cdots (t - x_n),$$

where C is a suitably chosen constant.

(3) Show that if $a > 1$, $p > 0$, and $0 < x < 1$, then

$$0 < \log_a (p + x) - \log_a p - x[\log_a (p + 1) - \log_a p] < \frac{x(1 - x)}{2p^2} \log_a e \leqslant \frac{\log_a e}{8p^2}.$$

AN APPLICATION: Tables of logarithms give the common logarithms of integers between 10^3 and 10^4. Give a maximum bound Δ for the error incurred in evaluating a logarithm by interpolation, that is, by taking $x[\log_{10} (p + 1) - \log_{10} p]$ for the value of $\log_{10} (p + x) - \log_{10} p$.

Solutions

CHAPTER 1

401

First question:

Proof by contradiction: If the denial of the conclusion is true, that is, if $a > b$, there exists at least one real number x_0 in the open interval (a, b) such that

$$x_0 > b \quad \text{and} \quad a > x_0,$$

which contradicts the hypothesis.

Second question:

The limit of a sequence is unchanged when we delete a finite number of terms. Therefore, if $u_n \leqslant x$ from some term on, it follows that $l \leqslant x$ (preservation of the conditional inequality). The inequality $l \leqslant x$ holds for every number $x > b$ and hence, from (1), implies that $l \leqslant b$.

402

Every upper bound of A is an upper bound of B. The set B, which is bounded above, has a least upper bound. The least upper bound of A, that is, sup A, is an upper bound of B. Therefore, it is at least equal to the least upper bound of B, that is, sup B.

403

Let us suppose that sup $B \leqslant A$, which implies

$$\text{sup (sup } A, \text{ sup } B) = \text{sup } A.$$

The number sup A is by definition an upper bound of A. It is also an upper bound of B since it is at least equal to sup B. Therefore, it is at least equal to every element in $A \cup B$. The set $A \cup B$, which is bounded above by sup A, has a least upper bound, which is the minimum of the set of upper bounds:

$$\sup (A \cup B) \leqslant \sup A.$$

But A is contained in $A \cup B$; hence (cf. exercise 402),

$$\sup A \leqslant \sup (A \cup B).$$

An examination of $\inf (A \cup B)$ can be made in an analogous manner. However, we can also consider the sets A' and B', consisting of the negatives of the elements of A and B respectively, and then use the fact that

$$\inf A = -\sup A' \cdots (A \cup B)' = A' \cup B'.$$

FIRST APPLICATION:

By the definition of the limit of a sequence, there exists a number n_0 such that

$$n > n_0 \Rightarrow |x_n - l| < 1.$$

Therefore, the set A of those x_n such that $n > n_0$ is bounded. The set B of those x_n such that $n \leqslant n_0$ is a finite set and hence is bounded. The set of all terms of the sequence is the set $A \cup B$. It is bounded since it is the union of two bounded sets.

SECOND APPLICATION:

By the definition of the limit of a function as its argument approaches $+\infty$, there exists a number x_0 such that

$$x > x_0 \Rightarrow |f(x) - l| < 1.$$

Therefore, the set

$$A = f(]x_0, +\infty[)$$

is bounded.

The function f is continuous on the closed interval $[0, x_0]$. Therefore, it is bounded in that interval (so-called theorem of the maximum; see PZ, Book III, Chapter III, section 2, theorem 1). The set $B = f([0, x_0])$ is bounded.

We know that

$$f([0, +\infty[) = f([0, x_0] \cup]x_0, +\infty[) = f([0, x_0]) \cup f(]x_0 + \infty[).$$

The set $f([0, +\infty[)$ is the set $A \cup B$. It is bounded and hence, by definition, the function f is bounded in the interval $[0, +\infty[$.

404

The set $A \cap B$, being contained in the bounded set A, is itself bounded. We shall prove the inequality indicated for $\inf (A \cap B)$ and leave to the reader the analogous proof for sup $(A \cap B)$. We shall use the result corresponding to the result of exercise 402 for sets that are bounded below.

$$A \cap B \subset A \Rightarrow \inf (A \cap B) \geqslant \inf A,$$
$$A \cap B \subset B \Rightarrow \inf (A \cap B) \geqslant \inf B.$$

The greater of the two numbers inf A and inf B is a lower bound of $A \cap B$. Therefore, it does not exceed the greatest lower bound of $A \cap B$, that is, inf $(A \cap B)$:

$$\sup (\inf A, \inf B) \leqslant \inf (A \cap B).$$

EXAMPLE:

$$A = \{0, 2, 3, 5\} \qquad B = \{1, 2, 3, 4\} \qquad A \cap B = \{2, 3\}$$
$$\sup (\inf A, \ \inf B) = \sup (0, 1) = 1, \qquad \inf (A \cap B) = 2,$$
$$\inf (\sup A, \sup B) = \ \inf (5, 4) = 4, \qquad \sup (A \cap B) = 3.$$

405

First question:

The sets A and B, which are bounded above, have least upper bounds sup A and sup B. From this, we conclude that $M = \sup A + \sup B$ is an upper bound of C:

$$x \leqslant \sup A \qquad \text{and} \qquad y \leqslant \sup B \Rightarrow z \leqslant \sup A + \sup B \qquad \forall z \in C.$$

If $M' < M$, we have, from the definition of the least upper bound of a set,

$$\exists x_0 \in A : \qquad \sup A - \frac{M - M'}{2} < x_0,$$

$$\exists y_0 \in B : \qquad \sup B - \frac{M - M'}{2} < y_0.$$

Therefore,

$$x_0 + y_0 \in C, \qquad M' = \sup A + \sup B - (M - M') < x_0 + y_0.$$

The number M' is not an upper bound of C. The least upper bound of C, that is, sup C is the number M.

Second question:

The study of (1) uses essentially the property that inequalities can be added termwise. To obtain an analogous result for products, it is necessary to confine ourselves to sets of nonnegative numbers. With this restriction, we can show just as in (1) that D is bounded above and that

$$\sup D = \sup A \sup B.$$

REMARKS: (a) This property is obviously false if we take arbitrary sets. For example, let us take

$A = B =$ set of negative reals, which is bounded above;

$D = AB =$ set of positive reals, which is not bounded above.

(b) The proof can be carried out by applying the property of (1) to the sets A' and B' defined by

$$x' \in A' \iff e^{x'} \in A, \qquad y' \in B' \iff e^{y'} \in B,$$

if we confine ourselves to sets of positive numbers.

Third question:

By hypothesis, the sets $f(X)$ and $g(X)$ are bounded above and therefore,

$$\sup[f(X)+g(X)] = \sup f(X) + \sup g(X) = \sup f + \sup g.$$

Also, we easily see that

$$(f+g)(X) \subset f(X) + g(X).$$

The set $(f+g)(X)$ is bounded above, and hence so is the function $f+g$. We have

$$\sup(f+g) = \sup[(f+g)(X)] \leqslant \sup[f(X)+g(X)] = \sup f + \sup g.$$

An example with strict inequality:

$$X = [0, 1], \qquad f(t) = t, \qquad g(t) = 1-t,$$
$$\sup f = \sup g = \sup(f+g) = 1.$$

406

First question:

To show that f is a bijective mapping of R onto itself, let us show that, for every real number y, the equation

$$f(x) = y$$

has one and only one solution:

If $y=0$,	$x=0$;
if y is a nonzero rational,	$x=1/y$;
if y is irrational,	$x=y$.

Second question:

The function f is discontinuous at 0. In every interval containing 0, there are rational numbers between -1 and 1 and hence numbers x such that

$$|f(x)-f(0)| = |f(x)| = \frac{1}{|x|} > 1.$$

If $x_0 \neq 0$, then, for every $\varepsilon > 0$, there exists a number $h > 0$ such that

$$|x-x_0| < h \Rightarrow \left|\frac{1}{x}-\frac{1}{x_0}\right| < \varepsilon$$

(because of the continuity of the reciprocal function, $x \rightarrow 1/x$).

For $x_0 = 1/x_0$, that is, for $x_0 = \pm 1$, one can easily show the continuity of f: for any value of x (rational or irrational),

$$|x-x_0| < \inf (h, \varepsilon) \Leftrightarrow |f(x)-f(x_0)| < \varepsilon.$$

For any other value x_0, the function f is discontinuous. Let us take ε so small that the intervals of length 2ε with centers at x_0 and $1/x_0$ have no common points, that is, so that $2\varepsilon < |x_0 - 1/x_0|$. Then, if $|x-x_0| < \inf (h, \varepsilon)$,

$$|f(x)-x_0 \ | < \varepsilon \Rightarrow |f(x)-1/x_0| > \varepsilon \qquad \text{if } x \text{ is irrational,}$$
$$|f(x)-1/x_0| < \varepsilon \Rightarrow |f(x)-x_0 \ | > \varepsilon \qquad \text{if } x \text{ is rational.}$$

Let $f(x_0)$ be equal to x_0 or to $1/x_0$. In every open interval containing x_0, there are points x such that

$$|x-x_0| < \inf (h, \varepsilon)$$

and hence points x such that

$$|f(x)-f(x_0)| > \varepsilon.$$

Therefore, it is impossible to find a number $\eta > 0$ such that

$$|x-x_0| < \eta \Rightarrow |f(x)-f(x_0)| < \varepsilon.$$

407

First question:

Let x_0 be an element of A, so that $f(x_0) < y$, and let u be an arbitrary number such that $u < f(x_0)$. From the definition of continuity of f at x_0, there exists an open interval $]x_0 - \eta, x_0 + \eta[$ the image of which under f is contained in $]u, y[$. This interval $]x_0 - \eta, x_0 + \eta[$ is therefore contained in A.

The proof is analogous for B.

Second question:

The set $f^{-1}(y)$ is nonempty because the function f, which is continuous on the interval $[a, b]$ assumes at least once every value between $f(a)$ and $f(b)$ (by the intermediate-value theorem; see PZ, Book III, Chapter III, section 2, theorem 2).

Let us show that $f^{-1}(y)$ has a greatest element, that is, that the set $f^{-1}(y)$ is bounded above and that its least upper bound belongs to the set. The set $f^{-1}(y)$ is bounded above because it is contained in $[a, b]$. Therefore, it has a least upper bound m (where $a \leqslant m \leqslant b$).

In every open interval containing m, there are points of the set $f^{-1}(y)$ (by the definition of a least upper bound). If m belongs to A, there is an open interval centered at m and contained in A, hence containing no points of $f^{-1}(y)$. Therefore, m cannot belong to A. For the same reason, it cannot belong to B. Therefore, it belongs to $f^{-1}(y)$.

The proof is the same for the smallest element.

408

FIRST PART

We show, for example, by induction on n, that

$$u_n = \sum_{k=1}^{n} \left(\frac{1}{k} - \frac{1}{k+1} \right) = 1 - \frac{1}{n+1}.$$

The limit of the sequence $\{u_n\}$ is therefore 1.

For the sequence $\{v_n\}$, we form

$$|v_m - v_n| = \left| \sum_{n+1}^{m} \left(\frac{1}{k} - \frac{1}{k+1} \right) \sin k\alpha \right| \leqslant \sum_{n+1}^{m} \left(\frac{1}{k} - \frac{1}{k+1} \right) |\sin k\alpha|$$

$$\leqslant \sum_{n+1}^{m} \left(\frac{1}{k} - \frac{1}{k+1} \right).$$

Therefore,

$$|v_m - v_n| \leqslant u_m - u_n.$$

The convergent sequence $\{u_n\}$ satisfies Cauchy's criterion and hence, so does the sequence $\{v_n\}$. This implies the convergence of the sequence $\{v_n\}$. (For $\{u_n\}$, we use the fact that satisfaction of Cauchy's criterion is a necessary condition for convergence; for $\{v_n\}$, we use the fact that satisfaction of the Cauchy criterion is a sufficient condition for convergence.)

SECOND PART

First question:

The continuity of f implies that the segments constituting the graphs of the linear functions representing the restrictions of f to adjacent intervals have in each case a common end-point.

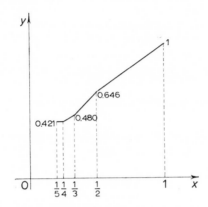

Second question:

We easily see that

$$f\left(\frac{1}{n}\right) = f(1) + \sum_{k=1}^{n-1}\left[f\left(\frac{1}{k+1}\right) - f\left(\frac{1}{k}\right)\right] = 1 - \sum_{k=1}^{n-1}\left[f\left(\frac{1}{k}\right) - f\left(\frac{1}{k+1}\right)\right].$$

In the interval $[1/(k+1), 1/k]$, the restriction of f is the linear function defined by condition c of the statement of the problem. Therefore,

$$f\left(\frac{1}{k}\right) - f\left(\frac{1}{k+1}\right) = \left(\frac{1}{k} - \frac{1}{k+1}\right)\sin k\alpha,$$

$$f\left(\frac{1}{n}\right) = 1 - \sum_{k=1}^{n-1}\left(\frac{1}{k} - \frac{1}{k+1}\right)\sin k\alpha = 1 - v_{n-1}.$$

The sequence $\{f(1/n)\}$ converges because the sequence $\{v_{n-1}\}$ converges.

Third question:

Let us denote by y_0 the limit of the sequence $\{f(1/n)\}$ and let us show that y_0 is the limit of the function f at the point 0. Let ε denote a given positive number.

By the definition of a limit, there exists an integer n_0 such that

$$n \geqslant n_0 \;\Rightarrow\; \left|f\left(\frac{1}{n}\right) - y_0\right| < \varepsilon.$$

If $0 < x < 1/n_0$, the number x belongs to an interval of the form $[1/(n+1), 1/n]$, where n is an integer $\geqslant n_0$. The restriction of f to this interval is a linear function and hence $f(x)$ lies between $f(1/n)$ and $f(1/(n+1))$). Therefore,

$$|f(x) - y_0| \leqslant \sup\left(\left|f\left(\frac{1}{n}\right) - y_0\right|, \left|f\left(\frac{1}{n+1}\right) - y_0\right|\right) < \varepsilon.$$

Thus, we have found a number $\eta = 1/n_0$ such that

$$0 < x < \eta \;\Rightarrow\; |f(x) - y_0| < \varepsilon.$$

This states that the limit of the function f as its argument approaches 0 is y_0.

Third Part

First question:

Suppose that x and x' are given numbers such that $x < x'$. There exist integers m and n such that

$$\frac{1}{m+1} \leqslant x < \frac{1}{m}, \qquad \frac{1}{n+1} < x' \leqslant \frac{1}{n}.$$

If $m = n$, then x and x' belong to the same interval and

$$|f(x') - f(x)| = |(x' - x)\sin n\alpha| \leqslant x' - x.$$

If $m > n$, let us bring in the intermediate points of the form $1/k$ as follows:

$$f(x') - f(x) = \left[f(x') - f\left(\frac{1}{n+1}\right)\right] + \left[f\left(\frac{1}{n+1}\right) - f\left(\frac{1}{n+2}\right)\right] + \cdots$$

$$+ \left[f\left(\frac{1}{m-1}\right) - f\left(\frac{1}{m}\right)\right] + \left[f\left(\frac{1}{m}\right) - f(x)\right]$$

$$= \left[f(x') - f\left(\frac{1}{n+1}\right)\right] + \sum_{k=n+1}^{m-1}\left[f\left(\frac{1}{k}\right) - f\left(\frac{1}{k+1}\right)\right] + \left[f\left(\frac{1}{m}\right) - f(x)\right].$$

The definition of f yields

$$|f(x')-f(x)| \leqslant \left|f(x')-f\left(\frac{1}{n+1}\right)\right|+\sum_{k=n+1}^{m-1}\left|f\left(\frac{1}{k}\right)-f\left(\frac{1}{k+1}\right)\right|+\left|f\left(\frac{1}{m}\right)-f(x)\right|,$$

$$\left|f(x')-f\left(\frac{1}{n+1}\right)\right| = \left|\left(x'-\frac{1}{n+1}\right)\sin n\alpha\right| \leqslant x'-\frac{1}{n+1},$$

$$\left|f\left(\frac{1}{k}\right)-f\left(\frac{1}{k+1}\right)\right| = \left|\left(\frac{1}{k}-\frac{1}{k+1}\right)\sin k\alpha\right| \leqslant \frac{1}{k}-\frac{1}{k+1},$$

$$\left|f\left(\frac{1}{m}\right)-f(x)\right| = \left|\left(\frac{1}{m}-x\right)\sin m\alpha\right| \leqslant \frac{1}{m}-x.$$

Therefore,

$$|f(x')-f(x)| \leqslant \left(x'-\frac{1}{n+1}\right)+\sum_{k=n+1}^{m}\left(\frac{1}{k}-\frac{1}{k+1}\right)+\left(\frac{1}{m}-x\right) = x'-x.$$

The constant M that satisfies for f the inequality written for g is 1.

Second question:

If $m>n$, then

$$\left|g\left(\frac{1}{m}\right)-g\left(\frac{1}{n}\right)\right| \leqslant M\left|\frac{1}{m}-\frac{1}{n}\right| \leqslant \frac{M}{n}.$$

The sequence $\{g(1/n)\}$ satisfies the Cauchy criterion: for any $\varepsilon>0$.

$$n > M/\varepsilon \quad \text{and} \quad m > M/\varepsilon \Rightarrow \left|g\left(\frac{1}{m}\right)-g\left(\frac{1}{n}\right)\right| < \varepsilon.$$

Therefore, the sequence $\{g(1/n)\}$ has a limit, which we denote by y_0. By considering the limit as $m\to\infty$ of the sequence $\{|g(1/m)-g(1/n)|\}$ for fixed n, we see that

$$\left|y_0-g\left(\frac{1}{n}\right)\right| \leqslant \frac{M}{n}.$$

For every x in $]0, 1[$, consider the integer n such that $1/n\leqslant x<1/(n-1)$. Then,

$$|g(x)-y_0| \leqslant \left|g(x)-g\left(\frac{1}{n}\right)\right|+\left|g\left(\frac{1}{n}\right)-y_0\right| \leqslant M\left(x-\frac{1}{n}\right)+\frac{M}{n},$$

$$|g(x)-y_0| \leqslant \frac{M}{n(n-1)}+\frac{M}{n} \leqslant \frac{2M}{n}.$$

Therefore, for given $\varepsilon > 0$, we have

$$0 < x < \varepsilon/2M \;\Rightarrow\; n \geqslant \frac{1}{x} > \frac{2M}{\varepsilon} \;\Rightarrow\; |g(x) - y_0| < \varepsilon.$$

The right-hand limit of the function g as its argument approaches 0 is y_0.

409

First question:

The hypotheses imply

$$u_0 \leqslant u_n \leqslant v_n \leqslant v_0.$$

The sequence $\{u_n\}$, which is nondecreasing and bounded above by v_0 has a limit u, which is the least upper bound of the set of values of the u_n. The sequence $\{v_n\}$, which is nonincreasing and bounded below by u_0 has a limit v, which is the greatest lower bound of the set of values of the v_n.

For arbitrary integers n and m,

$$u_n \leqslant u_{n+m} \leqslant v_{n+m} \leqslant v_m.$$

Here, u_n, being a lower bound of the set of the v_m, does not exceed the greatest lower bound v of the set of the v_m. The number v is an upper bound of the set of the u_n and is therefore at least equal to the least upper bound u of that set. Therefore,

$$u_n \leqslant u \leqslant v \leqslant v_n \;\Rightarrow\; 0 \leqslant v - u \leqslant v_n - u_n.$$

If $v - u \neq 0$ and if $\lim (v_n - u_n) = 0$, it is possible, by virtue of the definition of such a limit to find a number N such that

$$n > N \;\Rightarrow\; v_n - u_n < v - u.$$

This contradicts the hypothesis and hence $v - u = 0$ if $\lim (v_n - u_n) = 0$.

Second question:

We can easily verify by induction that $u_n > 0$ and $v_n > 0$.

$$v_n^2 - u_n^2 = \left(\frac{u_{n-1} + v_{n-1}}{2} \right)^2 - u_{n-1} v_{n-1} = \left(\frac{v_{n-1} - u_{n-1}}{2} \right)^2 \geqslant 0.$$

Since u_n and v_n are positive, we can write

$$v_n^2 - u_n^2 \geqslant 0 \;\Rightarrow\; v_n \geqslant u_n \;\Rightarrow\; \begin{cases} u_{n+1} = \sqrt{u_n v_n} \geqslant u_n \\[2mm] v_{n+1} = \dfrac{u_n + v_n}{2} \leqslant v_n. \end{cases}$$

The sequence $\{u_n\}$ is nondecreasing; the sequence $\{v_n\}$ is nonincreasing. Furthermore, $u_n \leqslant v_n$.
The preceding results imply that

$$u_n \leqslant u_{n+1} \leqslant v_{n+1} \leqslant v_n,$$

$$0 \leqslant v_{n+1} - u_{n+1} \leqslant v_{n+1} - u_n = \frac{v_n - u_n}{2}.$$

Therefore, by induction on n, we have

$$0 \leqslant v_n - u_n \leqslant \frac{v_0 - u_0}{2^n}.$$

The limit of the sequence $\{v_n - u_n\}$ is 0. All the hypotheses of (1) are verified and the sequences $\{u_n\}$ and $\{v_n\}$ converge to the same limit. This common limit of the two sequences is called the *arithmetico-geometric mean* of the two numbers u_0 and v_0.

410

First question:

Let us suppose that $u_1 \geqslant u_0$ and let us show by induction that the sequence is nondecreasing, that is, that $u_n \geqslant u_{n-1}$. This inequality is true by hypothesis for $n = 1$. Let us suppose it established for $u_{n-1} \geqslant u_{n-2}$. Since the function f is nondecreasing,

$$u_{n-2} \leqslant u_{n-1} \Rightarrow u_{n-1} = f(u_{n-2}) \leqslant f(u_{n-1}) = u_n.$$

The sequence $\{u_n\}$, which is nondecreasing, is, by hypothesis, bounded above by b. It converges and its limit u satisfies the inequalities

$$a \leqslant u_0 \leqslant u \leqslant b.$$

Similarly, we can show that the sequence $\{u_n\}$ is nonincreasing if $u_1 \leqslant u_0$. In this case, the sequence $\{u_n\}$, which is nonincreasing and bounded below by a, also converges and its limit u satisfies the inequalities

$$a \leqslant u \leqslant u_0 \leqslant b.$$

The function f is, by hypothesis, continuous at the value u in $[a,b]$. Therefore, the limit of the sequence $\{f(u_n)\}$ is $f(u)$ (see PZ, Book III, Chapter III, part 2, § 1, 3rd definition of continuity). The sequences $\{u_{n+1}\}$ and $\{f(u_n)\}$, which are term for term identical, have the same limit

$$u = f(u).$$

The preceding result can be presented on a graph, but this procedure cannot replace the proof made above.

The ordinate of the point on the graph of f whose abscissa is u_0 is u_1. The abscissa of the point of ordinate u_1 and on the line bisecting the angle between the axes is also u_1. The ordinate of the corresponding point of the graph of the function is $u_2 = f(u_1)$. We take the point of ordinate u_2 on the lines bisecting the angle and continue the process as indicated.

The graph on the right corresponds to the case of a nondecreasing sequence and that on the left corresponds to the case of a nonincreasing sequence.

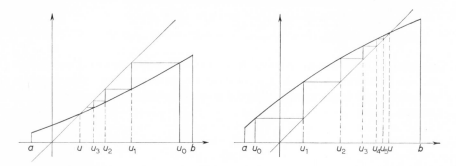

Second question:

Since the function f is continuous and nonincreasing, the inverse image of $[a, b]$ is an interval $[\alpha, \beta]$, which, by hypothesis, contains all the u_n. In the interval $[\alpha, \beta]$, the function $f \circ f$ is defined and nondecreasing:

$$x' < x \Rightarrow y' = f(x') \geqslant y = f(x) \Rightarrow (f \circ f)(x') = f(y') \leqslant f(y) = (f \circ f)(x).$$

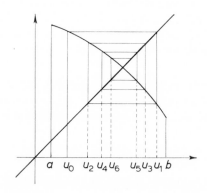

Graph constructed as in the case of (1) with the aid of the graph of f

The results of (1) are therefore applicable to sequences defined inductively with the aid of the function $f \circ f$ and, in particular, they are applicable to the sequences $\{u'_n\}$ and $\{u''_n\}$ defined by

$$u'_0 = u_0, \qquad u'_n = (f \circ f)(u'_{n-1}) = u_{2n} \qquad \text{(by induction)},$$

$$u''_0 = u_1, \qquad u''_n = (f \circ f)(u''_{n-1}) = u_{2n+1} \qquad \text{(by induction)}.$$

These two sequences are monotonic and bounded — hence convergent. Furthermore, it is easy to see that if one of these sequences is nondecreasing, the other is nonincreasing:

$$u_{2n} \geqslant u_{2n-2} \Rightarrow u_{2n+1} = f(u_{2n}) \leqslant f(u_{2n-2}) = u_{2n-1}.$$

Finally, we note that the sequence $\{u_n\}$ will converge if and only if the limits u' and u'' of the two sequences $\{u'_n\} = \{u_{2n}\}$ and $\{u''_n\} = \{u_{2n+1}\}$ are equal.

Third question:

EXAMINATION OF SEQUENCE $\{u_n\}$

We can see by induction that the u_n are positive and also that they do not exceed 1. Here, the function f is the homographic function defined by

$$f(x) = \frac{x+1}{x+2}.$$

It is defined, continuous, and increasing for x between 0 and 1. The hypotheses of (1) are satisfied. Furthermore, $u_1 = \frac{1}{2}$ is greater than $u_0 = 0$. The sequence $\{u_n\}$ converges, increasing toward a number u, which is a root of the equation

$$u = \frac{u+1}{u+2} \qquad \text{or} \qquad u^2 + u - 1 = 0.$$

This equation has one and only one positive root. Therefore, u is equal to that root, namely, $\frac{1}{2}(-1 + \sqrt{5})$.

EXAMINATION OF SEQUENCE $\{v_n\}$

We can show by induction that $0 \leqslant v_n \leqslant 1$. In the interval $[0, 1]$, the cosine function is decreasing. The hypotheses of (2) are satisfied; hence, we know that:
    The sequence $\{v_{2n}\}$ is increasing since $v_2 > v_0 = 0$, and it has a limit v';
    The sequence $\{v_{2n+1}\}$ is decreasing, and it has a limit v''.
    By writing that the sequences $\{\cos v_{2n}\}$ and $\{\cos v_{2n+1}\}$ converge to $\cos v'$ and $\cos v''$ (because of the continuity of the cosine function at v' and v''), we obtain the system of equations

$$\begin{cases} v' = \cos v'' \\ v'' = \cos v' \end{cases} \Rightarrow \begin{cases} v' = \cos v'' \\ v'' - v' = \cos v' - \cos v''. \end{cases}$$

It is impossible for $v'' - v'$ to be anything but 0 because

$$|\cos v' - \cos v''| < |v' - v''| \qquad \text{if} \qquad v' - v'' \neq 0.$$

The sequence $\{v_n\}$ converges and its limit v is the common limit of the two sequences $\{v_{2n}\}$ and $\{v_{2n+1}\}$ that is, it is a root of the equation

$$v = \cos v.$$

Examination of Sequence $\{w_n\}$

We can show by induction that $0 \leqslant w_n \leqslant 1$. The function f defined here by $f(x) = (1-x)^2$ is continuous and decreasing in that interval. The sequence $\{w_n'\}$, where $w_n' = w_{2n}$, is increasing since $w_2 = \frac{9}{16} > \frac{1}{2} = w_0$. Its limit is w', which verifies the double inequality $\frac{9}{16} \leqslant w' \leqslant 1$. Therefore, the sequence $\{w_n''\}$, where $w_n'' = w_{2n+1}$, is decreasing and its first term w_1 is $\frac{1}{4}$. It has a limit w'', which satisfies the double inequality $0 \leqslant w'' \leqslant \frac{1}{4}$. The two limits w' and w'' are not equal. Therefore, the sequence $\{w_n\}$ diverges. (If a sequence has a limit l, every subsequence of it converges to l.)

We can determine w' and w''. These two numbers constitute the solution of the system

$$\begin{cases} w' = (1-w'')^2 \\ w'' = (1-w')^2 \end{cases} \Rightarrow \begin{cases} w' - w'' = (w' - w'')(2 - w' - w'') \\ w' + w'' = 2 - 2(w' + w'') + w'^2 + w''^2. \end{cases}$$

We can divide the two members of the first equation by the nonzero factor $w' - w''$:

$$\begin{cases} 1 = 2 - (w' + w'') \\ w' + w'' = 2 - 2(w' + w'') + (w' + w'')^2 - 2w'w'' \end{cases} \Rightarrow \begin{cases} w' + w'' = 1 \\ w'w'' = 0. \end{cases}$$

Therefore, $w' = 1$ and $w'' = 0$.

411

First question:

The hypothesis made implies that the function f is uniformly continuous and therefore, it is *a fortiori* continuous at every point of $[a, b]$. For any $\varepsilon > 0$,

$$|u - v| < \varepsilon \Rightarrow |f(u) - f(v)| < \varepsilon.$$

The function F defined by $F(t) = f(t) - t$ is strictly decreasing:

$$F(u) - F(v) = [f(u) - u] - [f(v) - v] = [f(u) - f(v)] + [v - u].$$

From the hypothesis, the sign of this last sum is the same as that of the sum of $v - u$, which is the greater of the two terms in absolute value.

The values assumed by the function f belong to $[a, b]$. In particular,

$$a \leqslant f(a) \Rightarrow F(a) = f(a) - a \geqslant 0,$$
$$b \geqslant f(b) \Rightarrow F(b) = f(b) - b \leqslant 0.$$

The function F, which is continuous on $[a, b]$, assumes positive and negative values. Therefore, it assumes the value 0 somewhere in that interval (from the intermediate-value theorem, see PZ, Book III, Chapter III, § 2, theorem 2). Furthermore, F is strictly monotonic and hence injective. Therefore, it assumes the value 0 only once.

Second question:

We note first that the hypothesis $f([a, b]) \subset [a, b]$ implies, as we can see by induction on n, that all the terms x_n belong to $[a, b]$.

(a) By definition, $f(\theta) = \theta$. Then,

$$|x_n - \theta| = |f(x_{n-1}) - f(\theta)| < |x_{n-1} - \theta|.$$

The sequence $\{|x_n - \theta|\}$, which is nonincreasing and bounded below by zero converges to a limit $l \geqslant 0$.

(b) Let E be the set of indices n such that $x_n - \theta \geqslant 0$. If E is infinite and n_1, $n_2, \ldots, n_i, \ldots$ are the elements of E, the subsequence $\{x_{n_i}\}$ of the sequence $\{x_n\}$ is such that

$$x_{n_i} - \theta \geqslant 0 \Rightarrow x_{n_i} - \theta = |x_{n_i} - \theta|.$$

The subsequence $\{|x_{n_i} - \theta|\}$ of the sequence $\{|x_n - \theta|\}$ converges to l. Therefore, the sequence $\{x_{n_i}\}$ converges to $\theta + l$.

If E is finite and if n_0 is the greatest element of E, then

$$n > n_0 \Rightarrow x_n - \theta < 0 \Rightarrow x_n - \theta = -|x_n - \theta|.$$

Therefore, the sequence $\{x_n\}$ converges to $\theta - l$ and answers the question.

(c) The limit $\theta + \varepsilon l$ of the sequence $\{x_{n_i}\}$ of elements of the closed interval $[a, b]$ belongs to that interval. Therefore, the function f is continuous at that point and the sequence $\{f(x_{n_i})\}$ converges to $f(\theta + \varepsilon l)$ (see PZ, Book III, Chapter III, part 2, § 1, 3rd definition of continuity):

$$|f(\theta + \varepsilon l) - \theta| = \lim |f(x_{n_i}) - \theta| = \lim |x_{n_i + 1} - \theta| = l.$$

(The subsequence $\{|x_{n_i + 1} - \theta|\}$ of the sequence $\{|x_n - \theta|\}$ converges to l.) But, by virtue of the basis hypothesis, if $\theta + \varepsilon l \neq \theta$, that is, if $l \neq 0$,

$$l = |f(\theta + \varepsilon l) - \theta| = |f(\theta + \varepsilon l) - f(\theta)| < |\varepsilon l| = l,$$

which gives us the contradiction $l < l$. Therefore, the only possibility is $l = 0$.

REMARK: The preceding study can be applied to the function f defined in the interval $[0, \frac{1}{2}\pi]$ by

$$f(t) = \cos t.$$

(The inductively defined sequence associated with this function was studied in exercise 411.)

412

First question:

If $n < m$,

$$|S_m(t) - S_n(t)| = \left| \sum_{n+1}^{m} \frac{\sin kt}{k^2(k+1)} \right| \leqslant \sum_{n+1}^{m} \frac{1}{k^2(k+1)} \leqslant \sum_{n+1}^{m} \frac{1}{k(k+1)} \leqslant \frac{1}{n+1}.$$

Thus, the Cauchy criterion is satisfied and the sequence $\{S_n(t)\}$ has a limit.

Second question:

The familiar inequality

$$|\sin x' - \sin x| < |x' - x| \qquad \text{if} \qquad x' \neq x$$

enables us to write

$$|S_n(v) - S_n(u)| = \left| \sum_{1}^{n} \frac{\sin kv - \sin ku}{k^2(k+1)} \right| < \sum_{1}^{n} \frac{k|v-u|}{k^2(k+1)} = |v-u| \sum_{1}^{n} \frac{1}{k(k+1)}.$$

This inequality between the general terms of the two sequences holds also for the limits except that we must replace the strict inequality with a conditional inequality:

$$|S(v) - S(u)| \leqslant |v-u| \lim\left(\sum_{1}^{n} \frac{1}{k(k+1)} \right) = |v-u|.$$

The result obtained is therefore not the strict inequality required. To get this strict inequality, note that

$$\left| \sum_{2}^{n} \frac{\sin kv - \sin ku}{k^2(k+1)} \right| < |v-u| \sum_{2}^{n} \frac{1}{k(k+1)}.$$

Taking the limit in this inequality, we get

$$\left| S(v) - S(u) - \frac{\sin v - \sin u}{2} \right| \leqslant \frac{|v-u|}{2}.$$

Therefore,

$$|S(v) - S(u)| \leqslant \left| \frac{\sin v - \sin u}{2} \right| + \frac{|v - u|}{2}.$$

This, together with the strict inequality

$$|\sin v - \sin u| < v - u$$

yields

$$|S(v) - S(u)| < \frac{|v - u|}{2} + \frac{|v - u|}{2} = |v - u|.$$

To show that the function $S(t)$ for $-\pi \leqslant t \leqslant \pi$ satisfies the hypotheses of exercise 411, we need only show that the image of the interval $[-\pi, \pi]$ is contained in $[-\pi, \pi]$. This is obvious since

$$|S(t)| = |S(t) - S(o)| \leqslant |t| \leqslant \pi.$$

413

First question:

Let α denote an arbitrary positive number. By the definition of a limit, there exists a number N such that

$$n > N \Rightarrow |x_n - l| < \alpha.$$

The difference $y_n - l$ can be written

$$y_n - l = \frac{(x_1 + \cdots + x_N) + (x_{N+1} + \cdots + x_n) - nl}{n}$$

$$= \frac{x_1 + \cdots + x_N - Nl}{n} + \frac{(x_{N+1} - l) + \cdots + (x_n - l)}{n} = \xi_n + \eta_n.$$

By virtue of the choice of N, we have $|x_p - l| < \alpha$ whenever $p > N$. Therefore,

$$|\eta_n| = \left| \frac{(x_{N+1} - l) + \cdots + (x_n - l)}{n} \right| \leqslant \frac{|x_{N+1} - l| + \cdots + |x_n - l|}{n} < \frac{(n - N)\alpha}{n} < \alpha.$$

The sequence $\{\xi_n\}$ defined by

$$\xi_n = \frac{x_1 + \cdots + x_N - Nl}{n},$$

the numerators of which are independent of n, converges to 0. Therefore, there exists a number n_0 such that

$$n \geqslant n_0 \Rightarrow \left| \frac{x_1 + \cdots + x_N - Nl}{n} \right| < \alpha.$$

Finally, if $n \geqslant n_0$,

$$\begin{cases} |\xi_n| < \alpha \\ |\eta_n| < \alpha \end{cases} \Rightarrow |y_n - l| \leqslant |\xi_n| + |\eta_n| < 2\alpha.$$

The number α is an arbitrary positive number. Therefore, by definition, the sequence $\{y_n\}$ converges to l.

Second question:

Let us use the sequence $\{x_n\}$ to define a sequence $\{d_n\}$ by

$$d_1 = x_1 \quad \text{and} \quad d_n = x_n - x_{n-1} \quad \text{if} \quad n \geqslant 2.$$

The sequence $\{d_n\}$ converges to l by hypothesis. According to (1), the sequence whose general term is

$$\frac{d_1 + d_2 + \cdots + d_n}{n}$$

also converges to l, and we see immediately that

$$d_1 + d_2 + \cdots + d_n = x_n.$$

Third question:

We use the isomorphism between the multiplicative group of positive real numbers and the additive group of all real numbers:

$$p_n = \ln \sqrt[n]{x_1 \cdots x_n} = \frac{1}{n} \ln (x_1 \cdots x_n) = \frac{\ln x_1 + \cdots + \ln x_n}{n}.$$

By hypothesis, the sequence $\{x_n\}$ converges to l. Therefore, the sequence $\{\ln x_n\}$ converges to $\ln l$ (because of the continuity of the logarithm function), and the sequence $\{p_n\}$ also converges to $\ln l$ (result of (1)). The continuity of the exponential function then implies that

$$\sqrt[n]{x_1 \cdots x_n} = e^{p_n} \rightarrow e^{\ln l} = l.$$

Similarly, if the sequence $\{x_{n+1}/x_n\}$ converges to l, the sequence $\{\ln x_{n+1} - \ln x_n\}$ converges to $\ln l$. Then, the result of (2) yields

$$\ln \sqrt[n]{x_n} = \frac{1}{n} \ln x_n \to \ln l,$$

$$\sqrt[n]{x_n} \to e^{\ln l} = l.$$

Fourth questi on:

$$u_n = \frac{1}{n} \sum_{p=1}^{n} \frac{1}{p}.$$

The sequence whose general term is $x_p = 1/p$ converges to 0. Therefore, the same is true for the sequence $\{u_n\}$ defined by

$$u_n = \frac{x_1 + \cdots + x_n}{n}.$$

For the second sequence, set $x_n = n^3 + n^2 - 1$, so that the general term is $v_n = \sqrt[n]{x_n}$. The ratio x_{n+1}/x_n is the ratio of two polynomials in n with equal terms of highest degree. The limit of such a ratio is 1, so that the limit of the sequence $\{v_n\}$, where $v_n = \sqrt[n]{x_n}$ is also 1 in accordance with (3).

414

Every interval containing $x_0 = p_0/q_0$ contains irrational numbers x. For such an irrational number x,

$$|f(x) - f(x_0)| = \frac{1}{q_0}.$$

If $0 < \varepsilon < 1/q_0$, there is number η such that

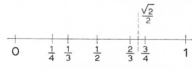

Graph corresponding to
$x_0 = \sqrt{2}/2,\ \varepsilon = 0.22$.

The numbers r are rational numbers
shown; $r_1 = \frac{2}{3}, r_2 = \frac{3}{4}$.

$$|x - x_0| < \eta \Rightarrow |f(x) - f(x_0)| \leqslant \varepsilon,$$

since the interval $]x_0 - \eta,\ x_0 + \eta[$ contains irrational numbers x for which the inequality is not satisfied.

If x_0 is irrational, let us take a number $\varepsilon > 0$ and show that the inequality $|f(x) - f(x_0)| < \varepsilon$ is verified by all numbers x in any open interval containing x_0.

Here, $f(x_0) = 0$ and $f(x) \geqslant 0$; therefore, the inequality of the preceding sentence can be replaced by $f(x) < \varepsilon$. The numbers r for which this inequality is not satisfied are the rational numbers whose denominators q do not exceed $1/\varepsilon$. There are finitely many of these, and x_0, which is irrational, is not one of them. Let us denote by r_1 the greatest such number r that is less than x_0 and let us denote by r_2 the smallest such number r that is greater than x_0. The interval $]r_1, r_2[$ contains none of these numbers r (by the definition of r_1 and r_2). Therefore, every number x in this interval satisfies the inequality $f(x) < \varepsilon$.

Finally, the above reasoning shows also that f approaches 0 as its argument approaches *any* value x_0. (We recall that the definition of the limit of a function f as its argument approaches a point x_0 does not involve the value of $f(x_0)$.) Just as before, choose from the set of numbers r, such that $f(r) \geqslant \varepsilon$, the numbers r_1 and r_2 closest to x_0 (from below and from above respectively), (exclude the number x_0 even if it belongs to the set of these numbers r). Every number x in one of the intervals $]r_1, x_0[$ or $]x_0, r_2[$ thus satisfies the inequality $f(x) < \varepsilon$. The function f has left- and right-hand limits at x_0 equal to 0. Therefore, its limit as its argument approaches x_0 is 0.

415

For any $\varepsilon > 0$, we need to study the inequality

$$|\sqrt{x'} - \sqrt{x}| < \varepsilon \Leftrightarrow \frac{|x' - x|}{\sqrt{x'} + \sqrt{x}} < \varepsilon,$$

where x and x' are two numbers in $[0,1]$.

CASE 1: If the two numbers x and x' are less than ε^2,

$$0 < \sqrt{x'} < \varepsilon \quad \text{and} \quad 0 < \sqrt{x} < \varepsilon \Rightarrow |\sqrt{x'} - \sqrt{x}| < \varepsilon.$$

CASE 2: If the conditions of case 1 are not satisfied, that is, if at least one of the numbers x, x' is at least equal to ε^2,

$$\sqrt{x'} + \sqrt{x} \geqslant \varepsilon \Rightarrow \frac{1}{\sqrt{x'} + \sqrt{x}} \leqslant \frac{1}{\varepsilon},$$

$$|x' - x| < \varepsilon^2 \Rightarrow \frac{|x' - x|}{\sqrt{x'} + \sqrt{x}} \leqslant \frac{|x' - x|}{\varepsilon} < \frac{\varepsilon^2}{\varepsilon} = \varepsilon.$$

Thus, for the number ε, we have chosen the number ε^2, which has the property that, for x and x' in $[0, 1]$,

$$|x' - x| < \varepsilon^2 \Rightarrow |\sqrt{x'} - \sqrt{x}| < \varepsilon.$$

(This result is valid whether the pair (x, x') satisfies the conditions of case 1 or of case 2.) Thus, by definition, the function is uniformly continuous on the interval $[0,1]$.

416

Since the function f is nondecreasing, it has, at every value x_0, right- and left-hand limits $f(x_0+0)$ and $f(x_0-0)$, and we know that

$$f(x_0-0) = \sup_{x<x_0} f(x) \leqslant f(x_0) \leqslant \inf_{x>x_0} f(x) = f(x_0+0).$$

Therefore, the function f is continuous at x_0 if and only if

$$f(x_0-0) = f(x_0+0).$$

Let us suppose that the function f is discontinuous at a value c in $[a, b]$. Then,

$$f(c-0) < f(c+0),$$

$$x < c \Rightarrow f(x) \leqslant \sup_{x<c} f(x) = f(c-0),$$

$$x > c \Rightarrow f(x) \geqslant \inf_{x>c} f(x) = f(c+0).$$

Except for the single value $f(c)$, the function f does not assume values in the interval $]f(c-0), f(c+0)[$. Since the definitions of $f(c-0)$ and $f(c+0)$ imply that

$$f(a) \leqslant f(c-0) < f(c+0) \leqslant f(b),$$

we have a contradiction with the hypothesis made.
    (If $c=a$ or $c=b$, we need to replace the interval in question with $]f(a), f(a+0)[$ or $]f(b-0), f(b)[$.)

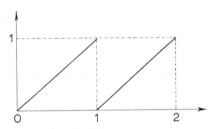

The function g defined on $[0,2]$ by

$$g(x) = x \qquad \text{if} \qquad 0 \leqslant x \leqslant 1,$$

$$g(x) = x-1 \qquad \text{if} \qquad 1 < x \leqslant 2,$$

assumes every value between $g(0)=0$ and $g(2)=1$, and it is discontinuous at $x=1$.

417

First question:

If $n=0$, we have $A_0=1$. Assume the result established for all integers up to and including n.

$$4 \cos n\theta \cos \theta = 2 \cos (n+1)\theta + 2 \cos (n-1)\theta \;\Rightarrow\; xA_n = A_{n+1} + A_{n-1}.$$

A_{n+1} is a polynomial whose term of highest degree is the term of highest degree of xA_n, namely x^{n+1}.

Second question:

If $-2 \leqslant x \leqslant 2$, we can set $x = 2 \cos \theta$ (where $0 \leqslant \theta \leqslant \pi$):

$$A_n(x) = 2 \;\Leftrightarrow\; \cos n\theta = 1 \;\Leftrightarrow\; n\theta = 0 \qquad \mathrm{mod}\ 2\pi.$$

The zeros of $A_n - 2$ are the numbers $u_k = 2 \cos (2k\pi/n)$, where $\leqslant k \leqslant \frac{1}{2}n$. In the interval $[-2,2]$, the polynomial $A_n - 2$ assumes only nonpositive values. If $-2 < u_k < 2$, this polynomial vanishes at u_k but is negative both to the left and to the right of this value. Therefore, the order of the zero u_k is even and at least equal to 2. Let us now evaluate the sum of the orders of the zeros u_k and treat the cases of even and odd n separately.

    (1) If $n = 2n'$, we have $u_0 = 2$ and $u_{n'} = -2$. There are $n' - 1$ numbers u_k in the interval $]-2, +2[$. The sum of the orders of these zeros is at least equal to

$$1 + 1 + 2(n' - 1) = 2n'.$$

Since the polynomial $A_n - 2$ is of degree $2n'$, we see that we have found all the zeros of this polynomial, that 2 and -2 are simple zeros, and that the other zeros are double zeros.

    (2) If $n = 2n' + 1$, $u_0 = 2$ and there are n' numbers u_k in $]-2, 2[$, where $k = 1, 2, \ldots, n'$. The sum of the orders of these zeros is at least equal to

$$1 + 2n' = 2n' + 1 = n.$$

We have all the zeros of $A_n - 2$; 2 is a simple zero and the others are double zeros.

    Study of the zeros of the polynomial $A_n + 2$ follows along identical lines. The zeros of this polynomial are all real; in particular, they are the numbers

$$v_h = 2 \cos \frac{(2h+1)\pi}{n} \qquad \left(0 \leqslant h \leqslant \frac{n-1}{2} \right),$$

and each of these is a double zero except possibly for -2.

Third question:

The numbers u_k and v_h defined above are both of the form

$$2 \cos \frac{m\pi}{n} ,$$

whether m is an even number $2k$ or an odd number $2h+1$. In either case, let us define

$$x_m = 2 \cos \frac{m\pi}{n} ,$$

so that $x_{2k}=u_k$ in the first case and $x_{2h+1}=v_h$ in the second. Then, from the definitions of u_h and v_k,

$$A(x_m) = 2(-1)^m.$$

By hypothesis, $|f(x_m)| < 2$. Therefore, the number $A(x_m)-f(x_m)$ has the same sign as does $A(x_m)$, which has the sign of $(-1)^m$. At two consecutive numbers x_m, the continuous function $A-f$ assumes values of opposite sign. Therefore, it assumes the value 0 at least once in the open interval $]x_m, x_{m+1}[$ (by the intermediate value theorem). There are n intervals $]x_m, x_{m+1}[$, all disjoint. Therefore, the function A_n-f vanishes at least n times in $[-2, 2]$.

Fourth question:

Proof by contradiction: If

$$|F(x)| < 2, \qquad \forall\, x \in [-2, 2],$$

the function F satisfies the hypotheses of (3) and the polynomial A_n-F has at least n zeros in $[-2, 2]$. In particular, if p denotes the degree of F, the polynomial A_p-F has at least p zeros in $[-2, 2]$. But the degree of this polynomial is at most $p-1$ (because the terms of highest degree in A_p and F are both equal to x^p). Therefore, it is the zero polynomial, so that we have a contradiction with the hypothesis that $|F(x)|<2$ since the polynomial A_p assumes the values -2 and 2.

Fifth question:

The substitution

$$x - \frac{a+b}{2} = \frac{b-a}{4} X$$

transforms the interval $[a, b]$ into the interval $[-2, 2]$ and the function G into the function Γ defined by

$$\Gamma(X) = G\left[\frac{a+b}{2} + \frac{b-a}{4} X\right].$$

The function

$$\left(\frac{4}{b-a}\right)^n \Gamma$$

is a polynomial whose term of highest degree is X^n. Let us apply to it the result obtained in (4). There exists at least one number X_0 in $[-2, 2]$ such that

$$\left|\frac{4}{(b-a)^n}\Gamma(X_0)\right| \geq 2.$$

Therefore, there exists at least one number x_0 in $[a, b]$, namely, the number

$$\frac{a+b}{2} + \frac{b-a}{4} X_0,$$

such that

$$|G(x_0)| \geq 2\left(\frac{b-a}{4}\right)^n.$$

The oscillation ω of the function G in $[a, b]$ is the difference $M-m$ between the greatest lower and least upper bound of G in $[a, b]$. Therefore, the greatest lower and least upper bounds of the function $G-\frac{1}{2}(M+m)$ in the interval $[a, b]$ are respectively $\frac{1}{2}\omega$ and $-\frac{1}{2}\omega$. The function $G-\frac{1}{2}(M+m)$ is a polynomial whose term of highest degree is x^n. The preceding result can be applied to it. Therefore, there exists at least one value x_0 in $[a, b]$ such that

$$2\left(\frac{b-a}{4}\right)^n \leq \left|G(x_0) - \frac{M+m}{2}\right| \leq \frac{1}{2}\omega.$$

418

First question:

We know that

$$(1+u)^n = 1 + nu + A(u),$$

where A is a polynomial with missing constant and first-degree terms. Each of the individual terms in $A(u)$ is $O(u^2)$, and so is their sum:

$$(1+u)^n = 1+nu+O(u^2).$$

Let us set $\sqrt[n]{1+u}=1+v$. We know that then

$$1+u = (1+v)^n = 1+nv+O(v^2) \Rightarrow u = nv+o(v).$$

The functions u and nv are therefore equivalent functions and, under these conditions,

$$v = O(u) \Rightarrow v^2 = O(u^2) \Rightarrow O(v^2) = O(u^2).$$

(For the properties of equivalent functions and the symbols o and O, see PZ, Book IV, Chapter I, §1.)
Thus,

$$v = \frac{u}{n}+O(u^2).$$

Second question:

Noting that $2x+3x^2 = O(x)$ and $x+x^2 = O(x)$, and using the above results, we can write

$$\sqrt[2]{1+2x+3x^2} = 1+\frac{2x+3x^2}{2}+O[(2x+3x^2)^2] = 1+\frac{2x+3x^2}{2}+O(x^2),$$

$$\sqrt[3]{1+x+x^2} = 1+\frac{x+x^2}{3}+O[(x+x^2)^2] = 1+\frac{x+x^2}{3}+O(x^2),$$

$$f(x) = 1+\frac{2x+3x^2}{2}+O(x^2)-[1+\frac{x+x^2}{3}+O(x^2)]$$

$$= \frac{2x}{3}+\frac{7x^2}{6}+O(x^2) = \frac{2x}{3}+O(x^2).$$

Since $O(x^2)$ is $o(x)$, we can also write

$$f(x) = \frac{2x}{3}+o(x) \Leftrightarrow f \sim \frac{2x}{3}.$$

Note that the same method applied to the function g produces no result: What we obtain is

$$g(x) = \frac{x^2}{4}+O(x^2) = O(x^2),$$

which is not sufficient to determine an equivalent function. Instead, let us write

$$g(x) = \sqrt[4]{1+x}\left(\sqrt[4]{1+\frac{x^2}{1+x}}-1\right) = \sqrt[4]{1+x}\left\{\frac{1}{4}\frac{x^2}{1+x}+O\left[\left(\frac{x^2}{1+x}\right)^2\right]\right\}$$

$$\frac{1}{1+x} = O(1) \Rightarrow O\left[\frac{x^4}{(1+x)^2}\right] = O[x^4]$$

$$\frac{g(x)}{x^2} = \frac{\sqrt[4]{1+x}}{4(1+x)}+\frac{O(x^4)}{x^2} = \frac{\sqrt[4]{1+x}}{4(1+x)}+O(x^2).$$

The limit of $g(x)/x^2$ as $x \to 1$ is $\frac{1}{4}$. Therefore, g is equivalent to the function $\frac{1}{4}x^2$.

419

First question:

We know that, for $x>0$,

$$\cot x < \frac{1}{x} < \frac{1}{\sin x} \Rightarrow \frac{\cos x}{x} < \cot x < \frac{1}{x},$$

$$\cos x = 1-2\sin^2\frac{x}{2} = 1-2[O(x)]^2 = 1-O(x^2),$$

so that

$$\frac{1-O(x^2)}{x} < \cot x < \frac{1}{x}.$$

Therefore, the difference $\cot x-1/x$ is $O(x^2)/x$, hence $O(x)$, and hence an infinitesimal that we can denote by $o(1)$.

Since the function $\cot x-1/x$ is an odd function, the result established for $x>0$ is also valid for $x<0$.

Second question:

The result of (1) enables us to write

$$f(x) = \sum_{k=1}^{n} a_k\left(\frac{1}{x}+o(1)\right) = \frac{1}{x}\left(\sum_{k=1}^{n} a_k\right)+o(1).$$

The function f approaches a limit as x approaches 0 if and only if

$$\sum_{k=1}^{n} a_k = 0.$$

Therefore, $f(x)=o(1)$, and the limit of f is 0.

(This last result was obvious from the start: The function f, being an odd function, cannot approach any limit other than zero as its argument approaches zero.)

420

First question:

The result is true for $n=1$. Let us assume it true for all values up to and including $n-1$:

$$x^n = x \times x^{n-1} \geqslant [1+(x-1)][1+(n-1)(x-1)]$$
$$= 1+n(x-1)+(n-1)(x-1)^2$$
$$(n-1)(x-1)^2 \geqslant 0 \Rightarrow x^n \geqslant 1+n(x-1)$$

with strict inequality holding for $n \geqslant 2$. If $x>1$, the sequence whose general term is $1+n(x-1)$ approaches $+\infty$ and so does the sequence whose general term is x^n.

If $x=1$, $x^n=1$, whence the sequence $\{x^n\}$ approaches 1.

If $x<1$, we have $x^n \times (1/x)^n = 1$. The sequence $\{(1/x)^n\}$ approaches $+\infty$, and the sequence $\{x^n\}$ approaches 0.

Second question:

For all x in $[0, h]$, the limit of the sequence $\{f_n(x)\}$ is 0. Therefore, the limit function f is the zero function (defined by $f(x)\equiv 0$). For each n, the function f_n is nondecreasing. Therefore,

$$|f_n(x)-f(x)| = f_n(x) \leqslant h^n, \qquad \forall\, x \in [0, h].$$

We have seen that the sequence $\{h^n\}$ converges to zero. For any number $\varepsilon>0$, there exists a number N such that

$$n > N \Rightarrow h^n < \varepsilon,$$

and hence

$$n > N \Rightarrow |f_n(x)-f(x)| < \varepsilon, \qquad \forall\, x \in [0, h].$$

This is the definition of uniform convergence in $[0, h]$.

REMARK: We have replaced the condition

$$|f_n(x) - f(x)| < \varepsilon, \qquad \forall\, x \in [0, h]$$

with the condition

$$\|f_n - f\| = \sup_{0 \leqslant x \leqslant h} |f_n(x) - f(x)| < \varepsilon.$$

The latter condition is easier to work with since the variable x no longer plays any role. It is often convenient to test for uniform convergence of sequences of functions in this manner.

Third question:

For no matter how large a number N, the function f_{N+1}, which is defined and continuous on $[0, 1]$, assumes values arbitrarily close to 1 in $[0, 1[$ and hence does not satisfy the condition

$$f_{N+1}(x) < \varepsilon, \qquad \forall\, x \in [0, 1[$$

if $\varepsilon < 1$.

The sequence $\{f_n(x)\}$ approaches zero if $0 \leqslant x < 1$ and approaches 1 if $x = 1$. The limit function f and the function $|f_n - f|$ are therefore defined by

$$f(x) = 0 \qquad \text{if} \qquad 0 \leqslant x < 1, \qquad f(1) = 1,$$

$$|f_n(x) - f(x)| = f_n(x) \qquad \text{if} \qquad 0 \leqslant x < 1, \qquad |f_n(1) - f(1)| = 0.$$

We have just shown that, for $0 < \varepsilon < 1$, there does not exist a number N such that

$$n > N \Rightarrow f_n(x) < \varepsilon, \qquad \forall\, x \in [0, 1[.$$

Therefore, there exists no number N such that

$$n > N \Rightarrow |f_n(x) - f(x)| < \varepsilon, \qquad \forall\, x \in [0, 1].$$

The sequence of the functions f_n does not converge uniformly in the interval $[0, 1]$.

In the interval $[0, 1]$, all the functions f_n are continuous. If the sequence of the functions f_n converges uniformly on $[0, 1]$, the limit function f is continuous on that interval. But f is discontinuous at 1.

Fourth question:

Let ε denote an arbitrary positive number. Then,

(a) $1 - \varepsilon < x \leqslant 1 \Rightarrow 1 - x < \varepsilon$ and $x^n < 1 \Rightarrow x^n(1 - x) < \varepsilon$;

(b) $0 \leqslant x \leqslant 1 - \varepsilon \Rightarrow 1 - x < 1 \Rightarrow g_n(x) \leqslant f_n(x)$.

The sequence $\{f_n\}$ converges uniformly in the interval $[0, 1-\varepsilon]$ (from the study made in (2)). Therefore, there exists a number N such that

$$n > N \Rightarrow f_n(x) < \varepsilon \Rightarrow g_n(x) < \varepsilon, \qquad \forall\, x \in [0,\, 1-\varepsilon].$$

In conclusion,

$$n > N \Rightarrow g_n(x) < \varepsilon, \qquad \forall\, x \in [0,\, 1].$$

This is true because, if $x \leqslant 1-\varepsilon$, this is the result (b) and, on the other hand, if $x > 1-\varepsilon$, the inequality on the right is true for every n.

Since $g_n(x) \geqslant 0$, we have thus expressed the fact that the sequence of the functions g_n converges uniformly to 0 on $[0, 1]$.

421

The functions u_n are odd functions (as can be verified immediately by induction). Therefore, we need to study only the interval $[0, \frac{1}{2}\pi]$ and, in what follows, we shall assume that x lies in that interval. If we fix x, the sequence $\{u_n(x)\}$ is an inductively defined numerical sequence, and we shall use the results of exercise 410. The function f of that exercise is now the sine function, which increases in the interval $[0, \frac{1}{2}\pi]$. The interval contains all the terms $u_n(x)$ (as can be verified by induction). For each x, the sequence $\{u_n(x)\}$ is nonincreasing since

$$u_1(x) = \sin x \leqslant x = u_0(x).$$

Since the terms $u_n(x)$ are bounded below by 0, the sequence of the $u_n(x)$ converges. The limit $u(x)$ of the sequence $\{u_n(x)\}$ satisfies the equation

$$u(x) = \sin u(x)$$

because of the continuity of the sine function. Therefore, $u(x)$ is equal to 0, the only root of the equation.

Just as in exercise 420, we shall prove that the sequence of functions u_n converges uniformly by showing that the numerical sequence

$$U_n = \sup_{-\pi/2 \,\leqslant\, x \,\leqslant\, \pi/2} |u_n(x) - u(x)| = \|u_n - u\|$$

converges to 0. I mentioned in the remark of the preceding exercise that this definition is easier to work with than the one in which x appears explicitly.

In the present problem $u(x) = 0$ and the function u_n is odd and takes positive values for $x \geqslant 0$:

$$U_n = \sup_{0 \,\leqslant\, x \,\leqslant\, \pi/2} u_n(x).$$

Let us show by induction on n that, for each n, the function u_n is a nondecreasing function:

The function u_0 defined by $u_0(x)=x$ is an increasing function.
Let us suppose that our result is established up to and inclusive of $n-1$:

$$x < x' \Rightarrow u_{n-1}(x) \leqslant u_{n-1}(x')$$

$$\Rightarrow u_n(x) = \sin\,[u_{n-1}(x)] \leqslant \sin\,[u_{n-1}(x')] = u_n(x'),$$

which means that it holds also for n.
Since u_n is nondecreasing, we see that

$$U_n = \sup_{0 \leqslant x \leqslant \pi/2} u_n(x) = u_n(\tfrac{1}{2}\pi).$$

We have shown that, for given x, the sequence $\{u_n(x)\}$ converges to 0. In particular, this result holds for $x=\tfrac{1}{2}\pi$. The sequence whose general term is $u_n\,(\tfrac{1}{2}\pi)=U_n$ converges to 0, and the sequence of the functions u_n converges uniformly to 0.

422

First question:

The denominator of f_n has no real zeros:

$$(x+1)^{2n+1} = (x-1)^{2n+1} \Leftrightarrow x+1 = x-1.$$

The function f_n is a rational fraction whose denominator has no real zeros. Therefore, it is a continuous function in R.

The numerator of f_n is an odd polynomial and its denominator is an even polynomial. Therefore f_n is an odd function. (One can see this either by replacing x with $-x$ or by studying the degrees of the monomials in the numerator and denominator.)

Therefore, we need study the variations of f_n only in the interval $[0, +\infty[$. We note that $x+1 \neq 0$ for $x \geqslant 0$. Therefore,

$$f_n(x) = \frac{1+\left(\dfrac{x-1}{x+1}\right)^{2n+1}}{1-\left(\dfrac{x-1}{x+1}\right)^{2n+1}},$$

so that the function f_n is the composite $w(v(u(\quad)))$, where u, v, and w are defined by

$$u = \frac{x-1}{x+1}, \qquad v = u^{2n+1}, \qquad w = \frac{1+v}{1-v}.$$

These three functions are all increasing functions. Therefore, the composite f_n is an increasing function and this result, which we have established for $x \geqslant 0$, remains valid for $x < 0$ since the function f_n is an odd function. Finally, u and v both approach 1 from below as x approaches $+\infty$, and w approaches $+\infty$ as x approaches $+\infty$. Therefore, $f_n(x)$ approaches $+\infty$:

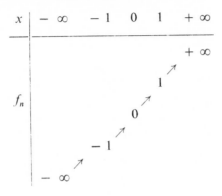

Second question:

The numerator of f_n is an odd polynomial and contains no term of degree $2n$. Similarly, the denominator is an even polynomial and contains no term of degree $2n-1$. This enables us to write

$$f_n(x) = \frac{x^{2n+1} + o(x^{2n})}{(2n+1)x^{2n} + o(x^{2n-1})} = \frac{x}{2n+1} \frac{1+o(1/x)}{1+o(1/x)}$$

as x approaches $\pm\infty$. If f is an infinitesimal, we may write

$$\frac{1}{1+f} = 1 - f + o(f).$$

Therefore,

$$\frac{1}{1+o(1/x)} = 1 - o\left(\frac{1}{x}\right) + o\left[o\left(\frac{1}{x}\right)\right] = 1 + o\left(\frac{1}{x}\right)$$

and

$$f_n(x) = \frac{x}{2n+1}\left[1 + o\left(\frac{1}{x}\right)\right] = \frac{x}{2n+1} + o(1).$$

The line representing the equation $y = x/(2n+1)$ is an asymptote of the graph of f_n. Elementary calculation yields

$$f_n(x) = \frac{2(2n+1)x + o(x)}{2} = (2n+1)x + o(x)$$

as x approaches 0. The line representing the equation $y=(2n+1)x$ is tangent to the graph at the origin. Finally, the graph of f_n is symmetric about the origin since the function f_n is an odd function.

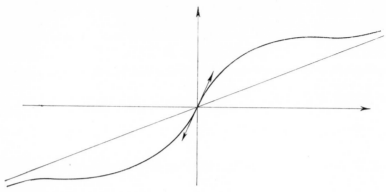

Third question:

Let us assume that $x>0$ and let us use the decomposition of f_n that we used above to study the variations:

$$x > 0 \Rightarrow -1 < u < 1 \Rightarrow v_n = u^{2n+1} \to 0$$

$$\Rightarrow f_n(x) = \frac{1+v_n}{1-v_n} \to 1.$$

If $x=0$, $f_n(0)=0$ and the limit is 0.

Finally, since the function f_n is an odd function, the sequence $\{f_n(x)\}$ approaches the constant -1 if $x<0$.

If the end-points a and b of the interval in question are of the same sign, we shall suppose that they are both nonnegative. The results that we obtain will remain valid for the interval $[-b, -a]$ since the functions f_n are odd functions. Then, the limit function f is the constant 1, and

$$f_n(x)-1 = \frac{2(x-1)^{2n+1}}{(x+1)^{2n+1}-(x-1)^{2n+1}} = 2\frac{\left(\dfrac{x-1}{x+1}\right)^{2n+1}}{1-\left(\dfrac{x-1}{x+1}\right)^{2n+1}}.$$

For a given $\varepsilon>0$, let us study the inequality

$$|f_n(x) - 1| < \varepsilon$$

or rather the two inequalities

(1) $$-\varepsilon < f_n(x) - 1 < \varepsilon.$$

These inequalities are equivalent to the following ones:

$$-\frac{\varepsilon}{2-\varepsilon} < \left(\frac{x-1}{x+1}\right)^{2n+1} < \frac{\varepsilon}{2+\varepsilon}.$$

The homographic function

$$\frac{x-1}{x+1} \quad \text{increases from} \quad \frac{a-1}{a+1} \quad \text{to} \quad \frac{b-1}{b+1}$$

in the interval $[a, b]$, and

$$\left(\frac{x-1}{x+1}\right)^{2n+1} \quad \text{increases from} \quad \left(\frac{a-1}{a+1}\right)^{2n+1} \quad \text{to} \quad \left(\frac{b-1}{b+1}\right)^{2n+1}.$$

Thus, the two inequalities (1) will be satisfied for every $x \in [a, b]$ if and only if

$$-\frac{\varepsilon}{2-\varepsilon} < \left(\frac{a-1}{a+1}\right)^{2n+1} < \left(\frac{b-1}{b+1}\right)^{2n+1} < \frac{\varepsilon}{2+\varepsilon}.$$

The two numbers

$$\frac{a-1}{a+1} \quad \text{and} \quad \frac{b-1}{b+1}$$

are less in absolute value than 1. Hence, the sequences whose general terms are

$$\left(\frac{a-1}{a+1}\right)^{n} \quad \text{and} \quad \left(\frac{b-1}{b+1}\right)^{n}$$

both converge to 0. By the definition of a limit, for the interval

$$\left]-\frac{\varepsilon}{2-\varepsilon}, \frac{\varepsilon}{2+\varepsilon}\right[,$$

which contains 0, there exists a number N such that

$$n > N \Rightarrow \left(\frac{a-1}{a+1}\right)^{n} \quad \text{and} \quad \left(\frac{b-1}{b+1}\right)^{n} \quad \text{belong to} \quad \left]-\frac{\varepsilon}{2-\varepsilon}, \frac{\varepsilon}{2+\varepsilon}\right[,$$

$$n > N \Rightarrow -\varepsilon < f_n(x)-1 < \varepsilon, \qquad \forall\, x \in [a, b].$$

Therefore, the sequence of the functions f_n converges uniformly in the interval $[a, b]$.

The functions f_n are continuous for all real values and, in particular, for the values in $[-a, a]$. If the sequence of the f_n were uniformly convergent in the interval $[-a, a]$, the limit function f would be continuous in $[-a, a]$, which is not the case since we know that it is discontinuous at zero.

423

First question:

(1) The solution of the problem is given by

$$l(x) = f(u) + (x-u)\frac{f(v)-f(u)}{v-u},$$

as can be verified immediately by assigning to x the values u and v.

(2) The image of the interval $[u, v]$ under the continuous monotonic function l is the interval $[f(u), f(v)]$. Since $l(x)$ is a value in this interval, it belongs to the set $f([u, v])$ of values assumed by f on $[u, v]$ (by the intermediate-value theorem). Therefore, there exists at least one number ξ in $[u, v]$ such that $l(x)=f(\xi)$.

(3) The preceding result implies that, for x in $[u, v]$,

$$|l(x)-f(x)| = |f(\xi)-f(x)| \leqslant \sup_{u \leqslant x \leqslant v} f(x) - \inf_{u \leqslant x \leqslant v} f(x) = \omega.$$

Thus, this ω is an upper bound of the set of numbers $|l(x)-f(x)|$. Therefore, it is either equal to or greater than the least upper bound of that set.

Second question:

The set $\varLambda$ is a subset of the vector space $\mathscr{C}$ of continuous real functions defined on $[a, b]$. We shall show that this is a vector space by showing that, for any two functions λ and μ in $\varLambda$ and any real number r, the difference $\lambda-\mu$ and the product $r\lambda$ are functions in $\varLambda$.

For $r\lambda$, this is obvious since the restriction of this function to each of the intervals resulting from the subdivision is the product of a first-degree polynomial multiplied by r and is itself therefore a first-degree polynomial.

Let us denote by $a, x_1, ..., x_{n-1}, b$ the end-points of the subintervals of the interval $[a, b]$ corresponding to λ and let us denote by $a, x'_1, ..., x'_{p-1}, b$ the end points of the subintervals corresponding to μ. Finally, let us denote by $y_1, ..., y_{q-1}$ the union of the two sets $\{x_i\}$ and $\{x'_j\}$ numbered in increasing order, so that the y_h, together with the numbers a and b, define a new partition of the interval $[a, b]$.

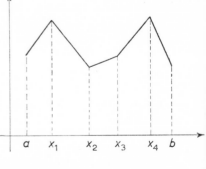

Between two consecutive y's, there is neither an x_i nor an x'_j since all these numbers are included in the set of y_h. Therefore, the interval $[y_k, y_{k+1}]$ is contained in an interval $[x_i, x_{i+1}]$ and in an interval $[x'_j, x'_{j+1}]$. The restrictions of λ and μ to the interval $[y_k, y_{k+1}]$ are polynomials of degree 1 at most, and the same is true for the restriction of $\lambda - u$ to that interval.

In each of the intervals $[x_i, x_{i+1}]$, the graph is a line segment. Any two consecutive such segments have a common end-point since the function λ is continuous.

Third question:

To show that a function f that is continuous on $[a, b]$ is the uniform limit of a sequence of functions belonging to Λ, we need show only that, for any $\varepsilon > 0$, there exists a function λ in Λ such that $|f(x) - \lambda(x)| < \varepsilon$ for every x in $[a, b]$. We then obtain a sequence of the type desired by taking for ε consecutive values ε_n of a sequence that approaches 0; for example, we may take $\varepsilon_n = 1/n$.

We know that, in each of the intervals $[x_i, x_{i+1}]$ of the partition defined by the function λ, this function coincides with a first-degree polynomial l. If l_i is the polynomial defined in the first part by $l_i(x_i) = f(x_i)$ and $l_i(x_{i+1}) = f(x_{i+1})$, we know that

$$x_i \leqslant x \leqslant x_{i+1} \Rightarrow |\lambda(x) - f(x)| = |l_i(x) - f(x)| \leqslant \omega_i(f),$$

where $\omega_i(f)$ denotes the oscillation of f in $[x_i, x_{i+1}]$.

For $|f(x) - \lambda(x)|$ to be less than ε for every x in $[a, b]$, it will be sufficient to have $\omega_i(f) < \varepsilon$ for all values of i. The function f, being continuous in the closed interval $[a, b]$ is uniformly continuous in that interval. Therefore, there exists a number η such that

$$|x' - x''| < \eta \Rightarrow |f(x') - f(x'')| < \varepsilon.$$

Then,

$$x_{i+1} - x_i < \eta \Rightarrow \omega_i(f) < \varepsilon$$

(cf. PZ, Book III, Chapter III, part 2, § 3).

Thus, we obtain a solution λ of the problem by making a partition of the interval $[a, b]$ every subinterval in which is less than η and by taking $\lambda(x_i) = f(x_i)$.

424

First question:

The number $[x]$ is the greatest integer that does not exceed x, and

$$[x + 1] = 1 + [x].$$

This bracket function, which is constant in every interval not containing an integer, is therefore continuous except at integral values of its argument, at which values it is continuous on the right but discontinuous on the left:

$$[n] = \lim_{x \to n^+} [x] = n; \qquad \lim_{x \to n^-} [x] = n - 1.$$

Finally, in every finite interval, this function has only a finite number of discontinuities and a finite number of values. It is a step-function.

The properties of the function D are immediately deduced:

$$D(x + 1) = x + 1 - [x + 1] = x - [x] = D(x).$$

The function D is continuous except at integral values of its argument, at which it is continuous on the right and discontinuous on the left:

$$D(+0) = D(0) = 0, \qquad D(-0) = 1.$$

Finally, D, being the difference between two regulated functions (namely, the linear identity function and the bracket function, which is a step-function) is a regulated function in every finite interval.

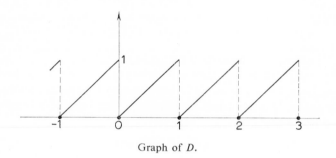

Graph of D.

Second question:

The functions u_n are periodic with period 1 since

$$D[k(x + 1)] = D[kx + k] = D[kx].$$

Therefore, we shall study the convergence of the sequence $\{u_n\}$ in an interval of length 1, for example, $I_0 = [-\frac{1}{2}, \frac{1}{2}]$. In this interval, the functions $D(kx)$ are regulated functions and hence so are the functions u_n.

If $p < q$, it is obvious, since $0 \leqslant D(kx) \leqslant 1$, that

$$0 \leqslant u_q(x) - u_p(x) = \sum_{k=p+1}^{q} \frac{D(kx)}{2^k} \leqslant \sum_{k=p+1}^{q} \frac{1}{2^k} < \frac{1}{2^p}.$$

If we make no assumption regarding the relative size of p and q, we have

$$\|u_q - u_p\| = \sup_{x \in I_0} |u_q(x) - u_p(x)| < \frac{1}{2^{\inf (p,q)}}.$$

The double sequence $\{\|u_p-u_q\|\}$ converges to 0. Therefore, the sequence of functions u_n converges uniformly in I_0 (cf. PZ, Book III, Chapter III, part 5, § 2, no. 3), and the limit function u is a regulated function in I_0 that is, the space of regulated functions is complete with respect to the norm

$$\|f\| = \sup_{x \in I_0} |f(x)|.$$

If x does not belong to I_0, there exists a number x_0 in I_0 such that $x=x_0+X$, where X is an integer. The functions u_n assume the same value at x and x_0. The two sequences $\{u_n(x)\}$ and $\{u_n(x_0)\}$ have the same limit. The limit function u is thus defined in R and it is periodic with period 1. Consequently,

$$\sup_{x \in R} |u_n(x)-u(x)| = \sup_{x \in I_0} |u_n(x)-u(x)|.$$

The uniform convergence of the sequence $\{u_n\}$ in I_0 implies the uniform convergence of the sequence $\{u_n\}$ in R.

Third question:

In this question and in the following one, we shall consider only values in the interval I_0. The general case is immediately obtained since the period of the function u is the length of the interval I_0.

If x_0 is irrational, the functions D_k defined by $D_k(x) = D(kx)$ are all continuous at that value (since the points of discontinuity are those at which kx is an integer, that is, rational numbers expressible with denominators that divide k). The functions u_n are continuous at x_0. The sequence $\{u_n\}$ converges uniformly in the interval I_0. Therefore, the limit u is continuous at x_0 (cf. PZ, Book III, Chapter III, part 5, § 2, no. 4, theorem 1).

Fourth question:

At every value x in I_0, the function u, being a regulated function, has a right-hand limit $u(x+0)$ and a left-hand limit $u(x-0)$. By the definition of uniform convergence, for every number $\varepsilon>0$, there corresponds a number N such that

$$n > N \Rightarrow |u_n(x) - u(x)| < \varepsilon, \qquad \forall x \in I_0.$$

The functions $u_n(x)$ and $u(x)$ approach right-hand limits $u_n(x_0+0)$ and $u(x_0+0)$ respectively, as $x \to 0$ from above.
The existence of these limits is known, and they satisfy the condition

$$n > N \Rightarrow |u_n(x_0 + 0) - u(x_0 + 0)| \leqslant \varepsilon,$$

which proves that $u(x_0+0)= \lim_{n \to \infty} u_n(x_0+0)$.
The same line of proof is used for left-hand limits.

The functions D_k are continuous on the right at every point, and hence so are the functions u_n:

$$u_n(x_0 + 0) = u_n(x_0) \Rightarrow u(x_0 + 0) = \lim u_n(x_0 + 0) = \lim u_n(x_0) = u(x_0).$$

The function u is continuous on the right everywhere.

If x_0 is a rational number represented, in its simplest form, by s/r, the functions D_k, where k is a multiple hr of r, are discontinuous at x_0 and

$$D_{hr}(x_0 - 0) = 1, \qquad D_{hr}(x_0) = 0.$$

The other functions D_k are continuous at x_0, and

$$D_k(x_0) = D_k(x_0 - 0).$$

Under these conditions, we see that

$$u_n(x_0 - 0) = u_n(x_0) + \sum_{hr \leqslant n} \frac{1}{2^{hr}}.$$

This result also implies that $u(x_0 - 0) \neq u(x_0)$ since

$$u_n(x_0 - 0) \geqslant u_n(x_0) + \frac{1}{2^r}$$

for $n \geqslant r$ and hence

$$\lim u_n(x_0 - 0) = u(x_0 - 0) \geqslant u(x_0) + \frac{1}{2} = \lim u_n(x_0) + \frac{1}{2^r}.$$

The function u is discontinuous on the left at x_0. More precisely, one can show that

$$u(x_0 - 0) = u(x_0) + \sum_h \frac{1}{2^{hr}} = u(x_0^{\cdot\cdot}) + \frac{1}{2^r - 1}.$$

425

First question:

Let us use the definition of continuity at 0 that is expressed in terms of sequences $\{x_n\}$ that converge to 0 (cf. PZ, Book III, Chapter III, part 2, § 1, definition 3).

If a sequence $\{x_n\}$ converges to 0, the sequence $\{f(x_n)\}$ also converges to 0:

$$\forall \, \varepsilon > 0, \qquad \exists \, N \qquad \text{such that} \qquad n > N \Rightarrow |x_n| < \varepsilon \Rightarrow |f(x_n)| < \varepsilon.$$

Then, by the hypothesis that f is continuous at 0, the sequence $\{f(x_n)\}$ converges to $f(0)$.

Second question:

The function g defined by

$$g(x) = \left| \frac{f(x)}{x} \right|$$

is defined and continuous at $x=0$. (We recall that the quotient of two non-vanishing continuous functions and the absolute value of a continuous function are continuous functions.)

The function g, which is defined and continuous in the two closed intervals $[-M, -\varepsilon]$ and $[\varepsilon, M]$, is bounded in these intervals, and there exist numbers x_1 and x_2 such that

$$-M \leqslant x_1 \leqslant -\varepsilon, \qquad g(x_1) = \sup_{-M \leqslant x \leqslant -\varepsilon} g(x),$$

$$\varepsilon \leqslant x_2 \leqslant M, \qquad g(x_2) = \sup_{\varepsilon \leqslant x \leqslant M} g(x).$$

(The extreme-value theorem [cf. PZ, Book III, Chapter III, part 2, § 2, corollary to theorem 1].)

By hypothesis,

$$g(x_1) = \left| \frac{f(x_1)}{x_1} \right| < 1 \qquad \text{and} \qquad g(x_2) = \left| \frac{f(x_2)}{x_2} \right| < 1.$$

The number sup $(g(x_1), g(x_2))$ is also less than 1 and serves as the required k:

$$\varepsilon \leqslant |x| \leqslant M \Rightarrow g(x) \leqslant k \Rightarrow |f(x)| \leqslant k |x|.$$

The function f defined by $f(x)=x/(|x|+1)$ provides the required counter-example:

$$|f(x)| < |x| \qquad \text{if} \qquad x \neq 0,$$

$$\sup_{\varepsilon \leqslant |x| \leqslant M} \left| \frac{f(x)}{x} \right| = \frac{1}{1+\varepsilon} = k.$$

The ratio $f(x)/x$ approaches 1 as x approaches 0. By the definition of a limit, it cannot be bounded above by any number $h<1$.

(Other possible examples are $\sin x$, $\tanh x$, and $\ln(|x|+1)$.)

Third question:

Proof by induction on n. First, let us take $n=0$. We then have

$$\left. \begin{array}{l} |x| < \varepsilon \Rightarrow |f_0(x)| < |x| < \varepsilon \\ \varepsilon \leqslant |x| \leqslant M \Rightarrow |f_0(x)| \leqslant k|x| \leqslant kM \end{array} \right\} \Rightarrow |f_0(x)| \leqslant \sup(\varepsilon, kM).$$

Suppose that the result holds for all numbers up to and including $n-1$. The proof for n is carried out just as was done above for the case of $n=0$ except that, here, $f_{n-1}(x)$ plays the role that x played above.

The hypothesis

$$|f_{n-1}(x)| \leqslant \sup(\varepsilon, k^n M) \implies |f_{n-1}(x)| < \varepsilon \qquad \text{or} \qquad \varepsilon \leqslant |f_{n-1}(x)| \leqslant k^n M,$$

$$|f_{n-1}(x)| < \varepsilon \implies |f[f_{n-1}(x)]| < |f_{n-1}(x)| < \varepsilon,$$

$$\varepsilon \leqslant f_{n-1}(x) \leqslant k^n M \implies \varepsilon \leqslant |f_{n-1}(x)| \leqslant M \qquad \text{since} \qquad k < 1.$$

Therefore, we can apply the result of (2) to the value $f_{n-1}(x)$ of the variable:

$$|f[f_{n-1}(x)]| \leqslant k\,|f_{n-1}(x)| \leqslant k \times k^n\, M = k^{n+1} M.$$

Combining the two results that we have obtained, we have

$$|f_n(x)| = |f[f_{n-1}(x)]| \leqslant \sup(\varepsilon, k^{n+1} M).$$

The preceding result can be expressed by

$$\|f_n - 0\| = \|f_n\| = \sup_{|x| \leqslant M} |f_n(x)| \leqslant \sup(\varepsilon, k^{n+1} M).$$

The limit of the sequence $\{k^{n+1} M\}$ is 0 since $k < 1$. Therefore, there exists an integer N such that

$$n > N \implies k^{n+1} M < \varepsilon \implies \|f_n - 0\| < \varepsilon.$$

In the preceding inequalities, ε is an arbitrary positive number. Therefore, we can interpret the results obtained by saying that the sequence $\{\|f_n - 0\|\}$ converges to 0, that is, that the sequence of the functions f_n converges uniformly to the zero function.

426

First question:

The function $1/x$ is continuous for $x \neq 0$, and the sine function is continuous at every real value. Therefore, the composite function f is continuous except possibly at $x=0$.

Let us show that the function f is not continuous at 0, no matter what the value of y_0 is that is, that f does not approach any limit as its argument approaches 0.

Let us define a sequence $\{x_n\}$ by

$$\frac{1}{x_n} = n\pi + \frac{\pi}{2}.$$

Obviously, the sequence $\{x_n\}$ converges to 0 and $f(x_n)=(-1)^n$. The sequence $\{f(x_n)\}$ does not converge to anything. Therefore, the function f has no right-hand limit as its argument approaches 0 (by virtue of the definition of a limit in terms of convergent sequences), and, in particular, it does not have the right-hand limit y_0.

Second question:

Let $0<x_1<x_2<...<1/\pi$ define a partition of the interval $[0, 1/\pi]$ into subintervals such that φ is constant in the interior of each subinterval of the partition.

The function f assumes every value of the interval $[-1, 1]$ as its argument ranges over $]0, x_1[$. We denote by φ_1 the value of φ in that interval. Then,

$$\sup |f(x)-\varphi(x)| = \sup (|1-\varphi_1|, |1+\varphi_1|),$$

$$|1-\varphi_1|+|1+\varphi_1| \geqslant |(1-\varphi_1)+(1+\varphi_1)| = 2,$$

and hence

$$\sup (|1 - \varphi_1|, |1 + \varphi_1|) \geqslant 1.$$

Consequently, it is impossible to have $|f(x)-\varphi(x)|<\frac{1}{2}$ for all values of x in $]0, x_1[$ or *a fortiori* $[0, 1/\pi]$.

In (1), we saw that the function f has no right-hand limit at 0. Therefore, this function is not a regulated function. Consequently, there exists a number ε_0 such that no step function φ satisfies the condition

$$|f(x)-\varphi(x)| < \varepsilon_0, \qquad \forall\, x \in \left[0, \frac{1}{\pi}\right].$$

The preceding result expresses this fact, taking ε_0 equal to $\frac{1}{2}$.

(The property cited to express the fact that f is not a regulated function is the negation of the property that we used to show that a function is a regulated function. In this connection, see exercise 423, (3) or PZ, Book III, Chapter III, part 5, § 1, final remark.)

The function f is continuous in $[a, x]$ and differentiable in $]a, x[$, where x is a number in $]a, b]$. Therefore, there exists a number c such that

$$a < c < x, \qquad \frac{f(x)-f(a)}{x-a} = f'(c)$$

(by the theorem of the mean).

By the definition of a limit, for any given $\varepsilon > 0$, there exists an α such that

$$a < u < \alpha \Rightarrow |f'(u)-l| < \varepsilon$$

and hence such that

$$a < x < \alpha \Rightarrow a < c < \alpha \Rightarrow |f'(c)-l| = \left| \frac{f(x)-f(a)}{x-a} - l \right| < \varepsilon,$$

which, by definition, means that l is the limit of the ratio

$$\frac{f(x)-f(a)}{x-a},$$

as x approaches a from above.

If the derivative f' has a right-hand limit at a, the function f is differentiable on the right at a and its right-hand derivative is the right-hand limit of f', which justifies the notation $f'(a+0)$ for the right-hand derivative at a.

EXAMPLE: The nonnegative-square-root function has a finite derivative in $(0, \infty)$. The sine function is everywhere differentiable. Therefore, the product is differentiable in $(0, \infty)$ and

$$f'(x) = \frac{\sin x}{2\sqrt{x}} + \sqrt{x} \cos x.$$

As $x \to 0$, we have

$$\frac{\sin x}{2\sqrt{x}} \sim \frac{x}{2\sqrt{x}} \to 0 \qquad \text{and} \qquad f'(x) \to 0.$$

The function f is right-differentiable at 0 and $f'(+0) = 0$.

(This result is obvious from observation of the fact that

$$\frac{f(x)}{x} = \frac{\sin x}{\sqrt{x}} \to 0 \quad \text{as} \quad x \to 0.)$$

428

The function f is differentiable if $x \neq 0$ since the function x^2 is differentiable and the function $\cos (1/x)$, being the composite of two differentiable functions, is differentiable.
Thus,

$$f'(x) = 2x \cos \frac{1}{x} + x^2 \left[\left(-\frac{1}{x^2} \right) \left(-\sin \frac{1}{x} \right) \right] = 2x \cos \frac{1}{x} + \sin \frac{1}{x}$$

for $x \neq 0$. As $x \to 0$, we have

$$\frac{f(x) - f(0)}{x} = x \cos \frac{1}{x} \to 0.$$

Thus, the derivative of f at 0 is 0; this function is differentiable everywhere in its domain of definition.

Since f is differentiable, it is continuous and hence is a regulated function. The function f' is not a regulated function. To see this, note that the function $2x \cos (1/x)$ is a regulated function (and is in fact continuous if we extend its definition by assigning it the value 0 at $x=0$). Therefore, if f' were a regulated function, the difference $f' - 2x \cos (1/x)$ would be a regulated function, but we have seen (see exercise 426) that the function $\sin (1/x)$ is not a regulated function on $[0, 1/\pi]$.

429

First question:

The function f is an even function that is defined and continuous for all x since the polynomial $x^4 - 2x^2 + 2$ assumes only values equal to or greater than 1. Being the ratio of composite functions of differentiable functions, f is differentiable and

$$f'(x) = x\sqrt{2}(x^4 - 2x^2 + 2)^{-\frac{1}{2}} + \frac{x^2\sqrt{2}}{2}\left(-\frac{1}{2} \right)(x^4 - 2x^2 + 2)^{-\frac{3}{2}}(4x^3 - 4x)$$

$$= \frac{x\sqrt{2}(2 - x^2)}{(x^4 - 2x^2 + 2)^{\frac{3}{2}}}.$$

The sign of $f'(x)$ is the same as the sign of the numerator, and we can immediately see the behavior of f:

As $x \to \pm\infty$, $f \sim x^2\sqrt{2}/2x^2$ and its limit is $\sqrt{2}/2$:

	$-\infty$		$-\sqrt{2}$		0		$\sqrt{2}$		$+\infty$
f	$\dfrac{\sqrt{2}}{2}$	$\nearrow$	1	$\searrow$	0	$\nearrow$	1	$\searrow$	$\dfrac{\sqrt{2}}{2}$

The composite function $g = \arcsin \circ f$ is defined since the values assumed by f are numbers in $[-1, 1]$. Being a function of the even function f, it is an even function. It increases when f increases since the arcsin is an increasing function.

	$-\infty$		$-\sqrt{2}$		0		$\sqrt{2}$		$+\infty$
g	$\tfrac{1}{4}\pi$	$\nearrow$	$\tfrac{1}{2}\pi$	$\searrow$	0	$\nearrow$	$\tfrac{1}{2}\pi$	$\searrow$	$\tfrac{1}{4}\pi$

Second question:

(1) If the functions f and arcsin have finite derivatives at x, the composite function g will be differentiable (as is the case except when $f(x) = 1$, that is, except when $x = -\sqrt{2}$ or $\sqrt{2}$). Therefore,

$$g'(x) = \frac{f'(x)}{\sqrt{1 - f^2(x)}} = \frac{x\sqrt{2}(2 - x^2)}{(x^4 - 2x^2 + 2)^{\frac{3}{2}}} \sqrt{\frac{2(x^4 - 2x^2 + 1)}{(x^2 - 2)^2}}$$

$$= \frac{2x(2 - x^2)}{(x^4 - 2x^2 + 2)\,|x^2 - 2|}.$$

Thus, we finally have

$$g'(x) = \frac{-2x}{x^4 - 2x^2 + 2} \qquad \text{if} \qquad |x| > \sqrt{2},$$

$$g'(x) = \frac{2x}{x^4 - 2x^2 + 2} \qquad \text{if} \qquad |x| < \sqrt{2}.$$

(2) Since g is an even function, we need examine only the behavior at $\sqrt{2}$:

$$g'(x) \to -\sqrt{2} \text{ as } x \to \sqrt{2} + 0 \text{ (i.e., as } x \to \sqrt{2} \text{ from above)}$$

and

$$g'(x) \to \sqrt{2} \text{ as } x \to \sqrt{2} - 0 \text{ (i.e., as } x \to \sqrt{2} \text{ from below).}$$

We know (cf. exercise 427) that g has right- and left-hand derivatives at $\sqrt{2}$:

$$g'(\sqrt{2}+0) = -\sqrt{2}, \qquad g'(\sqrt{2}-0) = \sqrt{2}.$$

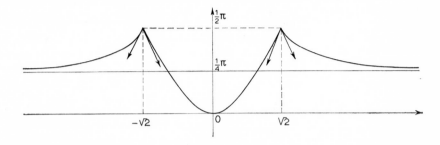

But g is not differentiable at that value since the right- and left-hand derivatives are unequal. (This is an example of a maximum that does not correspond to a zero of the derivative.)

The graph of g is symmetric about the y-axis since g is an even function. It has distinct right- and left-hand half-tangents at points with abscissas $-\sqrt{2}$ and $\sqrt{2}$.

430

First question:

The function f is defined, continuous, and differentiable everywhere except at $x = -1$.

The function $(1-x)/(1+x)$ approaches $+\infty$ (resp. $-\infty$) and $\arctan (1-x)/(1+x)$ approaches $\tfrac{1}{2}\pi$ (resp. $-\tfrac{1}{2}\pi$) as $x \to -1$ from the right (resp. left). The function f has right- and left-hand limits:

$$f(-1-0) = -\tfrac{1}{2}-\pi, \qquad f(-1+0) = -\tfrac{1}{2}+\pi.$$

For $x \neq -1$, the value of the derivative is

$$f'(x) = \frac{2(x^2+1)-2x(2x+1)}{(x^2+1)^2}+2\frac{1}{1+\left(\dfrac{1-x}{1+x}\right)^2}\times\frac{-2}{(1+x)^2}$$

$$= \frac{-2x(1+2x)}{(x^2+1)^2}.$$

The derivative $f'(x)$ approaches $-\tfrac{1}{2}$ as $x \to -1$. The function f_1, which is equal to f if $x > -1$ and equal to $f(-1+0)$ if $x = -1$, thus has a right-hand derivative equal to $-\tfrac{1}{2}$ at $x = -1$. Similarly, the function f_2, which is equal to f for $x < -1$ and equal to $f(-1-0)$ at $x = -1$, has a left-hand derivative equal to $-\tfrac{1}{2}$ at $x = -1$ (this result being obtained in exercise 427).

The sign of the derivative $f'(x)$ is the sign of the product $-x(1+2x)$. Thus, the behavior of f is immediately clear:

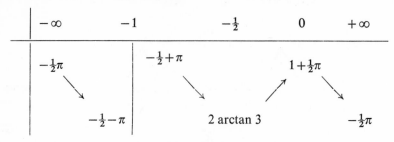

The value of $2 \arctan 3$ is approximately 2.5.

Construction of the graph presents no difficulties. We already know that the straight line representing the equation $y = -\frac{1}{2}\pi$ is an asymptote of this graph and that this graph has half-tangents, both right and left, of slope $-\frac{1}{2}$ at a point whose abscissa is -1.

The appearance of the graph can be made more precise, although this is not necessary here, by examining the sign of $f''(x)$ in order to determine the convexity of f:

$$f''(x) = 2\frac{4x^3 + 3x^2 - 4x - 1}{(x^2+1)^2} = 2\frac{N(x)}{(x^2+1)^2}$$

$$N(-\infty) = -\infty, \quad N(-2) = -13, \quad N(-1) = 2, \quad N(-\tfrac{1}{2}) = \tfrac{5}{4},$$

$$N(0) = -1, \quad N(1) = 2, \quad N(+\infty) = +\infty.$$

Graph of f

The function N thus has three zeros, which we denote by x_1, x_2, and x_3. These three zeros satisfy the inequalities

$$- 2 < x_1 < - 1 < - \tfrac{1}{2} < x_2 < 0 < x_3 > 1.$$

The graph is convex upward in the intervals $[x_1, -1[,]-1, x_2[$, and $[x_3, +\infty[$, and it is convex downward in the intervals $]-\infty, x_1]$ and $[x_2, x_3]$.

Finally, we note that the graph is rather flattened between -1 and 0 since the maxima of f are only slightly different from 2.64 and 2.57 and the minimum is only slightly different from 2.5.

Second question:

The result is established for $n=2$ (and, for that matter, for $n=3$) since

$$f'(x) = \frac{A_2(x)}{(x^2+1)^2},$$

where $A_2(x) = -2x(1+2x)$.

Suppose it is true for 2, 3, ..., n. Then,

$$f^{(n-1)}(x) = \frac{A_n(x)}{(x^2+1)^n} \Rightarrow f^{(n)}(x) = \frac{A_n'(x)(x^2+1)-2nxA_n(x)}{(x^2+1)^{n+1}}.$$

This is a rational function with denominator $(x^2+1)^{n+1}$ and with numerator

$$A_{n+1} = (x^2+1)A_n'-2nxA_n.$$

If we denote the term of highest degree in A_n by $a_n x^n$, the term of highest degree in A_{n+1} is

$$na_n x^{n+1} - 2na_n x^{n+1} = -na_n x^{n+1}.$$

Thus, A_{n+1} is of degree $n+1$.

Let us again establish by induction on n the required result regarding the zeros of A_n: The function A_2 has two real distinct zeros, namely, $-\tfrac{1}{2}$ and 0. Let us suppose that A_n has n real distinct zeros. Then, the derivative $f^{(n-1)}$ has n zeros x_1, x_2, ..., x_n, and between any of these zeros and the next there is at least one zero of the first derivative, $f^{(n)}$, of $f^{(n-1)}$ (by Rolle's theorem). Furthermore, $f^{(n-1)}$ approaches zero as x approaches $\pm\infty$. Between $-\infty$ and x_1 and again between x_n and $+\infty$, there is at least one extremum of $f^{(n-1)}$ and hence a zero of the derivative $f^{(n)}$.

In sum, the derivative $f^{(n)}$ has at least $n+1$ real distinct zeros (one in the interior of each of the intervals defined by $-\infty$, x_1, x_2, ..., x_n, $+\infty$). These zeros are the zeros of the polynomial A_{n+1}, which is of degree $n+1$. This polynomial can have no other zeros. Its zeros are all real and distinct.

431

First question:

Since 0 is not a root of the equation given, we can replace this equation with the equation

$$a^x x^{-\alpha} = 1.$$

To discuss this equation, let us study the behavior of the function f defined by $f(x) = a^x x^{-\alpha}$. This function is defined, continuous, and differentiable at every positive value of x and

$$f'(x) = x^{-\alpha} a^x \ln a - \alpha x^{-\alpha-1} a^x = a^x x^{-\alpha-1}(x \ln a - \alpha).$$

(1) If $\alpha \leqslant 0$, the derivative $f'(x)$ is positive and the function increases monotonically from 0 to $+\infty$. It therefore assumes the value 1 once and only once.
(2) If $\alpha > 0$, the derivative f' vanishes at $x = \alpha/(\ln a)$ and

$$x < \frac{\alpha}{\ln a} \Rightarrow f'(x) < 0, \qquad x > \frac{\alpha}{\ln a} \Rightarrow f'(x) > 0.$$

As $x \to 0$, we have $a^x \to 1$ and $x^{-\alpha} \to +\infty$, so that $f(x) \to +\infty$.

As $x \to +\infty$, the product $a^x x^{-\alpha}$ is an indeterminate form since a^x approaches $+\infty$ and $x^{-\alpha}$ approaches 0. But we know that this product approaches $+\infty$ because of the relative speed of increase of the exponential function and any power (cf., for example, PZ, Book III, Chapter III, part 7, near end of § 6).

Therefore, the general behavior of f is as shown by

where

$$m = a^{\alpha/\ln a} \left(\frac{\alpha}{\ln a} \right)^{-\alpha} = \left(\frac{e \ln a}{\alpha} \right)^{\alpha}.$$

The function f thus assumes the value 1 either twice or not at all according as m is less or greater than 1.

In sum,

$$\alpha > 0 \quad \text{and} \quad \frac{e \ln a}{\alpha} < 1 \quad \text{two solutions,}$$

$$\alpha > 0 \quad \text{and} \quad \frac{e \ln a}{\alpha} > 1 \quad \text{no solution,}$$

$$\alpha > 0 \quad \text{and} \quad e \ln a = \alpha \quad \text{one double solution,}$$

$$\alpha \leqslant 0 \qquad\qquad\qquad\qquad \text{one solution.}$$

Second question:

The result is due to the increase in the exponential function a^x, where $a > 1$:

$$a^x = x \Rightarrow a^{(a^x)} = x,$$
$$a^x > x \Rightarrow a^{(a^x)} > a^x > x,$$
$$a^x < x \Rightarrow a^{(a^x)} < a^x < x.$$

Therefore, for $a^{(a^x)}$ to be equal to x, it is necessary (and sufficient) that $a^x = x$. Since the roots x of this equation are positive numbers, we can use the results of (1). The given equation then has

$$\begin{array}{lll} \text{2 solutions} & \text{if} & e \ln a < 1, \\ \text{0 solution} & \text{if} & e \ln a > 1. \end{array}$$

432

We know that $f(x) = e^{x \ln x}$.

Thus, the function f is defined, continuous, and differentiable for $x > 0$, and

$$f'(x) = e^{x \ln x} (1 + \ln x).$$

This derivative has the same sign as does $1 + \ln x$ and hence the same as does $x - 1/e$:

$$x \to 0 \qquad \Rightarrow x \ln x \to 0 \qquad \Rightarrow f(x) \to 1 \quad ,$$
$$x \to +\infty \Rightarrow x \ln x \to +\infty \Rightarrow f(x) \to +\infty.$$

Regarding the tangent at the point $(0, 1)$, note that, as $x \to 0$,

$$\frac{f(x)-1}{x} = \frac{e^{x \ln x}-1}{x} \simeq \frac{x \ln x}{x} \to -\infty$$

or

$$f'(x) = (1+ \ln x)f(x) \to -\infty.$$

Thus, the y-axis is tangent to the graph.

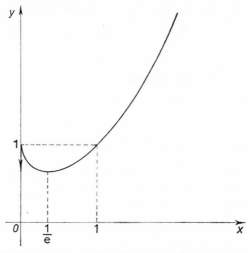

Graph of f

Regarding the behavior of the function as $x \to +\infty$, note that

$$\frac{f(x)}{x} = e^{(x-1) \ln x} \to +\infty.$$

Thus, at this end, the graph has no asymptote but it becomes more and more nearly vertical.

433

First question:

The system (S) is equivalent to the system (1) obtained by taking the logarithms of the numbers that appear in (S):

$$\begin{cases}(x+y) \ln x = n \ln y \\ (x+y) \ln y = m \ln x.\end{cases}$$

If either x or y is equal to 1, so is the other (trivial solution).

If neither x nor y is equal to 1, the two members of the first equation are nonzero, and we obtain a system equivalent to the system (1) when we replace the second equation with the equation obtained when we multiply the first of equations (1) by the second:

(2)
$$\begin{cases} (x+y) \ln x = n \ln y \\ (x+y)^2 \ln x \ln y = mn \ln x \ln y. \end{cases}$$

Then, if we divide both members of the second equation by the (nonzero) product $\ln x \ln y$, we have

(3)
$$\begin{cases} (x+y) \ln x = n \ln y \\ (x+y)^2 = mn. \end{cases}$$

For this system to have any solutions, it is necessary that mn be positive, as we shall assume to be the case. Since $x+y>0$, the system (3) is then equivalent to the system

(4)
$$\begin{cases} x+y = \sqrt{mn} \\ \ln y = \dfrac{m}{\sqrt{mn}} \ln x = \dfrac{m}{|m|} \sqrt{\dfrac{m}{n}} \ln x. \end{cases}$$

Second question:

Consider first the case in which $m>0$ (and hence $n>0$). Let us set $\sqrt{mn}=p$ and $\sqrt{m/n}=q$. The system (4) then becomes

(4) $\begin{cases} \ln y = q \ln x \\ x+y = p. \end{cases}$ $\Leftrightarrow$ (5) $\begin{cases} y = x^q \\ x+x^q = p. \end{cases}$

Define the function $\varphi(x)=x+x^q$. This function increases monotonically from 0 to $+\infty$. Thus, it assumes the value p exactly once, and the system (5) has a unique solution. This solution coincides with the trivial solution when $p=2$ (that is, when $mn=4$).

Third question:

Let us suppose now that $m<0$ (and hence $n<0$). In the same notations as in (2), the system (4) becomes

(4) $\begin{cases} \ln y = -q \ln x \\ x-y = p. \end{cases}$ $\Leftrightarrow$ (5) $\begin{cases} y = x^{-q} \\ x+x^{-q} = p. \end{cases}$

To find the number of solutions of the second equation, let us study the behavior of the function ψ defined by

$$\psi(x) = x + x^{-q}.$$

This function is defined, continuous, and differentiable for $x > 0$, and

$$\psi'(x) = 1 - qx^{-(q+1)}.$$

The derivative ψ' vanishes for $x_0 = q^{1/(q+1)}$ and has the same sign as does $x - x_0$. If we set $p_0 = \psi(x_0) = q^{1/(q+1)} + q^{-q/(q+1)}$, the behavior of ψ is shown by the table

	0	x_0	$+\infty$
ψ	$+\infty$		$+\infty$
		p_0	

If $p < p_0$, the system (5) has no solution.

If $p > p_0$, the system (5) has two solutions, one of them being the trivial solution when $p = 2$, that is, when $mn = 4$. If $p = p_0$, the system (5) has one solution, which we may regard as a double solution.

434

First question:

Proof by induction on n: For $n = 2$, we obtain the definition of a convex function. Suppose that the result is true for all integers not exceeding $n - 1$. Let us set

$$\sum_{i=1}^{n} p_i x_i = p_1 x_1 + \sum_{i=2}^{n} p_i x_i = p_1 x_1 + (1 - p_1) \sum_{i=2}^{n} \frac{p_i}{1 - p_1} x_i.$$

The sum

$$\sum_{i=2}^{n} \frac{p_i}{1 - p_1}$$

is equal to 1 since

$$\sum_{i=2}^{n} p_i = 1 - p_1.$$

We set

$$y_1 = \sum_{i=2}^{n} \frac{p_i}{1 - p_1} x_i.$$

The definition of a convex function enables us to write

$$f\left(\sum_{i=1}^{n} p_i x_i\right) = f[p_1 x_1 + (1-p_1)y_1] \leqslant p_1 f(x_1) + (1-p_1)f(y_1).$$

The result stated in the problem can be applied to a sum of $n-1$ terms:

$$f(y_1) = f\left(\sum_{i=2}^{n} \frac{p_i}{1-p_1} x_i\right) \leqslant \sum_{i=2}^{n} \frac{p_i}{1-p_1} f(x_i).$$

Finally, combining the two preceding results, we obtain

$$f\left(\sum_{i=1}^{n} p_i x_i\right) \leqslant p_1 f(x_1) + (1-p_1) \sum_{i=2}^{n} \frac{p_i}{1-p_1} f(x_i) \leqslant \sum_{i=1}^{n} p_i f(x_i).$$

Second question:

We know that

$$f'(x) = \alpha x^{\alpha-1}, \qquad f''(x) = \alpha(\alpha-1)x^{\alpha-2}.$$

The power $x^{\alpha-2}$ is nonnegative. The sign f'' is that of the product $\alpha(\alpha-1)$. Therefore, the function f'' is positive if $\alpha > 1$ and the function f is strictly convex (cf. PZ, Book III, Chapter III, part 6, § 4, corollary to theorem 2).

Let us write the formula of (1) and take all the p_i equal. Each of these is then equal to $1/n$ since their sum is

$$\left(\sum_{i=1}^{n} \frac{x_i}{n}\right)^{\alpha} \leqslant \sum_{i=1}^{n} \frac{1}{n} x_i^{\alpha},$$

and, when we multiply both sides by n^{α}, we have

$$\left(\sum_{i=1}^{n} x_i\right)^{\alpha} \leqslant n^{\alpha-1}\left(\sum_{i=1}^{n} x_i^{\alpha}\right).$$

We have used the formula established in the first part. Looking back at the proof of that formula, we see that the inequalities given are strict inequalities for a strictly convex function except in the single case in which the values x_1 and y_1 of the variable are equal.

Therefore, it is necessary that $x_1 = y_1$. If we take $x_1 = \inf x_i$, the condition $x_1 = y_1$ implies $x(i) = x_1$ for all values of i.

Equality will hold if and only if all the x_i are equal.

Third question:

Proof by induction on n: First, for $n=2$, let us study the behavior of the function φ defined for $t>0$ by $\varphi(t) = (x_1+t)^{\alpha} - t^{\alpha}$. This function is continuous and differentiable, and

$$\varphi'(t) = \alpha\,[(x_1+t)^{\alpha-1} - t^{\alpha-1}] > 0 \qquad \text{since} \qquad x_1 \neq 0.$$

Also, it is a strictly increasing function:

$$\varphi(x_2) = (x_1+x_2)^{\alpha} - x_2^{\alpha} > x_1^{\alpha} = \varphi(0).$$

Now, to go from $(n-1)$ numbers x_1 to n numbers, we use the above result:

$$\left[\sum_{i=1}^{n} x_i\right]^{\alpha} = \left[\left(\sum_{i=1}^{n-1} x_i\right)+x_n\right]^{\alpha} > \left(\sum_{i=1}^{n-1} x_i\right)^{\alpha}+x_n^{\alpha}.$$

Since the inequality required is assumed true for $(n-1)$ numbers,

$$\left(\sum_{i=1}^{n-1} x_i\right)^{\alpha} > \sum_{i=1}^{n-1} x_i^{\alpha}.$$

Combining these two inequalities, we obtain the required result.

435

First question:

The inequality in question is equivalent to the inequality

$$f\left(\frac{x}{y}\right) = \alpha\frac{x}{y}+\beta-\left(\frac{x}{y}\right)^{\alpha} \geqslant 0.$$

The function f defined for $u>0$ by $f(u)=\alpha u+\beta-u^{\alpha}$ is continuous and differentiable. Since $0<\alpha=1-\beta<1$, we have

$$f'(u) = \alpha(1-u^{\alpha-1}) \text{ is of the same sign as } u-1,$$
$$f(1) = \alpha+\beta-1 = 0.$$

The behavior of the function is indicated by the table

	0	1	$+\infty$
f		$\searrow 0 \nearrow$	

This table shows that f assumes only nonnegative values and that it assumes the value 0 only at $u=1$. (For the problem in question, we do not seek the limits of f as u tends to 0 or $+\infty$.)

If we assign to u the value x/y, we see that the conditional inequality in question is satisfied and that equality can hold only if $x=y$.

Second question:

The operations mentioned in the statement of the problem yield

$$x_i^\alpha y_i^\beta = \frac{a_i b_i}{a^\alpha b^\beta} \leqslant \alpha \frac{a_i^{1/\alpha}}{a} + \beta \frac{b_i^{1/\beta}}{b},$$

$$\sum_{i=1}^{n} a_i b_i \leqslant a^\alpha b^\beta \left[\frac{\alpha}{a} \left(\sum_{i=1}^{n} a_i^{1/\alpha} \right) + \frac{\beta}{b} \left(\sum_{i=1}^{n} b_i^{1/\beta} \right) \right].$$

If we assign to a and b the values

$$a = \sum_{i=1}^{n} a_i^{1/\alpha}, \qquad b = \sum_{i=1}^{n} b_i^{1/\beta},$$

the bracketed expression becomes $\alpha + \beta = 1$, and we obtain

$$\sum_{i=1}^{n} a_i b_i \leqslant \left(\sum_{i=1}^{n} a_i^{1/\alpha} \right)^\alpha \left(\sum_{i=1}^{n} b_i^{1/\beta} \right)^\beta.$$

436

First question:

In $]x_1, x_2]$, the denominator of φ is never 0, and, being the quotient of non-vanishing continuous functions, it is continuous. Since f is differentiable on the right at x_1, the function φ has a right-hand limit at x_1 equal to $f'(x_1)$. If we set $\varphi(x_1) = f'(x_1)$, the function φ is, by definition, continuous at x_1 and hence in the entire interval $[x_1, x_2]$.

The same is true of ψ if we set $\psi(x_2) = f'(x_2)$.

The continuous function φ assumes the values $\varphi(x_1) = f'(x_1)$, and

$$\varphi(x_2) = \frac{f(x_2) - f(x_1)}{x_2 - x_1} = r.$$

Thus, it assumes every value in the interval $[f'(x_1), r]$ at least once (cf. PZ, Book III, Chapter III, part 2, § 2, theorem 2). Similarly, the function ψ assumes every value in the interval $[r, f'(x_2)]$ at least once since $r = \psi(x_1)$ and $f'(x_2) = \psi(x_2)$.

Any value m between $f'(x_1)$ and $f'(x_2)$ belongs to one or the other of the two intervals $[f'(x_1), r]$ or $[r, f'(x_2)]$. Therefore, it is a value assumed either by the function φ or by the function ψ.

Second question:

By hypothesis, there exist numbers x_1 and x_2 in $[a, b]$ such that $f'(x_1) = \alpha$ and $f'(x_2) = \beta$. In the interval $[x_1, x_2]$, the function f satisfies the hypotheses of (1). Every value m lying between α and β is therefore assumed by one of the functions φ or ψ. Let us suppose that it is assumed by the function φ. Then, there exists a number x in $[x_1, x_2]$ such that

$$m = \frac{f(x) - f(x_1)}{x - x_1}.$$

The function f is continuous and differentiable in the interval $[x_1, x]$. Therefore' there exists, by the theorem of the mean, a number c in $]x_1, x[$ such that

$$m = \frac{f(x) - f(x_1)}{x - x_1} = f'(c).$$

Third question:

If the function f is convex, its derivative f' is nondecreasing. The derivative f' assumes the values $f'(a)$ and $f'(b)$ and hence every value between $f'(a)$ and $f'(b)$, as was shown in (2).

As a consequence of these two properties, the derivative f' is a continuous function. This is the result of exercise 416, to which we refer the reader for proof.

437

First question:

We assume that $h > 0$ (otherwise, we replace h with $-h$). By hypothesis, as $h \to 0$,

$$\frac{f(x_0 + h) - f(x_0)}{h} \to f'(x_0 + 0),$$

$$\frac{f(x_0 - h) - f(x_0)}{-h} \to f'(x_0 - 0).$$

Therefore, by the theorem on the limit of a sum,

$$\frac{f(x_0 + h) - f(x_0 - h)}{2h} = \frac{1}{2}\left[\frac{f(x_0 + h) - f(x_0)}{h} + \frac{f(x_0 - h) - f(x_0)}{-h}\right]$$

$$\to \tfrac{1}{2}[f'(x_0 + 0) + f'(x_0 - 0)].$$

The function f has a symmetric derivative $f_s'(x_0)$ and

$$f_s'(x_0) = \tfrac{1}{2}[f'(x_0+0)+f'(x_0-0)].$$

Second question:

The ratio

$$\frac{f(x)-f(0)}{x} = \sin\frac{1}{x}$$

has no limit as x tends to 0 (either from above or from below). Therefore, the function f has neither a right-hand nor a left-hand derivative at 0.

The function f is an even function. Hence, it has a symmetric derivative equal to 0 at 0:

$$\frac{f(0+h)-f(0-h)}{2h} = 0 \;\Rightarrow\; f_s'(0) \text{ exists and } f_s'(0) = 0.$$

Third question:

Let us take $h>0$. Then, by hypothesis,

$$f(x_0+h)-f(x_0-h) \geqslant 0,$$

$$f_s'(x_0) = \lim_{h\to 0}\frac{f(x_0+h)-f(x_0-h)}{2h} \geqslant 0.$$

Fourth question:

This study is analogous to that made regarding the derivative in the usual sense of the term.

$$\frac{(f+g)(x_0+h)-(f+g)(x_0-h)}{2h}$$

$$= \frac{f(x_0+h)+g(x_0+h)-f(x_0-h)-g(x_0-h)}{2h}$$

$$= \frac{f(x_0+h)-f(x_0-h)}{2h}+\frac{g(x_0+h)-g(x_0-h)}{2h}.$$

By hypothesis, as h tends to 0, each of the ratios on the right has a limit. Therefore, the sum has a limit, and the function $f+g$ has a symmetric derivative at x_0. The value of this limit is

$$(f+g)_s'(x_0) = f_s'(x_0)+g_s'(x_0).$$

Note that the continuity of f or g is not needed for this proof.

For the symmetric derivative of the product, let us use the identity

$$(fg)(x_0+h)-(fg)(x_0-h)$$
$$= f(x_0+h)g(x_0+h)-f(x_0-h)g(x_0-h)$$
$$= f(x_0+)[g(x_0+h)-g(x_0-h)]+g(x_0-h)[f(x_0+h)-f(x_0-h)].$$

Then, dividing both sides by $2h$, we get

$$\frac{(fg)(x_0+h)-(fg)(x_0-h)}{2h}$$

$$= f(x_0+h)\frac{g(x_0+h)-g(x_0-h)}{2h}+g(x_0-h)\frac{f(x_0+h)-f(x_0-h)}{2h}.$$

As h tends to 0, the following statements hold:

$$f(x_0+h) \to f(x_0) \quad \text{and} \quad g(x_0-h) \to g(x_0)$$

(because of the continuity of f and g); the two ratios approach $g'_s(x_0)$ and $f'_s(x_0)$ respectively; the left-hand member therefore approaches a limit; that is, the product function fg has a symmetric derivative, the value of which is given by

$$(fg)'_s(x_0) = f(x_0)g'_s(x_0) + g(x_0)f'_s(x_0).$$

438

First question:

The set E_λ is not empty (since x_1 belongs to it) and is bounded above by x_2. Therefore, it has a least upper bound ξ. By hypothesis, $\varphi(x_1) > \lambda$. Since φ is continuous at x_1, there exists an interval $[x_1, x_1 + \eta[$ throughout which $\varphi(x) > \lambda$. This interval is contained in E_λ, so that $\xi \geqslant x_1 + \eta > x_1$.

Similarly, $\varphi(x_2) < \lambda$, and φ is continuous at x_2. Therefore, there exists an interval $]x_2 - \eta', x_2]$ in which $\varphi(x) < \lambda$. This implies that $\xi \leqslant x_2 - \eta' < x_2$.

Let $]\alpha, \beta[$ be an interval containing ξ (that is, let α and β be two numbers such that $\alpha < \xi < \beta$). From the definition of a least upper bound, there is at least one number u in E_λ such that $\alpha < u \leqslant \xi$. Furthermore, all the numbers $v \geqslant \xi$ lie outside E_λ and hence verify the inequality $\varphi(v) \leqslant \lambda$.

Suppose that $\varphi(\xi)$ is not equal to λ. Then, there is an interval $]\alpha, \beta[$ containing ξ such that

$$|\varphi(x) - \varphi(\xi)| < |\varphi(\xi) - \lambda|$$

throughout that interval (because of the continuity of φ at ξ). Then, the difference $\varphi(x) - \lambda$ has the same sign as $\varphi(\xi) - \lambda$, and the function φ does not assume values less or greater than λ in $]\alpha, \beta[$. This is a contradiction of the above result.

The number ξ does not belong to E_λ because $\varphi(\xi) = \lambda$. Then, there exists an infinite sequence $\{u_i\}$ of numbers in E_λ that converges to ξ. Let us set $u_i = \xi - h_i$. The numbers $v_i = \xi + h_i$ are greater than ξ and do not belong to E_λ.

$$\varphi(u_i) = \varphi(\xi - h_i) > \lambda, \qquad \varphi(v_i) = \varphi(\xi + h_i) \leqslant \lambda.$$

By hypothesis, the function φ has a symmetric derivative at ξ, and the definition of the limit of a function in terms of convergent sequences implies that

$$\varphi'_s(\xi) = \lim \frac{\varphi(\xi + h_i) - \varphi(\xi - h_i)}{2h_i} \leqslant 0.$$

Second question:

If f is not a nondecreasing function, there exist numbers x_1 and x_2 in $[a, b]$ such that

$$x_1 < x_2 \qquad \text{but} \qquad f(x_1) > f(x_2).$$

The restriction φ of f to the interval $[x_1, x_2]$ satisfies the hypotheses of (1). Therefore, there exists a number ξ in $]x_1, x_2[$ and hence in $]a, b[$ such that

$$\varphi'_s(\xi) = f'_s(\xi) \leqslant 0,$$

contrary to the hypothesis.

Third question:

The function $kx - f$ has a positive symmetric derivative $k - f'_s$ in $]a, b[$. To see this, note that the symmetric derivative of kx is equal to its derivative (see exercise 437, (1)) and that the difference between two functions having symmetric derivatives has a symmetric derivative equal to the difference between the original two symmetric derivatives. (This is the result of 437, (4) slightly modified.)

The hypotheses of (2) are satisfied for the function $kx - f$. Therefore, this function is nondecreasing in $[a, b]$.

$$x_1 < x_2 \Rightarrow kx_1 - f(x_1) \leqslant kx_2 - f(x_2),$$
$$x_1 < x_2 \Rightarrow f(x_2) - f(x_1) \leqslant k(x_2 - x_1).$$

If the symmetric derivative f'_s has a least upper bound M in $]a, b[$, the preceding result can be applied to every number $k > M$. The number $f(x_2) - f(x_1)$ is a lower bound of the set of numbers $k(x_2 - x_1)$ where $k > M$. Therefore, it does not exceed the greatest lower bound of that set $M(x_2 - x_1)$:

$$f(x_2) - f(x_1) \leqslant M(x_2 - x_1).$$

Analogous reasoning gives us a greatest lower bound $m(x_2 - x_1)$. We consider the functions $f(x) - hx$, where h is an arbitrary number less than m.

If $f'_s(x) \geqslant 0$, the greatest lower bound m of f'_s is at least equal to zero, and the inequality obtained shows that

$$x_1 < x_2 \Rightarrow f(x_1) \leqslant f(x_2).$$

Thus, the function f is nondecreasing.

If f is not a strictly increasing function, there exist at least two numbers x_0 and x'_0 such that

$$x_0 < x'_0 \qquad \text{and} \qquad f(x_0) = f(x'_0).$$

Since the function f is nondecreasing, for every number x in $[x_0, x'_0]$,

$$x_0 \leqslant x \leqslant x'_0 \Rightarrow f(x_0) \leqslant f(x) \leqslant f(x'_0) = f(x_0).$$

Thus the function f is constant in the interval $[x_0, x'_0]$. In this interval, its derivative and symmetric derivative both exist and are equal to 0.

439

First question:

In the interval $]0, \tfrac{1}{2}\pi]$, the function f is the quotient of two continuous and differentiable functions. As the denominator, x, does not vanish in this interval, f too is continuous and differentiable, and

$$f'(x) = \frac{x \cos x - \sin x}{x^2}.$$

For the value 0, we obtain our result immediately if we write the Maclaurin expansion according to Taylor's formula with a remainder for the sine in the form

$$\sin x = x + o(x^2) \Rightarrow f(x) = 1 + o(x).$$

(The remainder, which may be written $\varphi_n(x)x^n/n!$ [cf. PZ, Book III, Chapter III, part 6, § 6] is by definition $o(x^n)$ since $\varphi_n(x)$ tends to 0 as x tends to 0.)

When $x \to 0$,

$$f(x) \to 1 \qquad \text{and} \qquad \frac{f(x) - 1}{x} = \frac{o(x)}{x} \to 0.$$

The function f is continuous at 0 and has a derivative equal to 0 at that point. (We can also find the derivative by seeking the limit of the derivative $f'(x)$, using Taylor's formula for the sine and cosine.)

Second question:

The derivative $f'(x)$ has the same sign as its numerator:

$$x \cos x - \sin x = \cos x(x - \operatorname{tg} x).$$

Therefore, it is negative in the interval $]0, \tfrac{1}{2}\pi]$, and the greatest lower and least upper bounds of the decreasing function f are the values of f at the end-points of the interval:

$$1 \geqslant \frac{\sin x}{x} \geqslant \frac{2}{\pi}.$$

440

First question:

The Maclaurin expansion, via the first two nonzero terms with remainder according to Taylor's formula, yields

$$\arctan x - \sin x = [x - \frac{x^3}{3} + o(x^3)] - [x - \frac{x^3}{6} + o(x^3)] = -\frac{x^3}{6} + o(x^3),$$

$$\operatorname{tg} x - \arcsin x = [x + \frac{x^3}{3} + o(x^3)] - [x + \frac{x^3}{6} + o(x^3)] = \frac{x^3}{6} + o(x^3).$$

As x approaches 0,

$$\frac{\arctan x - \sin x}{\operatorname{tg} x - \arcsin x} = \frac{-x^3/6 + o(x^3)}{x^3/6 + o(x^3)} = \frac{-\frac{1}{6} + o(1)}{\frac{1}{6} + o(1)} \to -1.$$

Second question:

The denominator, x, in the expression for f is of order 1. Therefore, to obtain the limit of f, we need determine only the first term in the expansion of $\ln[(e^x-1)/x]$. To get it, we need the first three terms of the expansion

$$\frac{e^x - 1}{x} = 1 + \frac{x}{2} + \frac{x^2}{6} + o(x^2)$$

(obtained from the expansion of e^x). The expansion of $\ln[(e^x-1)/x]$ is obtained by applying the result dealing with the composition of expansions (cf. PZ, Book IV, Chapter I, § 5):

$$\ln 1 + u = u + \frac{u^2}{2} + o(u^2),$$

$$\ln\left\{1 + \left[\frac{x}{2} + \frac{x^2}{6} + o(x^2)\right]\right\} = \frac{x}{2} + \frac{x^2}{6} - \frac{1}{2}\left(\frac{x}{2} + \frac{x^2}{6}\right)^2 + o(x^2) = \frac{x}{2} + \frac{x^2}{24} + o(x^2).$$

Therefore, as x tends to 0,

$$f(x) = \frac{1}{2} + \frac{x}{24} + o(x) \to \frac{1}{2}$$

and

$$\frac{1}{x}\left(f(x) - \frac{1}{2}\right) = \frac{1}{24} + o(1) \to \frac{1}{24}.$$

Third question:

The two functions $\cos x$ and $(1+ax^2)/(1+bx^2)$ have the same value at $x=0$. Obviously, by a suitable choice of a and b, we can make the second- and fourth-degree terms in the expansions of these functions equal, so that these terms disappear from the difference. Therefore, let us determine the first four nonzero terms in the expansions of these two functions:

$$\cos x = 1 - \frac{x^2}{2} + \frac{x^4}{24} - \frac{x^6}{720} + o(x^6).$$

If every term of a polynomial is of degree exceeding n, the polynomial is $o(x^n)$ as $x \to 0$. Division with powers arranged in increasing order therefore enables us to write

$$\frac{1+ax^2}{1+bx^2} = 1 + (a-b)x^2 - b(a-b)x^4 + b^2(a-b)x^6 + o(x^6),$$

$$g(x) = -[\tfrac{1}{2} + (a-b)]x^2 + [\tfrac{1}{24} + b(a-b)]x^4 - [\tfrac{1}{720} + b^2(a-b)]x^6 + o(x^6).$$

Let us take

$$a - b = -\tfrac{1}{2}, \qquad b(a-b) = -\tfrac{1}{24},$$

that is

$$b = \tfrac{1}{12}, \qquad a = -\tfrac{5}{12}.$$

Then,

$$g(x) = \frac{3x^6}{1440} + o(x^6).$$

Fourth question:

We set $x = 1/t$ to get a simpler case in which the variable t tends to 0:

$$H(t) = h\left(\frac{1}{t}\right) = \frac{1}{t}\left[\frac{1}{e} - (1+t)^{-1/t}\right].$$

Since the power of the denominator t of the first factor is 1, let us find the constant and first-degree term in the expansion for the bracketed expression:

$$(1+t)^{-1/t} = e^{-(1/t)\ln(1+t)} = e^{-1+t/2+o(t)} = \frac{1}{e} + \frac{1}{e}\frac{t}{2} + o(t),$$

$$H(t) = -\frac{1}{2e} + o(1) \rightarrow -\frac{1}{2e} \quad \text{as} \quad t \rightarrow 0.$$

441

First question:

From the hypotheses,

$$g = f + o(f) = f[1+o(1)],$$
$$\ln g = \ln f + \ln[1+o(1)] = \ln f + o(1).$$

Since f approaches 0 or $+\infty$, the function $|\ln f|$ approaches $+\infty$ under the conditions mentioned. If a function φ is $o(1)$, it is *a fortiori* $o(\ln f)$:

$$|\varphi(x)| \leqslant \varepsilon \Rightarrow |\varphi(x)| \leqslant \varepsilon|\ln f|.$$

Therefore,

$$\ln g = \ln f + o(\ln f),$$

which is the definition of two equivalent functions.

Second question:

As x tends to $+\infty$,

$$e^x - 1 \sim e^x \Rightarrow \ln(e^x-1) \simeq \ln e^x = x.$$

Let us apply this result to

$$f(x) = \frac{1}{x}[\ln(e^x-1)-\ln x] = \frac{\ln(e^x-1)}{x} - \frac{\ln x}{x}.$$

As x tends to $+\infty$,

$$\frac{\ln(e^x-1)}{x} \simeq \frac{\ln e^x}{x} = 1 \Rightarrow \frac{\ln(e^x-1)}{x} \rightarrow 1,$$

$$\frac{\ln x}{x} \rightarrow 0$$

(owing to the relative speed of increase of ln x and x^α [for any α])

$$f(x) \to 1.$$

442

Here, we have two applications of Taylor's formula with the remainder expressed in terms of the third-degree term for ln $(1+x)$ and $(1+x)^{\frac{3}{2}}$:

$$\ln (1+x) = x - \frac{x^2}{2} + \frac{x^3}{3} \frac{1}{(1+c)^3}, \qquad (0 < c < x),$$

$$(1+x)^{\frac{3}{2}} = 1 + \frac{3}{2}x + \frac{3}{8}x^2 - \frac{x^3}{16}(1+c)^{-\frac{3}{2}}, \qquad (0 < c < x).$$

Since c is a positive number,

$$1 + c > 1 \Rightarrow 0 < \frac{1}{(1+c)^3} < 1 \qquad \text{and} \qquad 0 < (1+c)^{-\frac{3}{2}} < 1.$$

From this, we conclude

$$0 < \frac{x^3}{3}(1+c)^3 = \ln (1+x) - x + \frac{x^2}{2} < \frac{x^3}{3},$$

$$0 > -\frac{x^3}{16}(1+c)^{-\frac{3}{2}} = (1+x)^{\frac{3}{2}} - (1 + \frac{3}{2}x + \frac{3}{8}x^2) > -\frac{x^3}{16}.$$

443

Let us define

$$v_n = \ln u_n = \sum_{p=1}^{n} \ln \left(1 + \frac{p}{n^2}\right).$$

Taylor's formula applied to the function ln $(1+x)$, involving the first- and second-degree terms, enables us to show, as in exercise 432, that, if $x > 0$,

$$x - \frac{x^2}{2} < \ln 1 + x < x.$$

When we write these inequalities for each of the terms in the sequence $\{v_n\}$, we have

$$\frac{p}{n^2} - \frac{p^2}{n^4} < \ln \left(1 + \frac{p}{n^2}\right) < \frac{p}{n^2},$$

$$\sum_{p=1}^{n} \frac{p}{n^2} - \sum_{p=1}^{n} \frac{p^2}{n^4} < \sum_{p=1}^{n} \ln \left(1 + \frac{p}{n^2}\right) < \sum_{p=1}^{n} \frac{p}{n^2}.$$

We know that

$$\sum_{p=1}^{n} p = \frac{n(n+1)}{2} \;\Rightarrow\; \sum_{p=1}^{n} \frac{p}{n^2} = \frac{n+1}{2n} \rightarrow \frac{1}{2},$$

$$\sum_{p=1}^{n} p^2 < n \times n^2 \;\Rightarrow\; \sum_{p=1}^{n} \frac{p^2}{n^4} < \frac{n^3}{n^4} \rightarrow 0.$$

Thus, the sequence $\{v_n\}$ being bounded below and above by sequences that converge to $\frac{1}{2}$, must itself converge to $\frac{1}{2}$. Therefore, the sequence $\{u_n\}=\{e^{v_n}\}$ converges to $e^{\frac{1}{2}}=\sqrt{e}$.

444

First question:

Since the functions g and h are composites of continuous and infinitely differentiable functions, they are themselves continuous and infinitely differentiable. Their first two derivatives are

$$g'(x) = -\frac{\sin\sqrt{x}}{2\sqrt{x}}, \qquad g''(x) = \frac{\sin\sqrt{x}-\sqrt{x}\cos\sqrt{x}}{4x\sqrt{x}},$$

$$h'(x) = -\frac{\sinh\sqrt{-x}}{2\sqrt{-x}}, \qquad h''(x) = \frac{\sinh\sqrt{-x}-\sqrt{-x}\cosh\sqrt{-x}}{4x\sqrt{-x}}.$$

Second question:

For $x>0$, the function f coincides with the function g. Therefore, it is con tinuous and twice differentiable when g is, and we can verify immediately that

$$4xf''(x)+2f'(x)+f(x) = 4xg''(x)+2g'(x)+g(x) = 0.$$

The same remark holds if $x<0$ since f coincides with h.

Therefore, we need determine only the continuity and differentiability of f at 0, by considering right- and left-hand limits separately.

The continuity of f is immediate:

$$f(+0) = g(+0) = 1, \qquad f(-0) = h(-0) = 1.$$

When we expand $\sin\sqrt{x}$ and $\sinh\sqrt{-x}$ in powers of $|x|^{\frac{1}{2}}$, we obtain

$$\sin(|x|^{\frac{1}{2}}) = |x|^{\frac{1}{2}}-\tfrac{1}{6}|x|^{\frac{3}{2}}+o(|x|^{\frac{3}{2}}) \;\Rightarrow\; g'(x) = -\tfrac{1}{2}+\frac{|x|}{12}+o(|x|),$$

$$\sinh(|x|^{\frac{1}{2}}) = |x|^{\frac{1}{2}}+\tfrac{1}{6}|x|^{\frac{3}{2}}+o(|x|^{\frac{3}{2}}) \;\Rightarrow\; h'(x) = -\tfrac{1}{2}-\frac{|x|}{12}+o(|x|).$$

As x approaches 0 from above, $f'(x)=g'(x)$ approaches $-\frac{1}{2}$. Therefore, the function f has a right-hand derivative $f'(+0)$ equal to $-\frac{1}{2}$ (see exercise 427).

Similarly, $f'(x) = h'(x)$ approaches $-\frac{1}{2}$ as x approaches 0 from below. Thus, the function f has a left-hand derivative $f'(-0) = -\frac{1}{2}$. Since the right- and left-hand derivatives are equal, the function f is differentiable at 0, and $f'(0) = -\frac{1}{2}$.

As x approaches 0 from above,

$$\frac{f'(x)+\frac{1}{2}}{x} = \frac{g'(x)+\frac{1}{2}}{x} = \left[\frac{1}{12}\frac{|x|}{x}+\frac{o(|x|)}{x}\right] \to \frac{1}{12},$$

and as x approaches 0 from below,

$$\frac{f'(x)+\frac{1}{2}}{x} = \frac{h'(x)+\frac{1}{2}}{x} = \left[-\frac{1}{12}\frac{|x|}{x}+\frac{o(|x|)}{x}\right] \to \frac{1}{12},$$

which proves that the derivative f' is itself differentiable at $x=0$, and its derivative is $f''(0)=\frac{1}{12}$. (We could have shown also that g'' and h'' approach limits at 0.)

Finally, at $x=0$,

$$4xf''(x)+2f'(x)+f(x) = 0+2(-\frac{1}{2})+1 = 0.$$

445

First question:

The two problematic points are 0 and 1. As x approaches 0 from above,

$$x \ln x \to 0 \;\Rightarrow\; f(x) \to 0 \;\Rightarrow\; f \text{ is continuous at 0 if } f(0) = 0,$$

and

$$\frac{f(x)-f(0)}{x} = \frac{\ln x}{x^2-1} \to +\infty \;\Rightarrow\; f$$

is differentiable at 0 and $f'(+0)= +\infty$.

As x approaches 1 (from either above or below), the function $u=x-1$ approaches 0 and

$$f(x) = f(1+u) = \frac{1+u}{2+u}\times\frac{\ln(1+u)}{u} = \left[\frac{1}{2}+\frac{u}{4}+o(u)\right]\left[1-\frac{u}{2}+o(u)\right]$$

$$= \tfrac{1}{2}+o(u) = \tfrac{1}{2}+o[(x-1)].$$

Thus, $f(x)$ approaches $\frac{1}{2}$ as x approaches 1 and the function f will be continuous if we define $f(1) = \frac{1}{2}$. Furthermore,

$$\frac{f(x) - f(1)}{x - 1} = \frac{f(x) - \frac{1}{2}}{x - 1} = \frac{o(x - 1)}{x - 1}.$$

The function f is differentiable, and its derivative vanishes at $x = 1$.

For values of x in the open intervals $]0, 1[$ and $]1, +\infty[$, the function f is continuous and differentiable since it is the quotient of continuous differentiable functions and the function representing the denominator does not vanish:

$$f'(x) = \frac{(1 + \ln x)(x^2 - 1) - 2x^2 \ln x}{(x^2 - 1)^2} = \frac{x^2 - 1 - (x^2 + 1) \ln x}{(x^2 - 1)^2}.$$

Second question:

To determine the sign of the derivative, let us replace the numerator with an expression proportional to it in which the logarithm is isolated (since the derivative of such an expression is rational and hence easy to study):

$$x^2 - 1 - (x^2 + 1) \ln x = (x^2 + 1) \left[\frac{x^2 - 1}{x^2 + 1} - \ln x \right] = (x^2 + 1) g(x).$$

The derivative has the same sign as its numerator, hence the sign of $g(x)$. Therefore, examination of the behavior of g will yield the desired result. The function g is continuous and differentiable for $x \geq 0$, and

$$g'(x) = \frac{4x}{(x^2 + 1)^2} - \frac{1}{x} = \frac{4x^2 - (x^2 + 1)^2}{x(x^2 + 1)^2} = \frac{-(x^2 - 1)^2}{x(x^2 + 1)^2} \leq 0.$$

Thus, g is nonincreasing, and $g(1) = 0$; also $g(x)$ has the same sign as does $(1 - x)$.

We know now the behavior of f and we need find only its limit as x approaches $+\infty$:

$$f(x) = \frac{x \ln x}{x^2 - 1} \simeq \frac{x \ln x}{x^2} \to 0.$$

Thus, we have the table

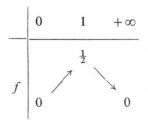

For the graph of f, we already know its behavior at these three points. Specifically, at the origin, it is tangent to the y-axis; at $x=1$, its tangent is horizontal; as $x \to \infty$, the graph approaches the x-axis as an asymptote.

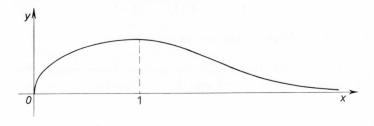

446

First question:

If $x \neq 0$, the classical theorems on composite functions and on the quotient of two functions enable us to assert that f is continuous and differentiable and that

$$f'(x) = \frac{1 + e^{1/x}(1 + 1/x)}{(1 + e^{1/x})^2}.$$

The inequality $|f(x)| < |x|$ implies the continuity of f at $x=0$,

$$x \to 0 \text{ from above} \Rightarrow \frac{1}{x} \to +\infty \Rightarrow e^{1/x} \to +\infty \Rightarrow \frac{f(x)}{x} \to 0.$$

Since $f(0)=0$, the limit of this last fraction is the right-hand derivative of $f(x)$ at $x=0$. Thus, f is differentiable on the right at the origin and $f'(+0)=0$. Similarly,

$$x \to 0 \text{ from the left} \Rightarrow \frac{1}{x} \to -\infty \Rightarrow e^{1/x} \to 0 \Rightarrow \frac{f(x)}{x} \to 1,$$

so that f is differentiable on the left at 0 and $f'(-0)=1$.

Second question:

The sign of the derivative $f'(x)$ is the sign of the numerator:

$$N(x) = 1 + e^{1/x}\left(1 + \frac{1}{x}\right).$$

The function N is continuous and differentiable except at $x=0$, and

$$N'(x) = -\frac{1}{x^2} e^{1/x} \frac{2x+1}{x}.$$

The function $N'(x)$ has the same sign as does the function $-x(2x+1)$ and from it we find the behavior of N:

	$-\infty$	$-\frac{1}{2}$	0	$+\infty$
N		$\searrow 1-e^{-2}\nearrow$	$\vert$	$\searrow 2$

In the two intervals $]-\infty, 0[$ and $]0, +\infty[$, in which N is continuous, its minima $1-e^{-2}$ and 2 are both positive. Therefore, $N(x)$ is positive for every x.

The function f is continuous in the intervals $]-\infty, 0]$ and $[0, +\infty[$ and differentiable, with positive derivative f', in the interiors of these intervals. Therefore, the function is increasing in the two intervals $]-\infty, 0]$ and $]0, +\infty]$ and hence over the entire real axis (these two intervals having a common point at 0).

Third question:

As x tends to $\pm\infty$, $1/x$ tends to 0 and

$$1+e^{1/x} = 2+\frac{1}{x}+\frac{1}{2x^2}+\frac{1}{6x^3}+o\left(\frac{1}{x^3}\right) = 2\left[1+\frac{1}{2x}+\frac{1}{4x^2}+\frac{1}{12x^3}+o\left(\frac{1}{x^3}\right)\right].$$

From this we get the expansion of the reciprocal (cf. PZ, Book IV, Chapter I,§ 6):

$$\frac{1}{1+e^{1/x}} = \frac{1}{2}\left[1-\left(\frac{1}{2x}+\frac{1}{4x^2}+\frac{1}{12x^3}\right)\right.$$

$$\left.+\left(\frac{1}{2x}+\frac{1}{4x^2}+\frac{1}{12x^3}\right)^2-\left(\frac{1}{2x}+\frac{1}{4x^2}+\frac{1}{12x^3}\right)^3+o\left(\frac{1}{x^3}\right)\right]$$

$$= \frac{1}{2}\left[1-\frac{1}{2x}+\frac{1}{24x^3}+o\left(\frac{1}{x^3}\right)\right] = \frac{1}{2}-\frac{1}{4x}+\frac{1}{48x^3}+o\left(\frac{1}{x^3}\right).$$

(We can also divide by the polynomial $2+u+u^2/2+u^3/6$ and arrange the powers in ascending order.)

Thus, the expansion of f is

$$f(x) = \frac{x}{2}-\frac{1}{4}+\frac{1}{48x^2}+o\left(\frac{1}{x^2}\right).$$

The line representing the equation $y=x/2-\frac{1}{4}$ is an asymptote of the graph of f, and it lies below the graph since the difference

$$f(x)-y = \frac{1}{x^2}\left[\frac{1}{48}+o(1)\right]$$

is positive for sufficiently small values of $1/x$.

At the origin, the values of the right-hand and left-hand derivatives, namely, $f'(+0)=0$ and $f'(-0)=1$, give the slopes of the half-tangents of the graph.

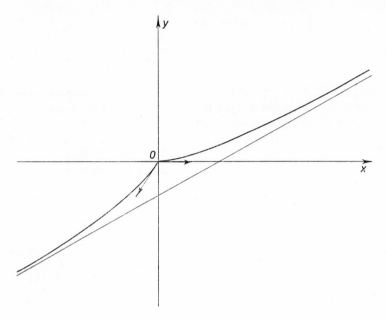

447

First question:

In the Maclaurin expansion, only the remainder term is nonzero:

$$|f(x)| = \left| \frac{x^n}{n!} f^{(n)}(c) \right| \leqslant \frac{|x|^n}{n!} n! \, k^n = (k|x|)^n.$$

If $k|x| < 1$, that is, if $|x| < 1/k$, then the sequence $\{(k\,|x|)^n\}$ converges to 0 and hence $f(x)=0$. The continuity of the function f enables us to extend this result to $x = \pm 1/k$. Therefore,

$$-\frac{1}{k} \leqslant x \leqslant \frac{1}{k} \Rightarrow f(x) = 0 \Rightarrow f^{(n)}(x) = 0.$$

If $[a, b]$ is contained in $[-1/k, 1/k]$, the proof is complete.

Otherwise, we write Taylor's formula beginning with the value $1/k$. The above reasoning then shows that the function f and its derivatives vanish in the interval $[0, 2/k]$. Proceeding successively in this way with $2/k$, $3/k$, ..., or with $-1/k$, $-2/k$, ..., we see that we eventually obtain all the numbers in $[a, b]$.

Second question:

The function is a composite of the function $-1/x^2$ and the exponential function. Therefore, it is differentiable except perhaps at $x=0$, and

$$g'(x) = \frac{2}{x^3}e^{-1/x^2} = G_1\left(\frac{1}{x}\right)e^{-1/x^2}.$$

Let us show by induction on n that g has an nth derivative of the form indicated. The result is established for $n=1$, and we assume it true for all integral values not exceeding n; in particular,

$$g^{(n)}(x) = G_n\left(\frac{1}{x}\right)e^{-1/x^2}.$$

The function g^n is still differentiable if $x\neq0$, and its derivative is

$$g^{(n+1)}(x) = \left[-\frac{1}{x^2}G'_n\left(\frac{1}{x}\right)+\frac{2}{x^3}G_n\left(\frac{1}{x}\right)\right]e^{-1/x^2} = G_{n+1}\left(\frac{1}{x}\right)e^{-1/x^2},$$

where G'_n is the polynomial representing the derivative of G_n. Thus, G_{n+1} is indeed a polynomial and we see that its degree is $3(n+1)$.

As x tends to 0, the ratio $1/x^2$ tends to $+\infty$. Therefore, for any exponent k, the product

$$\frac{1}{x^k}e^{-1/x^2}$$

approaches 0 as x approaches $+\infty$. (If we set $1/x^2=u$, this last assertion amounts to saying that $u^{k/2}e^{-u}$ approaches 0 as u approaches $+\infty$.) Thus, the product $G_n(1/x)e^{-1/x^2}$ also approaches 0.

We now show by induction on n that $g^{(n)}(0)$ exists and is equal to 0. Let us suppose that this is true for 1, 2, ..., $n-1$ and let us form the ratio

$$\frac{g^{(n-1)}(x)-g^{(n-1)}(0)}{x} = \frac{g^{(n-1)}(x)}{x}.$$

This ratio is the sum of terms of the form

$$\frac{1}{x^k}e^{-1/x^2},$$

and it approaches 0 as x approaches 0. Thus, the function $g^{(n-1)}$ is differentiable at $x=0$, and its derivative now is 0.

(We can also use the result of exercise 427. The function $g^{(n-1)}$ is continuous at 0 since $g^{(n-1)}(x)$ approaches $0=g^{(n-1)}(0)$, and its derivative $g^{(n)}(x)$ approaches 0 as x approaches 0.) Thus, the function g is infinitely differentiable at all real values. Its derivatives all vanish at 0 and yet this function is not the zero function.

448

First question:

To apply the results of exercise 410, we need to show that the terms in the sequence all belong to $[a, b]$.

The theorem of the mean enables us to write

$$x_0 - f(a) = f(x_0) - f(a) = (x_0 - a)f'(c), \qquad (a < c < x_0),$$

$$0 \leqslant f'(c) \leqslant 1 \Rightarrow 0 \leqslant x_0 - f(a) \leqslant x_0 - a \Rightarrow a \leqslant f(a) \leqslant x_0.$$

Similarly, we can show that $x_0 \leqslant f(b) \leqslant b$.

Since the function f is nondecreasing, the interval $f([a, b])$ is the interval $[f(a), f(b)]$ and is contained in $[a, b]$. From this we deduce by induction on n that $u_n \in [a, b]$ for every n.

Furthermore, by applying the result of exercise 410, (1), we conclude that the sequence $\{u_n\}$ is monotonic and that its limit is a number in $[a, b]$, this number being a solution of the equation $x = f(x)$. By hypothesis, this equation has only one root x_0 in $[a, b]$. Therefore, the limit of the sequence $\{u_n\}$ is x_0.

Second question:

Let us assume for definiteness that $u_0 < u_1$ (in the opposite case, we need only to change the direction of this and subsequent inequalities).

By hypothesis, u_0 and u_1 belong to $[a, b]$. Therefore, by the theorem of the mean,

$$u_1 - u_2 = f(u_0) - f(u_1) = (u_0 - u_1)f'(c), \qquad (u_0 < c < u_1),$$

$$-1 \leqslant f'(c) \leqslant 0 \Rightarrow 0 \leqslant u_1 - u_2 \leqslant u_1 - u_0 \Rightarrow u_0 \leqslant u_2 \leqslant u_1.$$

The function f is nonincreasing. Therefore, the interval $f([u_0, u_1])$ is the interval $[f(u_0), f(u_1)]$, which is $[u_2, u_1]$, and is contained in $[u_0, u_1]$. Therefore, the function $f \circ f$ is defined, continuous, and differentiable in $[u_0, u_1]$, and

$$(f \circ f)'(x) = f'[f(x)]f'(x).$$

In $[u_0, u_1]$ (and, in particular, for x and $f(x)$), the derivative f' assumes values between -1 and 0. Therefore,

$$0 \leqslant (f \circ f)'(x) \leqslant 1, \qquad \forall\, x \in [u_0, u_1].$$

The function $f \circ f$ satisfies the hypotheses of (1). Therefore, we can apply the results of (1) to the sequences defined by

$$v_0 = u_0, \qquad v_n = (f \circ f)(v_{n-1}); \qquad \text{that is,} \qquad v_n = u_{2n};$$

$$w_0 = u_1, \qquad w_n = (f \circ f)(w_{n-1}); \qquad \text{that is,} \qquad w_n = u_{2n+1}.$$

The terms v_n and w_n belong to $[a, b]$. All the u_n belong to $[a, b]$. The two sequences $\{v_n\}$ and $\{w_n\}$ are monotonic and convergent. Their limits v and w are roots of the equation $x = (f \circ f)(x)$.

The function $\varphi = x - f \circ f$ is nondecreasing since it has a nonnegative derivative. If it is not strictly increasing, it is constant in an interval $[\alpha, \beta]$:

$$\varphi(\alpha) = \varphi(\beta) \quad \text{and} \quad \alpha < x < \beta \Rightarrow \varphi(\alpha) \leqslant \varphi(x) \leqslant \varphi(\beta) = \varphi(\alpha).$$

Its derivative is 0 throughout that interval; hence,

$$f'[f(x)] \times f'(x) = 1.$$

Each of the two factors on the left is less than 1 in absolute value. Their product can be equal to 1 only if each of them is equal to -1.

The function φ is strictly increasing and cannot have more than one zero in the interval $[u_0, u_1]$. Obviously, x_0 is a zero of φ. Therefore, there is no other. The two sequences $\{v_n\}$ and $\{w_n\}$ converge to x_0. Therefore, the sequence $\{u_n\}$ converges to x_0.

Furthermore, if $u_0 < u_1$, the sequence $\{u_{2n}\}$ is nondecreasing since $u_0 < u_2$ and the sequence $\{u_{2n+1}\}$ is nonincreasing. The limit x_0 lies between every pair of consecutive terms. (If $u_0 > u_1$, the result is analogous; in this case, the sequence $\{u_{2n}\}$ is nonincreasing and the sequence $\{u_{2n+1}\}$ is nondecreasing.)

Third question:

Let us suppose that $|f'(x)| \geqslant 1$ in $[\alpha, \beta]$, where $\alpha < x_0 < \beta$. If the sequence $\{u_n\}$ converges to x_0, there exists an integer N such that

$$n > N \Rightarrow \alpha < u_n < \beta,$$

$$u_{n+1} - x_0 = f(u_n) - f(x_0) = (u_n - x_0)f'(c) \Rightarrow |u_{n+1} - x_0| \geqslant |u_n - x_0|.$$

Therefore, the sequence $\{u_n\}$ does not converge to x_0. The hypothesis made is therefore impossible.

(An obvious exception is the case in which $x_0 = u_0 = u_1 = \ldots = u_n = \ldots$.)

Fourth question:

Let us again apply the theorem of the mean:

$$u_n - x_0 = f(u_{n-1}) - f(x_0) = (u_{n-1} - x_0)f'(c) \Rightarrow |u_n - x_0| \leqslant k|u_{n-1} - x_0|.$$

Then, by induction on n,

$$|u_n - x_0| \leqslant k^n |u_0 - x_0| \leqslant k^n(b - a).$$

Fifth question:

We remind the reader that the derivative of the sine function is equal to the cosine function when x is measured in radians but not otherwise.

$$[\sin_{400} x]' = \left[\sin\left(\frac{\pi}{200}x\right)\right]' = \frac{\pi}{200}\cos\left(\frac{\pi}{200}x\right) = \frac{\pi}{200}\cos_{400} x.$$

The function $x - 32\sin_{400} x$ is defined and continuous for every x, and its derivative is

$$1 - 32\frac{\pi}{200}\cos_{400} x > 0.$$

It is strictly increasing and thus assumes the value 80 once and only once:

$$x = 100 \Rightarrow x - 32\sin_{400} x = 100 - 32 < 80,$$

$$x = \frac{400}{3} \Rightarrow x - 32\sin_{400} x = \frac{400}{3} - 32\frac{\sqrt{3}}{2} > 80.$$

The root x_0 of the equation lies between 100 and 400/3. In this interval, the derivative of the function

$$x \to 32\sin_{400} x + 80,$$

which is equal to

$$\frac{32\pi}{200}\cos_{400} x,$$

lies between -1 and 0. The root x_0 is the limit of the inductively defined sequence $\{u_n\}$ and it lies between any two consecutive terms of that sequence. In practice, one would determine not the sequence $\{u_n\}$ but the sequence $\{v_n\}$ defined by $v_n = f(v'_{n-1})$, where v'_{n-1} is an approximate value of v_{n-1}:

$$
\begin{aligned}
v_0 &= 100 \\
v_1 &= 32 + 80 & = 112 \\
v_2 &= 32\sin 112 + 80 & = 111.43^+ \\
v_3 &= 32\sin 111.43 + 80 & = 111.49^- \\
v_4 &= 32\sin 111.49 + 80 & = 111.480^+.
\end{aligned}
$$

The values given for v_2 and v_4, which are lower bounds for x_0 are less than the precise values. Therefore, these approximate values are *a fortiori* lower bounds for x_0. Similarly, the value given for v_3, which is an upper bound for x_0, is greater than the precise value; hence this approximate value is an upper bound of x_0.

The root x_0 lies between 111.480 and 111.49. In this interval, one can verify that

$$|f'(x)| = \frac{32\pi}{200} |\cos_{400} x| < 0{,}1,$$

$$|v_4 - x_0| < 0.1\,(111.49 - x_0) < 0.1\,(111.49 - 111.480) = 10^{-3}.$$

Since v_4 is a lower bound for x_0, it follows that x_0 lies between v_4 and $v_4 + 10^{-3}$:

$$111.480 \leqslant v_4 \leqslant 111.481 \Rightarrow 111.480 \leqslant x_0 \leqslant 111.482,$$

$$x_0 = 111.481,$$

with an error not exceeding 10^{-3}.

449

First question:

If $f''(x_0) > 0$, then in a neighborhood $]\alpha, \beta[$ of x_0, the continuous function f'' is positive and the function f' is strictly increasing. In this neighborhood,

$$x < x_0 \Rightarrow f'(x) < 1, \qquad x > x_0 \Rightarrow f'(x) > 1.$$

Let us take for u_0 a number in $]\alpha, x_0[$ sufficiently close to x_0 for the derivative f' to be nonnegative in the interval $[u_0, x_0]$. In that interval, the derivative f' verifies the double inequality $0 \leqslant f'(x) \leqslant 1$, and the function $x - f$, whose derivative $1 - f'$ is positive in $]u_0, x_0[$, is strictly increasing and it vanishes at the single value x_0.

The hypotheses of (1) of exercise 448 are satisfied. Thus we know that the sequence $\{u_n\}$ is nondecreasing and that it converges to x_0.

Let us suppose that $x_0 < u_0 < \beta$. We can then show that

$$u_1 - x_0 = f(u_0) - f(x_0) = (u_0 - x_0)f'(c) > u_0 - x_0,$$

and (by induction on n) that, for the elements $u_n < \beta$,

$$u_{n+1} - u_n \geqslant u_n - u_{n-1} \geqslant \dots \geqslant u_1 - u_0 \Rightarrow u_{n+1} - u_0 \geqslant (n+1)(u_1 - u_0).$$

Therefore, only finitely many terms of the sequence lie in $]x_0, \beta[$. Since no hypothesis has been made regarding the function for its derivative outside the interval $[\alpha, \beta]$, the study of the sequence cannot be carried out.

If $f''(x_0) < 0$, the sequence $\{u_n\}$ converges to x_0 if $u_0 > x_0$. If $u_0 < x_0$, we cannot study more than finitely many terms of the sequence.

Second question:

Let us suppose that $f'''(x_0) \neq 0$. Then, there exists a neighborhood $]\alpha, \beta[$ of x_0 in which $f'''(x)f'''(x_0) > 0$. In this interval, the variations in f'' and f' are governed by the sign of f''':

If $f'''(x_0) > 0$, the greatest lower bound of f' in $[\alpha, \beta]$ is equal to 1 and the sequence $\{u_n\}$ cannot converge to x_0 (see exercise 448, (3)).

If $f'''(x_0) < 0$, there exists an interval $]\alpha_1, \beta_1[\subset]\alpha, \beta[$ in which the derivative f' is positive. In this interval, $]\alpha_1, \beta_1[$,

$$0 \leqslant f'(x) \leqslant 1,$$

and the function $x - f$, which is strictly increasing, vanishes only at x_0.

The inductively defined sequence $\{u_n\}$ is monotonic and it converges to x_0.

Third question:

The composite function $F = f \circ f$ has continuous derivatives of the first three orders, and

$$F'(x) = (f' \circ f)(x)f'(x),$$

$$F''(x) = (f'' \circ f)(x)[f'(x)]^2 + (f' \circ f)(x)f''(x),$$

$$F'''(x_0) = (f''' \circ f)(x)[f'(x)]^3 + 3(f'' \circ f)(x)f'(x)f''(x) + (f' \circ f)(x)f'''(x).$$

The values of these derivatives at x_0 are

$$F'(x_0) = 1, \qquad F''(x_0) = 0, \qquad F'''(x_0) = -3[f''(x_0)]^2 - 2f'''(x_0).$$

Consider the inductively defined sequences associated with the function $f \circ f = F$:

$$v_n = (f \circ f)(v_{n-1}), \qquad v_0 = u_0 \Rightarrow v_n = u_{2n},$$

$$w_n = (f \circ f)(w_{n-1}), \qquad w_0 = u_1 \Rightarrow w_n = u_{2n+1}.$$

The numbers u_0 and u_1 lie on opposite sides of x_0 since $f'(x)$ is negative. However, the relative positions of these terms with respect to x_0 play no role in the present case since $F''(x_0) = 0$:

$$F'''(x_0) = -3[f''(x_0)]^2 - 2f'''(x_0) < 0 \Rightarrow v_n \to x_0 \quad \text{and} \quad w_n \to x_0.$$

Therefore, the sequence $\{u_n\}$ converges to x_0.

$$F'''(x_0) = -3[f''(x_0)]^2 - 2f'''(x_0) > 0 \Rightarrow v_n \nrightarrow x_0 \quad \text{and} \quad w_n \nrightarrow x_0;$$

and *a fortiori* the sequence $\{u_n\}$ does not converge to x_0.

Fourth question:

By applying Taylor's formula, we obtain

$$f(x) - f(x_1) = (x - x_1)f'(x_1)\left[1 + (x - x_1)\frac{f''(x_1)}{2f'(x_1)} + o(x - x_1)\right],$$

$$\frac{1}{f(x) - f(x_1)} = \frac{1}{(x - x_1)f'(x_1)}\left[1 - (x - x_1)\frac{f''(x_1)}{2f'(x_1)} + o(x - x_1)\right],$$

$$\frac{1}{f(x) - f(x_1)} - \frac{1}{(x - x_1)f'(x_1)} = -\frac{f''(x_1)}{2[f'(x_1)]^2} + o(1).$$

The desired limit is therefore

$$-\frac{1}{2}\frac{f''(x_1)}{[f'(x_1)]^2}.$$

The limit of the inductively defined sequence $\{u_n\}$ is x_0. We can apply the preceding result, by taking $x_1 = x_0$, $x = u_{n-1}$, and $f(x) = u_n$:

$$\lim\left(\frac{1}{u_n - x_0} - \frac{1}{u_{n-1} - x_0}\right) = -\frac{1}{2}f''(x_0).$$

We then know (see exercise 413) that

$$\lim \frac{1}{n(u_n - x_0)} = -\frac{1}{2}f''(x_0) \quad \text{or} \quad u_n - x_0 \simeq \frac{-2}{nf''(x_0)},$$

which is more precise than the result required.

One should understand that this method is not applicable to numerical calculation. For example, if $f''(x_0) = 1$, even if we evaluate the first 2000 terms of the sequence, we get an accuracy only of the order of 10^{-3}.

Fifth question:

We use the same method as in the fourth question:

$$f(x)-f(x_1) = (x-x_1)f'(x_1)\{1+\frac{(x-x_1)^2}{6}\frac{f'''(x_1)}{f'(x_1)}+o[(x-x_1)^2]\},$$

$$\frac{1}{f(x)-f(x_1)} = \frac{1}{(x-x_1)f'(x_1)}\{1-\frac{(x-x_1)^2}{6}\frac{f'''(x_1)}{f'(x_1)}+o[(x-x_1)^2]\},$$

$$\frac{1}{[f(x)-f(x_1)]^2} = \frac{1}{[(x-x_1)f'(x_1)]^2}\{1-2\frac{(x-x_1)^2}{6}\frac{f'''(x_1)}{f'(x_1)}+o[(x-x_1)^2]\}$$

$$= \frac{1}{[(x-x_1)f'(x_1)]^2} - \frac{f'''(x_1)}{3[f'(x_1)]^3}+o(1).$$

Therefore, the desired limit is

$$-\frac{f'''(x_1)}{3[f'(x_1)]^3}.$$

If $x_1 = x_0$, if $x = u_{n-1}$, and if $f(x) = u_n$, then

$$\lim\left[\frac{1}{(u_n-x_0)^2}-\frac{1}{(u_{n-1}-x_0)^2}\right] = -\frac{1}{3}f'''(x_0),$$

$$\lim\frac{1}{n(u_n-x_0)^2} = -\frac{1}{3}f'''(x_0), \qquad |u_n-x_0| \simeq \frac{1}{\sqrt{n}}\sqrt{\frac{-3}{f'''(x_0)}}.$$

Sixth question:

If $f'(x_0) = -1$, the preceding results can be applied to the sequences $\{v_n\}$ and $\{w_n\}$ defined in (3). If

$$3[f''(x_0)]^2+2f'''(x_0) > 0,$$

these sequences converge to x_0. Indeed, these sequences are defined by the function $F = f \circ f$, and we know that

$$F'(x_0) = 1, \qquad F''(x_0) = 0, \qquad F'''(x_0) = -3[f''(x_0)]^2-2f'''(x_0).$$

The conditions are those of (5), and

$$|u_{2n}-x_0| = |v_n-x_0| \simeq \frac{1}{\sqrt{n}}\sqrt{\frac{3}{3[f''(x_0)]^2+2f'''(x_0)}},$$

$$|u_{2n+1}-x_0| = |w_n-x_0| \simeq \frac{1}{\sqrt{n}}\sqrt{\frac{3}{3[f''(x_0)]^2+2f'''(x_0)}}$$

$$\Rightarrow |u_n-x_0| \simeq \frac{1}{\sqrt{n}}\sqrt{\frac{6}{3[f''(x_0)]^2+2f'''(x_0)}}.$$

450

First question:

Proof by induction on n: For $n=1$, we have Rolle's theorem. Suppose that the result is true for $n \leqslant k-1$. We denote by

$$u_1 < u_2 \ldots < u_{k+1}$$

the zeros of φ in $[a, b]$. In each of the k open intervals $]u_i, u_{i+1}[$ (for $i=1, 2, \ldots, k$), the derivative φ' has at least one zero v_i, and these zeros are distinct since the intervals $]u_i, u_{i+1}[$ are disjoint. The function φ' has at least k distinct zeros in $[a, b]$ and it has derivatives of the first $k-1$ orders in $[a, b]$. Its $(k-1)$st derivative, that is, the function $\varphi^{(k)}$ has at least one zero in $]a, b[$, a result we may assume to be valid (since we may apply it for $n=k-1$).

Second question:

Let us suppose that x is not one of the x_i (in the opposite case, the equation holds no matter what the value of c is since both sides are 0). The function φ vanishes at the values x_i for every C. Let us determine C so that φ will also vanish at the value x:

$$\varphi(x) = 0 \Leftrightarrow C = \frac{f(x) - A(x)}{(x - x_1) \cdots (x - x_n)}.$$

Now, the function φ vanishes at at least $(n+1)$ values in $[a, b]$. Its nth derivative vanishes at at least one value, which we denote by c, in the interval $]a, b[$ (result of the first question):

$$\varphi^{(n)}(c) = f^{(n)}(c) - A^{(n)}(c) - n! \, C = 0.$$

The nth derivative of the polynomial A, which is of degree not exceeding $n-1$, is 0. If we take $\varphi(x)=0$ and $C=f^{(n)}(c)/n!$, we get the required result.

Third question:

We use the preceding result and take for f the function defined by $f(x) = \log_a(p+x) - \log_a p$ and we take for A the polynomial that assumes the same value as f at $x_1=0$ and $x_2=1$, namely, the polynomial $x \, [\log_a(p+1) - \log_a p]$.

Then, we know that there exists a number c in $]0, 1[$ such that

$$f(x) - A(x) = \frac{(x-x_1)(x-x_2)}{2!} f''(c) = \frac{-x(1-x)}{2}\left(\frac{-\log_a e}{(p+c)^2}\right),$$

$$\left.\begin{array}{l} 0 < c < 1 \Rightarrow \dfrac{1}{(p+c)^2} < \dfrac{1}{p^2} \\[2mm] 0 < x < 1 \Rightarrow 0 < x(1-x) < \tfrac{1}{4} \end{array}\right\} \; 0 < f(x) - A(x) < \frac{\log_a e}{8p^2}.$$

APPLICATION: In the preceding notations, the error made is $f(x) - A(x)$, and we know that

$$0 < f(x) - A(x) < \frac{\log_{10} e}{8p^2} < \frac{0.44}{8p^2} \, .$$

For the numbers p in the table that exceed 10^{-3}, we can take

$$\Delta = 6 \times 10^{-8} > \frac{0.44}{8 \times 10^6} \, .$$

This error is very small relative to that arising from the omission of the sixth decimal place and beyond in the table, namely, 5×10^{-6}.

Part 5

INTEGRATION OF REAL REGULATED FUNCTIONS, INTEGRATION OVER A NONCOMPACT INTERVAL, EVALUATION OF INTEGRALS

Exercises related to the properties of Chapter IV of Book III
and Chapters II and V of Book IV of *Mathématiques générales*
by C. Pisot and M. Zamansky

Introduction

All the exercises dealing with the integration of functions from R into R are assembled in this part. My feeling was that the same majorization problems, for example, arise and that the same techniques for study are used whether we are dealing with integration of a regulated function over a closed interval or with integration over a noncompact interval. Finally, I did not wish to separate the technical problems of evaluating integrals with the aid of primitives from more theoretical problems regarding the properties of integrals.

Although there is no neat subdivision of this part, I can still point out that exercises 501-516 deal almost entirely with the properties of integrals of regulated functions over a closed interval. Exercises 502-504, 506, and 514 do not involve the properties of derivatives or associated properties (integration by parts and change of variables). It is only later (in exercises 517-528) that we come to the evaluation of integrals of simple functions. These are essentially exercises in techniques.

It is only when we get to exercise 531 that we encounter integrals over noncompact intervals. The properties of functions defined with the aid of integrals are also studied only in the last few exercises since they assume a knowledge of the properties of functions from R^2 into R, and these are studied in the following chapter.

Preliminary remarks

Step functions

The definition of the integral of a regulated (in particular, continuous) function f with the aid of step functions that approach f uniformly is often used in the problems. Sometimes, we use it to majorize the integral of f by choosing majorizing or minorizing step functions (for example,

$$f(x) \leqslant M \implies \int_a^b f(x)\,dx \leqslant M(b-a),$$

where we consider the step function φ that is constant and is equal to M). Sometimes, when a property of integrals of step functions is established, we assign to every number $\varepsilon > 0$ a step function φ such that

$$\left| \int_a^b f(x)\,dx - \int_a^b \varphi(x)\,dx \right| < \varepsilon.$$

Nonnegative functions

The integral of a nonnegative function f over an interval $[a, b]$, where $a<b$, is nonnegative. If f is also continuous, its integral can be zero only if f is the zero function. (This result is often used to show that a continuous function is the zero function.)

As a consequence of this last result in the vector space $\mathscr{C}$ of continuous mappings on $[a, b]$ into R, the mapping

$$f \rightarrow \int_a^b f^2(x)\,dx$$

is a positive-definite quadratic form. We can then define on $\mathscr{C}$ a Euclidean structure (see exercise 351 in Part 3).

Sequences of functions

(1) If the sequence $\{f_n\}$ converges uniformly to f in $[a, b]$, then

$$\lim \int_a^b f_n(t)\,dt = \int_a^b f(t)\,dt.$$

However, this result may not be true if the convergence is not uniform (see exercise 513).

(2) If the sequence $\{f_n\}$ converges uniformly to f in $[a, b]$, if F_n is a primitive of f_n, and if, for a particular value x_0 in $[a, b]$, the sequence $\{F_n(x_0)\}$ is convergent, then the sequence $\{F_n\}$ converges uniformly in $[a, b]$ to a primitive F of f.

(3) *In the case of differentiation, note that the conditions apply to the sequence of derivatives.* We apply the preceding result to a sequence of functions g_n and their continuous derivatives g_n' by taking $f_n = g_n'$ and $F_n = g_n$:

If the *sequence of derivatives* $\{g_n'\}$ converges uniformly in $[a, b]$ and if the sequence $\{g_n(x_0)\}$ is convergent, the sequence $\{g_n\}$ converges uniformly in $[a, b]$ to a function g. The function g is differentiable and $g' = \lim \{g_n'\}$.

Functions defined in terms of integrals

It is important to note that the hypotheses of the various theorems (continuity of the function, existence and continuity of partial derivatives, etc.) must always be satisfied in a *closed* rectangle, that is, the rectangle including its boundary.

Evaluation of integrals and primitives

For integration by parts, it is often useful to know the formula for n-fold integration by parts:

$$\int_a^b f^{(n)}(t)g(t)\,dt = \left[\sum_{k=0}^{n-1} (-1)^k f^{(n-k-1)}(t)g^{(k)}(t)\right]_a^b + (-1)^n \int_a^b f(t)g^{(n)}(t)\,dt.$$

In particular, if g is a polynomial of degree $< n$, the last integral vanishes and the integration is carried out in closed form.

In a change of variables, it is essential not to forget to determine the new limits of integration.

After a change of variables, we sometimes obtain an integral over a noncompact (for example, infinite) interval. However, in this case, it is certain that the integral obtained converges and there is no need to prove this in each individual case. (The reason is given in (2) of the following paragraph.)

Integration over a noncompact interval ($[a, +\infty]$ or $[a, b]$)

(1) *Absolute convergence* implies convergence and is often easier to determine. Therefore, we begin by testing whether a function f is of constant sign and, if not, whether its absolute value is integrable. Determination of absolute convergence is often made by finding an integrable function g that majorizes $|f|$.

(2) *A change of variable* does not affect the convergence or absolute convergence of an integral. Therefore, we may immediately perform a change of variable if this simplifies the determination of convergence of the integral.

(3) *Integration by parts* should be applied to the integral of a function over a closed interval. Then, we take the limit as one of the limits of integration approaches the desired value.

(4) *If the interval of integration contains neither of its end-points* (as, for example, with the intervals $]a, +\infty[$ and $]a, b[$) the simplest procedure is to study separately the integrability of the function over two intervals of the form $]a, x_0]$ and $[x_0, b[$, where x_0 is any value in the open interval $]a, b[$.

Exercises

501

Suppose that f is a regulated function on $[a, b]$ into R and that

$$f(a+b-x) = f(x).$$

Show that

$$\int_a^b x f(x) \, dx = \frac{a+b}{2} \int_a^b f(x) \, dx.$$

502

Let f denote a real function that is defined, continuous, and nonnegative in $[a, b]$. Show that the sequence $\{I_n\}$ defined by

$$I_n = \left\{ \int_a^b [f(x)]^n \, dx \right\}^{1/n}$$

converges to the least upper bound M of f in $[a, b]$. You may use very simple step functions φ and ψ that minorize and majorize f to show that, if n is sufficiently great,

$$M - \varepsilon < I_n < M + \varepsilon.$$

503

(1) Suppose that two functions f and g are defined and continuous on the interval $[a, b]$. Show that the equation

$$\Delta = \left(\int_a^b f(x) g(x) \, dx \right)^2 - \left(\int_a^b [f(x)]^2 \, dx \right) \left(\int_a^b [g(x)]^2 \, dx \right) = 0$$

implies the existence of a real number λ such that $f = \lambda g$.

(2) Use the Schwarz inequality to determine the minimum of the product

$$P_f = \left(\int_a^b f(x) \, dx \right) \left(\int_a^b \frac{1}{f(x)} \, dx \right),$$

defined on the set $\mathscr{C}'$ of functions f that are continuous and positive on $[a, b]$. (Note that the mapping that sends f into P_f is a mapping on $\mathscr{C}'$ into $\mathbf{R}$.)

For what functions f is the minimum attained?

(3) Show that the product P_f is not bounded above.

504

(1) Consider two regulated functions f and g defined on the intervals $[a, b]$. Denote by p and P the greatest lower and least upper bounds of f and denote by q and Q the greatest lower and least upper bounds of g on $[a, b]$. Suppose that p and q are positive numbers. Show that the function

$$(f - \frac{P}{q}g)(f - \frac{p}{Q}g)$$

is nonpositive and that the function

$$F_\lambda = f^2 + \lambda f g \left(\sqrt{\frac{PQ}{pq}} + \sqrt{\frac{pq}{PQ}} \right) + \lambda^2 g^2$$

is nonnegative for certain values of λ and nonpositive for others. From this, show that

$$\left(\int_a^b f(x)g(x)\,dx \right)^2 \left(\sqrt{\frac{PQ}{pq}} + \sqrt{\frac{pq}{PQ}} \right)^2 \geq 4 \left(\int_a^b [f(x)]^2\,dx \right) \left(\int_a^b [g(x)]^2\,dx \right).$$

(2) Show under what conditions equality may hold in the last conditional inequality under the assumption that the functions f and g are continuous in $[a, b]$.

(3) Consider $2n$ nonnegative numbers p_i and q_i (for $i = 1, 2, ..., n$). Assume that

$$0 < r \leqslant p_i \leqslant R, \qquad 0 < s \leqslant q_i \leqslant S.$$

Show that

$$4 \left(\sum_{i=1}^n p_i^2 \right) \left(\sum_{i=1}^n q_i^2 \right) \leqslant \left(\sqrt{\frac{RS}{rs}} + \sqrt{\frac{rs}{RS}} \right)^2 \left(\sum_{i=1}^n p_i q_i \right)^2.$$

505

(1) Let f denote a function that is defined and continuous on the interval $[a, b]$. Let $\{\sigma_n\}$ denote a sequence of partitions of the interval $[a, b]$. Suppose that the sequence $\{l_n\}$ converges to 0, where l_n denotes the maximum length of the intervals of σ_n.

In each interval
$$[x_{i-1}^{(n)}, \quad x_i^{(n)}]$$
of the partition σ_n, let us choose a value $\xi_i^{(n)}$. Show that the sequence of sums
$$S_n = \sum_i (x_i^{(n)} - x_{i-1}^{(n)}) f(\xi_i^{(n)})$$

converges to

$$\int_a^b f(x)\,dx.$$

What in particular do we obtain when σ_n is the partition resulting from dividing $[a, b]$ into n equal intervals and

$$\xi_i^{(n)} = x_i^{(n)}?$$

(2) If a is a real number $\geqslant 1$, show that the sequence $n(a^{1/n} - 1)$ converges to

$$\int_1^a \frac{dx}{x}.$$

Consider the partitions σ_n defined by the values $a^{i/n}$.

(3) Show that the preceding result and the monotonic increase of the function that maps x into a^x implies the differentiability of that function at 0.

506

Find the limits of the sequences $\{x_n\}$, $\{y_n\}$, and $\{z_n\}$ defined by

$$x_n = \sum_{p=1}^{n} \frac{n}{n^2 + p^2},$$

$$y_n = \frac{1}{n} \sqrt[n]{(n+1)(n+2)\cdots(n+p)\cdots 2n} = \frac{1}{n}\sqrt[n]{\prod_{p=1}^{n}(n+p)},$$

$$z_n = \sqrt[n]{\frac{(a+1)\cdots(a+n)}{n!}} = \sqrt[n]{\frac{1}{n!}\prod_{p=1}^{n}(a+p)}.$$

For some of these, one can use the result of (1) of exercise 505.

507

Let p and q denote two numbers greater than 1 such that

$$\frac{1}{p} + \frac{1}{q} = 1.$$

(1) If u and v are nonnegative numbers, show that

$$uv \leqslant \frac{u^p}{p} + \frac{v^q}{q}.$$

Suggestion: Examine the behavior of the function that maps u into $u^p/p - uv$.

(2) Define mappings N_p and N_q that assign to any regulated functions f and g defined on the interval $[a, b]$ (where $a < b$) the numbers

$$N_p(f) = \left(\int_a^b |f(t)|^p \, dt \right)^{1/p}, \qquad N_q(g) = \left(\int_a^b |g(t)|^q \, dt \right)^{1/q}.$$

By applying the inequality of (1) to the numbers

$$u = \frac{|f(t)|}{N_p(f)}, \qquad v = \frac{|g(t)|}{N_q(g)},$$

show that

$$\left| \int_a^b f(t)g(t) \, dt \right| \leqslant N_p(f)N_q(g).$$

(3) Consider two nonnegative regulated functions φ and ψ defined on the interval $[a, b]$. Let α and β denote two positive numbers whose sum is equal to 1. Show that

$$\int_a^b [\varphi(t)]^\alpha [\psi(t)]^\beta \, dt \leqslant \left(\int_a^b \varphi(t) \, dt \right)^\alpha \left(\int_a^b \psi(t) \, dt \right)^\beta.$$

508

Let f denote a continuous strictly increasing function f defined on the interval $[0, a]$ such that $f(0) = 0$. Let g denote the inverse function of f.

(1) Define a function F on the interval $[0, a]$ by

$$F(x) = \int_0^x f(t) \, dt + \int_0^{f(x)} g(t) \, dt - xf(x).$$

Show that the function F is differentiable and that its derivative is everywhere 0. First assume that the function f is differentiable. Then, in the general case, use the definition of a derivative. What is the value of $F(x)$?

(2) Show that

$$0 \leqslant u \leqslant a \quad \text{and} \quad 0 \leqslant v \leqslant f(a) \Rightarrow uv \leqslant \int_0^u f(t) \, dt + \int_0^v g(t) \, dt.$$

Use this result to find again the inequality proven in (1) of exercise 507:

$$uv \leqslant \frac{u^p}{p} + \frac{u^q}{q} \quad \text{if} \quad \frac{1}{p} + \frac{1}{q} = 1.$$

509

Let f and g denote two functions that are defined and continuous on the interval $[a, b]$, where $a < b$. Suppose that f is nondecreasing and that g and $1-g$ are positive. Finally, let us define

$$l = \int_a^b g(t)\,dt.$$

(1) Study the behavior of the function G defined by

$$G(x) = \int_a^x g(t)\,dt$$

and show that $a + G(x) \leqslant x$.

(2) Compare the functions φ and ψ defined by

$$\varphi(x) = \int_a^x f(t)g(t)\,dt, \qquad \psi(x) = \int_a^{a+G(x)} f(t)\,dt$$

and show that

$$\int_a^b f(t)g(t)\,dt \geqslant \int_a^{a+l} f(t)\,dt.$$

(3) Can equality ever hold in this conditional inequality?

510

A second theorem of the mean for continuous functions

(1) Let f denote a function continuous and monotonic on the interval $[a, b]$ and possessing a continuous and bounded derivative except possibly at a finite number of values at which f' has right- and left-hand limits.

Show by integration by parts that, if g is a continuous function on $[a, b]$, there exists a number ξ such that

$$\int_a^b f(t)g(t)\,dt = f(a)\int_a^\xi g(t)\,dt + f(b)\int_\xi^b g(t)\,dt.$$

(2) Show that the preceding result holds if the function f is the uniform limit of a sequence of functions f_n verifying the hypotheses of (1).

Use the results of exercise 423 of Part 4 to show that all continuous monotonic functions belong to the set of functions f for which the result is valid.

511

Proof of the irrationality of π and e

(1) Let a and b denote positive integers and let P_n denote the polynomial

$$\frac{x^n(bx-a)^n}{n!}.$$

Show that the polynomial P_n and all its derivatives assume integral values at $x=0$ and $x=a/b$. You may, for example, use the Maclaurin expansion of P_n to examine the value 0.

(2) Show that the sequence whose general term is

$$I_n = \int_0^\pi P_n(x) \sin x \, dx$$

converges to 0.

(3) Show that if π is rational and if $\pi=a/b$ where a and b are integers, then the number I_n is a noninteger in contradiction with the result of (2).

(4) Show in a similar way, by examining the sequence whose general term is

$$J_n = \int_0^r P_n(x) e^x \, dx,$$

that the number e^r, where r is any positive rational number, is irrational.

512

Let f denote a regulated function defined on a closed interval $[a, b]$. To every number h such that $0 < h < (b-a)/2$, let us assign the function g_h defined on the interval $[a+h, b-h]$ by

$$g_h(x) = \frac{1}{2h} \int_{x-h}^{x+h} f(t) \, dt = \frac{1}{2h} \int_{-h}^h f(x+u) \, du.$$

(1) Show that the function g_h is continuous and differentiable if f is continuous and that it has derivatives of the first $n+1$ orders if f has derivatives of the first n orders.

(2) Show that g_h is a convex function if the function f is convex.

(3) Show that, as h tends to 0, $g_h(x)$ tends to

$$\tfrac{1}{2}[f(x+0)+f(x-0)].$$

(4) Show that, if f is continuous, it is the uniform limit of a sequence of functions of the type g_h in every closed subinterval of the open interval $]a, b[$.

(5) Let f denote a convex function defined in an open interval $]a_1, b_1[$ and let $[\alpha, \beta]$ denote a closed subinterval of $]a_1, b_1[$. Show that the function f is the uniform limit on $[\alpha, \beta]$ of convex functions that are n times differentiable, where n is any given integer.

513

Evaluate the integral

$$I_n = \int_0^1 f_n(x)\,dx,$$

where

$$f_n(x) = \frac{nx-1}{(x \ln n+1)(1+nx^2 \ln n)}.$$

Find the limit of the sequence $\{I_n\}$ and the limit $f(x)$ of the sequence $\{f_n(x)\}$. Why is there no contradiction between these two results?

514

(1) Show that the function s_n defined on $[0, \pi/2]$ by

$$s_n(0) = n^2, \qquad s_n(t) = \frac{\sin^2 nt}{\sin^2 t} \qquad \text{if} \qquad t \neq 0$$

is continuous and find a minorant of s_n on $[0,\pi/2n]$. Show that the sequence of the integrals I_n converges to $+\infty$ if

$$I_n = \int_0^{\pi/2} \frac{\sin^2 nt}{\sin^2 t}\,dt.$$

(2) Let φ denote a regulated function on $[0, \pi/2]$ the right-hand limit of which is 0 at 0, that is, $\varphi(+0)=0$. Show that the sequence $\{u_n\}$ defined by

$$u_n = \frac{1}{I_n}\int_0^{\pi/2} \varphi(t)\frac{\sin^2 nt}{\sin^2 t}\,dt$$

converges to 0. This can be done by decomposing the interval $[0, \pi/2]$ into two subintervals $[0, \eta]$ and $[\eta, \pi/2]$ and choosing first η, then n, in such a way as to make u_n less than any given positive number ε.

(3) Show that, if f is a regulated function on $[0, \pi/2]$, the sequence $\{v_n\}$ defined by

$$v_n = \frac{1}{I_n} \int_0^{\pi/2} f(t) \frac{\sin^2 nt}{\sin^2 t} \, dt$$

converges to $f(+0)$.

515

Consider functions f_n and g_n defined on $[-1, 1]$ by

$$f_n(x) = \frac{\int_0^x (1-t^2)^n \, dt}{\int_0^1 (1-t^2)^n \, dt}, \qquad g_n(x) = \int_0^x f_n(t) \, dt.$$

(1) Show that f_n and g_n are polynomials. Check whether they are even or odd functions or neither.

(2) Use the inequalities

$$t(1-t^2)^n \leqslant (1-t^2)^n \qquad \text{if} \qquad 0 \leqslant t \leqslant 1,$$

$$(1-t^2)^n \leqslant \frac{t}{x}(1-t^2)^n \qquad \text{if} \qquad 0 \leqslant x \leqslant t \leqslant 1,$$

to show that $\lim f_n(x) = 1$ if $x > 0$. Find the limit of $f_n(x)$ if $x \leqslant 0$.

(3) Show that the sequence f_n converges uniformly on the interval $[\alpha, 1]$ whatever the value of $\alpha > 0$ but that this sequence does not converge uniformly on the interval $[-\alpha, \alpha]$.

(4) Show that the sequence $\{g_n\}$ converges uniformly to the absolute-value function on the interval $[-1, 1]$. If x is positive, one may majorize the difference $x - g_n(x)$ by treating separately the cases: $x < h$ and $x \geqslant h$, where h is an arbitrarily chosen number.

516

Weierstrass' theorem: Any continuous function on a closed interval $[a, b]$ is the uniform limit of a sequence of polynomials.

This exercise assumes knowledge of the results of the preceding exercise and of exercise 423 of Part 4.

(1) Consider $n+1$ distinct real numbers $x_0 < x_1 < ... < x_n$. Then, for any system of $n+1$ real numbers $c_0, c_1, ..., c_n$, define a function μ by

$$\mu(x) = \sum_{i=0}^{n+1} c_i |x - x_i|.$$

Show that the function μ is a continuous linear function in each of the intervals $[x_{i-1}, x_i]$ and that it has a derivative everywhere except possibly at the values x_i. Show that if the function φ vanishes at $x_0, x_1, \ldots, x_n$, it vanishes for all x and that then the coefficients c_i are all 0. Conclude from this that the system of equations

$$(S) \qquad \mu(x_j) = \sum_{i=0}^{n+1} c_i |x_j - x_i| = y_j \qquad (j = 0, 1, \ldots, n),$$

where the c_i are the unknowns, has a unique solution whatever the numbers y_j.

(2) In the notations of exercise 423 of Part 4, show that every function λ defined on the interval $[a, b]$ can be defined as the function μ of (1), that is, that there exist distinct numbers $a = x_0 < x_1 < \ldots < x_n = b$ and coefficients c_i such that

$$\lambda(x) = \sum_{i=0}^{n} c_i |x - x_i|.$$

(3) Show that every function that is defined and continuous on a closed interval $[a, b]$ is the uniform limit on that interval of a sequence of polynomials. Consider first the case $a = 0$, $b = 1$ and use the result of the preceding exercise.

517

Consider the set E of real continuous differentiable functions defined on $\mathbf{R}$ such that

$$f'(x) = \tfrac{1}{2}[f(x+a) - f(x-a)],$$

where a is a given positive constant.

(1) Show that E is a vector space.
(2) See whether E contains the functions e^{rx}, $\cos \omega x$, and $\sin \omega x$.
(3) Let f denote a real function defined on $\mathbf{R}$. Suppose that, for every closed interval $[x_1, x_2]$, there exists a sequence $\{f_n\}$ of functions belonging to E that converges uniformly to f on $[x_1, x_2]$. Show that the function f is an element of the set E.

518

Evaluate the integral

$$I_m = \int_0^1 \frac{t^2 + 2t - 1}{(t^2 + 1)^m} \, dt$$

for $m = 1, 2$, and $\tfrac{3}{2}$.

Indicate a method of evaluation that is applicable to the case in which m is an integer and to the case when $m-\frac{1}{2}$ is an integer. (Do not carry out the calculations, however.)

519

Evaluate the integral

$$I_x = \int_1^x \frac{2t \ln t}{(1+t^2)^2}\,dt.$$

Does this integral have a limit as x tends to $+\infty$?

520

(1) Evaluate the integral

$$I = \int_0^1 \frac{dx}{x^4+1}.$$

(2) Integrate the above expression by parts to obtain the value of

$$J = \int_0^1 \frac{dx}{(x^4+1)^2}.$$

(3) Evaluate the integral

$$K = \int_0^1 \frac{(x+1)dx}{(x^4+1)^2}.$$

521

Evaluate

$$\int_a^b \sqrt{(x-a)(b-x)}\,dx \qquad (a < b).$$

522

Evaluate the integral

$$I_x = \int_x^3 \frac{1+\sqrt{1+t}}{\sqrt{1+t}-1}\,dt \qquad (x > 0).$$

What is the derivative of I_x? Show that I_x tends to $+\infty$ as x tends to 0 (this result can be predicted by anyone knowing the theory of integration over a noncompact interval).

523

Evaluate the integral

$$I = \int_0^a \frac{dx}{x - 2\sqrt[3]{x} + 4} \qquad (a > 0).$$

524

Evaluate the integral

$$I = \int_{\pi/3}^{\pi/2} \frac{dx}{\sin^3 x},$$

(1) by setting $\tan (x/2) = t$,
(2) by setting $\cos x = u$.

525

Evaluate the integral

$$I = \int_0^{\pi/3} \frac{\sin x \sin 2x}{\sin^4 x + \cos^4 x + 1} \, dx.$$

526

Denote by $I(a, b)$ the integral

$$\int_a^b \frac{(1 - x^2) \, dx}{(1 + x^2)\sqrt{1 + x^4}}.$$

(1) Show that

$$I(a, b) = I(-b, -a)$$

and that, if a and b have the same sign,

$$I(a, b) = I\left(\frac{1}{a}, \frac{1}{b}\right).$$

Finally, show that

$$I\left(a, \frac{1}{a}\right) = 0.$$

(2) Evaluate $I(a, b)$. Treat first the case in which a and b are at least equal to 1 and set $t = x + (1/x)$.

527

Denote by I_n the integral

$$I_n = \int_0^1 x^n \sin(\pi x) \, dx.$$

(1) Evaluate I_0 and I_1. Establish a relationship between I_n and I_{n+2}, and use this relationship to evaluate I_4.

(2) Show that the sequence $\{I_n\}$ is nonincreasing and that it converges to 0. From this, show that I_n is equivalent to π/n^2. More precisely, show that

$$\frac{\pi}{(n+1)(n+2)}\left[1 - \frac{\pi^2}{(n+3)(n+4)}\right] < I_n < \frac{\pi}{(n+1)(n+2)}.$$

528

Evaluate the integral

$$I = \int_0^1 x\sqrt{x^2 - 2x + 2}\,dx.$$

529

Consider the integral

$$I_\alpha = \int_0^{\pi/2} \sin^\alpha x \, dx,$$

where α is a nonnegative real number.

(1) By integrating by parts, establish a relationship between I_α and $I_{\alpha+2}$, and show that the function f defined for $\alpha \geqslant 0$ by

$$f(\alpha) = (\alpha+1)I_\alpha I_{\alpha+1}$$

is periodic with period 1. Evaluate $f(0)$.

(2) Show that the function that maps α into I_α is nonincreasing and hence that

$$p \leqslant \alpha < p+1 \Rightarrow \frac{p+1}{p+2}f(0) < f(\alpha) < \frac{p+2}{p+1}f(0).$$

Find the limit of the sequence whose general term is $f(\alpha+n)$ and show that the function f is constant.

(3) Show that the ratio $I_{\alpha+1}/I_\alpha$ tends to 1 as α tends to $+\infty$ and find a simple function that is equivalent to I_α.

530

Denote by E the vector space over $\mathbf{R}$ of polynomials with real coefficients and of degree not exceeding n. We recall that the dimension of E is $n+1$.

(1) For every real function f that is defined and continuous on $[a, b]$, define a mapping F on E into $\mathbf{R}$ by

$$F(A) = \int_a^b A(x)f(x)\,\mathrm{d}x.$$

Show that F is a linear form on E.

(2) If the function f is a polynomial of degree not exceeding n, show that the linear form F cannot be the zero form unless f is the zero polynomial. If $f_1, f_2, ..., f_p$ are p polynomials of degree not exceeding n, show that a necessary and sufficient condition for the forms $F_1, F_2, ..., F_p$ to be independent is that the polynomials $f_1, f_2, ..., f_p$ be independent.

(3) Conversely, show that, for every linear form G on E, there exists a polynomial g of degree not exceeding n such that

$$G(A) = \int_a^b A(x)g(x)\,\mathrm{d}x.$$

531

This exercise assumes knowledge of the results of the preceding exercise. Finding of the solution will be facilitated if one refers to exercises 315 and 352 of Part 3, although the solution can also be found by a straightforward procedure.

Consider the vector space E of polynomials of degree not exceeding $2k-1$ with real coefficients. Let D denote the polynomial

$$\frac{\mathrm{d}^k}{\mathrm{d}x^k}(x^2-1)^k.$$

(1) Show that if A is a polynomial of degree less than k, then

$$\int_{-1}^{1} A(x)D(x)\,dx = 0.$$

(2) Show that the zeros of the polynomial D, which we denote by $x_1, \ldots, x_k$, are all real and distinct and that they lie between -1 and 1.

(3) Show that the polynomials which belong to E, and which are multiples of D constitute a k-dimensional vector subspace V. Show that the linear forms l and l_i (for $i=1, 2, \ldots, k$) defined by

$$l(A) = \int_{-1}^{1} A(x)\,dx, \qquad l_i(A) = A(x_i),$$

belong to a vector subspace W of E^* of zero forms on the subspace V and that the forms l_i constitute a basis for W.

(4) Show that there exist constants m_i (for $i=1, 2, \ldots, k$) such that, for every polynomial A of degree not exceeding $2k-1$,

$$\int_{-1}^{1} A(x)\,dx = \sum_{i=1}^{k} m_i A(x_i).$$

Show finally that

$$m_i = \frac{1}{D'(x_i)} \int_{-1}^{1} \frac{D(x)}{x-x_i}\,dx.$$

532

Show that the function f defined for $|x| < 1$ by

$$f(x) = \frac{x}{(1+x^2)\sqrt{1-x^4}}$$

is integrable over the interval $[0, 1[$ and evaluate

$$I = \int_{0}^{1-0} f(x)\,dx.$$

533

Let p denote a positive integer. Show that the integral

$$I = \int_{0}^{1} (\ln x)^p\,dx$$

converges and calculate its value.

534

Show that

$$\lim_{n \to \infty} \frac{1}{n} \sum_{p=1}^{n} \ln \frac{p}{n} = \int_0^1 \ln x \, dx$$

and deduce from the preceding result that

$$\lim_{n \to \infty} \frac{\sqrt[n]{n!}}{n} = \frac{1}{e}.$$

535

(1) Denote by P and Q two polynomials and denote by a a real number exceeding all of the real zeros of Q. Suppose that deg $P \leqslant$ deg $Q - 2$. Show that the integrals

$$\int_a^{+\infty} \frac{P(t)}{Q(t)} \sin t \, dt, \qquad \int_a^{+\infty} \cos t \frac{P(t)}{Q(t)} dt,$$

are absolutely convergent.

(2) Show by integrating by parts that the integral

$$I = \int_0^{+\infty} \frac{t \sin t}{(1+t^2)} dt$$

converges.

(3) Show that the procedure of (2) can be applied to integrals of the form

$$\int_a^{+\infty} \sin t \frac{A(t)}{B(t)} dt,$$

where A and B are polynomials such that deg $A =$ deg $B - 1$ and where a is a real number exceeding all the real zeros of B.

536

Let α and β denote real numbers such that

$$0 < \alpha < 1, \qquad 1 - \alpha < \beta \leqslant 1,$$

and let λ denote a positive parameter.

(1) Show that the integrals

$$I(\lambda) = \int_1^{+\infty} \frac{dx}{x^\alpha(1+\lambda x)}, \qquad J(\lambda) = \int_1^{+\infty} \frac{dx}{(1+x^\alpha)(1+\lambda x)}$$

converge.

(2) Show that

$$I(\lambda) - \lambda^{\alpha-1} \int_{+0}^{+\infty} \frac{dt}{t^\alpha(1+t)}$$

approaches a finite limit as λ approaches 0.

(3) Show that the integral

$$K(\lambda) = \int_1^{+\infty} \frac{dx}{x^\beta(1+x^\alpha)(1+\lambda x)}$$

converges and approaches

$$\int_1^{+\infty} \frac{dx}{x^\beta(1+x^\alpha)},$$

as λ tends to 0. The result of (2) can be used to majorize the difference.

(4) Show that, if $1/(n+1) < \alpha < 1/n$ (where n is a positive integer), there exist constants $C_1, C_2, ..., C_n$, such that

$$J(\lambda) - C_1\lambda^{\alpha-1} - C_2\lambda^{2\alpha-1} - \cdots - C_n\lambda^{n\alpha-1}$$

tends to a finite limit as λ tends to 0.

(5) What modification do we need to make if $\alpha = 1/n$?

537

Let a and b denote two numbers such that $0 < a < b$ and let f denote a function that is continuous on $[0, +\infty[$. Suppose finally that the function $f(t)/t$ is integrable over $1 \leqslant t < \infty$.

(1) Show that the function $f(kx)/x$ is integrable over $u \leqslant x < \infty$, where u and k are positive numbers. Show also that

$$\int_u^{+\infty} \frac{f(bx)-f(ax)}{x}\,dx = \int_{bu}^{au} \frac{f(t)}{t}\,dt.$$

(2) Show that the function

$$\frac{f(bx)-f(ax)}{x}$$

is integrable over $0 < x < \infty$ and that

$$\int_{+0}^{+\infty} \frac{f(bx) - f(ax)}{x} \, dx = f(0) \ln \frac{a}{b}.$$

(3) Show that the integral

$$\int_{+0}^{+\infty} \frac{e^{-bt} - e^{-at}}{t} \, dt$$

is convergent and find its value.

538

Denote by a and b two real numbers such that $0 < a < b$.

(1) Consider the function f defined in $]0, +\infty[$ by

$$f(u) = v = \frac{ab + u^2}{2u}.$$

Examine the behavior of f. How must we restrict the domain of definition of f for the inverse function to be defined? Give an expression for the inverse function once such a restriction has been made.

(2) Consider the integral

$$I(a, b) = \int_0^{\pi/2} \frac{d\theta}{\sqrt{a^2 \cos^2 \theta + b^2 \sin^2 \theta}}.$$

By making the substitutions

$$u = \sqrt{a^2 \cos^2 \theta + b^2 \sin^2 \theta}$$

and

$$v = \frac{ab + u^2}{2u},$$

find the new expressions for I. (Note that the integrand, when the variable v is used, is not given by a single formula throughout the interval of integration.) From this, show that

$$I(a, b) = I\left(\sqrt{ab}, \frac{a+b}{2}\right).$$

(3) We recall that the sequences $\{a_n\}$ and $\{b_n\}$ defined by

$$a_0 = a, \qquad b_0 = b, \qquad a_{n+1} = \sqrt{a_n b_n}, \qquad b_{n+1} = \frac{a_n + b_n}{2},$$

have a common limit l and that $a_n \leqslant l \leqslant b_n$ (cf. exercise 409 of Part 4). Show that

$$I(a, b) = \frac{\pi}{2l}.$$

539

(1) Suppose that $1 < \alpha < 3$. Show that the function $t^{-\alpha} \sin^2 t$ is integrable over the interval $]0, +\infty[$.

(2) Let φ denote a real function defined and continuous in the interval $]0, +\infty[$. Suppose that there exist positive constants M and K and a number β satisfying the inequalities $-1 < \beta \leqslant 1$ such that

$$0 < t \leqslant 1 \;\Rightarrow\; |\varphi(t)| \leqslant Mt^\beta,$$

$$1 \leqslant t \;\Rightarrow\; |\varphi(t)| \leqslant K.$$

Show that the integral

$$I_n = \int_{+0}^{+\infty} \varphi\left(\frac{t}{n}\right) \frac{\sin^2 t}{t^2} \, dt$$

converges for every positive integer n. Show also that, if $\beta \neq 1$,

$$|I_n| \leqslant \frac{A}{n^\beta},$$

where A is a number independent of n, and that, if $\beta = 1$,

$$|I_n| \leqslant \frac{B \ln n}{n},$$

where B is a number independent of n. (*Suggestion:* Decompose the interval $]0, +\infty[$ into the three intervals $]0, 1]$, $[1, n]$, and $[n, +\infty[$.)

(3) Let f denote a real function that is defined, continuous, and bounded on the real line. Suppose that there exist numbers H and γ satisfying the inequalities $H \geqslant 0$ and $0 < \gamma \leqslant 1$ such that

$$|f(x') - f(x)| \leqslant H|x' - x|^\gamma.$$

Set

$$F_n(x) = \int_0^{+\infty} \left[f\left(x + \frac{t}{n}\right) + f\left(x - \frac{t}{n}\right) \right] \frac{\sin^2 t}{t^2} \, dt,$$

$$C = 2 \int_0^{+\infty} \frac{\sin^2 t}{t^2} \, dt.$$

Show that the sequence $\{F_n\}$ converges uniformly to Cf.

540

Let n denote a positive integer.

(1) Show that

$$\frac{t^2}{n} - \frac{t^4}{2n^2} < \ln\left(1 + \frac{t^2}{n}\right) < \frac{t^2}{n},$$

and from this show that the sequence $\{(1+t^2/n)^{-n}\}$ converges uniformly to the function e^{-t^2} over every finite interval $[a, b]$. Show also that

$$e^{-t^2} < \left(1 + \frac{t^2}{n}\right)^{-n} < (1+t^2)^{-1}.$$

(2) Show that the functions e^{-t^2} and $(1+t^2/n)^{-n}$ are integrable over $[0, +\infty]$ and that

$$\lim_{n \to \infty} \int_0^{+\infty} \left(1 + \frac{t^2}{n}\right)^{-n} dt = \int_0^{+\infty} e^{-t^2} dt.$$

Consider separately the intervals $[0, a]$ and $[a, +\infty[$. Show that in each of these two intervals, the difference between the integrals can be majorized by $\varepsilon/2$ if the number a is suitably chosen.

(3) Show that the evaluation of the integral

$$\int_0^{+\infty} \left(1 + \frac{t^2}{n}\right)^{-n} dt$$

can be reduced by a change of variable to evaluation of one of the integrals

$$I_\alpha = \int_0^{\pi/2} \sin^\alpha x \, dx.$$

From the results of question (3) of exercise 529, evaluate

$$\int_0^{+\infty} e^{-t^2} dt.$$

541

Evaluate the integral

$$f_x(y) = \int_0^x \frac{dt}{t^2 + y},$$

where y denotes a positive number. Show that the function $f_x(y)$ is defined, continuous, and infinitely differentiable in the interval $]0, +\infty[$, and that

$$f_x^{(n)}(y) = (-1)^n(n\,!) \int_0^x \frac{dt}{(t^2+y)^{n+1}}.$$

APPLICATION: Evaluate the integral

$$I = \int_0^1 \frac{dt}{(t^2+1)^3}.$$

542

Let α denote a given real number.
  (1) Show that, if $|\alpha| \neq 1$,

$$P_n(\alpha) = \prod_0^{n-1} (1-2\alpha \cos \frac{k\pi}{n}+\alpha^2) = \frac{\alpha-1}{\alpha+1}(\alpha^{2n}-1).$$

  (2) Use this result to show that

$$I(\alpha) = \int_0^\pi \ln (1-2\alpha \cos x+\alpha^2)dx = \begin{cases} 0 & \text{if} & |\alpha| < 1 \\ \pi \ln \alpha^2 & \text{if} & |\alpha| > 1. \end{cases}$$

To do this, consider the mean value of the function

$$\ln (1-2\alpha \cos x+\alpha^2).$$

  (3) Use this result to evaluate the integral

$$J(\alpha) = \int_0^\pi \frac{\alpha-\cos x}{1-2\alpha \cos x+\alpha^2}dx,$$

when $|\alpha| \neq 1$.
  (4) Make the substitution $t=\tan (x/2)$ and evaluate $J(\alpha)$ by using the expression obtained.

543

Use the result obtained in the preceding exercise:

$$I(\alpha) = \int_0^\pi \ln (1-2\alpha \cos x+\alpha^2)dx = \begin{cases} 0 & \text{if} & |\alpha| < 1 \\ \pi \ln \alpha^2 & \text{if} & |\alpha| > 1. \end{cases}$$

(1) Suppose that $0 \leqslant \alpha \leqslant 1$ and $0 \leqslant x \leqslant \pi/3$. Show that

$$\sin^2 x \leqslant 1 - 2\alpha \cos x + \alpha^2 \leqslant 1.$$

If in addition $x \neq 0$, show that

$$\frac{3}{4} \leqslant \frac{1 - 2\alpha \cos x + \alpha^2}{2 - 2 \cos x} \leqslant \frac{1}{\sin^2 x}.$$

(2) Show that the following integrals are convergent:

$$K = \int_0^{\pi/3} \ln (\sin x) \, \mathrm{d}x, \qquad I(1) = \int_0^{\pi} \ln (2 - 2 \cos x) \, \mathrm{d}x.$$

(3) Show that $I(\alpha) - I(1)$ tends to 0 as α tends to 1. To majorize this difference, partition the interval $]0, \pi]$ into two intervals of the forms $]0, X]$ and $[X, \pi]$, and choose X and then α so that the corresponding integrals will be arbitrarily small.

(4) Use the above result to evaluate the integrals

$$\int_0^{\pi} \ln \left(\sin \frac{x}{2} \right) \mathrm{d}x, \qquad \int_0^{\pi/2} \ln (\sin x) \, \mathrm{d}x, \qquad \int_0^{\pi/2} \ln (\cos x) \, \mathrm{d}x.$$

544

Consider the sequence of functions f_n defined in $]0, +\infty[$ by

$$f_n(x) = \int_0^n e^{-xt} \frac{\sin t}{t} \, \mathrm{d}t.$$

(1) Show that, for every number $x > 0$, the sequence $\{f_n(x)\}$ converges. Show that the function f defined by

$$f(x) = \lim_{n \to \infty} f_n(x)$$

approaches 0 as x approaches $+\infty$.

(2) Show that the function f is continuous and differentiable and evaluate its derivative $f'(x)$ at an arbitrary value x. From this, find an expression for $f(x)$ in terms of elementary functions.

Solutions

501

Let us make the change of variable $a+b-x=u$. Then, when $x=a$, $u=b$, and when $x=b$, $u=a$; $dx=-du$. Thus,

$$\int_a^b xf(x)dx = \int_b^a (a+b-u)f(a+b-u)[-du]$$

$$= \int_a^b (a+b-u)f(a+b-u)du$$

$$= \int_a^b (a+b-u)f(u)du = (a+b)\int_a^b f(u)du - \int_a^b uf(u)du.$$

The symbol u on the right can be replaced with any other symbol, including x, so that

$$\int_a^b xf(x)dx = (a+b)\int_a^b f(x)dx - \int_a^b xf(x)dx,$$

from which we get the required result.

502

To majorize f, let us take the function ψ equal to the constant M:

$$\int_a^b [f(x)]^n\,dx \le \int_a^b [\psi(x)]^n dx = M^n(b-a) \Rightarrow I_n \le M(b-a)^{1/n}.$$

The sequence whose general term is

$$(b-a)^{1/n} = e^{(1/n)\ln(b-a)}$$

converges to 1. For any $\varepsilon>0$, there exists a number N_1 such that

$$n > N_1 \Rightarrow (b-a)^{1/n} < 1+\frac{\varepsilon}{M} \Rightarrow I_n < M+\varepsilon.$$

The function f, being continuous on a closed interval, attains its least upper bound M for at least one value x_0. Since the function is continuous at x_0, there exists an interval $[\alpha, \beta]$ containing x_0 such that

$$M_1 = \inf_{\alpha \leqslant x \leqslant \beta} f(x) > M - \varepsilon.$$

To minorize f, let us take the function φ defined by

$$\varphi(x) = M_1 \quad \text{if} \quad \alpha \leqslant x \leqslant \beta \quad \text{and}$$

$$\varphi(x) = 0 \quad \text{if} \quad x < \alpha \quad \text{or} \quad x > \beta.$$

Then,

$$I_n^n = \int_a^b [f(x)]^n \, dx \geqslant \int_a^b [\varphi(x)]^n \, dx = M_1^n(\beta - \alpha).$$

The limit of the sequence $\{(\beta - \alpha)^{1/n}\}$ is 1 and the limit of the sequence $\{M_1(\beta - \alpha)^{1/n}\}$ is the number $M_1 > M - \varepsilon$. Therefore, there exists a number N_2 such that

$$n > N_2 \Rightarrow M_1(\beta - \alpha)^{1/n} > M - \varepsilon \Rightarrow I_n > M - \varepsilon.$$

The two results we have obtained show that

$$n \geqslant \sup(N_1, N_2) \Rightarrow M - \varepsilon < I_n < M + \varepsilon.$$

Here we have the definition of a limit.

503

First question:

We know that Δ is the discriminant of the trinomial in k,

$$\int_a^b [f(x) + kg(x)]^2 \, dx = k^2 \int_a^b [g(x)]^2 \, dx + 2k \int_a^b f(x)g(x) \, dx + \int_a^b [f(x)]^2 \, dx.$$

If $\Delta = 0$, this trinomial has a double root $-\lambda$, and

$$\int_a^b [f(x) - \lambda g(x)]^2 \, dx = 0.$$

The function $(f - \lambda g)^2$ is continuous and nonnegative in $[a, b]$. Since its integral is 0, it is the zero function (cf. PZ, Book III, Chapter IV, Part III, § 1, remark 2) and $f = \lambda g$.

Second question:

Since the function f is continuous and positive on $[a, b]$, the functions $1/f$, $\varphi = \sqrt{f}$, and $\psi = 1/\varphi$, are defined and continuous on $[a, b]$.

When we apply the Schwarz inequality to the functions φ and ψ, we obtain

$$\left(\int_a^b \varphi(x)\psi(x)\,dx\right)^2 \leqslant \left(\int_a^b [\varphi(x)]^2\,dx\right)\left(\int_a^b [\psi(x)]^2\,dx\right),$$

so that

$$\left(\int_a^b dx\right)^2 = (b-a)^2 \leqslant \left(\int_a^b f(x)\,dx\right)\left(\int_a^b \frac{1}{f(x)}\,dx\right) = P_f.$$

The product P_f is minorized by $(b-a)^2$.

The result of the first question implies that P_f is equal to $(b-a)^2$ if and only if $\varphi = \lambda\psi$, that is,

$$f = \lambda^2 \frac{1}{f}.$$

It is necessary and sufficient that the function f be a constant.

Third question:

Let us define the functions f and $1/f$ by

$$a \leqslant x \leqslant a + \frac{b-a}{3} \Rightarrow f(x) = M \Rightarrow \frac{1}{f(x)} = \frac{1}{M},$$

$$a + 2\frac{b-a}{3} \leqslant x \leqslant b \Rightarrow f(x) = \frac{1}{M} \Rightarrow \frac{1}{f(x)} = M.$$

In the interval

$$\left[a + \frac{b-a}{3}, a + 2\frac{b-a}{3}\right],$$

f is an arbitrary continuous nonincreasing function, for example, the linear function defined by

$$f(x) = M - \frac{3}{b-a}\left(M - \frac{1}{M}\right)\left[x - a - \frac{b-a}{3}\right],$$

and $1/f$ is the inverse of that function.

Obviously, since the functions f and $1/f$ are nonnegative,

$$\left.\begin{array}{l} \displaystyle\int_a^b f(x)\,dx \geqslant \int_a^{a+(b-a)/3} f(x)\,dx = \frac{b-a}{3}M \\[3mm] \displaystyle\int_a^b \frac{1}{f(x)}\,dx \geqslant \int_{a+2(b-a)/3}^b \frac{1}{f(x)}\,dx = \frac{b-a}{3}M \end{array}\right\} \Rightarrow P_f \geqslant \frac{(b-a)^2}{9}M^2.$$

Since M is an arbitrary number, P_f can therefore take values arbitrarily large.

504

First question:

From the definitions of p, P, q, and Q,

$$g \geqslant q \;\Rightarrow\; \frac{P}{q}g \geqslant P \;\Rightarrow\; f - \frac{P}{q}g \leqslant 0,$$

$$g \leqslant Q \;\Rightarrow\; \frac{P}{Q}g \leqslant p \;\Rightarrow\; f - \frac{P}{Q}g \geqslant 0,$$

and therefore,

$$\left(f - \frac{P}{q}g\right)\left(f - \frac{P}{Q}g\right) \leqslant 0.$$

This product is the function F_λ relative to

$$\lambda = -\sqrt{\frac{Pp}{Qq}}.$$

For this value of λ, the function F_λ is thus negative.

If $\lambda \geqslant 0$, the function F_λ, being the sum of three nonnegative functions, is nonnegative.

The integral

$$\int_a^b F_\lambda(x)\,\mathrm{d}x$$

is a trinomial in λ that assumes nonnegative values for $\lambda \geqslant 0$ and that assumes nonpositive values, for example, at $\lambda = -\sqrt{Pp/Qq}$. Its discriminant Δ is nonnegative. Thus,

$$\int_a^b F_\lambda(x)\,\mathrm{d}x = \int_a^b [f(x)]^2\,\mathrm{d}x + \lambda\left(\sqrt{\frac{PQ}{pq}} + \sqrt{\frac{pq}{PQ}}\right)\int_a^b f(x)g(x)\,\mathrm{d}x + \lambda^2 \int_a^b [g(x)]^2\,\mathrm{d}x,$$

$$\Delta = \left(\sqrt{\frac{PQ}{pq}} + \sqrt{\frac{pq}{PQ}}\right)^2 \left(\int_a^b f(x)g(x)\,\mathrm{d}x\right)^2 - 4\left(\int_a^b [f(x)]^2\,\mathrm{d}x\right)\left(\int_a^b [g(x)]^2\,\mathrm{d}x\right) \geqslant 0.$$

Second question:

We shall have equality if $\Delta = 0$, that is, if the trinomial in λ

$$\int_a^b F_\lambda(x)\,\mathrm{d}x$$

does not change sign and in this case does not become negative. For $\lambda = -\sqrt{Pp/Qq}$, the function F_λ is nonpositive and so is its integral. Since its integral is 0, if f and g and hence F_λ are continuous functions, it follows that F_λ is the zero function (cf. PZ, Book III, Chapter IV, Part III, § 1, remark 2). For that value of λ,

$$F_\lambda = (f - \frac{P}{q}g)(f - \frac{p}{Q}g).$$

The first of these two factors is 0. Therefore, the functions f and g are proportional and the ratio between them is given by

$$\frac{P}{Q} = \frac{p}{q} = \begin{cases} \dfrac{P}{q} \\ \dfrac{p}{Q} \end{cases}$$

depending on which factor of F_λ is 0. In both cases, we have $p = P$ and $q = Q$. The two functions f and g must both be constants.

This necessary condition is also obviously sufficient.

Third question :

Consider the two step functions f and g defined by

$$i - 1 < x < i \Rightarrow f(x) = p_i \quad \text{and} \quad g(x) = q_i.$$

We know that, from the definition of an integral,

$$\int_0^n [f(x)]^2 \, dx = \sum_{i=1}^n p_i^2, \qquad \int_0^n [g(x)]^2 \, dx = \sum_{i=1}^n q_i^2$$

$$\int_0^n f(x)g(x) \, dx = \sum_{i=1}^n p_i q_i.$$

The numbers $p = \inf p_i$, $P = \sup p_i$, $q = \inf q_i$, and $Q = \sup q_i$ are the greatest lower and least upper bounds of the functions f and g. Therefore, the result of (1) can be written

$$4\left(\sum_{i=1}^n p_i^2\right)\left(\sum_{i=1}^n q_i^2\right) \leqslant \left(\sqrt{\frac{PQ}{pq}} + \sqrt{\frac{pq}{PQ}}\right)^2 \left(\sum_{i=1}^n p_i q_i\right)^2.$$

The numbers r, s, R, and S satisfy the inequalities

$$\left. \begin{array}{l} r \leqslant p \leqslant P \leqslant R \Rightarrow \dfrac{R}{r} \geqslant \dfrac{P}{p} \\[2mm] s \leqslant q \leqslant Q \leqslant S \Rightarrow \dfrac{S}{s} \geqslant \dfrac{Q}{q} \end{array} \right\} \Rightarrow \sqrt{\frac{RS}{rs}} \geqslant \sqrt{\frac{PQ}{pq}}.$$

The monotonic increase of the function $u + 1/u$ for $u > 1$ then implies

$$\sqrt{\frac{RS}{rs}} + \sqrt{\frac{rs}{RS}} \geqslant \sqrt{\frac{PQ}{pq}} + \sqrt{\frac{pq}{PQ}},$$

which proves the assertion made.

505

First question:

The sum S_n is the integral over $[a, b]$ of the step function φ_n defined by

$$\varphi_n(x) = f(\xi_i^{(n)}) \quad \text{if} \quad x_{i-1}^{(n)} < x < x_i^{(n)} \quad \text{and} \quad \varphi(x_i^{(n)}) = f(x_i^{(n)}).$$

The function f is uniformly continuous on $[a, b]$. Therefore,

$$\forall \, \varepsilon > 0, \, \exists \, \eta > 0 \quad \text{such that} \quad |x' - x| < \eta \Rightarrow |f(x') - f(x)| < \varepsilon.$$

Let us now determine N such that $l_n < \eta$ whenever $n > N$:

$$n > N \Rightarrow x_i^{(n)} - x_i^{(n-1)} \leqslant l_n < \eta$$
$$\Rightarrow |f(x) - f(\xi_i^{(n)})| < \varepsilon \quad \text{if} \quad x_{i-1}^{(n)} < x < x_i^{(n)}.$$

Every x in $[a, b]$ either belongs to an interval of the form $]x_{i-1}^{(n)}, x_i^{(n)}[$ or is one of the $x_i^{(n)}$. Therefore,

$$n > N \Rightarrow |f(x) - \varphi_n(x)| < \varepsilon, \quad \forall \, x \in [a, b].$$

The sequence of the functions φ_n converges uniformly to f over $[a, b]$ and, by the definition of an integral,

$$\int_a^b f(x)\,dx = \lim \int_a^b \varphi_n(x)\,dx = \lim S_n.$$

In the particular case in question,

$$S_n = \sum_{i=1}^n \left[\frac{b-a}{n} f(x_i^{(n)}) \right] = \frac{b-a}{n} \sum_{i=1}^n f\left(a + i\frac{b-a}{n} \right),$$

$$\lim \frac{1}{n} \sum_{i=1}^n f\left(a + i\frac{b-a}{n} \right) = \frac{1}{b-a} \int_a^b f(x)\,dx.$$

Second question:

Note that

$$a^{1/n} - 1 = \frac{a^{i/n} - a^{(i-1)/n}}{a^{(i-1)/n}}.$$

If we take for f the function defined by $f(x) = 1/x$, if we take for the set of points of subdivision defining the partition σ_n of $[1, a]$ the sequence of the $a^{i/n}$, and if we take for $\xi_i^{(n)}$ the number $x_{i-1}^{(n)}$, then the sum S_n defined in (1) is

$$S_n = \sum_{i=1}^{n} (x_i^{(n)} - x_{i-1}^{(n)}) f(x_{i-1}^{(n)}) = n(a^{1/n} - 1).$$

The length of each of the intervals is

$$a^{i/n} - a^{(i-1)/n} < a(a^{1/n} - 1).$$

The maximum l_n of these lengths is no greater than $a(a^{1/n} - 1)$, and its limit is 0. Now, we can use the result of (1):

$$\lim n(a^{1/n} - 1) = \int_1^a \frac{dx}{x}.$$

Third question:

Let us examine the matter of right-hand differentiability. For every number x between 0 and 1 exclusively, there corresponds an integer n defined by

$$\frac{1}{n} \leqslant x < \frac{1}{n-1}.$$

The monotonic increase of the exponential function then implies

$$a^{1/n} - 1 \leqslant a^x - 1 < a^{1/(n-1)} - 1,$$

so that

$$(n-1)(a^{1/n} - 1) < \frac{a^x - 1}{x} < n(a^{1/(n-1)} - 1).$$

As x tends to 0, n tends to $+\infty$, so that

$$(n-1)(a^{1/n} - 1) = \frac{n-1}{n} [n(a^{1/n} - 1)] \to \int_1^a \frac{dx}{x},$$

$$n(a^{1/(n-1)} - 1) = \frac{n}{n-1} [(n-1)(a^{1/(n-1)} - 1)] \to \int_1^a \frac{dx}{x},$$

and hence

$$\frac{a^x - 1}{x} \to \int_1^a \frac{dx}{x}.$$

This is the right-hand derivative.

If $x < 0$, let us write $x = -|x|$. Then,

$$\frac{a^x - 1}{x} = \frac{a^{-|x|} - 1}{-|x|} = a^{-|x|}\frac{1 - a^{|x|}}{-|x|} = a^{-|x|}\frac{a^{|x|} - 1}{|x|}.$$

As x tends to 0, $a^{-|x|}$ tends to 1 and $(a^{|x|} - 1)/|x|$ tends to $\displaystyle\int_1^a \frac{dx}{x}$.

506

$$x_n = \sum_{p=1}^{n} \frac{1}{n} \frac{1}{1 + p^2/n^2} = \sum_{p=1}^{n} (x_p^{(n)} - x_{p-1}^{(n)}) f(x_p^{(n)}),$$

where we set $x_p^{(n)} = p/n$ (thus defining a partition σ_n of $[0, 1]$ into n equal subintervals) and $f(x) = 1/(1 + x^2)$.

Thus, the sequence $\{x_n\}$ converges to

$$\int_0^1 \frac{dx}{1 + x^2} = [\arctan x]_1^0 = \frac{\pi}{4}$$

(cf. exercise 505, (1)).

For y_n, let us take the logarithm, so as to replace the product with a sum:

$$\ln y_n = \ln \sqrt[n]{\prod_{p=1}^{n}\left(1 + \frac{p}{n}\right)} = \frac{1}{n}\sum_{p=1}^{n} \ln\left(1 + \frac{p}{n}\right).$$

Thus, $\ln y_n$ is the sum S_n relative to the logarithm function and the partition σ_n of the interval $[1, 2]$ into n equal subintervals if we choose

$$\xi_p^{(n)} = x_p^{(n)} = 1 + p/n$$

(in the notations of (1) of exercise 505). Therefore, the sequence $\{\ln y_n\}$ converges to

$$\int_1^2 \ln x \, dx = [x \ln x - x]_1^2 = 2 \ln 2 - 1.$$

The sequence $\{z_n\}$ is studied in a different way, and the resemblance between the expressions for z_n and y_n is only superficial.

The sequence whose general term is $u_n = 1 + a/n$ converges to 1. We have shown (Part 4, exercise 413, (3)) that the sequence $\{z_n\}$ defined by

$$z_n = \sqrt[n]{u_1 \cdots u_n}$$

has the same limit as the sequence $\{u_n\}$. Thus, the sequence $\{z_n\}$ converges to 1.

507

First question:

Let h denote the function defined by

$$h(u) = \frac{u^p}{p} - uv.$$

This function is continuous and differentiable for all $u \geqslant 0$.

$$h'(u) = u^{p-1} - v \Rightarrow h'(u) = 0 \quad \text{if} \quad u = u_0 = v^{1/(p-1)}.$$

The derivative $h'(u)$ has the same sign as does $u - u_0$. Therefore, the function h has a minimum at u_0 equal to

$$h(u_0) = \frac{u_0^p}{p} - u_0 v = v^{p/(p-1)} \left(\frac{1}{p} - 1 \right) = -\frac{v^q}{q}$$

$$h(u) \geqslant h(u_0) \Leftrightarrow \frac{u^p}{p} - uv \geqslant -\frac{v^q}{q}.$$

Second question:

When we assign to u and v the values indicated, we obtain

$$\frac{|f(t)g(t)|}{N_p(f)N_q(g)} \leqslant \frac{|f(t)|^p}{p[N_p(f)]^p} + \frac{|g(t)|^q}{q[N^q(g)]^q}.$$

This inequality implies the corresponding inequality for the integrals:

$$\frac{\int_a^b |f(t)g(t)| \, dt}{N_p(f)N_q(f)} \leqslant \frac{\int_a^b |f(t)|^p \, dt}{p[N_p(f)]^p} + \frac{\int_a^b |g(t)|^q \, dt}{q[N_q(f)]^q} = \frac{1}{p} + \frac{1}{q} = 1.$$

The inequality asserted is immediately obtained if we note that

$$\left| \int_a^b f(t)g(t) \, dt \right| \leqslant \int_a^b |f(t)g(t)| \, dt.$$

Third question:

Let us apply the result of (2) by taking

$$\frac{1}{p} = \alpha, \qquad \frac{1}{q} = \beta \Rightarrow \frac{1}{p} + \frac{1}{q} = 1,$$

$$f = \varphi^\alpha, \qquad g = \psi^\beta.$$

All these functions are nonnegative, therefore,

$$\int_a^b [\varphi(t)]^\alpha [\psi(t)]^\beta \, dt \leqslant \left\{ \int_a^b [\varphi(t)]^{\alpha p} \, dt \right\}^\alpha \left\{ \int_a^b [\varphi(t)]^{\beta q} \, dt \right\}^\beta,$$

and since $\alpha p = \beta q = 1$, the asserted inequality follows.

508

First question:

1. Examination under the assumption that f is differentiable

Since the function f is continuous, the function that maps x into

$$\int_a^x f(t) \, dt$$

is differentiable and its derivative is $f(x)$.

The inverse g of a strictly monotonic continuous function is continuous. Therefore, the function that maps y into

$$\int_a^y g(t) \, dt$$

has a derivative equal to $g(y)$. The function that maps x into

$$\int_a^{f(x)} g(t) \, dt$$

is a composite of the function f (assumed to be differentiable) and the preceding function. Therefore, it has a derivative equal to

$$g[f(x)]f'(x) = xf'(x).$$

(By the definition of an inverse function, $x = g[f(x)]$.)

Therefore, the function F is differentiable and

$$F'(x) = f(x) + xf'(x) - [xf(x)]' = 0.$$

2. A direct examination in the general case

Let us consider the values x and $x + \Delta x$ of the independent variable and the corresponding values $f(x)$ and $f(x + \Delta x) = f(x) + \Delta f$ of the function. In our proof, we shall assume that $\Delta x > 0$ in order to specify the direction of the inequalities that we shall use. The result, however, remains the same for $\Delta x < 0$.

Then,

$$\Delta F = F(x+\Delta x)-F(x) = \int_x^{x+\Delta x} f(t)\,dt$$

$$+\int_{f(x)}^{f(x+\Delta x)} g(t)\,dt - [(x+\Delta x)(f(x)+\Delta f)-xf(x)].$$

Since f and g are strictly increasing functions,

$$\Delta xf(x) \leqslant \int_x^{x+\Delta x} f(t)\,dt \leqslant \Delta xf(x+\Delta x) = \Delta x[f(x)+\Delta f],$$

$$x\Delta f = g[f(x)]\Delta f \leqslant \int_{f(x)}^{f(x+\Delta x)} g(t)\,dt \leqslant g[f(x+\Delta x)]\Delta f = (x+\Delta x)\Delta f.$$

If we add these two inequalities and compare the results with the expression for ΔF (expanding the terms in the brackets in that expression). We obtain

$$-\Delta x\Delta f \leqslant \Delta F \leqslant \Delta x\Delta f \Leftrightarrow -\Delta f \leqslant \frac{\Delta F}{\Delta x} \leqslant \Delta f.$$

As Δx tends to 0, so does Δf (because of the continuity of f). Therefore, the function F has a derivative that is everywhere 0 and hence F is a constant:

$$F(x) = F(0) = 0.$$

Second question:

If the inequality $v \geqslant f(u)$ is not satisfied, we have $f(u) > v$ and hence $u > g(v)$. Let us suppose that $v \geqslant f(u)$, remaining the functions f and g, which play the same role, if necessary.

We know from (1) that

$$uf(u) = \int_0^u f(t)\,dt + \int_0^{f(u)} g(t)\,dt.$$

In the interval $[f(u), v]$, the function g is minorized by $g[f(u)] = u$ and

$$u[v-f(u)] \leqslant \int_{f(u)}^v g(t)\,dt.$$

If we add $u\,f(u)$ to both members of this inequality, we obtain

$$uv \leqslant \int_0^u f(t)\,dt + \int_0^{f(u)} g(t)\,dt + \int_{f(u)}^v g(t)\,dt = \int_0^u f(t)\,dt + \int_0^v g(t)\,dt.$$

Let us set $f(u)=u^{p-1}$. The function f is continuous and strictly increasing since $p-1>0$. The inverse function g is defined by

$$g(v) = u \Leftrightarrow v = f(u) = u^{p-1} \Leftrightarrow u = v^{1/(p-1)} = v^{q-1}.$$

For these particular functions, the inequality becomes

$$uv \leqslant \int_0^u t^{p-1}\,\mathrm{d}t + \int_0^v t^{q-1}\,\mathrm{d}t = \frac{u^p}{p} + \frac{v^q}{q}.$$

509

First question:

The function G is a primitive of the function g. It is nondecreasing and its maximum and minimum values are respectively $G(a)=0$ and $G(b)=l$.

The derivative of the function $x-a-G(x)$ is $1-g(x)$. It is nondecreasing and everywhere at least as great as its value at $x=a$ (which is 0).

Second question:

The functions f and g are continuous. Therefore, the function φ is differentiable and its derivative is $f(x)g(x)$. The derivative of the function that maps u into

$$\int_a^u f(t)\,\mathrm{d}t$$

is $f(u)$. The function ψ, being a composite of the function $a+G$ and the preceding function, is differentiable and its derivative is

$$\psi'(x) = f[a+G(x)]G'(x) = f[a+G(x)]g(x).$$

The function f is nondecreasing and the function g is nonnegative. Therefore,

$$a+G(x) \leqslant x \Rightarrow f[a+G(x)] \leqslant f(x) \Rightarrow \psi'(x) \leqslant \varphi'(x).$$

The function $\varphi-\psi$ is nondecreasing and its value at $x=a$ is 0. Therefore, it is nonnegative in $[a, b]$ and $\varphi \geqslant \psi$. In particular, $\varphi(b) \geqslant \psi(b)$ and

$$\int_a^b f(t)g(t)\,\mathrm{d}t \geqslant \int_a^{a+l} f(t)\,\mathrm{d}t.$$

Third question:

Equality will hold in the last conditional inequality if $\varphi(b)-\psi(b)=0$. The nondecreasing function $\varphi-\psi$, which vanishes at a and at b, is then the constant zero function and its derivative $\varphi' - \psi'$ is 0 (a necessary and sufficient condition):

$$\varphi'(x) - \psi'(x) = \{f(x) - f[a + G(x)]\}\, g(x) = 0.$$

By hypothesis, g is nonzero. Therefore, it is necessary and sufficient that

$$f(x) = f[a + G(x)].$$

By hypothesis, the function $x - (a + G)$ has a positive derivative. Therefore, it is strictly increasing and positive if $x > a$. The nondecreasing function f assumes the same value at $a + G(x)$ and x. It is constant in the interval $[a + G(x), x]$.

In particular, the function f is constant and equal to $f(b)$ in the interval $[a + G(b), b]$. Therefore, let us consider the greatest lower bound x_0 of the set of those x such that $f(x) \equiv f(b)$ in the interval $[x, b]$.

In every interval of the form $[x_0, x_0 + h[$, there exists x such that $f(x) = f(b)$. The continuity of the function at x_0 then implies that $f(x_0) = f(b)$ and that the function is constant in the interval $[x_0, b]$.

If $x_0 > a$, the function f is constant on the interval $[a + G(x_0), x_0]$. Its value on that interval is $f(x_0) = f(b)$, and the function f is constant and equal to $f(b)$ on the interval $[a + G(x_0), b]$ in contradiction with the hypothesis since

$$a + G(x_0) < x_0 = \inf\{ x\colon\ f(t) = f(b) \quad \text{for} \quad t \in [x, b]\}.$$

In sum, equality will hold in the conditional inequality in question if and only if the function f is constant on $[a, b]$.

510

First question:

Let us denote by G the primitive of g that vanishes at a. If, for every value x, the function f has a derivative $f'(x)$ and if f' is continuous, we know that

$$(1) \qquad \int_a^b f(t)g(t)\,dt = [f(t)G(t)]_a^b - \int_a^b f'(t)G(t)\,dt.$$

If for a finite number of values $x_1 < x_2 < \ldots < x_n$, the function f is not differentiable or if the derivative f' is discontinuous, we write the above formula for each of the subintervals $[a, x_1], [x_1, x_2], \ldots [x_n, b]$ and add the equations obtained. This gives us the formula (1). By hypothesis, the function f is monotonic. We shall assume that f is nondecreasing. (In the opposite case, we can replace f with its negative.) The derivative f' of the nondecreasing function f is nonnegative and the function G is continuous. The hypotheses of the first theorem of the mean are satisfied for the functions f' and G:

$$\int_a^b f'(t)G(t)\,dt = G(\xi) \int_a^b f'(t)\,dt = G(\xi)[f(b) - f(a)].$$

(If f' is continuous, f is a primitive of f' and we know that in this case

$$\int_a^b f'(t)\mathrm{d}t = f(b) - f(a).$$

Otherwise, we apply the above result to the intervals $[x_i, x_{i+1}]$ defined by the points of discontinuity and then add the equations obtained.)

Then, eq. (1) becomes

$$\int_a^b f(t)g(t)\mathrm{d}t = f(b)G(b) - G(\xi)[f(b) - f(a)] = f(a)G(\xi) + f(b)[G(b) - G(\xi)]$$

$$= f(a)\int_a^\xi g(t)\mathrm{d}t + f(b)\int_\xi^b g(t)\mathrm{d}t.$$

Second question:

For each of the functions f_n, we can write

$$\int_a^b f_n(t)G(t)\mathrm{d}t = f_n(b)G(b) - G(\xi_n)[f_n(b) - f_n(a)].$$

By hypothesis, the sequence $\{f_n\}$ converges uniformly to f. The function G is bounded. Therefore, the sequence $\{f_n G\}$ converges uniformly to fG, and

$$\lim \int_a^b f_n(t)G(t)\mathrm{d}t = \int_a^b f(t)G(t)\mathrm{d}t.$$

The sequence $\{\xi_n\}$ contains a convergent subsequence $\{\xi_{n_i}\}$ (by a corollary of the Bolzano-Weierstrass theorem), and the limit of this subsequence, which we denote by ξ, belongs to $[a, b]$. The continuity of the function G implies that the sequence $G(\xi_{n_i})$ converges to $G(\xi)$. The subsequence whose general term is

$$\int_a^b f_{n_i}(t)G(t)\mathrm{d}t,$$

taken from the convergent sequence whose general term is

$$\int_a^b f_n(t)G(t)\mathrm{d}t,$$

has the same limit as does the latter, and therefore,

$$\int_a^b f(t)G(t)\mathrm{d}t = \lim \int_a^b f_{n_i}(t)G(t)\mathrm{d}t = f(b)G(b) - G(\xi)[f(b) - f(a)].$$

Thus, we again arrive at the result of (1).

The result of exercise 423 can be stated as follows: A continuous function f defined on an interval $[a, b]$ is the uniform limit of a sequence of functions λ. Each of these functions is defined by a partition $a < x_1 < x_2 < \ldots < x_n < b$ of $[a, b]$ and the requirements

$$\lambda \text{ is a linear function on } [x_i, x_{i+1}];$$
$$\lambda(x_i) = f(x_i).$$

It is then evident that λ is a continuous function differentiable except at the values x_i. This derivative is constant in each of the intervals $]x_i, x_{i+1}[$ and hence has right- and left-hand limits at x_i. Furthermore, λ is monotonic in each of the subintervals $[x_i, x_{i+1}]$. If the function f is nondecreasing, the conditions $f(x_i) \leqslant f(x_{i+1})$ imply that λ is nondecreasing in each of the subintervals $[x_i, x_{i+1}]$ and hence nondecreasing in $[a, b]$. (If f is nonincreasing, λ is nonincreasing.)

Every continuous monotonic function is the uniform limit of a sequence $\{\lambda_n\}$ of monotonic functions λ, that is, of functions satisfying the hypotheses of (1).

511

First question:

The polynomial P_n can be expanded by the binomial formula on the one hand and by Maclaurin's formula on the other:

$$P_n(x) = \sum_{k=0}^{n} (-1)^{n-k} \frac{\binom{n}{k}}{n!} b^k a^{n-k} x^{n+k} = \sum_h \frac{P_n^{(h)}(0)}{h!} x^h.$$

From this, we get

$$P_n^{(h)}(0) = 0 \quad \text{if} \quad h < n \quad \text{or} \quad h > 2n$$

$$P_n^{(n+k)}(0) = (-1)^{n-k}(n+k)! \frac{\binom{n}{k}}{n!} b^k a^{n-k} \quad \text{if} \quad h = n+k.$$

The numbers $\binom{n}{k}$ and $(n+k)!/n!$ are both integers. Therefore, $P_n^{(n+k)}(0)$ is an integer.

To study the values of the derivatives at $x = a/b$, let us set $x - a/b = u$:

$$P_n(x) = \left(\frac{a}{b} + u\right)^n \frac{b^n u^n}{n!} = \frac{u^n(a + bu)^n}{n!} = \Pi_n(u).$$

The derivatives of Π_n, for $u = 0$, are integers. They are equal to the derivatives of P_n for $x = a/b$ since the Taylor expansion of P_n about the number a/b is the Maclaurin expansion of Π_n.

Second question:

Define

$$M = \sup_{0 \leqslant x \leqslant \pi} |x(bx-a)|.$$

Then,

$$|P_n(x) \sin x| \leqslant |P_n(x)| \leqslant \frac{M^n}{n!} \quad \text{if} \quad 0 \leqslant x \leqslant \pi,$$

$$|I_n| \leqslant \int_0^\pi |P_n(x) \sin x| \, dx \leqslant \int_0^\pi \frac{M^n}{n!} \, dx = \pi \frac{M^n}{n!}.$$

The sequence of ratios M_n of consecutive terms of the sequence $\{\pi M^n/n!\}$ converges to 0. Therefore, the sequence $\{\pi M^n/n!\}$ converges to 0 and so does the sequence $\{I_n\}$.

Third question:

Let us write the formula for $(2n+1)$-fold integration by parts:

$$\int_{x_1}^{x_2} f(x)g^{(2n+1)}(x)\,dx$$

$$= [f(x)g^{(2n)}(x) - f'(x)g^{(2n-1)}(x) + \cdots + (-1)^{2n}f^{(2n)}(x)g(x)]_{x_1}^{x_2}$$

$$+ (-1)^{2n+1} \int_{x_1}^{x_2} f^{(2n+1)}(x)g(x)\,dx$$

(cf. PZ, Book III, Chapter IV, Part 3, § 4).

Let us apply this formula by taking for f the polynomial P_n and taking for $g^{(2n+1)}$ the sine function. The $(2n+1)$st derivative of P_n is the zero function, and the successive primitives $g^{(2n)}, \ldots, g$ are $\pm \sin x$ or $\pm \cos x$. Therefore, taking $x_1 = 0$ and $x_2 = a/b$, we obtain

$$I_n = \int_0^{a/b} P_n(x) \sin x \, dx$$

$$= [-P_n(x) \cos x + \cdots + (-1)^{n+1} P_n^{(2n)}(x) \cos x]_0^{a/b}.$$

If $a/b = \pi$, the sine and cosine are equal respectively to 0 and ± 1 at 0 and at a/b. All the derivatives of the polynomial P_n assume integral values at these same points. Therefore, the integral I_n is an integer.

I_n is the integral over $[0, \pi]$ of the function $P_n \sin x$, which is *continuous* and nonnegative in that interval and assumes somewhere in that interval nonzero values. Therefore, I_n cannot be equal to 0 (cf. PZ, Book III, Chapter IV, part 3, § 1, remark 2).

In conclusion, if $\pi = a/b$, I_n is a nonzero integer. Then, the sequence $\{I_n\}$ cannot converge to 0, in contradiction with the result of (2). Therefore, the number π is irrational.

Fourth question :

Let us show as in (2) that the limit of the sequence $\{I_n\}$ is 0 (e^x is bounded above by e^r and J_n by $re^r M^n/n!$). Applying the formula given above for $(2n+1)$-fold integration by parts to the functions $f = P_n$ and $g = e^x$, we get

$$ J_n = \int_0^r P_n(x) e^x \, dx = [P_n(x) e^x + \cdots + P_n^{(2n)}(x) e^x]_0^r. $$

If we choose for a and b two integers such that $a/b = r$, the values of P_n and its derivatives at 0 and at a/b are integers, and

$$ J_n = e^r N_1 - N_2, $$

where N_1 and N_2 are integers. Finally, the integral of the *continuous* nonnegative function $P_n e^x$ is a positive number. If e^r is a rational number p/q, the nonzero number

$$ J_n = \frac{p N_1 - q N_2}{q} $$

is at least equal to $1/q$, and the sequence $\{J_n\}$ cannot converge to 0. Therefore, the numbers e^r and, in particular, $e^1 = e$ are irrational.

512

First question :

Let us form the difference

$$ 2h[g_h(x') - g_h(x)] = \left[\int_{x-h}^{x'+h} f(t) \, dt - \int_{x-h}^{x'-h} f(t) \, dt \right] - \int_{x-h}^{x+h} f(t) \, dt $$

$$ = \int_{x+h}^{x'+h} f(t) \, dt - \int_{x-h}^{x'-h} f(t) \, dt. $$

The regulated function $|f|$ is bounded above on the interval $[a, b]$ by the number M. Therefore,

$$ \left| \int_{x+h}^{x'+h} f(t) \, dt \right| \leq \left| \int_{x+h}^{x'+h} |f(t)| \, dt \right| \leq \left| \int_{x+h}^{x'+h} M \, dt \right| = M|x' - x|. $$

We have the same result for the second integral. Therefore,

$$2h|g_h(x')-g_h(x)| \leqslant \left| \int_{x+h}^{x'+h} f(t)dt \right| + \left| \int_{x-h}^{x'-h} f(t)dt \right| \leqslant 2M|x'-x|.$$

Obviously, this inequality implies continuity of the function g_h.

If f is continuous and if $x_0 \in [a, b]$, we can write

$$g_h(x) = \frac{1}{2h} \left[\int_{x_0}^{x+h} f(t)dt - \int_{x_0}^{x-h} f(t)dt \right].$$

Each of the two integrals, being a composite function of x through the function $x+h$ or $x-h$, is differentiable and so is g_h:

$$g_h'(x) = \frac{1}{2h}[f(x+h)-f(x-h)].$$

Therefore, it is obvious that, if f is n times differentiable on $[a, b]$, the function g_h' is n times differentiable on $[a+h, b-h]$. Therefore, g_h is $n+1$ times differentiable.

Second question:

Let us apply the definition of convex functions. Let λ denote a number such that $0 < \lambda < 1$. Then,

$$g_h[\lambda x + (1-\lambda)x'] = \frac{1}{2h} \int_{-h}^{h} f[\lambda x + (1-\lambda)x' + u]du.$$

From the definition of a convex function,

$$f[\lambda x + (1-\lambda)x' + u]$$

$$= f[\lambda(x+u) + (1-\lambda)(x'+u)] \leqslant \lambda f[x+u] + (1-\lambda)f[x'+u]$$

$$g_h[\lambda x + (1-\lambda)x']$$

$$\leqslant \frac{1}{2h} \int_{-h}^{h} \{\lambda f[x+u] + (1-\lambda)f[x'+u]\}du = \lambda g_h(x) + (1-\lambda)g_h(x').$$

Therefore, the function g_h is convex.

Third question:

The function f has right- and left-hand limits at every value of x since it is a regulated function. Therefore, let us examine

$$g_h(x) - \frac{f(x+0)+f(x-0)}{2}$$

$$= \frac{1}{2h}\int_x^{x+h} [f(t)-f(x+0)]\,dt + \frac{1}{2h}\int_{x-h}^x [f(t)-f(x-0)]\,dt.$$

From the definition of a limit, for any $\varepsilon>0$, there exists an $\eta>0$ such that

$$x < t < x+\eta \Rightarrow |f(t)-f(x+0)| < \varepsilon,$$

and

$$x-\eta < t < x \Rightarrow |f(t)-f(x-0)| < \varepsilon.$$

If we take for h a number less than η, we obtain

$$]x, x+h[\subset]x, x+\eta[\Rightarrow \left|\int_x^{x+h} [f(t)-f(x+0)]\,dt\right| < \varepsilon h,$$

$$]x-h, x[\subset]x-\eta, x[\Rightarrow \left|\int_{x-h}^x [f(t)-f(x-0)]\,dt\right| < \varepsilon h,$$

$$\left|g_h(x) - \frac{f(x+0)+f(x-0)}{2}\right|$$

$$\leqslant \frac{1}{2h}\left[\left|\int_x^{x+h} [f(t)-f(x+0)]\,dt\right| + \left|\int_{x-h}^x [f(t)-f(x-0)]\,dt\right|\right] < \varepsilon.$$

The limit as $h\to 0$ of the function that maps h into $g_h(x)$ is $\frac{1}{2}[f(x+0)+f(x-0)]$.

Fourth question:

If f is continuous, the limit of $g_h(x)$ as $h\to 0$ is $f(x)$. Let us take values of h so small that $a+h < \alpha < \beta < b-h$. To show that f is the uniform limit of functions g_h in $[\alpha, \beta]$, we need show only that f can be approximated uniformly by the functions g_h; that is, for given $\varepsilon>0$, the number η in the preceding proof can be the same for all x in $[\alpha, \beta]$. This result is a consequence of the uniform continuity of the function f in $[a, b]$.

$$\forall \varepsilon, \exists \eta \quad \text{such that} \quad |t-x| < \eta \Rightarrow |f(t)-f(x)| < \varepsilon \quad \forall x \in [a, b],$$

$$h < \inf(\eta, \alpha-a, b-\beta) \Rightarrow |g_h(x)-f(x)| < \varepsilon \quad \forall x \in [\alpha, \beta].$$

(Here, we need choose only a sequence $\{h_n\}$ that converges to 0. Then, the corresponding sequence $[g_{h_n}]$ converges uniformly to f in $[\alpha, \beta]$.)

Fifth question:

Proof by induction on n

In the two parts of the proof, the numbers a and b will be numbers satisfying the inequalities

$$a_1 < a < \alpha < \beta < b < b_1.$$

If $n=1$, the function f is continuous on $[a, b]$ since it is convex in that interval. Therefore, it is the uniform limit in $[\alpha, \beta] \subset]a, b[$ of functions g_h. These functions are convex (cf. (2)) and differentiable since f is continuous (cf. (1)).

Suppose the result is true for $n=k-1$. Let us apply it to the interval $[a, b]$. The function f is the uniform limit in $[a, b]$ of convex functions that are $k-1$ times differentiable. Therefore, for any $\varepsilon > 0$, there exists a convex function φ that is $k-1$ times differentiable such that

$$\sup_{a \leqslant x \leqslant b} |f(x) - \varphi(x)| < \frac{\varepsilon}{2}.$$

The function φ is the uniform limit in $[\alpha, \beta] \subset]a, b[$ of functions γ_h defined in terms of φ in the same way that the functions g_h were defined in terms of f. Therefore, there exists a function γ_{h_0} that is convex (since φ is convex) and k times differentiable (since φ is $k-1$ times differentiable) and that satisfies the inequality

$$\sup_{\alpha \leqslant x \leqslant \beta} |\varphi(x) - \gamma_{h_0}(x)| < \frac{\varepsilon}{2}.$$

The two inequalities obtained imply that

$$\sup_{\alpha \leqslant x \leqslant \beta} |f(x) - \gamma_{h_0}(x)| \leqslant \sup_{\alpha \leqslant x \leqslant \beta} |f(x) - \varphi(x)| + \sup_{\alpha \leqslant x \leqslant \beta} |\varphi(x) - \gamma_{h_0}(x)| < \varepsilon.$$

The function f can be approximated uniformly by a function γ_{h_0} that is convex and k times differentiable. Therefore, it is the uniform limit of such functions.

513

The function f_n can be decomposed easily as follows:

$$f_n(x) = -\frac{1}{\ln n} \cdot \frac{1}{x + \dfrac{1}{\ln n}} + \frac{nx}{1 + nx^2 \ln n}.$$

Then, F_n a primitive of f_n, is given by

$$F_n(x) = -\frac{1}{\ln n}\ln(x\ln n + 1) + \frac{1}{2\ln n}\ln(1 + nx^2\ln n).$$

With the aid of F_n, we evaluate I_n:

$$I_n = F_n(1) - F_n(0) = -\frac{1}{\ln n}\ln(1 + \ln n) + \frac{1}{2\ln n}\ln(1 + n\ln n).$$

We saw in exercise 441, (4) that, if f and g are two equivalent functions both tending to $+\infty$, the functions $\ln f$ and $\ln g$ are also equivalent functions. Therefore, as $n \to +\infty$,

$$\frac{\ln(1 + \ln n)}{\ln n} \simeq \frac{in(\ln n)}{\ln n} \to 0$$

(since $t = \ln n \to +\infty$ and we know that $(\ln t)/t \to 0$ as $t \to 0$).

$$\frac{\ln(1 + n\ln n)}{\ln n} \simeq \frac{\ln(n\ln n)}{\ln n} = \frac{\ln n}{\ln n} + \frac{\ln(\ln n)}{\ln n} \to 1.$$

The sequence $\{I_n\}$ converges to $\frac{1}{2}$.

$$\text{If} \quad x \neq 0, \quad f_n(x) \simeq \frac{nx}{x^3n(\ln n)^2} \to \quad 0 = f(x);$$

$$\text{if} \quad x = 0, \quad f_n(0) = -1 \quad \to -1 = f(0).$$

The integral of the step function f is 0 and consequently,

$$\lim \int_0^1 f_n(x)\,dx \neq \int_0^1 [\lim f_n(x)]\,dx.$$

This result is not particularly surprising. We have proven that these two numbers are equal when the convergence of the sequence $\{f_n\}$ is uniform, which is not the case here since

$$f_n\left(\frac{1}{\sqrt{n}}\right) = \frac{\sqrt{n-1}}{\left(\frac{in\,n}{\sqrt{n}} + 1\right)(1 + \ln n)} \simeq \frac{\sqrt{n}}{\ln n} \to +\infty \quad \text{as} \quad n \to +\infty,$$

$$\|f_n - f\| = \sup_{0 \leqslant x \leqslant 1} |f_n(x) - f(x)| \geqslant f_n\left(\frac{1}{\sqrt{n}}\right) - f\left(\frac{1}{\sqrt{n}}\right) = f_n\left(\frac{1}{\sqrt{n}}\right).$$

The limit of the sequence $\{\|f_n - f\|\}$ is $+\infty$ and not 0. Therefore, the sequence $\{f_n\}$ does not converge to f uniformly.

514

First question:

As t tends to 0, $\sin nt \simeq nt$ and $\sin t \simeq t$ and

$$\frac{\sin^2 nt}{\sin^2 t} \simeq \frac{n^2 t^2}{t^2} = n^2 \implies \frac{\sin^2 nt}{\sin^2 t} \to n^2.$$

We know (exercise 439 of part 4) that

$$0 \leqslant t \leqslant \frac{\pi}{2} \implies \frac{2t}{\pi} \leqslant \sin t \leqslant t,$$

$$0 \leqslant t \leqslant \frac{\pi}{2n} \implies 0 \leqslant nt \leqslant \frac{\pi}{2} \implies \frac{2nt}{\pi} \leqslant \sin nt,$$

$$0 < t \leqslant \frac{\pi}{2n} \implies \frac{\sin nt}{\sin t} \geqslant \frac{2nt}{\pi} \frac{1}{t} = \frac{2n}{\pi}.$$

The function s_n is therefore minorized by $4n^2/\pi^2$ on $[0, \pi/2n]$. This is a consequence of the preceding result for $]0, \pi/2n]$ and the fact that $4n^2/\pi^2$ is a minorant of $n^2 = s_n(0)$.

The function s_n is nonnegative. Consequently,

$$I_n = \int_0^{\pi/2n} s_n(t)\,dt + \int_{\pi/2n}^{\pi/2} s_n(t)\,dt \geqslant \int_0^{\pi/2n} s_n(t)\,dt \geqslant \int_0^{\pi/2n} \frac{4n^2}{\pi^2}\,dt = \frac{2n}{\pi}.$$

The sequence $\{I_n\}$, which is minorized by the sequence $\{2n/\pi\}$, approaches $+\infty$.

Second question:

Let us set

$$\varphi_\eta = \sup_{0 < t < \eta} |\varphi(t)| \qquad \text{and} \qquad \Phi = \sup_{0 < t < \pi/2} |\varphi(t)|.$$

We can write the following inequalities regarding the integrals:

$$\left| \int_0^\eta \varphi(t) \frac{\sin^2 nt}{\sin^2 t}\,dt \right| \leqslant \int_0^\eta |\varphi(t)| \frac{\sin^2 nt}{\sin^2 t}\,dt \leqslant \varphi_\eta \int_0^\eta \frac{\sin^2 nt}{\sin^2 t}\,dt \leqslant \varphi_\eta I_n.$$

Since $\Phi/\sin^2\eta$ is a majorant of

$$\left| \varphi(t) \frac{\sin^2 nt}{\sin^2 t} \right|$$

on $]\eta, \pi/2[$, we have

$$\left| \int_\eta^{\pi/2} \varphi(t) \frac{\sin^2 nt}{\sin^2 t} dt \right| \leq \int_\eta^{\pi/2} \frac{\Phi}{\sin^2 \eta} dt = \left(\frac{\pi}{2} - \eta\right) \frac{\Phi}{\sin^2 \eta} \leq \frac{\pi}{2} \frac{\Phi}{\sin^2 \eta}.$$

These inequalities yield for $|u_n|$

$$|u_n| \leq \frac{1}{I_n} \left(\varphi_n I_n + \frac{\pi}{2} \frac{\Phi}{\sin^2 \eta}\right) = \varphi_n + \frac{1}{I_n} \frac{\pi}{2} \frac{\Phi}{\sin^2 \eta}.$$

Since $\varphi(+0) = 0$, for every $\varepsilon > 0$, there exists an η such that

$$t < \eta \Rightarrow |\varphi(t)| < \frac{\varepsilon}{2} \Rightarrow \varphi_\eta \leq \frac{\varepsilon}{2}.$$

Suppose that we have chosen such a number η. Then, the expression

$$\frac{\pi}{2} \frac{\Phi}{\sin^2 \eta}$$

is defined and is independent of n. Its product with $1/I_n$ approaches 0 since I_n approaches $+\infty$. Therefore, there exists a number N such that

$$n > N \Rightarrow \frac{1}{I_n} \frac{\pi}{2} \frac{\Phi}{\sin^2 \eta} < \frac{\varepsilon}{2}.$$

In conclusion, for given ε, we have determined a number N (in terms of η, but the manner of determination is immaterial) such that

$$n > N \Rightarrow |u_n| < \varepsilon.$$

Then, by definition, the sequence $\{u_n\}$ converges to 0.

Third question:

Obviously,

$$f(+0) = \frac{I_n}{I_n} f(+0) = \frac{1}{I_n} \int_0^{\pi/2} f(+0) \frac{\sin^2 nt}{\sin^2 t} dt,$$

$$v_n - f(+0) = \frac{1}{I_n} \int_0^{\pi/2} [f(t) - f(+0)] \frac{\sin^2 nt}{\sin^2 t} dt.$$

The function $f - f(+0)$ is a regulated function and its right-hand limit at 0 is zero. It satisfies the same hypotheses as does the function φ in (2). Therefore,

$$\lim (v_n - f(+0)) = 0 \Rightarrow \lim v_n = f(+0).$$

515

First question:

The primitive that vanishes at 0 of the even polynomial $(1-t^2)^n$ is an odd polynomial. Since f_n is the quotient obtained by dividing this polynomial by a constant, f_n too is an odd polynomial.

The primitive g_n of the odd polynomial f_n is an even polynomial.

Second question:

The inequalities mentioned are easily obtained by multiplying the inequalities that t satisfies by the nonnegative number $(1-t^2)^n$. If $0 < x \leqslant 1$,

$$1 - f_n(x) = \frac{\displaystyle\int_x^1 (1-t^2)^n dt}{\displaystyle\int_0^1 (1-t^2)^n dt},$$

being the ratio of two nonnegative numbers, is bounded below by 0. We obtain an upper bound for this ratio by majorizing the numerator and minorizing the denominator:

$$\int_x^1 (1-t^2)^n dt \leqslant \int_x^1 \frac{t}{x}(1-t^2)^n dt = \frac{(1-x^2)^{n+1}}{2(n+1)x},$$

$$\int_0^1 (1-t^2)^n dt \geqslant \int_0^1 t(1-t^2)^n dt = \frac{1}{2(n+1)},$$

$$0 \leqslant 1 - f_n(x) \leqslant \frac{(1-x^2)^{n+1}}{2(n+1)x} : \frac{1}{2(n+1)} = \frac{(1-x^2)^{n+1}}{x},$$

$$0 \leqslant 1 - x^2 < 1 \Rightarrow \lim_{n \to \infty} (1-x^2)^{n+1} = 0 \Rightarrow \lim_{n \to \infty} 1 - f_n(x) = 0,$$

$$f_n(0) = 0 \Rightarrow \lim_{n \to \infty} f_n(0) = 0,$$

$$f_n(x) = -f_n(-x) \Rightarrow \lim_{n \to \infty} f_n(x) = -\lim_{n \to \infty} f_n(-x) = -1 \quad \text{if} \quad -1 \leqslant x < 0.$$

Third question:

In the interval $[\alpha, 1]$, the limit f of the sequence $\{f_n\}$ is the function whose constant value is equal to 1. If we denote by v_α the norm relative to this interval, we have

$$v_\alpha(f-f_n) = \sup_{\alpha \leqslant x \leqslant 1} (1-f_n(x)) \leqslant \frac{(1-\alpha^2)^{n+1}}{\alpha}.$$

The numerical sequence $(1-\alpha^2)^{n+1}/\alpha$ converges to 0 and hence so does the sequence $\{v_\alpha(f-f_n)\}$. The sequence $\{f_n\}$ converges uniformly to f on $[\alpha, 1]$.

The functions f_n are continuous on $[-\alpha, \alpha]$. If the sequence $\{f_n\}$ is uniformly convergent, the limit function f is continuous on $[-\alpha, \alpha]$. Now, the left-hand and right-hand limits of the function f at 0 are respectively

$$f(-0) = -1, \qquad f(+0) = 1.$$

The function f is not continuous on $[-\alpha, \alpha]$. Therefore, the convergence of the sequence $\{f_n\}$ is not uniform.

We might also note that, because of the continuity of the function f_n,

$$\sup_{0<x<\alpha} [1-f_n(x)] = \sup_{0 \leqslant x \leqslant \alpha} [1-f_n(x)] = 1-f_n(0) = 1.$$

Therefore, $\sup\limits_{-\alpha \leqslant x \leqslant \alpha} |f(x)-f_n(x)| = 1$ does not tend to 0 and the sequence $\{f_n\}$ does not converge uniformly on $[-\alpha, \alpha]$.

Fourth question:

The functions g_n and the function $|x|$ are even functions. Therefore, using the customary notation $\| \quad \|$ for the norm of functions defined on $[-1, 1]$, we have

$$\| |x|-g_n \| = \sup_{-1 \leqslant x \leqslant 1} | |x|-g_n(x)| = \sup_{0 \leqslant x \leqslant 1} | |x|-g_n(x)|.$$

For nonnegative values of x, we know that

$$0 \leqslant 1-f_n(x) \leqslant 1 \Rightarrow 0 \leqslant x-g_n(x) \leqslant x \Rightarrow \| |x|-g_n \| = \sup_{0 \leqslant x \leqslant 1} [x-g_n(x)].$$

These inequalities imply that

$$\sup_{0 \leqslant x \leqslant h} [x-g_n(x)] \leqslant \sup_{0 \leqslant x \leqslant h} x = h.$$

The uniform convergence, proven in (3), of the sequence $\{f_n\}$ on $[h, 1]$, enables us to write

$$\int_h^x [1-f_n(t)]dt = x-g_n(x)-[h-g_n(h)],$$

$$\int_h^x [1-f_n(t)]dt \leqslant \int_h^x v_h(1-f_n)dt = (x-h)v_h(1-f_n) \leqslant v_h(1-f_n) \leqslant \frac{(1-h^2)^{n+1}}{h},$$

so that, on $[h, 1]$,

$$x - g_n(x) = h - g_n(h) + \int_h^x [1 - f_n(t)] \, dt \leqslant h + \frac{(1 - h^2)^{n+1}}{h}.$$

This majorization holds also for values in the interval $[0, h]$. Therefore,

$$\| |x| - g_n \| = \sup_{0 \leqslant x \leqslant 1} [x - g_n(x)] \leqslant h + \frac{(1 - h^2)^{n+1}}{h},$$

where h is an arbitrary positive number.

For an arbitrary positive number ε, we can take $h = \varepsilon/2$:

$$\| |x| - g_n \| \leqslant \frac{\varepsilon}{2} + 2\frac{(1 - \varepsilon^2/4)^{n+1}}{\varepsilon}.$$

The sequence whose general term is

$$\frac{2}{\varepsilon}\left(1 - \frac{\varepsilon^2}{4}\right)^{n+1}$$

converges to 0. Therefore, there exists an N such that

$$n > N \;\Rightarrow\; \frac{2}{\varepsilon}\left(1 - \frac{\varepsilon^2}{4}\right)^{n+1} < \frac{\varepsilon}{2} \;\Rightarrow\; \| |x| - g_n \| < \varepsilon.$$

The sequence $\{g_n\}$ converges uniformly to the function $|x|$.

516

First question:

Since the functions $|x - x_j|$ are continuous, μ is continuous. In the interval $[x_{i-1}, x_i]$, each of the functions $|x - x_j|$ is a linear function. Specifically, the function $|x - x_j|$ is $-(x - x_j)$ if $i \leqslant j$ and $x - x_j$ if $i > j$. Therefore, the function μ is a linear function.

In each of the open intervals $]x_{i-1}, x_i[$, the functions $|x - x_i|$ and hence the function μ are differentiable. At x_i, the functions $|x - x_j|$ of index $j \leqslant i$ are differentiable but the function $|x - x_i|$ is not. Therefore, the function μ is differentiable at x_i if and only if $c_i = 0$.

If $\mu(x_i) = 0$ for every i, the function μ, which is a linear function in each of the intervals $[x_{i-1}, x_i]$ vanishes for every x. Therefore, it is differentiable at every value of x and, in particular, at all the values x_i. The coefficients c_i are all zero.

We have just seen that the homogeneous system corresponding to the system

$$(S) \qquad\qquad \mu(x_j) = 0, \qquad (j = 0, 1, ..., n)$$

has no solution other than the zero solution. Since the number of unknowns is equal to the number of equations, this property implies that the system (S) is a Cramer system (that is, one in which the number of unknowns equals the number of equations and the set of row vectors composed of the coefficients is a linearly independent set) and hence has a unique solution whatever the values of the y_j (theorem of the alternative, PZ, Book II, Chapter VII, part 6, § 2).

Second question:

The functions λ are continuous functions characterized by the following fact: There exists a partition $a=x_0<x_1<x_2<...<x_n=b$ of $[a, b]$ such that the function λ is a linear function in each subinterval $[x_{i-1}, x_i]$.

The function μ defined by

$$\mu(x) = \sum_{i=0}^{n} c_i|x-x_i|,$$

also has this property. It will coincide with the function λ if

(S') $\qquad\qquad \mu(x_j) = \lambda(x_j), \qquad (j = 0, 1, ..., n),$

since in each of the intervals $[x_{i-1}, x_i]$, the linear functions λ and μ will then be identical.

The system (S') is simply the system (S) studied in (1), where we set $\lambda(x_j)=y_j$. We know that this system has one and only one solution.

Third question:

We know (cf. exercise 423, Part 4) that every function f that is continuous on $[0, 1]$ is the uniform limit of functions λ, that is, of functions μ as defined in (2). Therefore, for an arbitrary positive number α, there exists a function μ such that

$$\|f-\mu\| = \sup_{0\leqslant x\leqslant 1} |f(x)-\mu(x)| < \alpha.$$

We saw in the preceding exercise that the function $|x|$ is the uniform limit in $[-1, 1]$ of a sequence of polynomials. The same is true for the function $|x-x_i|$ in $[x_i-1, x_i+1]$. The function μ being a finite linear combination of functions $|x-x_i|$, where $0\leqslant x_i\leqslant 1$, is therefore the uniform limit in the interval $[0, 1]$ (which is contained in all the intervals $[x_i-1, x_i+1]$) of a sequence of polynomials. Thus, there exists a polynomial A such that

$$\|\mu-A\| = \sup_{0\leqslant x\leqslant 1} |\mu(x)-A(x)| < \alpha.$$

The two results obtained imply that

$$\|f-A\| \leqslant \|f-\mu\| + \|\mu-A\| < 2\alpha.$$

Since α is an arbitrary number, this inequality shows that the function f can be approximated uniformly by a polynomial. Therefore, we can find a sequence of polynomials that converges uniformly to f. We need only take a sequence $\{\alpha_n\}$ that converges to 0 and then consider the sequence of polynomials A_n such that

$$\|f - A_n\| > 2\alpha_n.$$

Finally, if the function F is defined and continuous on $[a, b]$, we make a change of variables transforming $[a, b]$ into $[0, 1]$:

$$x = \frac{t-a}{b-a}, \qquad f(x) = F(t) = F[(b-a)x+a].$$

The function f is the uniform limit in $[0, 1]$ of a sequence of polynomials A_n. Therefore, the function F is the uniform limit in $[a, b]$ of the polynomials B_n, i.e., of the images of the polynomials A_n under the transformation, that is, of the polynomials defined by

$$B_n(t) = A_n\left(\frac{t-a}{b-a}\right).$$

517

First question:

If to every function f we assign the function g defined by

$$g(x) = \tfrac{1}{2}[f(x+a) - f(x-a)] - f'(x),$$

we define a linear mapping of the vector space of real continuous differentiable functions on R into the vector space of continuous functions. The kernel of this mapping, the set E, is therefore a vector space.

Second question:

(1) The function e^{rx} belongs to E if

$$re^{rx} = \tfrac{1}{2}(e^{r(x+a)} - e^{r(x-a)}) \iff r = \sinh ra.$$

If r is a solution, so is $-r$. Let us find the nonnegative solutions by studying the behavior of the function φ defined for $r \geqslant 0$ by

$$\varphi(r) = \sinh ra - r \implies \varphi'(r) = a \cosh ra - 1.$$

If $a \geqslant 1$, the derivative φ' is nonnegative, and the increasing function φ vanishes only at 0.

If $a < 1$, the derivative φ' is negative at 0. It is increasing and it vanishes only at r_0.

	0	r_0	$+\infty$
φ	0		$+\infty$
	$\searrow$	$\nearrow$	
		φ_0	

Therefore, the function φ vanishes only at 0 and at a value $\rho > r_0$.

In conclusion, E contains two functions $e^{\rho x}$ and $e^{-\rho x}$ if $a < 1$ and it contains no exponentials if $a \geqslant 1$. In both cases, the constants (corresponding to $r=0$) are functions in E.

(2) In a similar way, we can verify that the functions $\cos \omega x$ and $\sin \omega x$ belong to E if

$$\omega = \sin \omega a.$$

If ω is a solution of this equation, so is $-\omega$. Let us seek nonnegative solutions by studying the behavior of the function ψ defined for $\omega \geqslant 0$ by

$$\psi(\omega) = \sin \omega a - \omega \Rightarrow \psi'(\omega) = a \cos \omega a - 1.$$

If $a \leqslant 1$, the derivative ψ' is nonpositive, and the decreasing function ψ vanishes only at 0.

If $a > 1$, the derivative ψ' vanishes at a value ω_0 lying between 0 and $\pi/2a$ and again at the values $\omega_0 + 2k\pi/a$ and $2k\pi/a - \omega_0$.

	0	ω_0	$2\pi/a - \omega_0$	$\omega_0 + 2\pi/a$	$4\pi/a - \omega_0$
		M_0		M_1	
ψ		$\nearrow$ $\searrow$	$\nearrow$	$\searrow$	$\nearrow$
	0	m_1		m_2	

The minima m_h are negative since

$$\sin a(2h\pi/a - \omega_0) = -\sin a\omega_0 < 0.$$

The number of values at which ψ vanishes is therefore twice the number of positive maxima M_h. The sequence of maxima is a decreasing one since

$$M_h = \sin(a\omega_0 + 2h\pi) - (\omega_0 + 2h\pi/a) = M_0 - 2h\pi/a,$$

and the last positive maximum M_n is defined by

$$2n\pi/a \leqslant \sin a\omega_0 - \omega_0 < 2(n+1)\pi/a.$$

Thus, denoting by $[u]$ the greatest integer not exceeding u, we have

$$n = \left[\frac{a(\sin a\omega_0 - \omega_0)}{2\pi} \right] = \left[\frac{a\,M_0}{2\pi} \right].$$

Therefore, the number of nonnegative maxima is $n+1$ and the number of values ω at which ψ vanishes is $2(n+1)$. Excluding the value 0, we finally obtain the following result:

The number of positive solutions of the equation

$$\omega = \sin \omega a$$

is

$$2n+1 = 1 + 2\left(\frac{a(\sin a\omega_0 - \omega_0)}{2\pi} \right)$$

if

$$a > 1 \quad \text{and} \quad \omega_0 = (1/a)\arccos(1/a).$$

REMARK: Drawing the graphs of the functions $\sin \omega a$ and the identity function (mapping ω into ω) in a single coordinate system makes some of these results intuitively clear, but this does not in itself constitute a proof.

Third question:

Consider an interval $[x_1,\ x_2]$ such that $x_2 - x_1 > 2a$. In this interval, the function f is the uniform limit of a sequence of functions f_n in E, that is, of functions such that

$$f'_n(x) = \tfrac{1}{2}[f_n(x+a) - f_n(x-a)].$$

If $x_1 + a \leqslant x \leqslant x_2 - a$, the two numbers $x+a$ and $x-a$ belong to $[x_1,\ x_2]$. Therefore, in the interval $[x_1 + a,\ x_2 - a]$, we have uniform convergence of

the sequence of functions $\{f_n(x+a)\}$ to the function $f(x+a)$,
the sequence of functions $\{f_n(x-a)\}$ to the function $f(x-a)$,
the sequence of functions $\{f'_n(x)\}$ to the function $\tfrac{1}{2}[f(x+a) - f(x-a)]$.

Since the sequence of the derivatives f'_n converges uniformly in the interval $[x_1 + a,\ x_2 - a]$, the limit f of the sequence $\{f_n\}$ is differentiable in that interval and

$$f'(x) = \lim f'_n(x) = \tfrac{1}{2}[f(x+a) - f(x-a)]. \quad \cdot$$

But every value x belongs to an interval of length exceeding $2a$, for example, the interval $[x-2a,\ x+2a]$, and this interval can play the role of the interval $[x_1,\ x_2]$ in the preceding discussion. (On every closed interval, the function f is the uniform limit of a sequence of functions in E.)

Therefore, the function f is differentiable at every value x, and its derivative is given by

$$f'(x) = \tfrac{1}{2}[f(x+a)-f(x-a)],$$

which proves that the function f belongs to the set E.

518

If $m=1$, we have

$$\frac{t^2+2t-1}{t^2+1} = 1+\frac{2t}{t^2+1}-\frac{2}{t^2+1},$$

$$I_1 = \int_0^1 \frac{t^2+2t-1}{t^2+1}dt = [t+\ln(t^2+1)-2\arctan t]_0^1 = 1+\ln 2-\frac{\pi}{2}.$$

For other values of m,

$$\int_0^1 2t(t^2+1)^{-m}dt = \left[\frac{(t^2+1)^{1-m}}{1-m}\right]_0^1 = \frac{1-2^{1-m}}{m-1}.$$

If $m=2$, we obtain by setting $\varphi=\arctan t$,

$$\int_0^1 \frac{t^2-1}{(t^2+1)^2}dt = -\int_0^{\pi/4} \cos 2\varphi\,d\varphi = -\tfrac{1}{2}[\sin 2\varphi]_0^{\pi/4} = -\tfrac{1}{2},$$

$$I_2 = -\frac{1}{2}+\frac{1-2^{-1}}{2-1} = 0.$$

If $m=\tfrac{3}{2}$, we again set $\varphi=\arctan t$:

$$\int_0^1 \frac{t^2-1}{(t^2+1)^{\frac{3}{2}}}dt = \int_0^{\pi/4}\left[\frac{1}{\cos\varphi}-2\cos\varphi\right]d\varphi$$

$$= \left[\ln\left(\operatorname{tg}\frac{\varphi}{2}+\frac{\pi}{4}\right)-2\sin\varphi\right]_0^{\pi/4} = \ln(1+\sqrt{2})-\sqrt{2}$$

$$I_{\frac{3}{2}} = \ln(1+\sqrt{2})-\sqrt{2}+\frac{1-2^{-\frac{1}{2}}}{1/2}$$

$$= 2-2\sqrt{2}+\ln(1+\sqrt{2}).$$

In the general case, we can write

$$\int_0^1 \frac{t^2-1}{(t^2+1)^m}dt = \int_0^1 \frac{dt}{(t^2+1)^{m-1}}-2\int_0^1 \frac{dt}{(t^2+1)^m} = J_{m-1}-2J_m.$$

We find a recursion relationship between J_m and J_{m-1} that enables us to reduce the calculation of J_m to the calculation of J_1 if m is an integer and to the calculation of $J_{\frac{1}{2}}$ if $m - \frac{1}{2}$ is an integer.

The relation between J_m and J_{m-1} is obtained by integration by parts:

$$J_{m-1} = \int_0^1 \frac{dt}{(t^2+1)^{m-1}} = \left[\frac{t}{(t^2+1)^{m-1}} \right]_0^1 + \int_0^1 \frac{2(m-1)t^2\,dt}{(t^2+1)^m}$$

$$= \frac{1}{2^{m-1}} + 2(m-1)(J_{m-1} - J_m),$$

$$2(m-1)J_m = \frac{1}{2^{m-1}} + (2m-3)J_{m-1}.$$

We can also obtain the value of J_m by differentiating under the integral sign (cf. exercise 41).

519

We integrate by parts to replace the logarithm with its derivative, which is a rational function:

$$I_x = \left[\frac{-\ln t}{1+t^2} \right]_1^x + \int_1^x \frac{dt}{t(1+t^2)} = \frac{-\ln x}{1+x^2} + \int_1^x \left(\frac{1}{t} - \frac{t}{t^2+1} \right) dt$$

$$= \frac{-\ln x}{1+x^2} + \ln x - \frac{1}{2} [\ln(x^2+1) - \ln 2]$$

$$= \frac{-\ln x}{1+x^2} + \frac{1}{2} \ln \left(\frac{2x^2}{x^2+1} \right).$$

As $x \to \infty$,

$$\frac{\ln x}{1+x^2} \to 0 \quad \text{and} \quad \ln \frac{2x^2}{x^2+1} \to \ln 2,$$

so that

$$\lim_{x \to \infty} I_x = \tfrac{1}{2} \ln 2.$$

This is a prime example of an integral over a noncompact interval. By definition,

$$\lim_{x \to +\infty} I_x = \int_1^{+\infty} \frac{2t \ln t}{(1+t^2)^2} dt = \tfrac{1}{2} \ln 2.$$

520

First question:

The poles are imaginary and simple. Therefore, let us decompose the fraction into real constituent partial fractions:

$$x^4+1 = (x^2+1)^2-2x^2 = (x^2+x\sqrt{2}+1)(x^2-x\sqrt{2}+1),$$

$$\frac{1}{x^4+1} = \frac{Ax+B}{x^2+x\sqrt{2}+1}+\frac{-Ax+B}{x^2-x\sqrt{2}+1}.$$

(Like the original fraction, the decomposition into partial fractions is invariant when we replace x with $-x$ since the decomposition is unique.)

We determine A and B by assigning convenient values to x:

$$x = 0 \Rightarrow 1 = 2B \Rightarrow B = \tfrac{1}{2},$$

$$x = i \Rightarrow \frac{1}{2} = \frac{Ai+B}{i\sqrt{2}}+\frac{-Ai+B}{-i\sqrt{2}} \Rightarrow A = \frac{\sqrt{2}}{4},$$

from which we get

$$Ax+B = \frac{\sqrt{2}}{4}x+\frac{1}{2} = \frac{\sqrt{2}}{8}(2x+\sqrt{2})+\frac{1}{4};$$

$(2x+\sqrt{2}$ is the derivative of the denominator $x^2+x\sqrt{2}+1)$:

$$\int_0^1 \frac{2x+\sqrt{2}}{x^2+x\sqrt{2}+1}\,dx = [\ln(x^2+x\sqrt{2}+1)]_0^1 = \ln(2+\sqrt{2}),$$

$$\int_0^1 \frac{dx}{x^2+x\sqrt{2}+1} = \int_0^1 \frac{dx}{\left(x+\frac{\sqrt{2}}{2}\right)^2+\left(\frac{\sqrt{2}}{2}\right)^2}$$

$$= \sqrt{2}[\arctan(x\sqrt{2}+1)]_0^1 = \sqrt{2}\left(\frac{3\pi}{8}-\frac{\pi}{4}\right).$$

The calculations are identical for the other fraction, so that

$$I = \frac{\sqrt{2}}{8}[\ln(2+\sqrt{2})-\ln(2-\sqrt{2})]+\frac{\sqrt{2}}{4}\left[\frac{\pi}{8}+\frac{3\pi}{8}\right]$$

$$= \frac{\sqrt{2}}{8}\ln(3+2\sqrt{2})+\frac{\pi\sqrt{2}}{8}.$$

Second question:

We obtain

$$\int_0^1 \frac{dx}{x^4+1} = \left[\frac{x}{x^4+1}\right]_0^1 + \int_0^1 \frac{4x^4 dx}{(x^4+1)^2} = \frac{1}{2}+4(I-J),$$

$$J = \frac{1}{8}+\frac{3}{4}I = \frac{3\pi\sqrt{2}+4}{32}+\frac{3\sqrt{2}}{32}\ln\,(3+2\sqrt{2}).$$

Third question:

We note that

$$K = J+ \int_0^1 \frac{x\,dx}{(x^4+1)^2} = J+\frac{1}{2}\int_0^1 \frac{dt}{(t^2+1)^2}\,.$$

The second integral can be evaluated by setting $\varphi=\arctan t$ or by using the recursion formula in the solution of exercise 517 (it is the integral called J_2 in that exercise):

$$\frac{1}{2}\int_0^1 \frac{dt}{(t^2+1)^2} = \frac{1}{2}\int_0^{\pi/4} \cos^2\varphi\,d\varphi = \frac{1}{8}[2\varphi+\sin 2\varphi]_0^{\pi/4} = \frac{\pi}{16}+\frac{1}{8}\,.$$

521

We use a parametric representation of the relation

$$y = \sqrt{(x-a)(b-x)} \iff y \geqslant 0 \qquad \text{and} \qquad y^2 = \left(\frac{a-b}{2}\right)^2 - \left(x-\frac{a+b}{2}\right)^2,$$

from which, by making the change of variables

$$x = \frac{a+b}{2}+\frac{b-a}{2}\cos t, \qquad 0 \leqslant t \leqslant \pi,$$

$$dx = -\frac{b-a}{2}\sin t\,dt, \qquad y = \frac{b-a}{2}\sin t \geqslant 0,$$

we get

$$\int_a^b \sqrt{(x-a)(b-x)}\,dx = \int_\pi^0 -\frac{(b-a)^2}{4}\sin^2 t\,dt$$

$$= \int_0^\pi \frac{(b-a)^2}{4}\frac{(1-\cos 2t)}{2}dt = \pi\frac{(b-a)^2}{8}\,.$$

522

Let us set $u=\sqrt{1+t}-1$:

$$I_x = \int_{\sqrt{1+x}-1}^{1} \frac{u+2}{u} 2(u+1)\,du = [u^2+6u+4\ln u]_{\sqrt{1+x}-1}^{1}$$

$$= 11-x-4\sqrt{1+x}-4\ln(\sqrt{1+x}-1).$$

We know that

$$\frac{dI_x}{dx} = -\frac{d}{dx}\left(\int_3^x \frac{1+\sqrt{1+t}}{\sqrt{1+t}-1}dt\right) = -\frac{1+\sqrt{1+x}}{\sqrt{1+x}-1},$$

which is easily verified directly.

As x tends to 0, the quantity $\ln(\sqrt{1+x}-1)$ tends to $-\infty$ and the other terms tend to finite limits. Therefore, I_x tends to $+\infty$:

$$1 < \sqrt{1+t} < 1+\frac{t}{2} \Rightarrow \frac{1+\sqrt{1+t}}{\sqrt{1+t}-1} > \frac{2}{1+\frac{1}{2}t-1} = \frac{4}{t}.$$

The function $4/t$ is not integrable over $[0, 3]$ and hence neither is the function in question.

523

Let us set $t=\sqrt[3]{x}$. Then,

$$I = \int_0^{\sqrt[3]{a}} \frac{3t^2\,dt}{t^3-2t+4},$$

$$\frac{3t^2}{t^3-2t+4} = \frac{3}{5}\left[\frac{2}{t+2}+\frac{3t-2}{t^2-2t+2}\right] = \frac{3}{5}\left[\frac{2}{t+2}+\frac{3(2t-2)+2}{2(t^2-2t+2)}\right],$$

$$I = \left[\frac{6}{5}\ln(t+2)+\frac{9}{10}\ln(t^2-2t+2)+\frac{3}{5}\arctan(t-1)\right]_0^{\sqrt[3]{a}}$$

$$= \frac{6}{5}\ln(\sqrt[3]{a}+2)+\frac{9}{10}\ln(\sqrt[3]{a^2}-2\sqrt[3]{a}+2)$$

$$+\frac{3}{5}\arctan(\sqrt[3]{a}-1)-\frac{21}{10}\ln 2+\frac{3\pi}{20}.$$

524

$$dx = \frac{2\,dt}{1+t^2}, \qquad \mathrm{tg}\,\frac{\pi}{6} = \frac{\sqrt{3}}{3}, \qquad \mathrm{tg}\,\frac{\pi}{4} = 1\,; \tag{1}$$

$$I = \int_{\sqrt{3}/3}^{1} \frac{(1+t^2)^2}{4t^3}\,dt = \left[-\frac{1}{8t^2} + \frac{1}{2}\ln t + \frac{t^2}{8} \right]_{\sqrt{3}/3}^{1} = \frac{1}{3} + \frac{1}{4}\ln 3.$$

$$du = -\sin x\,dx, \qquad \cos\frac{\pi}{3} = \frac{1}{2}, \qquad \cos\frac{\pi}{2} = 0\,; \tag{2}$$

$$I = \int_{\pi/3}^{\pi/2} \frac{\sin x\,dx}{(1-\cos^2 x)^2} = \int_{0}^{1/2} \frac{du}{(1-u^2)^2}\,.$$

The rational fraction has double poles 1 and -1. Since the fraction is an even function, we can deduce without calculation the elements in the partial-fraction decomposition corresponding to the pole -1 from the elements corresponding to the pole 1:

$$\frac{1}{(1-u^2)^2} = \frac{1}{4}\left[\frac{1}{(u-1)^2} - \frac{1}{u-1} + \frac{1}{(u+1)^2} + \frac{1}{u+1} \right]$$

$$I = \frac{1}{4}\left[-\frac{1}{u-1} - \frac{1}{u+1} + \ln\left|\frac{u+1}{u-1}\right| \right]_{0}^{1/2} = \frac{1}{3} + \frac{1}{4}\ln 3.$$

525

To check whether we obtain the integral of a rational function by taking as the new variable sin x, cos x, *or* tg x *(the most advantageous choice is very often* tg(x/2))*, we can see whether the differential element (including the differential* dx*) is invariant under one of the following transformations:*

$x \to -x$ when the new variable is cos x,
$x \to \pi - x$ when the new variable is sin x,
$x \to \pi + x$ when the new variable is tg x.

In the particular case at hand, we can take sin $x = u$ since

$$\frac{\sin(\pi-x)\sin(2\pi-2x)}{\sin^4(\pi-x)+\cos^4(\pi-x)+1}(-dx) = \frac{\sin x \sin 2x}{\sin^4 x + \cos^4 x + 1}\,dx,$$

$$I = \int_{0}^{\pi/3} \frac{2\sin^2 x \cos x\,dx}{\sin^4 x + \cos^4 x + 1} = \int_{0}^{\sqrt{3}/2} \frac{u^2\,du}{u^4 - u^2 + 1}\,.$$

The biquadratic trinomial $u^4 - u^2 + 1$ has no real zeros. Therefore, we can represent it in the form

$$u^4 - u^2 + 1 = (u^2 + 1)^2 - 3u^2 = (u^2 - u\sqrt{3} + 1)(u^2 + u\sqrt{3} + 1).$$

Then, the fact that the rational fraction in question is an even function enables us to write

$$\frac{u^2}{u^4 - u^2 + 1} = \frac{Au + B}{u^2 - u\sqrt{3} + 1} + \frac{-Au + B}{u^2 + u\sqrt{3} + 1}.$$

If we assign to u numerical values, for example, 0 and i, we obtain

$$A = \frac{\sqrt{3}}{6}, \qquad B = 0.$$

If we make the change of variables $v = -u$, we transform the integral as follows:

$$\int_0^{\sqrt{3/2}} \frac{-u\,du}{u^2 + u\sqrt{3} + 1} = \int_0^{-\sqrt{3/2}} \frac{-v\,dv}{v^2 - v\sqrt{3} + 1} = \int_{-\sqrt{3/2}}^0 \frac{v\,dv}{v^2 - v\sqrt{3} + 1},$$

$$I = \frac{\sqrt{3}}{6}\left[\int_0^{\sqrt{3/2}} \frac{u\,du}{u^2 - u\sqrt{3} + 1} + \int_0^{\sqrt{3/2}} \frac{-u\,du}{u^2 + u\sqrt{3} + 1}\right]$$

$$= \frac{\sqrt{3}}{6}\int_{-\sqrt{3/2}}^{\sqrt{3/2}} \frac{u\,du}{u^2 - u\sqrt{3} + 1},$$

$$I = \frac{\sqrt{3}}{12}\int_{-\sqrt{3/2}}^{\sqrt{3/2}} \frac{(2u - \sqrt{3}) + \sqrt{3}}{u^2 - u\sqrt{3} + 1}\,du$$

$$= \frac{\sqrt{3}}{12}[\ln(u^2 - u\sqrt{3} + 1) + 2\sqrt{3}\arctan(2u - \sqrt{3})]_{-\sqrt{3/2}}^{\sqrt{3/2}},$$

$$I = \frac{1}{2}\arctan 2\sqrt{3} - \frac{\sqrt{3}}{12}\ln 13.$$

526

First question:

If we set $u = -x$, we obtain

$$I(a, b) = \int_{-a}^{-b} \frac{(1 - u^2)}{(1 + u^2)\sqrt{1 + u^4}}(-du) = -I(-a, -b).$$

If a and b are two numbers of the same sign, $x=1/v$ defines a continuous change of variable on $[a, b]$, and

$$I(a, b) = \int_{1/a}^{1/b} \frac{(1-v^{-2})(-v^{-2}dv)}{(1+v^{-2})\sqrt{1+v^{-4}}} = \int_{1/a}^{1/b} \frac{(1-v^2)dv}{(1+v^2)\sqrt{1+v^4}} = I\left(\frac{1}{a}, \frac{1}{b}\right).$$

In particular, if $b=1/a$, we see that

$$I\left(a, \frac{1}{a}\right) = I\left(\frac{1}{a}, a\right) = -I\left(a, \frac{1}{a}\right) = 0.$$

Second question:

The function that maps x into $t=x+1/x$ is continuous and differentiable if $x \neq 0$, and it is a strictly increasing function if $x \geqslant 1$ since

$$dt = \frac{x^2-1}{x^2}dx.$$

If a and b are at least equal to 1, this function therefore defines a differentiable homeomorphism between $[a, b]$ and $[\alpha, \beta]$ (we set $\alpha=a+1/a$ and $\beta=b+1/b$):

$$I(a, b) = \int_{\alpha}^{\beta} \frac{-x^2 dt}{(1+x^2)\sqrt{1+x^4}} = \int_{\alpha}^{\beta} \frac{-dt}{t\sqrt{t^2-2}} = \left[\frac{\sqrt{2}}{2}\arcsin\frac{\sqrt{2}}{t}\right]_{\alpha}^{\beta},$$

$$I(a, b) = \frac{\sqrt{2}}{2}\left[\arcsin\frac{b\sqrt{2}}{b^2+1} - \arcsin\frac{a\sqrt{2}}{a^2+1}\right] = f(a, b).$$

If a and b both lie strictly between 0 and 1, their reciprocals are both greater than 1 and

$$I(a, b) = I\left(\frac{1}{a}, \frac{1}{b}\right) = f\left(\frac{1}{a}, \frac{1}{b}\right) = f(a, b).$$

If $a \leqslant 1 \leqslant b$, we obtain similarly

$$I(a, b) = I\left(a, \frac{1}{a}\right) + I\left(\frac{1}{a}, b\right) = I\left(\frac{1}{a}, b\right) = f\left(\frac{1}{a}, b\right) = f(a, b).$$

The value of $I(a, b)$ is therefore known if a and b are both positive. If a and b are both negative, we know that

$$I(a, b) = -I(-a, -b) = -f(-a, -b) = f(a, b).$$

Finally, the functions that map a into $I(a, b)$ and b into $I(a, b)$ are continuous functions since I is the integral of a continuous function. Therefore, if a and b

have the same sign, we have

$$I(0, b) = \lim_{a \to 0} I(a, b) = \lim_{a \to 0} f(a, b) = f(0, b),$$

from which we get the general formula

$$I(a, b) = I(0, b) - I(0, a) = f(0, b) - f(0, a) = f(a, b)$$

$$= \frac{\sqrt{2}}{2}\left[\arcsin \frac{b\sqrt{2}}{b^2 + 1} - \arcsin \frac{a\sqrt{2}}{a^2 + 1} \right].$$

527

First question:

$$I_0 = \int_0^1 \sin(\pi x)\,dx = \left[-\frac{1}{\pi}\cos \pi x \right]_0^1 = \frac{2}{\pi},$$

$$I_1 = \int_0^1 x \sin(\pi x)\,dx = \left[\frac{-x\cos(\pi x)}{\pi} \right]_0^1 + \int_0^1 \frac{\cos(\pi x)}{\pi}\,dx = \frac{1}{\pi}.$$

If we integrate twice by parts, we obtain

$$I_{n+2} = \left[\frac{-x^{n+2}\cos(\pi x)}{\pi} + \frac{(n+2)x^{n+1}\sin(\pi x)}{\pi^2} \right]_0^1 - \int_0^1 \frac{(n+2)(n+1)x^n \sin(\pi x)}{\pi^2}\,dx,$$

$$I_{n+2} = \frac{1}{\pi} - \frac{(n+1)(n+2)}{\pi_2}I_n,$$

and if we assign to n the values 0 and 2, we obtain

$$I_2 = \frac{1}{\pi} - \frac{2}{\pi^2}I_0 = \frac{1}{\pi} - \frac{4}{\pi^3},$$

$$I_4 = \frac{1}{\pi} - \frac{12}{\pi^2}I_2 = \frac{1}{\pi} - \frac{12}{\pi^3} + \frac{48}{\pi^5}.$$

Second question:

If $0 < x < 1$, the sequence $\{x^n\}$ is strictly decreasing:

$$x^n \sin(\pi x) > x^{n+1}\sin(\pi x) \Rightarrow \int_0^1 x^n \sin(\pi x)\,dx > \int_0^1 x^{n+1}\sin(\pi x)\,dx.$$

Furthermore, for $0 < x < 1$,

$$0 < x^n \sin \pi x < x^n \Rightarrow 0 < I_n < \int_0^1 x^n \, dx = \frac{1}{n+1}.$$

The sequence $\{I_n\}$ is strictly decreasing and it converges to 0. The relationship between I_n and I_{n+2} enables us to write

$$(R) \qquad I_n = \frac{\pi}{(n+1)(n+2)} - \frac{\pi^2 I_{n+2}}{(n+1)(n+2)} = \frac{\pi}{(n+1)(n+2)} + o\left(\frac{1}{n^2}\right),$$

and, since $\{I_{n+2}\}$ tends to 0 as x tends to $+\infty$,

$$I_n = \frac{\pi}{n^2} + o\left(\frac{1}{n^2}\right) \Rightarrow I_n \sim \frac{\pi}{n^2}.$$

More precisely, we conclude from the relation (R) that

$$I_{n+2} > 0 \Rightarrow I_n < \frac{\pi}{(n+1)(n+2)}.$$

Using this inequality for the value $n+2$ instead of n, we have

$$I_{n+2} < \frac{\pi}{(n+3)(n+4)} \Rightarrow I_n > \frac{\pi}{(n+1)(n+2)} - \frac{\pi^2}{(n+1)(n+2)} \frac{\pi}{(n+3)(n+4)}.$$

528

$$x^2 - 2x + 2 = (x-1)^2 + 1.$$

Let us set $x - 1 = \sinh \varphi$. Then, if we denote by α the solution of the equation $-1 = \sinh \alpha$, we have

$$I = \int_\alpha^0 (\sinh \varphi + 1) \cosh^2 \varphi \, d\varphi = \left[\frac{\cosh^3 \varphi}{3} + \frac{\varphi}{2} + \frac{\sinh 2\varphi}{4} \right]_\alpha^0$$

$$= \frac{1}{3} - \left(\frac{\cosh^3 \alpha}{3} + \frac{\alpha}{2} + \frac{\sinh 2\alpha}{4} \right),$$

$$\sinh \alpha = -1 \Rightarrow \cosh \alpha = \sqrt{1 + \sinh^2 \alpha} = \sqrt{2} \Rightarrow e^\alpha = \sinh \alpha + \cosh \alpha = \sqrt{2} - 1,$$

$$I = \frac{1}{2} \log(\sqrt{2} + 1) + \frac{\sqrt{2}}{2} - \frac{2\sqrt{2} - 1}{3}.$$

529

First question:

Integrating by parts, we have

$$I_{\alpha+2} = \int_0^{\pi/2} \sin^{\alpha+1} x \,(\sin x \, dx)$$

$$= -[\sin^{\alpha+1} x \cos x]_0^{\pi/2} + (\alpha+1)\int_0^{\pi/2} \sin^\alpha x \cos^2 x \, dx$$

$$I_{\alpha+2} = (\alpha+1)(I_\alpha - I_{\alpha+2}) \;\Rightarrow\; (\alpha+2)I_{\alpha+2} = (\alpha+1)I_\alpha.$$

This relation implies that

$$f(\alpha+1) = (\alpha+2)I_{\alpha+1}I_{\alpha+2} = (\alpha+1)I_\alpha I_{\alpha+1} = f(\alpha).$$

The function f has period 1, and, if p is an integer, we have

$$f(p) = f(0) = I_0 I_1 = \frac{\pi}{2}.$$

Second question:

$$0 < x < \frac{\pi}{2} \;\Rightarrow\; 0 < \sin x < 1,$$

$$\alpha < \alpha' \;\Rightarrow\; \sin^\alpha x > \sin^{\alpha'} x \;\Rightarrow\; \int_0^{\pi/2} \sin^\alpha x \, dx > \int_0^{\pi/2} \sin^{\alpha'} x \, dx.$$

Therefore, the inequalities $p \leqslant \alpha < p+1$ imply

$$I_p \geqslant I_\alpha > I_{p+1}, \qquad I_{p+1} \geqslant I_{\alpha+1} > I_{p+2},$$

$$\frac{p+2}{p+1}f(p) = (p+2)I_p I_{p+1} > (\alpha+1)I_\alpha I_{\alpha+1} > (p+1)I_{p+1}I_{p+2} = \frac{p+1}{p+2}f(p+1).$$

Since

$$f(p+1) = f(p) = f(0) = \frac{\pi}{2},$$

we have

$$\frac{p+2}{p+1}\frac{\pi}{2} > (\alpha+1)I_\alpha I_{\alpha+1} > \frac{p+1}{p+2}\frac{\pi}{2}.$$

The above double inequality applied to the value $\alpha+n$ of the parameter yields

$$\frac{n+p+2}{n+p+1}\,\frac{\pi}{2} > f(\alpha+n) > \frac{n+p+1}{n+p+2}\,\frac{\pi}{2}.$$

Therefore, the limit of the sequence $\{f(\alpha+n)\}$ is $\pi/2$. But the function f has period 1, and $f(\alpha+n)$ is equal to $f(\alpha)$:

$$\lim_{n\to\infty} f(\alpha+n) = f(\alpha) = \frac{\pi}{2}.$$

Third question:

The function I_α is a decreasing function of α:

$$I_\alpha > I_{\alpha+1} > I_{\alpha+2} \;\Rightarrow\; 1 > \frac{I_{\alpha+1}}{I_\alpha} > \frac{I_{\alpha+2}}{I_\alpha} = \frac{\alpha+2}{\alpha+1}.$$

The ratio $I_{\alpha+1}/I_\alpha$, which lies between 1 and $(\alpha+2)/(\alpha+1)$, tends to 1 as α tends to $+\infty$:

$$f(\alpha) = \alpha I_\alpha^2 \frac{I_{\alpha+1}}{I_\alpha} = \frac{\pi}{2} \;\Rightarrow\; \alpha I_\alpha^2 \to \frac{\pi}{2} \;\Rightarrow\; I_\alpha \simeq \sqrt{\frac{\pi}{2\alpha}}.$$

530

First question:

F is a mapping on E into $\mathbf{R}$. Therefore, we need only show that, for arbitrary real numbers λ and μ,

$$F(\lambda A+\mu B) = \lambda F(A)+\mu F(B),$$

which is a well-known property of integrals.

Second question:

Since the degree of f does not exceed n, consider

$$F(f) = \int_a^b f^2(x)\,dx.$$

This integral cannot be 0 unless the function f^2, which (being a polynomial) is a continuous function, is the zero function. Therefore, if F is the zero form, f is the zero polynomial.

The mapping on E into its dual E^* (mapping f into F) is linear (since $\lambda f_1 + \mu f_2$ is mapped into $\lambda F_1 + \mu F_2$) and therefore

$$\sum_{i=1}^{p} \lambda_i F_i = 0 \iff \sum_{1}^{p} \lambda_i f_i = 0.$$

Third question:

The mapping on E into E^* (mapping f into F) is a linear mapping the kernel of which is 0 in accordance with (2). The image of E under this mapping is therefore a subspace E' of E^*, the dimension of which is equal to the dimension of E. We know that the dimension of the dual E^* and of E are equal. Since E' is a subspace of E^* and has the same dimension as that of E^*, it must be E^*, and every linear form G on E is an element of E'; that is, there exists a polynomial g of degree not exceeding n such that

$$G(A) = \int_a^b g(x)A(x)\,dx.$$

531

First question:

The proof is made by integrating by parts k times:

$$\int_{-1}^{1} A(x)\left\{\frac{d^k}{dx^k}(x^2-1)^k\right\}dx$$

$$= \left[\sum_{p=0}^{k-1}(-1)^p A^{(p)}(x)\left\{\frac{d^{k-p-1}}{dx^{k-p-1}}(x^2-1)^k\right\}\right]_{-1}^{1} + (-1)^k \int_{-1}^{1} A^{(k)}(x)(x^2-1)^k\,dx.$$

For -1 and 1, the derivatives of $(x^2-1)^k$ of order less than k are 0. Therefore, the terms outside the integral sign are equal to 0. On the other hand, $A^{(k)}(x)=0$ since the degree of A is less than k.

Second question:

The result is obtained by studying the zeros of the successive derivatives

$$\frac{d^h}{dx^h}(x^2-1)^k.$$

(See (4) of exercise 352 in Part 3. The polynomial U_p in that question is proportional to D.)

Third question:

We can easily see that, if A and B belong to V and if λ and μ are arbitrary real numbers, then $\lambda A + \mu B$ belongs to V. Furthermore, $D, Dx, ..., Dx^{k-1}$ constitute a basis for V.

If A is a multiple of D, then, by definition, $A = DQ$, where the degree of Q does not exceed $2k - 1 - k = k - 1$, and we have seen that

$$\int_{-1}^{1} D(x)Q(x)\,dx = 0 \qquad \text{if} \qquad \deg Q < k.$$

Similarly,

$$l_i(A) = A(x_i) = D(x_i)Q(x_i) = 0.$$

Let us denote by Δ_i the polynomial

$$\frac{D}{x - x_i} = \prod_{j \neq i} (x - x_j).$$

The Δ_i are independent and they constitute a basis for the vector subspace V' of polynomials of degree less than the degree of D. This space V' is a space complementary to the space V:

$$l_j(\Delta_i) = 0 \qquad \text{if} \qquad j \neq i; \qquad l_i(\Delta_i) = \prod_{j \neq i} (x_i - x_j) \neq 0.$$

Let us suppose now that the forms l_j satisfy the relation

$$\sum_j \lambda_j l_j = 0 \Leftrightarrow \sum_j \lambda_j l_j(A) = 0, \qquad \forall A.$$

If we take for A the polynomial Δ_i, we obtain

$$\sum_j \lambda_j l_j(\Delta_i) = \lambda_i \prod_{j \neq i} (x_i - x_j) = 0.$$

The coefficients λ_j are all 0 and the forms l_j are independent.

If $\mathcal{L}$ is a form in W, we can determine the coefficients μ_i from the conditions

$$(\mathcal{L} - \sum_i \mu_i l_i)(\Delta_j) = 0, \qquad (j = 1, 2, ..., k).$$

Specifically, we have

$$\mathcal{L}(\Delta_j) - \mu_j l_j(\Delta_j) = 0,$$

which determines the coefficients μ_j.

By hypothesis, the form

$$\mathcal{L} - \sum_i \mu_i l_i$$

is the zero form on the space V. From the choice of the μ_i, it is 0 on the space V' complementary to V (since V' is generated by the Δ_j). Therefore, it is 0 throughout the entire space.

Every form in W is a linear combination of the independent l_i, and the l_i constitute a basis for W.

Fourth question:

The form l belongs to W. Therefore, it is a linear combination of the forms l_i of the basis for W. That is, there exist numbers m_i such that

$$l = \sum_{i=1}^{k} m_i l_i \iff \int_{-1}^{1} A(x)\,dx = \sum_{i=1}^{k} m_i A(x_i)$$

for every polynomial A in E. In other words, the polynomial A is of degree not exceeding $2k-1$.

To determine the m_i, let us take for A successively all the polynomials Δ_j:

$$\int_{-1}^{1} \Delta_j(x)\,dx = \sum_{i=1}^{k} m_i \Delta_j(x_i) = m_j \Delta_j(x_j).$$

By definition,

$$D(x) = (x - x_j)\Delta_j(x),$$

$$D'(x) = \Delta_j(x) + (x - x_j)\Delta_j'(x) \implies D'(x_j) = \Delta(x_j)$$

$$\int_{-1}^{1} \frac{D(x)}{x - x_j}\,dx = m_j D'(x_j).$$

532

If $0 \leqslant x < 1$, the function f is nonnegative. Also, in this interval,

$$\frac{x}{1 + x^2} < 1 \quad \text{and} \quad \frac{1}{\sqrt{1 - x^4}} < \frac{1}{\sqrt{1 - x}},$$

so that

$$f(x) < \frac{1}{\sqrt{1 - x}}.$$

Thus, the nonnegative function f is majorized by the integrable function $(1 - x)^{-\frac{1}{2}}$. Hence, it too is integrable over that interval.

We know that a change of variable transforms a convergent integral into an equal convergent integral (cf. PZ, Book IV, Chapter II, § 3). Let us set $x^2 = t$:

$$I = \frac{1}{2} \int_0^{1-0} \frac{dt}{(1+t)\sqrt{1-t^2}}.$$

Then, let us set $\varphi = \arccos t$:

$$I = \frac{1}{2} \int_{\pi/2}^{+0} \frac{-\sin\varphi\, d\varphi}{(1+\cos\varphi)\sin\varphi} = \frac{1}{2} \int_0^{\pi/2} \frac{d\varphi}{2\cos^2 \varphi/2} = \frac{1}{2}\left[\operatorname{tg}\frac{\varphi}{2}\right]_0^{\pi/2} = \frac{1}{2}.$$

(We note that, after the second change of variable, the integral I can be defined on the closed (that is, compact) interval $[0, \pi/2]$.)

533

We know that, for every $\alpha > 0$, as x tends to 0,

$$x^\alpha \ln x \to 0 \;\Rightarrow\; x^{\alpha p} (\ln x)^p \to 0.$$

If we take $\alpha = \frac{1}{2}p$, there exists, by the definition of a limit, an x_0 such that

$$x < x_0 \;\Rightarrow\; |x^{\frac{1}{2}}(\ln x)^p| < 1 \;\Rightarrow\; |(\ln x)^p| < x^{-\frac{1}{2}}.$$

The function $|(\ln x)^p|$ is majorized by the integrable function $x^{-\frac{1}{2}}$. This function and the function $(\ln x)^p$ are integrable over $]0, 1]$.

Let us set $t = \ln x$ (hence $x = e^t$):

$$I = \int_{-\infty}^0 t^p e^t\, dt.$$

We shall calculate the above integral by $p+1$ integrations by parts. However, this procedure has to be applied to a compact interval $[u, 0]$, after which we pass to the limit:

$$\int_u^0 t^p e^t\, dt = [e^t(t^p - pt^{p-1} + \cdots + (-1)^p p!)]_u^0$$

$$= (-1)^p p! - e^u(u^p - pu^{p-1} + \cdots + (-1)^p p!).$$

We know that, as u tends to $-\infty$, the product of the exponential e^u and any polynomial tends to 0, so that

$$I = \lim_{u \to -\infty} \int_u^0 t^p e^t\, dt = (-1)^p p!.$$

534

The logarithm function is increasing:

$$\frac{p}{n} < x \;\Rightarrow\; \ln\frac{p}{n} < \ln x \;\Rightarrow\; \frac{1}{n}\ln\frac{p}{n} < \int_{p/n}^{(p+1)/n} \ln x \, dx$$

$$\Rightarrow \frac{1}{n}\sum_{p=1}^{n-1} \ln\frac{p}{n} < \sum_{p=1}^{n-1}\int_{p/n}^{(p+1)/n} \ln x \, dx = \int_{1/n}^{1} \ln x \, dx.$$

Similarly,

$$x < \frac{p+1}{n} \;\Rightarrow\; \ln x < \ln\frac{p+1}{n} \;\Rightarrow\; \int_{p/n}^{(p+1)/n} \ln x \, dx < \frac{1}{n}\ln\frac{p+1}{n}.$$

These last inequalities are valid for $p=0$ because the logarithm function is integrable in the interval $]0, 1/n]$. From this, we conclude that

$$\frac{1}{n}\sum_{p=0}^{n-1} \ln\frac{p+1}{n} > \sum_{p=0}^{n-1}\int_{p/n}^{(p+1)/n} \ln x \, dx = \int_{+0}^{1} \ln x \, dx.$$

Noting that $\ln(n/n)=0$, we see that the sum above and the sum originally given are identical and that

$$\int_{+0}^{1} \ln x \, dx < \frac{1}{n}\sum_{p=1}^{n} \ln\frac{p}{n} < \int_{1/n}^{1} \ln x \, dx.$$

As $1/n$ tends to 0, the integral

$$\int_{1/n}^{1} \ln x \, dx$$

converges to

$$\int_{+0}^{1} \ln x \, dx$$

by the definition of this last integral, and

$$\lim_{n\to\infty} \frac{1}{n}\sum_{p=1}^{n} \ln\frac{p}{n} = \int_{+0}^{1} \ln x \, dx = [x \ln x - x]_{+0}^{1} = -1.$$

The expression in question can be written

$$u_n = \frac{\sqrt[n]{n!}}{n} = \sqrt[n]{\frac{n!}{n^n}} = \sqrt[n]{\prod_{1}^{n}\frac{p}{n}} \;\Rightarrow\; \ln u_n = \frac{1}{n}\sum_{1}^{n} \ln\frac{p}{n}.$$

We know that the sequence $\{\ln u_n\}$ converges to -1 and hence (by the continuity of the exponential) that the sequence $\{u_n\}$ converges to $e^{-1} = 1/e$.

535

First question:

The function $(t^2+1)P/Q$ is continuous for $t \geqslant a$ and has a finite limit as t tends to $+\infty$. Therefore, this function is bounded on the interval $[a, +\infty[$, and there exists a number k such that

$$\left| \sin t \, \frac{P(t)}{Q(t)} \right| \leqslant \left| \frac{P(t)}{Q(t)} \right| \leqslant \frac{k}{t^2+1}, \qquad \forall t \geqslant a.$$

The function $k/(t^2+1)$ is integrable over $[a, +\infty[$ and, hence, the function

$$\sin t \, \frac{P(t)}{Q(t)}$$

is absolutely integrable.

Second question:

Let us integrate the integral over the compact interval $[0, x]$ by parts:

$$\int_0^x \frac{t}{t^2+1} (\sin t \, dt) = \frac{-x \cos x}{x^2+1} + \int_0^x \frac{1-t^2}{(1+t^2)^2} \cos t \, dt.$$

The integral

$$\int_0^{+\infty} \frac{1-t^2}{(1+t^2)^2} \cos t \, dt$$

is convergent by virtue of (1). The expression

$$\frac{-x \cos x}{x^2+1}$$

tends to 0 as x tends to $+\infty$. Therefore, as $x \to +\infty$,

$$\int_0^x \frac{t \sin t}{1+t^2} \, dt \to \int_0^{+\infty} \frac{1-t^2}{(1+t^2)^2} \cos t \, dt.$$

By definition, the integral I is convergent.

Third question:

Integrating by parts, we obtain

$$\int_a^x \frac{A(t)}{B(t)}(\sin t \, dt) = \frac{-\cos x \, A(x)}{B(x)} + \frac{\cos a \, A(a)}{B(a)} + \int_a^x \cos t \frac{A'(t)B(t) - A(t)B'(t)}{B^2(t)} \, dt.$$

The last integral is absolutely convergent since

$$\deg(A'B - AB') = \deg A + \deg B - 1 = 2 \deg B - 2 = \deg B^2 - 2.$$

This integral has a limit as x tends to $+\infty$. The function

$$\frac{-\cos x \, A(x)}{B(x)}$$

approaches 0. Therefore, the integral in question is convergent.

REMARK: The convergence of this integral can also be established by applying Cauchy's criterion and the second theorem of the mean (cf. exercise 510): There exists a t_0 such that the function $A(t)/B(t)$ is monotonic in the interval $[t_0, +\infty[$. If x and x' are equal to or greater than t_0, we have

$$\int_x^{x'} \sin t \frac{A(t)}{B(t)} dt = \frac{A(x)}{B(x)} \int_x^\xi \sin t \, dt + \frac{A(x')}{B(x')} \int_\xi^{x'} \sin t \, dt.$$

The absolute value of each of the integrals on the right does not exceed 2. Therefore,

$$\left| \int_x^{x'} \sin t \frac{A(t)}{B(t)} dt \right| \leq 2 \left| \frac{A(x)}{B(x)} \right| + 2 \left| \frac{A(x')}{B(x')} \right| \Rightarrow \lim_{\substack{x \to +\infty \\ x' \to +\infty}} \int_x^{x'} \sin t \frac{A(t)}{B(t)} dt = 0,$$

which proves that the integral converges.

536

First question:

Obviously, in the interval $[1, +\infty[$,

$$0 < \frac{1}{(1+x^\alpha)(1+\lambda x)} < \frac{1}{x^\alpha(1+\lambda x)} < \frac{1}{\lambda x^{\alpha+1}}.$$

The functions

$$\frac{1}{(1+x^\alpha)(1+\lambda x)} \quad \text{and} \quad \frac{1}{x^\alpha(1+\lambda x)}$$

are positive and majorized by the function

$$\frac{1}{\lambda x^{\alpha+1}},$$

which is integrable in $[1, +\infty[$ since $\alpha+1>1$. Therefore, they are integrable in $[1, +\infty[$.

Second question:

Since $\alpha<1$, the inequality

$$0 < \frac{1}{x^{\alpha}(1+\lambda x)} < \frac{1}{x^{\alpha}}$$

implies that the function

$$\frac{1}{x^{\alpha}(1+\lambda x)}$$

is integrable over $]0,1]$ and hence over $]0, +\infty[$. Furthermore, if we set $t=\lambda x$, we obtain

$$\int_{+0}^{+\infty} \frac{dx}{x^{\alpha}(1+\lambda x)} = \lambda^{\alpha-1}\int_{+0}^{+\infty} \frac{dt}{t^{\alpha}(1+t)}$$

$$\Rightarrow I(\lambda)-\lambda^{\alpha-1}\int_{+0}^{+\infty} \frac{dt}{t^{\alpha}(1+t)} = -\int_{+0}^{1} \frac{dx}{x^{\alpha}(1+\lambda x)}.$$

For values of x in the interval $]0, 1]$,

$$\frac{1}{1+\lambda} \leqslant \frac{1}{1+\lambda x} \leqslant 1 \Rightarrow \frac{1}{1+\lambda}\int_{+0}^{1} \frac{dx}{x^{\alpha}} \leqslant \int_{+0}^{1} \frac{dx}{x^{\alpha}(1+\lambda x)} \leqslant \int_{0}^{1} \frac{dx}{x^{\alpha}}.$$

As λ tends to 0, the integral

$$\int_{+0}^{1} \frac{dx}{x^{\alpha}(1+\lambda x)}$$

tends to $1/(1-\alpha)$, which is the common limit of the majorant and the minorant. Therefore,

$$I(\lambda)-\lambda^{\alpha-1}\int_{+0}^{+\infty} \frac{dt}{t^{\alpha}(1+t)} \rightarrow \frac{-1}{1-\alpha}.$$

Third question:

The inequalities $\beta > 0$ and $x > 1$ imply

$$0 < \frac{1}{x^\beta(1+x^\alpha)(1+\lambda x)} < \frac{1}{(1+x^\alpha)(1+\lambda x)}.$$

Therefore, the convergence of $K(\lambda)$ follows from the convergence of $J(\lambda)$. Furthermore, the positive function

$$\frac{1}{x^\beta(1+x^\alpha)},$$

which is majorized by the function

$$\frac{1}{x^{\alpha+\beta}},$$

is integrable over the interval $[1, +\infty[$ if the latter function is integrable over that interval (since $\alpha + \beta > 1$), and we have

$$0 < K(\lambda) - \int_1^{+\infty} \frac{dx}{x^\beta(1+x^\alpha)} = \int_1^{+\infty} \frac{\lambda x\, dx}{x^\beta(1+x^\alpha)(1+\lambda x)} < \int_1^{+\infty} \frac{\lambda\, dx}{x^{\alpha+\beta-1}(1+\lambda x)}.$$

The majorant obtained is an integral $I(\lambda)$ depending on the exponent $\alpha + \beta - 1$, which, from the hypotheses made, lies between 0 and 1. Therefore, from (2), we know that, as λ tends to 0,

$$\lambda \int_1^{+\infty} \frac{dx}{x^{\alpha+\beta-1}(1+\lambda x)} \simeq \lambda \times \lambda^{\alpha+\beta-2} \int_0^{+\infty} \frac{dx}{x^{\alpha+\beta-1}(1+x)}.$$

The limit of this majorant, like that of $\lambda^{\alpha+\beta-1}$ is 0. Thus, we have established that

$$\lim_{\lambda=0} K(\lambda) - \int_1^{+\infty} \frac{dx}{x^\beta(1+x^\alpha)} = 0.$$

Fourth question:

The expression for the sum of the terms in a geometric progression enables us to write

$$\frac{1}{1+x^\alpha} = \frac{1}{x^\alpha(1+x^{-\alpha})} = \sum_{k=1}^n (-1)^{k-1} x^{-k\alpha} + (-1)^n \frac{x^{-n\alpha}}{1+x^\alpha},$$

$$J(\lambda) = \sum_{k=1}^n (-1)^{k-1} \int_1^{+\infty} \frac{x^{-k\alpha}dx}{1+\lambda x} + (-1)^n \int_1^{+\infty} \frac{x^{-n\alpha}}{(1+x^\alpha)(1+\lambda x)}dx.$$

The first integral on the right was studied in (1) (for $0 < k\alpha < 1$) and the second was studied in (3) (for $0 < \beta = n\alpha \leqslant 1$). Therefore, we know that, as λ tends to 0,

$$(-1)^{k-1}\int_1^{+\infty}\frac{x^{-k\alpha}}{1+\lambda x}-C_k\lambda^{k\alpha-1} \to \frac{(-1)^k}{1-k\alpha},$$

where

$$C_k = (-1)^{k-1}\int_0^{+\infty}\frac{dt}{t^{k\alpha}(1+t)},$$

$$\int_1^{+\infty}\frac{x^{-n\alpha}dx}{(1+x^\alpha)(1+\lambda x)} \to \int_1^{+\infty}\frac{dx}{x^{n\alpha}(1+x^\alpha)},$$

$$J(\lambda)-\sum_{k=1}^n C_k\lambda^{k\alpha-1} \to \sum_{k=1}^n\frac{(-1)^k}{1-k\alpha}+(-1)^n\int_1^{+\infty}\frac{dx}{x^{n\alpha}(1+x^\alpha)}.$$

Fifth question:

If $\alpha = 1/n$, the only result in (4) that needs to be changed is that regarding the integral:

$$\int_1^{+\infty}\frac{x^{-n\alpha}dx}{1+\lambda x} = \int_1^{+\infty}\frac{dx}{x(1+\lambda x)} = -\ln\lambda+\ln(1+\lambda).$$

As λ tends to zero, the sum

$$\int_1^{+\infty}\frac{x^{-n\alpha}dx}{1+\lambda x}+\ln\lambda$$

tends to zero, and therefore

$$J(\lambda)-\sum_{k=1}^{n-1}C_k\lambda^{k/(n-1)}-(-1)^n\ln\lambda \to \sum_{k=1}^{n-1}\frac{(-1)^k}{1-k/n}.$$

537

First question:

We do not affect the convergence of an integral by making a change of variable, in the present case $kx = t$:

$$\int_u^{+\infty}\frac{f(kx)}{x}dx = \int_{ku}^{+\infty}\frac{f(t)}{t}dt = \int_{ku}^1\frac{f(t)}{t}dt+\int_1^{+\infty}\frac{f(t)}{t}dt.$$

The function $f(t)/t$ is integrable over $[ku, 1]$ since it is continuous on that interval, and it is integrable over $[1, +\infty[$ by hypothesis. Therefore, it is integrable over $[ku, +\infty[$.

If we apply the preceding result with $k=b$ and $k=a$, we obtain

$$\int_u^{+\infty} \frac{f(bx)-f(ax)}{x} dx = \int_{bu}^{+\infty} \frac{f(t)}{t} dt - \int_{au}^{+\infty} \frac{f(t)}{t} dt = \int_{bu}^{au} \frac{f(t)}{t} dt.$$

Second question:

The function

$$\frac{f(bx)-f(ax)}{x}$$

is integrable over $[0, +\infty[$ if the integral of this function over $[u, +\infty[$ has a limit as u tends to 0, that is, if

$$\int_{bu}^{au} \frac{f(t)}{t} dt$$

has a limit.

As u tends to 0,

$$S(u) = \sup_{t \in [au,\, bu]} |f(t)-f(0)|$$

tends to 0 (because of the continuity of f at 0) and also,

$$\left| \int_{bu}^{au} \frac{f(t)-f(0)}{t} dt \right| \leqslant S(u) \left| \int_{bu}^{au} \frac{dt}{t} \right| = S(u) \ln \frac{b}{a}.$$

Under these conditions, as u tends to 0,

$$\int_{bu}^{au} \frac{f(t)}{t} dt = \int_{bu}^{au} f(0)\frac{dt}{t} + \int_{bu}^{au} \frac{f(t)-f(0)}{t} dt \to f(0) \ln \frac{a}{b},$$

$$\int_0^{+\infty} \frac{f(bx)-f(ax)}{x} dx = \lim_{u=0} \int_u^{+\infty} \frac{f(bx)-f(ax)}{x} dx = f(0) \ln \frac{a}{b}.$$

Third question:

The exponential is continuous in $\mathbf{R}$. As t tends to $+\infty$, the product te^{-t} approaches 0, and there exists a value t_0 such that

$$t \geqslant t_0 \Rightarrow te^{-t} < 1 \Rightarrow \frac{e^{-t}}{t} < \frac{1}{t^2}.$$

Therefore, the function e^{-t}/t is integrable over $[1, +\infty[$, and the hypotheses on the function f are satisfied by the function e^{-t}. Consequently, the function

$$\frac{e^{-bt} - e^{-at}}{t}$$

is integrable over $]0, +\infty[$ and

$$\int_{+0}^{+\infty} \frac{e^{-bt} - e^{-at}}{t} dt = e^0 \ln \frac{a}{b} = \ln \frac{a}{b}.$$

538

First question:

The function f is continuous and differentiable, and

$$f'(u) = \frac{u^2 - ab}{2u^2}.$$

Its behavior is indicated by

The inverse of a function f can be defined only if f is injective. If f is continuous, this means the inverse of f can be defined only if f is strictly monotonic (cf. PZ, Book III, Chapter III, part 3, section 2, theorem 3). Here, we need to restrict the definition of f to the interval $]0, \sqrt{ab}]$ or to the interval $[\sqrt{ab}, +\infty[$.

The restriction f_1 of f to the interval $]0, \sqrt{ab}]$ has an inverse φ_1 defined in the interval $[\sqrt{ab}, +\infty[$ by

$$u = \varphi_1(v) = v - \sqrt{v^2 - ab}.$$

The restriction f_2 of f to the interval $[\sqrt{ab}, +\infty[$ has an inverse φ_2 defined in the same interval $[\sqrt{ab}, +\infty[$ as φ_1 by

$$u = \varphi_2(v) = v + \sqrt{v^2 - ab}.$$

Second question:

The function that maps θ into u is continuous and strictly increasing on $[0, \pi/2]$. Its derivative is

$$\frac{du}{d\theta} = \frac{(b^2 - a^2)\cos\theta\sin\theta}{\sqrt{a^2\cos^2\theta + b^2\sin^2\theta}}.$$

The inverse function that maps u into θ is continuous and strictly increasing on $[a, b]$. It has a finite derivative in the open interval $]a, b[$ and an infinite derivative at a and at b. The change from the variable θ to the variable u thus transforms I into an integral over the noncompact interval $]a, b[$. However, this integral is convergent since it is obtained by a change of variable from the integral of a continuous function over the compact interval $[0, \pi/2]$ and since a change of variable preserves convergence.

We obtain easily

$$\cos\theta = \sqrt{\frac{b^2 - u^2}{b^2 - a^2}}, \qquad \sin\theta = \sqrt{\frac{u^2 - a^2}{b^2 - a^2}},$$

$$I(a, b) = \int_a^b \frac{du}{\sqrt{(b^2 - u^2)(u^2 - a^2)}}.$$

If we take as our variable of integration the variable v defined in (1), we need to distinguish two cases according as $u \leqslant \sqrt{ab}$ or $u \geqslant \sqrt{ab}$ since the inverse functions φ_1 and φ_2 of f do not have the same expression in the two cases. We know that

$$u = v + \varepsilon\sqrt{v^2 - ab}, \qquad \text{where} \qquad \varepsilon = \frac{u - \sqrt{ab}}{|u - \sqrt{ab}|},$$

$$du = \frac{\varepsilon u\, dv}{\sqrt{v^2 - ab}}, \qquad (b^2 - u^2)(u^2 - a^2) = -4u^2\left[\left(\frac{a+b}{2}\right)^2 - v^2\right],$$

$$\int_a^{\sqrt{ab}} \frac{du}{\sqrt{(b^2 - u^2)(u^2 - a^2)}} = \int_{(a+b)/2}^{\sqrt{ab}} \frac{-dv}{2\sqrt{v^2 - ab}\sqrt{\left(\frac{a+b}{2}\right)^2 - v^2}}$$

$$(\varepsilon = -1),$$

$$\int_{\sqrt{ab}}^b \frac{du}{\sqrt{(b^2 - u^2)(u^2 - a^2)}} = \int_{\sqrt{ab}}^{(a+b)/2} \frac{dv}{2\sqrt{v^2 - ab}\sqrt{\left(\frac{a+b}{2}\right)^2 - v^2}}$$

$$(\varepsilon = 1).$$

Finally, adding the two integrals above, we have

$$I(a, b) = \int_{\sqrt{ab}}^{(a+b)/2} \frac{dv}{\sqrt{[v^2 - ab]\left[\left(\dfrac{a+b}{2}\right)^2 - v^2\right]}} = I\left(\sqrt{ab}, \frac{a+b}{2}\right).$$

(Changing from the variable u to the variable v transforms I into an integral of the same type in which a and b are replaced with $\sqrt{ab}$ and $(a+b)/2$ both in the limits of integration and in the integrand.)

Third question:

We have shown that $I(a, b) = I(a_1, b_1)$. We now show by induction on n that $I(a, b) = I(a_n, b_n)$:

$$I(a, b) = \int_0^{\pi/2} \frac{d\theta}{\sqrt{a_n^2 \cos^2 \theta + b_n^2 \sin^2 \theta}},$$

$$a_n \leqslant \sqrt{a_n^2 \cos^2 \theta + b_n^2 \sin^2 \theta} \leqslant b_n \Rightarrow \int_0^{\pi/2} \frac{d\theta}{b_n} \leqslant I(a, b) \leqslant \int_0^{\pi/2} \frac{d\theta}{a_n}.$$

The number $I(a, b)$ lies between two variable numbers both of which approach $\pi/2l$. Therefore, $I(a, b)$ is equal to this limiting value.

539

First question:

Simple majorizations of the sine function show that

$$0 < t \leqslant 1 \Rightarrow \frac{\sin^2 t}{t^\alpha} \leqslant \frac{t^2}{t^\alpha} = t^{2-\alpha}, \quad \text{where} \quad 2 - \alpha > -1,$$

$$1 \leqslant t \Rightarrow \frac{\sin^2 t}{t^\alpha} \leqslant \frac{1}{t^\alpha}, \quad \text{where} \quad \alpha > 1.$$

The nonnegative function

$$\frac{\sin^2 t}{t^\alpha},$$

which is majorized in $]0, 1]$ and $[1, +\infty[$ by integrable functions, is integrable in each of these two intervals and hence in the interval $]0, +\infty[$.

Second question:

We note first that the inequalities satisfied by φ correspond to $t/n \leqslant 1$ or $t/n \geqslant 1$ according as $t \leqslant n$ or $t \geqslant n$.

$$0 < t \leqslant n \Rightarrow \left| \varphi\left(\frac{t}{n}\right) \frac{\sin^2 t}{t^2} \right| \leqslant \frac{Mt^\beta}{n^\beta} \frac{t^2}{t^2} = M \frac{t^\beta}{n^\beta}, \qquad \text{where} \qquad \beta > -1,$$

$$n \leqslant t \Rightarrow \left| \varphi\left(\frac{t}{n}\right) \frac{\sin^2 t}{t^2} \right| \leqslant \frac{K}{t^2}.$$

The functions Mt^β/n^β and K/t^2 are integrable respectively in the intervals $]0, n]$ and $[n, +\infty[$. Therefore, the integral I_n is absolutely convergent.

Using the majorants of the function $|\varphi|$ referred to in the statement of the problem and majorizing the ratio

$$\frac{\sin^2 t}{t^2}$$

by 1 if $t \leqslant 1$ and by $1/t^2$ for other values of t, we obtain

$$\left| \int_0^1 \varphi\left(\frac{t}{n}\right) \frac{\sin^2 t}{t^2} \, dt \right| \leqslant \int_0^1 \frac{Mt^\beta}{n^\beta} \, dt = \frac{M}{n^\beta} \frac{1}{1+\beta},$$

$$\left| \int_1^n \varphi\left(\frac{t}{n}\right) \frac{\sin^2 t}{t^2} \, dt \right| \leqslant \int_1^n \frac{Mt^{\beta-2}}{n^\beta} \, dt = \begin{cases} \dfrac{M}{n^\beta} \dfrac{1-n^{\beta-1}}{1-\beta} < \dfrac{M}{n^\beta} \dfrac{1}{1-\beta} & \text{if} \quad \beta < 1 \\[2ex] \dfrac{M}{n} \ln n & \text{if} \quad \beta = 1, \end{cases}$$

$$\left| \int_n^{+\infty} \varphi\left(\frac{t}{n}\right) \frac{\sin^2 t}{t^2} \, dt \right| \leqslant \int_n^{+\infty} \frac{K}{t^2} \, dt = \frac{K}{n} \leqslant \frac{K}{n^\beta} \qquad \text{since} \qquad \beta \leqslant 1.$$

Adding these inequalities, we obtain

$$\text{if} \qquad \beta < 1, \qquad |I_n| < \frac{1}{n^\beta}\left(\frac{M}{1+\beta} + \frac{M}{1-\beta} + K\right),$$

$$\text{if} \qquad \beta = 1, \qquad |I_n| < \frac{M}{n} \frac{1}{2} + \frac{M}{n} \ln n + \frac{K}{n},$$

and since $\ln n > 1$ whenever $n \geqslant 3$, we have

$$|I_n| < \frac{\ln n}{n}\left(\frac{3M}{2} + K\right).$$

Third question :

Let us define a function φ_x for $t \geq 0$ by

$$\varphi_x(t) = f(x+t) + f(x-t) - 2f(x).$$

This function satisfies the hypotheses made in (2):

$$|\varphi_x(t)| \leq |f(x+t) - f(x)| + |f(x-t) - f(x)| \leq 2Ht^\gamma,$$

$$|\varphi_x(t)| \leq |f(x+t)| + |f(x-t)| + 2|f(x)| \leq 4 \sup_{u \in R} |f(u)|.$$

These conditions imply those assumed in (2) if we set

$$\beta = \gamma, \qquad M = 2H, \qquad K = 4 \sup_{u \in R} |f(u)|.$$

We then know that

$$\left| \int_0^{+\infty} \varphi_x\left(\frac{t}{n}\right) \frac{\sin^2 t}{t^2} dt \right| \leq \begin{cases} \dfrac{4}{n^\gamma}\left(\dfrac{H}{1-\gamma^2} + \sup_{u \in R}|f(u)|\right) & \text{if} \quad \gamma < 1 \\[3ex] \dfrac{\ln n}{n}(3H + 4 \sup_{u \in R}|f(u)|) & \text{if} \quad \gamma = 1, \end{cases}$$

and we can see immediately that

$$\int_0^{+\infty} \varphi_x\left(\frac{t}{n}\right) \frac{\sin^2 t}{t^2} dt = F_n(x) - Cf(x).$$

The majorizations obtained for this quantity do not depend on x. If we now define

$$\|g\| = \sup_{x \in R} |g(x)|,$$

we obtain

$$\|F_n - Cf\| \leq \frac{4}{n^\gamma}\left(\frac{H}{1-\gamma^2} + \|f\|\right) \qquad \text{if} \quad \gamma < 1,$$

$$\|F_n - Cf\| \leq \frac{\ln n}{n}(3H + 4\|f\|) \qquad \text{if} \quad \gamma = 1.$$

In both cases, the sequence $\{\|F_n - Cf\|\}$ converges to 0 and hence the sequence $\{F_n\}$ converges uniformly to Cf on the real line.

540

First question:

Let us write Maclaurin's formula with first-degree term plus remainder:

$$\ln(1+u) = u - \frac{u^2}{2}\frac{1}{1+c}, \qquad \text{where} \qquad c \in]0, u[,$$

$$u \geqslant 0 \Rightarrow c \geqslant 0 \Rightarrow 0 < \frac{1}{1+c} < 1,$$

$$u - \frac{u^2}{2} < \ln(1+u) < u \Rightarrow \frac{t^2}{n} - \frac{t^4}{2n^2} < \ln\left(1+\frac{t^2}{n}\right) < \frac{t^2}{n}.$$

From this, we easily deduce

$$-t^2 + \frac{t^4}{2n} > -n\ln\left(1+\frac{t^2}{n}\right) > -t^2 \Rightarrow e^{-t^2+t^4/2n} > \left(1+\frac{t^2}{n}\right)^{-n} > e^{-t^2}$$

$$\Rightarrow 0 < \left(1+\frac{t^2}{n}\right)^{-n} - e^{-t^2} < e^{-t^2}(e^{t^4/2n}-1) \leqslant e^{t^4/2n}-1.$$

In the interval $[a, b]$, the function $|t|$ is bounded above by the number $\theta = \sup(|a|, |b|)$, and

$$\sup_{t\in[a, b]}\left|\left(1+\frac{t^2}{n}\right)^{-n} - e^{-t^2}\right| \leqslant e^{\theta^4/2n} - 1$$

approaches 0 as n approaches $+\infty$. The sequence of the functions

$$\left(1+\frac{t^2}{n}\right)^{-n}$$

converges uniformly to the function e^{-t^2} on the interval $[a, b]$.
The binomial expansion enables us to write

$$\left(1+\frac{t^2}{n}\right)^n > 1 + n\frac{t^2}{n} = 1+t^2,$$

where

$$\left(1+\frac{t^2}{n}\right)^{-n} < (1+t^2)^{-1},$$

and we showed above that

$$e^{-t^2} < \left(1 + \frac{t^2}{n}\right)^{-n}.$$

Second question:

The function $(1+t^2)^{-1}$ is integrable in $[0, +\infty[$ and

$$0 < e^{-t^2} < \left(1 + \frac{t^2}{n}\right)^{-n} < (1+t^2)^{-1},$$

which ensures integrability in the interval $[0, +\infty[$ of the other two functions.
Furthermore, the preceding inequalities imply that

$$0 < \int_a^{+\infty} \left(1 + \frac{t^2}{n}\right)^{-n} dt - \int_a^{+\infty} e^{-t^2} dt < \int_a^{+\infty} (1+t^2)^{-1} dt = \arctan \frac{1}{a}.$$

For given $\varepsilon > 0$, this expression will be less than $\varepsilon/2$ if

$$a > \left(\operatorname{tg} \frac{\varepsilon}{2}\right)^{-1}.$$

Let a_0 denote a number satisfying this condition. In the finite interval $[0, a_0]$,
we know that the sequence of the functions $(1 + t^2/n)^{-n}$ converges uniformly to
the function e^{-t^2} and hence that

$$\int_0^{a_0} e^{-t^2} dt = \lim_{n \to \infty} \int_0^{a_0} \left(1 + \frac{t^2}{n}\right)^{-n} dt.$$

This means that we know how to find a number N such that

$$n > N \Rightarrow \int_0^{a_0} \left(1 + \frac{t^2}{n}\right)^{-n} dt - \int_0^{a_0} e^{-t^2} dt < \frac{\varepsilon}{2}.$$

When we add the two inequalities that we have obtained, we see that

$$n > N \Rightarrow \int_0^{+\infty} \left(1 + \frac{t^2}{n}\right)^{-n} dt - \int_0^{+\infty} e^{-t^2} dt < \varepsilon.$$

REMARKS: 1. The number a_0 does not appear in the final inequality. It serves
only to determine the number N.

2. The value of N can be made precise by noting that, in accordance with (1),

$$\int_0^{a_0} \left[\left(1 + \frac{t^2}{n}\right)^{-n} - e^{-t^2}\right] dt < a_0 \sup_{0 \leqslant t \leqslant a_0} \left[\left(1 + \frac{t^2}{n}\right)^{-n} - e^{-t^2}\right] < a_0(e^{a_0^4/2n} - 1).$$

Therefore, we can take for N the root of the equation

$$a_0(e^{a_0^4/2n} - 1) = \frac{\varepsilon}{2},$$

namely,

$$N = \frac{a_0^4}{2 \ln (1 + \varepsilon/2a_0)};$$

but this has nothing to do with the proof.

Third question:

Let us use the variable φ defined by $t = \sqrt{n} \cot \varphi$:

$$\int_0^{+\infty} \left(1 + \frac{t^2}{n}\right)^{-n} dt = -\sqrt{n} \int_{\pi/2}^0 \sin^{2n-2} \varphi \, d\varphi = \sqrt{n} I_{2n-2}.$$

We know (exercise 529, (3)) that

$$I_\alpha \simeq \sqrt{\frac{\pi}{2\alpha}},$$

as α tends to $+\infty$ and hence that

$$\lim_{n \to \infty} \sqrt{n} I_{2n-2} = \lim_{n \to \infty} \sqrt{n} \sqrt{\frac{\pi}{2(2n-2)}} = \frac{\sqrt{\pi}}{2} = \int_0^{+\infty} e^{-t^2} dt.$$

541

$$f_x(y) = \frac{1}{\sqrt{y}} \arctan\left(\frac{x}{\sqrt{y}}\right).$$

The mapping on R^2 into R that maps (t, y) into $(t^2 + y)^{-1}$ has derivatives of all orders with respect to y, the nth derivative being

$$(-1)^n(n!) (t^2 + y)^{-n-1}.$$

This derivative is a continuous function of (t, y) in the closed rectangle

$$0 \leqslant t \leqslant x, \qquad 0 \leqslant y_1 \leqslant y \leqslant y_2.$$

Thus, the function f_x is infinitely differentiable in every interval $[y_1, y_2]$ (as can be shown by induction on n) and hence (since y_1 and y_2 are arbitrary positive numbers) for every positive value of y; also,

$$f_x^{(n)}(y) = (-1)^n(n!) \int_0^x (t^2+y)^{-n-1} dt.$$

APPLICATION:

$$f_x''(y) = 2 \int_0^x \frac{dt}{(t^2+y)^3} \Rightarrow I = \frac{1}{2} f_1''(1),$$

$$f_1(y) = y^{-1/2} \arctan(y^{-1/2}),$$

$$f_1'(y) = -\frac{1}{2} y^{-3/2} \arctan(y^{-1/2}) - \frac{1}{2} \frac{1}{y(y+1)},$$

$$f_1''(y) = \frac{3}{4} y^{-5/2} \arctan(y^{-1/2}) + \frac{1}{4} \frac{1}{y^2(y+1)} + \frac{1}{2} \frac{2y+1}{y^2(y+1)^2},$$

$$I = \frac{1}{2} f_1''(1) = \frac{1}{2}\left[\frac{3}{4}\frac{\pi}{4}+\frac{1}{8}+\frac{3}{8}\right] = \frac{3\pi}{32}+\frac{1}{4}.$$

542

First question:

$$P_n(\alpha) = \prod_{k=0}^{n-1} \left(1 - 2\alpha \cos\frac{k\pi}{n} + \alpha^2\right) = \prod_{k=0}^{n-1} (\alpha - e^{ik\pi/n})(\alpha - e^{-ik\pi/n}).$$

$P_n(\alpha)$ is a polynomial in α the zeros of which are the $(2n)$th roots of unity other than -1 (which is not a zero of $P_n(\alpha)$). Each of these zeros is of order 1 with the exception of 1 (the order of which is 2). Thus, we have

$$P_n(\alpha) = (\alpha^{2n}-1)\frac{\alpha-1}{\alpha+1}.$$

Second question:

The mean value relative to the partition of $[0, \pi]$ into n equal subintervals is

$$M_n = \frac{1}{n}\sum_{k=0}^{n-1} \ln\left(1-2\alpha\cos\frac{k\pi}{n}+\alpha^2\right) = \frac{1}{n}\ln\left[\prod_{k=0}^{n-1}\left(1-2\alpha\cos\frac{k\pi}{n}+\alpha^2\right)\right],$$

and we know that

$$\frac{1}{\pi} I(\alpha) = \lim_{n \to \infty} M_n = \lim_{n \to \infty} \frac{1}{n} \ln \left[\frac{\alpha - 1}{\alpha + 1} (\alpha^{2n} + 1) \right],$$

$$\frac{1}{\pi} I(\alpha) = \lim_{n \to \infty} \frac{1}{n} \ln \left| \frac{\alpha - 1}{\alpha + 1} \right| + \lim_{n \to \infty} \frac{1}{n} \ln |\alpha^{2n} - 1|.$$

$$\ln \left| \frac{\alpha - 1}{\alpha + 1} \right| \text{ is independent of } n \Rightarrow \lim_{n \to \infty} \frac{1}{n} \ln \left| \frac{\alpha - 1}{\alpha + 1} \right| = 0.$$

$$|\alpha| < 1 \Rightarrow \lim_{n \to \infty} |\alpha^{2n} - 1| = 1 \Rightarrow \lim \frac{1}{n} \ln |\alpha^{2n} - 1| = 0,$$

$$|\alpha| > 1 \Rightarrow |\alpha^{2n} - 1| \simeq \alpha^{2n} \Rightarrow \ln |\alpha^{2n} - 1| \simeq \ln (\alpha^{2n})$$

$$\Rightarrow \lim_{n \to \infty} \frac{1}{n} \ln |\alpha^{2n} - 1| = \lim \frac{1}{n} \ln \alpha^{2n} = \ln (\alpha^2).$$

Therefore,

$$I(\alpha) = 0 \qquad \text{if} \qquad |\alpha| < 1,$$

$$I(\alpha) = \pi \log \alpha^2 \qquad \text{if} \qquad |\alpha| > 1.$$

Third question:

$$\frac{\partial}{\partial \alpha} \ln (1 - 2\alpha \cos x + \alpha^2) = \frac{2(\alpha - \cos x)}{1 - 2\alpha \cos x + \alpha^2}.$$

The derivative is a continuous function of (x, α) in the closed rectangle

$$0 \leqslant x \leqslant \pi, \qquad \alpha_1 \leqslant \alpha \leqslant \alpha_2,$$

if the interval $[\alpha_1, \alpha_2]$ contains neither -1 nor 1.

Thus, we know that the function $I(\alpha)$ is differentiable for every value of α other than 1 and -1 and that

$$\frac{dI}{d\alpha} = \int_0^\pi \frac{\partial}{\partial \alpha} \{\ln (1 - 2\alpha \cos x + \alpha^2)\} \, dx = 2J(\alpha).$$

Thus,

$$J(\alpha) = 0 \qquad \text{if} \qquad |\alpha| < 1,$$

$$J(\alpha) = \frac{\pi}{\alpha} \qquad \text{if} \qquad |\alpha| > 1.$$

Fourth question:

If we set $\tan(x/2) = t$, we have

$$J(\alpha) = \int_0^{+\infty} \frac{\alpha(1+t^2)-(1-t^2)}{(1+\alpha^2)(1+t^2)-2\alpha(1-t^2)} \frac{2\,dt}{1+t^2}.$$

The integrand can be decomposed into real partial fractions (we treat t^2 as the variable)

$$\frac{1}{\alpha(1+t^2)} + \frac{1}{\alpha}\frac{\alpha-1}{\alpha+1} \frac{1}{t^2 + \left(\dfrac{1-\alpha}{1+\alpha}\right)^2},$$

from which, by integrating, we get

$$J(\alpha) = \frac{1}{\alpha}\left[\arctan t - \arctan\left(t\frac{1+\alpha}{1-\alpha}\right)\right]_0^{+\infty},$$

$$|\alpha| < 1 \Rightarrow \frac{1+\alpha}{1-\alpha} > 0 \Rightarrow \left[\arctan\left(t\frac{1+\alpha}{1-\alpha}\right)\right]_0^{+\infty} = \frac{\pi}{2} \Rightarrow J(\alpha) = 0,$$

$$|\alpha| > 1 \Rightarrow \frac{1+\alpha}{1-\alpha} < 0 \Rightarrow \left[\arctan\left(t\frac{1+\alpha}{1-\alpha}\right)\right]_0^{+\infty} = -\frac{\pi}{2} \Rightarrow J(\alpha) = \frac{\pi}{\alpha}.$$

543

First question:

Let us study the behavior of the function T_x of the single variable α defined in $[0, 1]$ by

$$T_x(\alpha) = 1 - 2\alpha \cos x + \alpha^2.$$

For each x, the trinomial $T_x(\alpha)$ is minimum at $\alpha = \cos x$ and equal at that point to $\sin^2 x$. Its maximum is at one of the end-points of the interval $0 \leqslant \alpha \leqslant 1$:

$$T_x(0) = 1, \qquad T_x(1) = 2-2\cos x = 4\sin^2\frac{x}{2} \leqslant 1 \qquad \text{if} \qquad x \leqslant \frac{\pi}{3}.$$

Thus, for every (α, x) in the rectangle $0 \leqslant \alpha \leqslant 1$, $0 \leqslant x \leqslant \pi/3$, we have

$$\sin^2 x \leqslant 1 - 2\alpha \cos x + \alpha^2 \leqslant 1.$$

Since $2 - 2\cos x > 0$, these inequalities imply

$$\frac{\sin^2 x}{2-2\cos x} \leqslant \frac{1-2\alpha \cos x + \alpha^2}{2-2\cos x} \leqslant \frac{1}{2-2\cos x}.$$

Furthermore,

$$2 - 2 \cos x = T_x(1) \geqslant \sin^2 x \Rightarrow \frac{1}{2 - 2 \cos x} \leqslant \frac{1}{\sin^2 x},$$

$$\frac{\sin^2 x}{2 - 2 \cos x} = \cos^2 \frac{x}{2} \geqslant \frac{3}{4}, \quad \text{since} \quad 0 \leqslant x \leqslant \frac{\pi}{3}.$$

Second question:

The integrands are negative. Therefore, let us study their absolute values, which, in the intervals in question, are equal to the functions

$$\ln \frac{1}{\sin x} \quad \text{and} \quad \ln \frac{1}{2 - 2 \cos x}.$$

In the interval $]0, \pi/3]$,

$$\frac{1}{2} \ln \frac{1}{2 - 2 \cos x} \leqslant \ln \frac{1}{\sin x} \leqslant \ln \left(\frac{\pi}{2x} \right).$$

(For the last inequality, see exercise 439 of Part 4.)

The function $\ln (\pi/2x) = -\ln (2x/\pi)$ is integrable in $]0, \pi/3]$. One can evaluate the primitive of this function or note that, for $0 < \gamma < 1$, we have $\ln (2x/\pi) = o(x^{-\gamma})$ as x tends to 0. The two functions $\ln \sin x$ and $\ln (2 - 2 \cos x)$ are therefore integrable in the interval $]0, \pi/3]$. Since the second of these is continuous on $[\pi/3, \pi]$, it is integrable over the interval $]0, \pi]$.

Third question:

We can write

$$I(\alpha) - I(1) = \int_{+0}^{X} \ln \left(\frac{1 - 2\alpha \cos x + a^2}{2 - 2 \cos x} \right) dx + I_X(\alpha) - I_X(1),$$

setting

$$I_X(\alpha) = \int_{X}^{\pi} \ln (1 - 2\alpha \cos x + \alpha^2) dx.$$

By virtue of (1), we can majorize the first integral if $X < \pi/3$:

$$\left| \int_{+0}^{X} \ln \left(\frac{1 - 2\alpha \cos x + \alpha^2}{2 - 2 \cos x} \right) dx \right| \leqslant \int_{+0}^{X} \ln \left(\frac{1}{\sin^2 x} \right) dx = 2 \left| \int_{+0}^{X} \ln (\sin x) \, dx \right|.$$

The integral K is convergent. Therefore, by definition,

$$\lim_{X \to 0} \int_{+0}^{X} \ln(\sin x) dx = \lim_{X \to 0} \left[\int_{+0}^{\pi/3} \ln(\sin x) dx - \int_{X}^{\pi/3} \ln(\sin x) dx \right] = 0.$$

For given $\varepsilon > 0$, there exists, by the definition of a limit, a particular value X_0 for which the first integral has an absolute value less than $\varepsilon/2$.

In the rectangle defined by

$$\tfrac{1}{2} \leqslant \alpha \leqslant \tfrac{3}{2}, \qquad X_0 \leqslant x \leqslant \pi,$$

the function $\ln(1 - 2\alpha \cos x + \alpha^2)$ is a continuous function of α and x. The function $I_{X_0}(\alpha)$ is therefore a continuous function of α (cf. PZ, Book III, Chapter IV, part 4, § 2). Consequently, there exists a number $\eta > 0$ such that

$$|\alpha - 1| < \eta \implies |I_{X_0}(\alpha) - I_{X_0}(1)| < \frac{\varepsilon}{2}.$$

Returning to the expression for $I(\alpha) - I(1)$, we see that

$$|\alpha - 1| < \eta \implies$$

$$|I(\alpha) - I(1)| \leqslant \left| \int_{+0}^{X_0} \ln\left(\frac{1 - 2\alpha \cos x + \alpha^2}{2 - 2\cos x} \right) dx \right| + |I_{X_0}(\alpha) - I_{X_0}(1)| < \varepsilon;$$

that is, the function $I(\alpha)$ is continuous at 1. (We note again that the choice of X_0 was an intermediary step in the determination of $\eta = \sup |\alpha - 1|$ but that X_0 does not enter into the conclusion.) Thus, we have

$$I(1) = \lim_{\alpha \to 1} I(\alpha) = 0.$$

Fourth question:

From the definition of $I(1)$,

$$I(1) = \int_{+0}^{\pi} \ln\left(4 \sin^2 \frac{x}{2} \right) dx = \int_{+0}^{\pi} 2 \left[\ln 2 + \ln\left(\sin \frac{x}{2} \right) \right] dx = 0,$$

$$\int_{+0}^{\pi} \ln\left(\sin \frac{x}{2} \right) dx = -\pi \ln 2.$$

If we first set $x/2 = t$ and then $\pi/2 - t = u$, we obtain

$$\int_{+0}^{\pi} \ln\left(\sin \frac{x}{2} \right) dx = 2 \int_{+0}^{\pi/2} \ln(\sin t) dt = -\pi \ln 2,$$

$$\int_{+0}^{\pi/2} \ln(\sin t) dt = \int_{0}^{\pi/2 - 0} \ln(\cos u) du = -\frac{\pi}{2} \ln 2.$$

544

First question:

The function

$$e^{-xt}\frac{\sin t}{t}$$

tends to 1 as t tends to 0. If we assign to this function the value 1 at 0, we obtain a function that is continuous in $[0, +\infty[$. Thus, the functions f_n are well defined. The function

$$e^{-xt}\frac{\sin t}{t}$$

is integrable since its absolute value is majorized by e^{-xt}, which is integrable, and

$$\left|\int_0^{+\infty} e^{-xt}\frac{\sin t}{t}dt\right| \leqslant \int_0^{+\infty} e^{-xt}dt = \frac{1}{x}.$$

The definition of an integral over the interval $[0, +\infty[$ implies that

$$\lim_{n\to\infty} f_n(x) = \lim_{n\to\infty}\int_0^n e^{-xt}\frac{\sin t}{t}dt = \int_0^{+\infty} e^{-xt}\frac{\sin t}{t}dt = f(x).$$

Finally, we have shown that $|f(x)| \leqslant 1/x$ and hence that the function f approaches 0 as x approaches $+\infty$.

Second question:

We shall show that the function f is differentiable, which will prove that it is continuous, by studying the sequence of the derivatives f_n' of the functions f_n. The function

$$\frac{\partial}{\partial x}\left[e^{-xt}\frac{\sin t}{t}\right] = -e^{-xt}\sin t$$

is continuous on the closed rectangle

$$0 \leqslant t \leqslant n, \qquad x_1 \leqslant x \leqslant x_2$$

for any values of x_1 and x_2. Therefore, the function f_n is differentiable in the interval $]0, +\infty[$ and its derivative is

$$f_n'(x) = \int_0^n -e^{-xt}\sin t\, dt.$$

The integral is easily evaluated with the aid of a primitive:

$$-e^{-xt}e^{it} = -\left(\frac{e^{-xt}e^{it}}{i-x}\right)' = \left[e^{-xt}(\cos t + i\sin t)\frac{x+i}{x^2+1}\right]',$$

$$-e^{-xt}\sin t = \left[e^{-xt}\frac{x\sin t + \cos t}{x^2+1}\right]',$$

$$f_n'(x) = \left[e^{-xt}\frac{x\sin t + \cos t}{x^2+1}\right]_0^n = e^{-nx}\frac{x\sin n + \cos n}{x^2+1} - \frac{1}{x^2+1}.$$

To study the convergence of the sequence $\{f_n'\}$, we note that

$$\left|e^{-nx}\frac{x\sin n + \cos n}{x^2+1}\right| \leqslant e^{-nx}\frac{x+1}{x^2+1} \leqslant 2e^{-nx}.$$

The function e^{-nx} is decreasing. Therefore, if $x_0 > 0$,

$$\left|f_n'(x) + \frac{1}{1+x^2}\right| \leqslant 2e^{-nx_0}, \qquad \forall\, x \in [x_0, +\infty[.$$

The sequence of the functions f_n' converges uniformly in the interval $[x_0, +\infty[$ to the function

$$-\frac{1}{x^2+1}.$$

Thus, we know that the function f, which is the limit of the sequence $\{f_n\}$, is differentiable in the interval $[x_0, +\infty[$ and that its derivative is given by

$$f'(x) = \lim f_n'(x) = -\frac{1}{x^2+1}.$$

For every $x > 0$, there exists a number x_0 such that $0 < x_0 < x$. Therefore, the preceding result is valid. The function f is differentiable at all $x > 0$ and

$$f'(x) = -\frac{1}{x^2+1}.$$

The function f is the primitive of f', which approaches 0 as x approaches $+\infty$. Therefore,

$$f(x) = \frac{\pi}{2} - \arctan x.$$

Part 6

MAPPINGS OF R INTO R^p, OF R^p INTO R OR R^q, AND OF C INTO C; LINE INTEGRALS; DOUBLE AND TRIPLE INTEGRALS

Exercises dealing with the properties of Chapters v-ix of Book iii of *Mathématiques générales* by C. Pisot and M. Zamansky.

Introduction

Although the theory of vector-valued functions or functions from R^p into R or R^q has numerous geometrical applications, in this portion of the book I give only exercises in analysis. This is not because I consider geometric problems as being of no interest. However, the importance given to these problems in courses in general mathematics is quite variable and if I had wished to treat geometry carefully, I would have needed to enlarge this book in a way that would not have been of use to many students.

Finally, the elementary problems in geometry lead almost immediately to analysis problems and will present no difficulties for students who have mastered the algebra and analysis program, which constitutes the basis of the course in general mathematics.

Review

Before listing particular properties, it is of interest to emphasize the analogies that exist between the theories regarding the space R and the space R^p.

Some of the essential topological properties of R^p are the same as those of R (for example, the Borel–Lebesgue, and Bolzano–Weierstrass theorems). The properties of limits and of continuous functions are also almost identical (the definition of limits in terms of convergent sequences for example, uniform continuity of a function that is continuous on a closed bounded set).

There is however one essential difference. Specifically, ordering plays an important part in the properties of R (for example, monotonic functions), but the space R^p is not an ordered set (and of course, the field C of complex numbers is not ordered).

Norms

The existence of three norms in R^p is no complication since these norms are equivalent. Therefore, we can use indifferently any one of these norms to study a limit or to write that a function is uniformly continuous.

In R, the three norms reduce to absolute value. Thus, a norm in R^p is a generalization of absolute value in R and it is often used in the same way.

Jordan arcs

To ascertain that an arc is a rectifiable Jordan arc, we shall often use the following theorem:

A differentiable and injective vector-valued function whose derivative is a bounded function defines a rectifiable Jordan arc.

Rectifiable Jordan arcs play an important role in the theory of double integrals since they are plane sets of measure 0.

Differentials

For functions from R^p into R or R^q, the concept of a differentiable function is very important, and the properties of a differential are frequently used.

To evaluate differentials, we use the classical rules of calculus:

$$d(\lambda f + \mu g) = \lambda\,df + \mu\,dg, \qquad d(fg) = g\,df + f\,dg,$$

$$d\left(\frac{f}{g}\right) = \frac{g\,df - f\,dg}{g^2}.$$

The fundamental theorem has to do with composite functions: The composite of two differentiable functions is differentiable, and its differential is the composite of the two differentials (this is the invariance of a differential under a change of variable):

Let $\varphi: R^p \to R^q$ be a mapping. Then, for every mapping of the form

$$L_p^i: \qquad f \to \sum_{i=1}^{p} a_i(\bar{x}) \frac{\partial f(\bar{x})_j}{\partial x_i} \qquad (\bar{x} = (x_1, x_2, ..., x_p)),$$

and for f and $a_i(\bar{x})$ infinitely differentiable on R^p, there is a corresponding mapping (if φ is suitably restricted)

$$L_q: \qquad g \to \sum_{i=1}^{p} a_i(\bar{x}) \frac{\partial g(\varphi(\bar{x}))}{\partial x_i}$$

for g infinitely differentiable on R^q. The *mapping* associating L_q to L_p is called the *differential* of φ:

$$d\varphi: L_p \to L_q.$$

If $\psi: R^s \to R^p$ is another mapping, then the differential of ψ, $d\psi$, is well defined, and thus the composition of $d\psi$ and $d\varphi$ is meaningful:

$$d\varphi \circ d\psi : L_s \to L_q.$$

This theorem also enables us to evaluate the partial derivatives of a composite function since they are the coefficients in the expression for the differential.

We also use the following result: If the partial derivatives of a function are continuous functions in a neighborhood of the point in question, the function is differentiable at that point.

Taylors' formula

To study the local properties of a function (limits, relative maxima, etc.), it is convenient to write Taylor's formula in the following form:

If a function f has continuous partial derivatives of the first n orders, then

$$f(X+H)-f(X) = \frac{1}{1!}\left(\sum_p \frac{\partial f(X)}{\partial x_k} h_k\right) + \cdots$$

$$+ \frac{1}{n!}\left(\sum_p \frac{\partial f(X)}{\partial x_k} h_k\right)^{[n]} + g_n(X, H)[V(H)]^n,$$

where the function $g_n(X, H)$ approaches 0 as H approaches 0. (The notation $V(H)$ denotes anyone of the three norms.)

The complex exponential

The essential result is that the function e^{it} on R into C defines an isomorphism of the additive group of real numbers modulo 2π and the multiplicative group of complex numbers of absolute value 1.

In general, it is more convenient to use the exponential e^{it} than the trigonometric functions. (For example, differentiation is easier, as is operation with the addition formulas.)

Integration of differentials

If U is a differentiable function and Γ is a Jordan arc with end-points at X_1 and X_2, then

$$\int_{\Gamma^+} dU = U(X_2) - U(X_1).$$

In particular, if Γ is a closed curve, the value of the integral is zero.

In the plane, Riemann's formula often enables us to show that the line integral

$$\int_\Gamma (P\,dx + Q\,dy)$$

over every closed rectifiable Jordan curve Γ is 0. However, it is essential that in the region $\bar{D}$ consisting of Γ and the points enclosed by Γ, the functions P and Q and their derivatives $\partial P/\partial y$ and $\partial Q/\partial x$ be continuous.

Evaluation of double integrals

When we evaluate a double integral by converting it into an iteration of two simple integrations, for example, by integrating first with respect to y and then with respect to x, it is necessary in each case to determine the limits of the integration; the inequalities that define the region of integration D usually provide the result quite simply.

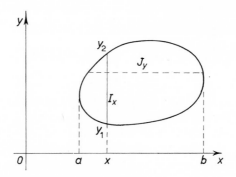

However, to check, I advise the student to make a quick drawing of the bounddary of the region of integration. In the integration with respect to y, the variable x is treated as a constant and the limits of integration are the (variable) ordinates y_1 and y_2 of the end-points of the segment I_x of abscissa x which lies in D. (If the boundary of D is intersected at more than two points by a line parallel to the y-axis, we shall have not one but several such segments.) The limits a and b in the integration with respect to x are also clear from the drawing. These are the end-points of the projection of the region D onto the x-axis.

If we integrate first with respect to x, the procedure is analogous; in the drawing, the segment J_y (shown by the horizontal dashed line) is considered instead of I_x.

Change of variables

In a double integral, we sometimes need to simplify not so much the function to be integrated as the region over which the integration is made, and the simplest such region of integration is undoubtedly a rectangle.

In making such a change of variables, one must not forget to multiply by the *absolute value* of the Jacobian of the transformation.

Exercises

601

Let f be the mapping on R^2 into R defined by

$$f(0, 0) = 0,$$

$$f(x, y) = xy\,\frac{x^2 - y^2}{x^2 + y^2} \quad \text{if} \quad (x, y) \neq (0, 0).$$

(1) Show that the function f is differentiable except possibly at the point $(0, 0)$ and find its differential.

(2) Show that the function f is also differentiable at the point $(0, 0)$ and that its differential is 0.

(3) Show that the function f has second partial derivatives f''_{xy} and f''_{yx} everywhere and evaluate these second derivatives at the point $(0, 0)$. Explain why the result obtained implies that at least one of the partial derivatives f''_{xy} and f''_{yx} is not continuous at the point $(0, 0)$.

602

This problem deals with a study of real functions f that are defined, continuous, and differentiable in the half-plane $x > 0$ of R^2 and that verify the equation

$$(E) \qquad x\,\frac{\partial f}{\partial x} + y\,\frac{\partial f}{\partial y} = 0.$$

(1) Show that the following are solutions of E: (a) the function φ defined by $\varphi(x, y) = y/x$; (b) the functions $g \circ \varphi$, where g is a continuous and differentiable function from R into R.

(2) Show that these are the only solutions of this equation.

603

This is a study of functions f on R^2 into R that have first and second continuous derivatives and that satisfy the equation

$$(e) \qquad a f''_{x^2} + 2b f''_{xy} + c f''_{y^2} = 0$$

(1) Let us make the change of variables

$$u = x+my, \qquad v = x+ny,$$

where m and n are distinct real numbers. This defines a transformation that maps the function f into a function F such that

$$F(u, v) = F(x+my, x+ny) = f(x, y).$$

Show that, if f has continuous first and second partial derivatives, so does the function F. Evaluate the partial derivatives of f in terms of the partial derivatives of F and the constants m and n. Find the equation (E) satisfied by the partial derivatives of the function F.

(2) Show that, if $b^2 - ac > 0$, we can choose m and n in such a way that the equation (E) is the equation $F''_{uv} = 0$. Characterize the functions f that are solutions of the equation (e).

(3) Show that, if $b^2 - ac = 0$, it is possible to choose m and n in such a way that the equation (E) will be the equation F''_{v^2}. Characterize the functions f that are solutions of equation (e).

(4) Show that, if $b^2 - ac < 0$, it is possible to choose m and n in such a way that equation (E) becomes the equation $F''_{u^2} + F''_{v^2} = 0$. (In this case, you are not asked to characterize the functions f.)

604

(1) For every real function φ that is defined and differentiable in a region D of $\mathbf{R}^2$ that does not contain the point $(0, 0)$, we define a function Φ by

$$\Phi(r, \theta) = \varphi(r \cos \theta, r \sin \theta)$$

in the region Δ such that

$$(r, \theta) \in \Delta \Rightarrow (r \cos \theta, r \sin \theta) \in D.$$

Show that the function Φ is differentiable and express Φ'_r and Φ'_θ as a function of φ'_x, φ'_y, r, and θ and express φ'_x and φ'_y as a function of Φ'_r, Φ'_θ, r, and θ.

(2) Let f denote a function that is defined and twice continuously differentiable in a region D in $\mathbf{R}^2$ that does not contain the point $(0, 0)$. As in (1), define a function F in the region Δ corresponding to D by

$$F(r, \theta) = f(r \cos \theta, r \sin \theta).$$

Show that the function F has continuous first and second derivatives and use the formulas in (1) to show that

$$\Delta f = f''_{x^2} + f''_{y^2} = F''_{r^2} + \frac{1}{r^2} F''_{\theta^2} + \frac{1}{r} F'_r.$$

Does there exist a function f, depending only on x^2+y^2, that satisfies the equation

$$\Delta f = f''_{x^2}+f''_{y^2} = 0.$$

605

This exercise assumes knowledge of the results of the preceding one.

In R^3, for a point $X=(x, y, z)$, the numbers ρ, θ, φ such that

$$x = \rho \cos \varphi \cos \theta, \qquad y = \rho \cos \varphi \sin \theta, \qquad z = \rho \sin \varphi$$

are called the spherical coordinates of X.

(1) Show that if X belongs to the complement ω (with respect to R^3) of the set defined by $x=y=0$ (that is, if X is not on the z-axis), there exists a unique system of spherical coordinates satisfying the inequalities

$$\rho > 0, \qquad -\frac{\pi}{2} < \varphi < \frac{\pi}{2}, \qquad 0 \leqslant \theta < 2\pi.$$

Denote by P the point (ρ, θ, φ) in R^3 thus defined, denote by Ω the set of points P, and denote by T the mapping on Ω into ω defined by $X=T(P)$.

(2) Show that the mapping T is the composite $T_1 \circ T_2$ of mappings T_1 and T_2 defined respectively by

$$T_1 \qquad x = r \cos \theta, \qquad y = r \sin \theta, \qquad z = z\,;$$
$$T_2 \qquad r = \rho \cos \varphi, \qquad \theta = \theta, \qquad z = \rho \sin \varphi.$$

Use the above remark and the results of exercise 604 to express

$$\Delta f = f''_{x^2}+f''_{y^2}+f''_{z^2}$$

as a function of ρ, θ, φ and the partial derivatives of the function $F=f \circ T$. (Assume that the function f is defined and twice continuously differentiable.)

Find the functions f depending on the single variable $x^2+y^2+z^2$ that satisfy the equation

$$\Delta f = f''_{x^2}+f''_{y^2}+f''_{z^2} = 0.$$

606

Let f denote a real function defined in a region D of R^n. Every vector X in R^n can be given in terms of its coordinates $(x_1, x_2, ..., x_n)$ relative to a basis α for the vector space R^n. Then, to every basis α, we assign a function f_α defined by

$$f_\alpha(x_1, ..., x_n) = f(X).$$

Relative to any other basis β for $\boldsymbol{R}^n$, the vector X has coordinates $(y_1, ..., y_n)$. We know that there exists a nonsingular matrix $P = (p_{ij})$ such that

$$\begin{pmatrix} x_1 \\ \vdots \\ x_n \end{pmatrix} = P \begin{pmatrix} y_1 \\ \vdots \\ y_n \end{pmatrix}.$$

(1) Show that, if the function f_α is differentiable at the point X, the functions f_β relative to other bases are also differentiable at the point X and

$$\begin{pmatrix} \dfrac{\partial f_\beta}{\partial y_1} \\ \vdots \\ \dfrac{\partial f_\beta}{\partial y_n} \end{pmatrix} = P^T \begin{pmatrix} \dfrac{\partial f_\alpha}{\partial x_1} \\ \vdots \\ \dfrac{\partial f_\alpha}{\partial x_n} \end{pmatrix}.$$

(2) Show that, if the matrix P is orthogonal, the vector F_α, whose coordinates are the $\partial f_\alpha/\partial x_i$ relative to the basis α, coincides with the vector F_β whose coordinates are the $\partial f/\partial y_i$ relative to the basis β.

We thus define a mapping that maps the function f at every point at which f is differentiable into a definite vector-valued function (independent of the orthonormal basis in $\boldsymbol{R}^n$ [treated as a Euclidean space]). This vector-valued function is called the *gradient* of f.

(3) Show that the differentials df_α, df_β, ... are expressions in the bases α, β, ..., of the same linear form df on $\boldsymbol{R}^n$. Use the dual of $\boldsymbol{R}^n$ in $\boldsymbol{R}^n$ (treated as a Euclidean space) and the scalar product to find again the result of (2).

607

Consider a field of vectors in $\boldsymbol{R}^3$. At every point M defined by its coordinates x, y, z relative to a basis α, we assign the vector S_M with coordinates X, Y, Z relative to the same basis α. Suppose that the functions X, Y, and Z are differentiable.

Relative to another basis β, the point M and the vector S_M have coordinates u, v, w and U, V, W respectively, defined in terms of the nonsingular matrix P by

$$\begin{pmatrix} x \\ y \\ z \end{pmatrix} = P \begin{pmatrix} u \\ v \\ w \end{pmatrix}, \qquad \begin{pmatrix} X \\ Y \\ Z \end{pmatrix} = P \begin{pmatrix} U \\ V \\ W \end{pmatrix}.$$

(1) Show that, if f is a function of (x, y, z) and $\bar{f}$ is the function of u, v, and w that is a composite of the function f and the linear mapping p associated with P,

then

$$(\bar{f}'_u \bar{f}'_v \bar{f}'_w) = (f'_x f'_y f'_z) P.$$

From this show that

$$B(S_M) = \begin{pmatrix} U'_u & U'_v & U'_w \\ V'_u & V'_v & V'_w \\ W'_u & W'_v & W'_w \end{pmatrix} = P^{-1} \begin{pmatrix} X'_x & X'_y & X'_z \\ Y'_x & Y'_y & Y'_z \\ Z'_x & Z'_y & Z'_z \end{pmatrix} P = P^{-1} A(S_M) P.$$

(2) From the above result, show that

$$X'_x + Y'_y + Z'_z = U'_u + V'_v + W'_w.$$

Can one obtain other invariants without calculation?

608

Let Γ denote an arc of a circle with end-points A and B. Suppose that the points M on Γ are defined by a parameter t, which may be the length of the arc AM.

(1) Show that the area of the triangle AMB (where M denotes an arbitrary point on Γ) has a maximum when M is equidistant from A and B.

(2) Inscribe in the arc Γ a polygonal convex line with $n+1$ sides: $AM_1 M_2 ... M_n B$. Admit the possibility that consecutive vertices coincide. Show that, for fixed n but variable points M_1, M_2, ..., M_n, the area of the convex polygon $AM_1 M_2 ... M_n B$ has a maximum attained by at least one polygonal line $AP_1 ... P_n B$.

(3) Use the result of (1) to show that the polygonal line $AP_1 P_2 ... P_n B$ is regular.

609

Let f denote a real function defined on a subset E of R^p. This function f is said to have a maximum (minimum) at the point X_0 if there exists a sphere B with center at X_0 such that

$$X \in B \cap E \Rightarrow f(X) \leq f(X_0) \qquad (f(X) \geq f(X_0)).$$

(Either a maximum or a minimum at X_0 is called an *extremum* at that point.)

(1) Suppose that the function f has an extremum at a point X_0, that f has partial derivatives at that point, and that the set E contains a sphere with center at X_0. Show that the partial derivatives of f are all 0 at the point X_0. Show that these conditions are not sufficient by considering the function f defined on R^p by

$$\varphi(x_1, ..., x_p) = x_1^3 + \cdots + x_p^3.$$

(2) Consider the function f defined on the set E of points (x, y) in R^2 such that $-a\sqrt{6} \leqslant x \leqslant a\sqrt{6}$, where a is a positive constant, by

$$f(x, y) = \frac{x^2}{4a} + \sqrt{6a^2 - x^2} \cos y.$$

Show that f has continuous partial derivatives of all orders in the interior of the set E and find the points at which the first partial derivatives vanish.

Show by writing Taylor's formula for the second-degree terms plus a remainder for the function in a neighborhood of one of these points, which we denote by X_0, that f has or does not have an extremum according as the quantity

$$\delta = [f''_{xy}(X_0)]^2 - f''_{x^2}(X_0) f''_{y^2}(X_0)$$

is negative or positive (do not treat the case in which this expression is zero) and apply this result to the various points obtained.

(3) Ascertain whether the function f has an extremum at points not satisfying the conditions of (2). Suggestion: set $t = \sqrt{6a^2 - x^2}$.

610

Consider two real functions U and V defined in a region D in R^2 and assume that the function U is positive. To these functions, let us assign the real functions P and Q and the complex function f defined by

$$f(z) = f(x + iy) = U(x, y)e^{iV(x,y)} = P(x, y) + iQ(x, y).$$

(1) A necessary condition for the function f to be a differentiable function of z is that the functions P and Q be differentiable and that

$$P'_x = Q'_y, \qquad P'_y = -Q'_x,$$

or, equivalently, that the functions U and V be differentiable and that

$$U'_x = UV'_y, \qquad U'_y = -UV'_x.$$

From this, show that the functions P, Q, $\ln U$, and V satisfy the equation

$$\varphi''_{x^2} + \varphi''_{y^2} = 0,$$

if they have continuous second partial derivatives.

(2) Use the result of exercise 604 to determine the functions f such that $|f(z)|$ is a function of $|z|$ and not of arg z.

611

Let z denote a given complex number.
(1) Test the sequence $\{z^{(2^n)}\}$ if $|z| \neq 1$.

(2) Show that, if $|z| = 1$, a necessary and sufficient condition for convergence of the sequence $\{z^{(2^n)}\}$ is

$$\arg z = \frac{r\pi}{2^s},$$

where r and s are integers.

612

We recall that the homographic transformations H that leave invariant the set E of numbers of absolute value not exceeding 1 are of the form

$$Z = e^{i\alpha}\frac{z - z_0}{1 - \bar{z}_0 z}, \qquad (\alpha \text{ real}; |z_0| < 1).$$

(See exercise 217 in Part 2.) Let us denote by $H(\alpha, z_0)$ the homographic function thus defined.

(1) Determine the image of z_0 under $H(\alpha, z_0)$ and use this result to show that the transformations H in which z_0 is homologous to Z_0 can be defined by

$$H = [H(0, Z_0)]^{-1} \circ H(\alpha, z_0) \quad \text{or} \quad \frac{Z - Z_0}{1 - \bar{Z}_0 Z} = e^{i\alpha}\frac{z - z_0}{1 - \bar{z}_0 z}.$$

(2) Show that the function $H(\alpha, z_0)$ is differentiable and evaluate its derivative dZ/dz. What do we obtain in particular if $z = z_0$? From the results of (1) and (2), show that

$$\frac{|dZ|}{1 - |Z|^2} = \frac{|dz|}{1 - |z|^2}.$$

(3) Assign to every point (x, y) in the plane $\mathbf{R}^2$ such that $x^2 + y^2 < 1$ the complex number $z = x + iy$ and define a metric on that set by taking as length of an arc Γ the value of the line integral

$$\int_\Gamma \frac{|dz|}{1 - |z|^2}.$$

Evaluate the length l of the line segment joining the origin to the image of the number $z = re^{i\theta}$ (where $r < 1$). Show that the length of every arc Γ with the same end-points is equal to or greater than l.

Let M_1 and M_2 denote the images (in the complex plane) of the complex numbers z_1 and z_2, where $|z_1| < 1$ and $|z_2| < 1$. Show that of all arcs having these two points as end-points, there is exactly one of minimum length.

613

Let B denote the closed region in R^2 defined by

$$(s, t) \in B \quad \text{if} \quad |t| \leqslant \pi.$$

(1) Let f denote the mapping on B into the field C of complex numbers defined by

$$f(s, t) = \frac{s+it}{e^{s+it}-1} \quad \text{if} \quad (s, t) \neq (0, 0),$$

$$f(0, 0) = 1.$$

Show that f is continuous everywhere in B.

(2) Let g be a mapping on R into C defined by

$$g(s) = \int_0^\pi f(s, t)\, dt.$$

Show that the function g is continuous and that it approaches 0 as s approaches $+\infty$.

(3) Show that the function g is differentiable everywhere except possibly at the value 0. Give an expression for $g'(s)$ in closed form (not containing any integrals). (Note that f depends on s and t only through $s+it$.)

(4) Show that the function g is differentiable at 0 and that its derivative g' is continuous. Show that

$$\int_0^{+\infty} g'(s)\, ds = -g(0)$$

and, by comparing this with the definition of $g(0)$, find the values of the two integrals

$$\int_0^{+\infty} \frac{se^s}{e^{2s}-1}\, ds, \quad \int_0^{\pi/2} \ln(\sin t)\, dt.$$

614

Let us set

$$\omega = P(x, y)dx + Q(x, y)dy,$$

where P and Q are functions defined by

$$P(x, y) = \frac{(3x^2 - y^2)(x^2 + y^2)}{x^2 y}, \quad Q(x, y) = \frac{(3y^2 - x^2)(x^2 + y^2)}{xy^2}.$$

(1) Show that, in the region D defined by $x>0$ and $y>0$, the two functions P and Q are continuously differentiable and that $\partial P/\partial y=\partial Q/\partial x$.

(2) In the region D, define a function U whose differential dU is equal to ω.

(3) Evaluate the line integral $\int_\Gamma \omega$, where the curve Γ is defined by

$$0 \leqslant t \leqslant \frac{\pi}{2}, \qquad x = t+\cos^2 t, \qquad y = 1+\sin^2 t.$$

615

Let φ denote a mapping on $\mathbf{R}^2$ into $\mathbf{R}$. We define P and Q by

$$P(x, y) = y\left[1 + \frac{\varphi(x, y)}{x^2+y^2}\right], \qquad Q(x, y) = -x\left[1 + \frac{\varphi(x, y)}{x^2+y^2}\right].$$

(1) For $P(x, y)dx+Q(x, y)dy$ to be the differential, in a neighborhood of a given point $(x_0, y_0)\neq(0, 0)$, of a function U possessing continuous first and second partial derivatives, it is necessary that the function φ have continuous partial derivatives satisfying the equation (E):

$$(E) \qquad x\frac{\partial\varphi}{\partial x}(x, y)+y\frac{\partial\varphi}{\partial y}(x, y)+2(x^2+y^2) = 0.$$

(2) We define r, θ, and Φ by

$$x = r\cos\theta, \qquad y = r\sin\theta, \qquad \Phi(r, \theta) = \varphi(r\cos\theta, r\sin\theta).$$

Find the functions Φ associated with the functions φ that satisfy equation (E). Do this either directly or by writing the equation (E') denoting a transformation of equation (E).

(3) Take for φ the function defined in the complement Ω of $(0, 0)$ with respect to $\mathbf{R}^2$ by

$$\varphi(x, y) = -(x^2+y^2) + \frac{4x^2}{x^2+y^2}.$$

Find a function U whose differential is $P(x, y)dx+Q(x, y)dy$ in a neighborhood of the point $(1, 0)$. Is it possible to define a function U_0 whose differential is $P(x,y)dx+Q(x,y)dy$ in the entire set Ω?

616

Evaluate the double integral

$$I = \iint_D (x^2+y^2)\,dx\,dy$$

in the region D defined by

$$x \geqslant 0, \qquad y \geqslant 0, \qquad \frac{x}{a} + \frac{y}{b} - 1 \leqslant 0,$$

where a and b are positive constants.

REMARK: This integral gives the value of the moment of inertia of a homogeneous triangular plate of density 1 about a vertex.

617

Evaluate the integral

$$I = \iint_C x \, dx \, dy,$$

where C is the circle $x^2 + y^2 - 2x \leqslant 0$.

618

Evaluate the integral

$$I = \iint_D x \sqrt{1 - x^2 - y^2} \, dx \, dy$$

in the region D defined by

$$x \geqslant 0, \qquad y \geqslant 0, \qquad x^2 + y^2 \leqslant 1.$$

619

Consider three given numbers $a \leqslant x_0 \leqslant b$. Denote by f a function that is defined and continuous on the closed interval $[a, b]$.

(1) By evaluating a double integral in two different ways, show that

$$\int_{x_0}^x du \left(\int_{x_0}^u (u - v)^n f(v) \, dv \right) = \int_{x_0}^x \frac{(x - v)^{n+1}}{n + 1} f(v) \, dv.$$

(2) Show that the function of x

$$\frac{1}{(n - 1)!} \int_{x_0}^x (x - v)^{n-1} f(v) \, dv$$

is a primitive of order n of f:

(a) by using the result of (1),

(b) by evaluating the derivative of this function directly.

620

By evaluating the integral

$$I = \iint_D e^{-(x^2+y^2)} dx \, dy$$

over a convenient region D in two different ways, show that

$$\left(\int_0^a e^{-x^2} dx \right)^2 = \int_0^{\pi/4} (1 - e^{-a^2/\cos^2\theta}) \, d\theta.$$

621

Evaluate the volume μ of the region D defined by

$$\begin{cases} 0 \leqslant z \leqslant \dfrac{1}{2}\left(\dfrac{x^2}{a} + \dfrac{y^2}{b}\right) \\ \dfrac{x^2}{a^2} + \dfrac{y^2}{b^2} - 1 \leqslant 0, \end{cases}$$

where a and b are two positive numbers.

622

(1) Evaluate the integral

$$I = \iiint_S (x^2 + y^2 + z^2) \, dx \, dy \, dz$$

in the region S defined by

$$x \geqslant 0, \qquad y \geqslant 0, \qquad z \geqslant 0, \qquad x^2 + y^2 + z^2 \geqslant 1.$$

Suggestion: Make the change of variables

$$x = \rho \cos \varphi \cos \theta, \qquad y = \rho \cos \varphi \sin \theta, \qquad z = \rho \sin \varphi.$$

(2) Evaluate the integral

$$J = \iiint_E (x^2 + y^2 + z^2) \, dx \, dy \, dz$$

in the region E defined by

$$x \geqslant 0, \qquad y \geqslant 0, \qquad z \geqslant 0, \qquad \frac{x^2}{a^2} + \frac{y^2}{b^2} + \frac{z^2}{c^2} - 1 \leqslant 0,$$

where a, b, and c are positive numbers.

623

(1) Evaluate the integral

$$I_R = \iint_D e^{-(x^2+y^2)} \, dx \, dy$$

in the region D defined by

$$x \geqslant 0, \qquad y \geqslant 0, \qquad x^2 + y^2 \leqslant R^2.$$

Express the double integral

$$J_a = \iint_\Delta e^{-(x^2+y^2)} \, dx \, dy$$

over the region Δ defined by

$$0 \leqslant x \leqslant a, \qquad 0 \leqslant y \leqslant a,$$

in terms of the simple integral

$$K(a) = \int_0^a e^{-t^2} \, dt.$$

(2) By comparing I_R and J_a, obtain a minorant and a majorant of $K(a)$. Show that the integral

$$\int_0^{+\infty} e^{-t^2} \, dt$$

converges and evaluate it.

624

(1) Show that the line integral

$$\int_\Gamma e^{-x^2+y^2} \, [\cos(2xy)\,dx + \sin(2xy)\,dy]$$

over a closed rectifiable Jordan curve Γ is 0.

(2) Express the above result by taking for Γ the boundary of the rectangle whose sides are the line segments

$$x = 0, \qquad x = a, \qquad y = 0, \qquad y = b,$$

where a and b are positive numbers.

(3) Show that the integral

$$\int_0^{+\infty} e^{-x^2} \cos 2bx \, dx$$

is convergent and evaluate it. Use the fact, shown in exercise 623, (2) that

$$\int_0^{+\infty} e^{-t^2} \, dt = \frac{\sqrt{\pi}}{2}.$$

•

625

Let I_Γ denote the line integral

$$\int_\Gamma P(x, y) \, dx + Q(x, y) \, dy$$

over the Jordan arc Γ, where P and Q are defined by

$$P(x, y) = \frac{e^{-y}}{x^2 + y^2} (x \sin x - y \cos x), \quad Q(x, y) = \frac{e^{-y}}{x^2 + y^2} (x \cos x + y \sin x),$$

in the complement of the point $(0, 0)$ with respect to $\mathbf{R}^2$.

(1) Show that if Γ is a closed curve not encircling or touching the origin, the value of I_Γ is 0.

(2) Take for Γ the semicircle C of radius R with center at the origin and lying in the half-plane $y \geqslant 0$. Show that

$$I_C = \int_0^\pi e^{-R \sin \theta} \cos (R \cos \theta) \, d\theta,$$

where the direction around the contour is counterclockwise, and that I_C approaches 0 and π as R tends to $+\infty$ and 0 respectively.

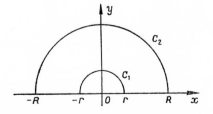

(3) Consider the contour Γ consisting of two semicircles C_1 and C_2 of radius r and R with centers at the origin and lying in the half-plane $y \geqslant 0$ and of the two segments $[-R, -r]$ and $[r, R]$ joining the end-points of these two semicircles. From the above results, show that the integral

$$\int_0^{+\infty} \frac{\sin t}{t} \, dt$$

converges and evaluate it.

Solutions

First question:

In the complement Ω of $(0, 0)$ with respect to R^2, the function f is a rational fraction the denominator of which is nonzero. Therefore, it is continuous and has partial derivatives:

$$f_x'(x, y) = y\frac{x^4+4x^2y^2-y^4}{(x^2+y^2)^2}, \qquad f_y'(x, y) = x\frac{x^4-4x^2y^2-y^4}{(x^2+y^2)^2}.$$

These partial derivatives define two continuous mappings on Ω into R^2. Therefore, the function f is differentiable in Ω (cf PZ, Book III, Chapter VII, part 2, § 1) and

$$df = y\frac{x^4+4x^2y^2-y^4}{(x^2+y^2)^2}dx+x\frac{x^4-4x^2y^2-y^4}{(x^2+y^2)^2}dy.$$

Second question:

$$f(0, y) = 0 \Rightarrow f_y'(0, 0) = 0,$$
$$f(x, 0) = 0 \Rightarrow f_x'(0, 0) = 0.$$

If f is differentiable at 0, its differential is zero. To prove the existence of the differential, it is therefore necessary, in accordance with the definition, to show that there exists a function φ that approaches 0 as its argument approaches $(0, 0)$ with the property that

$$|f(x, y)-f(0, 0)| = |f(x, y)| = \varphi(x, y)(|x|+|y|).$$

This result is immediate since

$$|f(x, y)| \leqslant |x|\,|y| \leqslant \frac{(|x|+|y|)^2}{2} \Rightarrow |\varphi(x, y)| \leqslant \frac{|x|+|y|}{2}.$$

Third question:

In Ω, the functions f_x' and f_y' are rational fractions. Therefore, they are continuous and differentiable. In particular, the derivatives f_{xy}'' and f_{yx}'' exist. These are rational fractions and hence continuous functions. They are equal.

At the point $(0, 0)$ we must make a direct examination:

$$f'_x(0, y) = -y \Rightarrow f''_{xy}(0, 0) = -1,$$
$$f'_y(0, x) = x \quad \Rightarrow f''_{yx}(0, 0) = 1.$$

At the point $(0, 0)$, the two derivatives f''_{xy} and f''_{yx} are not equal. If these two derivatives are continuous at the point $(0, 0)$, they are equal (cf. PZ, Book III, Chapter VII, part 1, § 4). Since this hypothesis is not satisfied, at least one of the derivatives is not continuous at that point.

602

First question:

In the region $x > 0$, the rational fraction φ is continuous and differentiable since the denominator is nonzero:

$$\frac{\partial \varphi}{\partial x}(x, y) = -\frac{y}{x^2}, \qquad \frac{\partial \varphi}{\partial y}(x, y) = \frac{1}{x} \Rightarrow x\frac{\partial \varphi}{\partial x} + y\frac{\partial \varphi}{\partial y} = 0.$$

Similarly, the function $g \circ \varphi$, being the composite of two differentiable functions, is differentiable, and

$$d(g \circ \varphi) = (g' \circ \varphi)d\varphi.$$

Therefore, the partial derivatives

$$(g' \circ \varphi)\frac{\partial \varphi}{\partial x} \qquad \text{and} \qquad (g' \circ \varphi)\frac{\partial \varphi}{\partial y}$$

satisfy the given equation.

Second question:

Let us make the change of variables

$$\begin{cases} u = x \\ v = \dfrac{y}{x} \end{cases} \Leftrightarrow \begin{cases} x = u \\ y = uv. \end{cases}$$

(Thus, we define a bijective mapping of the region $x > 0$ into the region $u > 0$.) The functions x and y are differentiable functions of u and v. Therefore, the same is true of the function F defined by

$$F(u, v) = f(u, uv).$$

Thus,

$$\frac{\partial F}{\partial u} = \frac{\partial f}{\partial x}\frac{\partial x}{\partial u} + \frac{\partial f}{\partial y}\frac{\partial y}{\partial u} = \frac{\partial f}{\partial x} + \frac{y}{x}\frac{\partial f}{\partial y}.$$

If f is a solution of (E), the derivative $\partial F/\partial u$ is 0 and F is a differentiable function of the single variable v:

$$f(x, y) = F\left(\frac{y}{x}\right) = (F \circ \varphi)(x, y).$$

Thus, we obtain the only functions considered in (1).

603

First question:

The change of variables in question defines an injective linear transformation T of $\boldsymbol{R}^2$ into itself. The functions x and y, being linear functions of u and v, have continuous partial derivatives of all orders with respect to u and v. The theorem on the differentiation of composite functions then enables us to assert that $F = f \circ T^{-1}$ has continuous partial derivatives of the same order as f and, in the present case, of orders 1 and 2 in particular.

Let us therefore differentiate the equation

$$f(x, y) = F(u, v).$$

We get

$$f'_x = F'_u u'_x + F'_v v'_x = F'_u + F'_v,$$
$$f'_y = F'_u u'_y + F'_v v'_y = mF'_u + nF'_v.$$

Then, by applying the rules of calculus to the two derivatives, we get

$$f''_{x^2} = \frac{\partial}{\partial u}(F'_u + F'_v) + \frac{\partial}{\partial v}(F'_u + F'_v) = F''_{u^2} + 2F''_{uv} + F''_{v^2},$$

$$f''_{xy} = m\frac{\partial}{\partial u}(F'_u + F'_v) + n\frac{\partial}{\partial v}(F'_u + F'_v) = mF''_{u^2} + (m+n)F''_{uv} + nF''_{v^2},$$

$$f''_{y^2} = m\frac{\partial}{\partial u}(mF'_u + nF'_v) + n\frac{\partial}{\partial v}(mF'_u + nF'_v) = m^2 F''_{u^2} + 2mn F''_{uv} + n^2 F''_{v^2}.$$

Therefore, the equation (E) which is (e) transformed, is

$$[a + 2bm + cm^2] F''_{u^2} + 2[a + b(m+n) + cmn] F''_{uv} + [a + 2bn + cn^2] F''_{v^2} = 0.$$

Second question:

If $b^2 - ac > 0$ and $c \neq 0$, we can take for m and n the two zeros of the trinomial $a + 2b\lambda + c\lambda^2$. In (E), the coefficients of F''_{u^2} and F''_{v^2} are therefore 0 and the coefficient of F''_{uv} is nonzero:

$$a + b(m+n) + cmn = a + b\left(-\frac{2b}{c}\right) + c\frac{a}{c} = 2\frac{ac - b^2}{c} \neq 0.$$

With this choice of m and n, equation (E) reduces to $F''_{uv} = 0$, and the solution of this equation is given by

$$F(u, v) = g(u) + h(v)$$

or

$$f(x, y) = g(x + my) + h(x + ny),$$

where g and h are two arbitrary differentiable functions.

If $c = 0$ and $a \neq 0$, we need to reverse the roles of x and y so as to have the preceding result remain valid (we set $u = mx + y$ and $v = nx + y$).

If $a = c = 0$, equation (e) is the equation $f''_{xy} = 0$, and the solution is of the type indicated above with x and y replacing u and v.

Third question:

If $b^2 - ac = 0$, we can take for m the double zero of the trinomial $a + 2b\lambda + c\lambda^2$. Then,

$$a + 2bm + cm^2 = 0, \qquad a + bm = 0, \qquad b + cm = 0,$$
$$\forall n \neq m, \qquad a + 2bn + cn^2 \neq 0.$$

Equation (E) is then reduced to $F''_{v^2} = 0$ and the solution of this equation is given by

$$F(u, v) = g(u) + vh(u),$$

or, if $n = 0$,

$$f(x, y) = g(x + my) + xh(x + my),$$

where g and h are two arbitrary functions possessing continuous first and second derivatives.

The only exceptional case: $b = c = 0$. Then, the trinomial $a + 2b\lambda + c\lambda^2$ has no zero. But equation (e) is $f''_{x^2} = 0$ and

$$f(x, y) = g(y) + xh(y).$$

Fourth question:

For equation (E) to be of the type indicated, it is necessary and sufficient that

$$\begin{cases} a+2bm+cm^2 = a+2bn+cn^2 \\ \\ a+b(m+n)+cmn = 0 \end{cases} \Leftrightarrow \begin{cases} m+n = -\dfrac{2b}{c} \\ \\ mn = \dfrac{2b^2-ac}{c^2} \end{cases}$$

since $m-n\neq0$ and $c\neq0$ (because, otherwise, b^2-ac is nonnegative).
    The system thus obtained has a solution since

$$\frac{(m+n)^2}{4} - mn = \frac{b^2-(2b^2-ac)}{c^2} = \frac{ac-b^2}{c^2} > 0.$$

604

First question:

We denote by $X=(x, y)$ and $U=(r, \theta)$ the elements of R^2 and we note by g the mapping on Δ into D defined by

$$(x, y) = (r \cos \theta, r \sin \theta) = g(r, \theta) \quad \text{or} \quad X = g(U).$$

The mapping g is differentiable. The function $\Phi = \varphi \circ g$, being the composite of two differentiable functions, is differentiable, and

$$d\Phi = \varphi'_x(X)[dr \cos \theta - r \sin \theta\, d\theta] + \varphi'_y(X)[dr \sin \theta + r \cos \theta\, d\theta]$$

$$= [\varphi'_x(X) \cos \theta + \varphi'_y(X) \sin \theta] dr + r[-\varphi'_x(X) \sin \theta + \varphi'_y(X) \cos \theta] d\theta.$$

By definition, the coefficients of dr and $d\theta$ are $\Phi'_r(U)$ and $\Phi'_\theta(U)$:

$$\begin{cases} \Phi'_r(U) = \varphi'_x(X) \cos \theta + \varphi'_y(X) \sin \theta \\ \Phi'_\theta(U) = r[-\varphi'_x(X) \sin \theta + \varphi'_y(X) \cos \theta] \end{cases}$$

$$\Leftrightarrow \begin{cases} \varphi'_x(X) = \Phi'_r(U) \cos \theta - \dfrac{\sin \theta}{r} \Phi'_\theta(U) \\ \\ \varphi'_y(X) = \Phi'_r(U) \sin \theta + \dfrac{\cos \theta}{r} \Phi'_\theta(U). \end{cases}$$

Second question:

The function f and its first derivatives (which have continuous partial derivatives) are differentiable. The functions $F=f \circ g$, $f'_x \circ g$ and $f'_y \circ g$ are therefore differentiable (g being as defined in (1)), and we know (on the basis of the formulas (1) with X replaced by $g(U)$) that

$$F_r'(U) = (f' \circ g)(U) \cos \theta + (f_y' \circ g)(U) \sin \theta,$$

$$F_\theta'(U) = r\left[-(f_x' \circ g)(U) \sin \theta + (f_y' \circ g)(U) \cos \theta\right].$$

All the functions that appear in F_r' and F_θ' are differentiable. Therefore, the functions F_r' and F_θ' are differentiable. This enables us to apply the result of (1) to evaluate the second derivatives. We already know that

$$f_x'(X) = F_r'(U) \cos \theta - \frac{\sin \theta}{r} F_\theta'(U),$$

$$f_y'(X) = F_r'(U) \sin \theta + \frac{\cos \theta}{r} F_\theta'(U).$$

To obtain the derivative $f_{x^2}''(X)$ let us apply the result of (1) to the two functions

$$\varphi(X) = f_x'(X), \qquad \Phi(U) = F_r'(U) \cos \theta - \frac{\sin \theta}{r} F_\theta'(U),$$

$$f_{x^2}''(X) = \varphi_x'(X) = \cos \theta \Phi_r'(U) - \frac{\sin \theta}{r} \Phi_\theta'(U)$$

$$= F_{r^2}''(U) \cos^2 \theta - 2\frac{\sin \theta \cos \theta}{r} F_{r\theta}''(U) + \frac{\sin^2 \theta}{r^2} F_{\theta^2}''(U)$$

$$+ \frac{\sin^2 \theta}{r} F_r'(U) + 2\frac{\sin \theta \cos \theta}{r^2} F_\theta'(U).$$

In the same way, we obtain

$$f_{y^2}''(X) = \frac{\partial}{\partial y}(f_y')(X) = \sin \theta \frac{\partial}{\partial r}\left[F_r' \sin \theta + \frac{\cos \theta}{r} F_\theta'\right](U)$$

$$+ \frac{\cos \theta}{r}\frac{\partial}{\partial \theta}\left[F_r' \sin \theta + \frac{\cos \theta}{r} F_\theta'\right](U)$$

$$= F_{r^2}''(U) \sin^2 \theta + 2\frac{\sin \theta \cos \theta}{r} F_{r\theta}''(U) + \frac{\cos^2 \theta}{r^2} F_{\theta^2}''(U)$$

$$+ \frac{\cos^2 \theta}{r} F_r'(U) - 2\frac{\sin \theta \cos \theta}{r^2} F_\theta'(U),$$

$$\Delta f = f_{x^2}''(X) + f_{y^2}''(X) = F_{r^2}''(U) + \frac{1}{r^2} F_{\theta^2}''(U) + \frac{1}{r} F_r'(U).$$

The function $F = f \circ g$ must depend on the single variable r and hence it satisfies the differential equation

$$\frac{d^2F}{dr^2} + \frac{1}{r}\frac{dF}{dr} = \frac{1}{r}\frac{d}{dr}\left(r\frac{dF}{dr}\right) = 0,$$

$$r\frac{dF}{dr} = \alpha \;\Rightarrow\; F(r) = \alpha \log|r| + \beta,$$

$$f(x, y) = \alpha \log\left(\sqrt{x^2 + y^2}\right) + \beta,$$

where α and β are arbitrary constants.

605

First question:

We obtain immediately

$$\rho = \sqrt{x^2 + y^2 + z^2}, \qquad \varphi = \arcsin\frac{z}{\rho}.$$

Finally, θ is defined by the two equations

$$\cos\theta = \frac{x}{\rho\cos\varphi}, \qquad \sin\theta = \frac{y}{\rho\cos\varphi}.$$

These equations are compatible since

$$\cos^2\theta + \sin^2\theta = \frac{x^2 + y^2}{\rho^2\cos^2\varphi} = \frac{x^2 + y^2}{\rho^2 - z^2} = 1.$$

Therefore, there exists one and only one number θ up to a multiple of 2π that satisfies the two equations, that is, one and only one θ in $[0, 2\pi[$ that satisfies them.

Second question:

Replacement of r and z with their expressions as functions of θ and φ in the equations that define T_1 shows immediately that $T = T_1 \circ T_2$.

Let us consider first the function $h = f \circ T_1$ defined by

$$h(r, \theta, z) = f(r\cos\theta, r\sin\theta, z).$$

In the differentiations with respect to z, the functions h and f are considered as functions of the variable z alone. Therefore, the transformation T_1 is the identity transformation. The function h is twice differentiable with respect to z and

$$h'_z = f'_z \circ T_1, \qquad h''_{z^2} = f''_{z^2} \circ T_1.$$

Therefore, if z is constant, the function f is a mapping on $\mathbf{R}^2$ into $\mathbf{R}$ and the mapping T_1 is the mapping studied in the preceding exercise. Thus, denoting by U the point (r, θ, z) in $\mathbf{R}^3$, we have

$$f''_{x^2}(X) + f''_{y^2}(X) = h''_{r^2}(U) + \frac{1}{r^2} h''_{\theta^2}(U) + \frac{1}{r} h'_r(U).$$

By adding $f''_{z^2}(X) = h''_{z^2}(U)$, we have

$$\Delta f(X) = f''_{x^2}(X) + f''_{y^2}(X) + f''_{z^2}(X) = h''_{r^2}(U) + \frac{1}{r^2} h''_{\theta^2}(U) + \frac{1}{r} h'_r(U) + h''_{z^2}(U).$$

The function F is defined by

$$F = f \circ T = f \circ (T_1 \circ T_2) = (f \circ T_1) \circ T_2 = h \circ T_2$$

or

$$F(P) = F(\rho, \theta, \varphi) = h(\rho \cos \varphi, \theta, \rho \sin \varphi) = h(U).$$

In the differentiations with respect to θ, we treat h or F as a function only of that variable. Then, T_2 is reduced to the identity mapping, and

$$F'_\theta(P) = h'_\theta(U), \qquad F''_{\theta^2}(P) = h''_{\theta^2}(U).$$

Thus, if we assign to θ a fixed value, the transformation T_2 is the transformation of $\mathbf{R}^2$ into $\mathbf{R}^2$ studied in exercise 604, and

$$h''_{r^2}(U) + h''_{z^2}(U) = F''_{\rho^2}(P) + \frac{1}{\rho^2} F''_{\varphi^2}(P) + \frac{1}{\rho} F'_\rho(P),$$

$$h'_r(U) = F'_\rho(P) \cos \varphi - \frac{\sin \varphi}{\rho} F'_\varphi(P).$$

Finally, by adding the results obtained, we have

$$\Delta f(X) = F''_{\rho^2}(P) + \frac{1}{\rho^2 \cos^2 \varphi} F''_{\theta^2}(P) + \frac{1}{\rho^2} F''_{\varphi^2}(P) + \frac{2}{\rho} F'_\rho(P) - \frac{\operatorname{tg} \varphi}{\rho^2} F'_\varphi(P).$$

If f is a function only of $x^2 + y^2 + z^2$, the function F is a function only of ρ and it is a solution of the differential equation

$$\frac{d^2 F}{d\rho^2} + \frac{2}{\rho}\frac{dF}{d\rho} = \frac{1}{\rho^2}\frac{d}{d\rho}\left(\rho^2\frac{dF}{d\rho}\right) = 0,$$

$$\rho^2 \frac{dF}{d\rho} = \alpha \Rightarrow F(\rho) = -\frac{\alpha}{\rho} + \beta,$$

$$f(x, y, z) = \frac{-\alpha}{\sqrt{x^2 + y^2 + z^2}} + \beta,$$

where α and β are arbitrary constants.

606

First question:

Let us denote by p the linear transformation corresponding to the matrix P. The mapping p on R^p into R^p is differentiable, and the function $f_\beta = f_\alpha \circ p$, being the composite of two differentiable mappings, is differentiable. The chain rule for the differentiation of composite functions enables us to write

$$\frac{\partial f_\beta}{\partial y_i} = \sum \frac{\partial f_\alpha}{\partial x_j}\frac{\partial x_j}{\partial y_i} = \sum_j p_{ji}\frac{\partial f_\alpha}{\partial x_j}.$$

Thus, the derivatives $\partial f_\beta/\partial y_i$ are linear combinations of the derivatives $\partial f_\alpha/\partial x_j$ and the general term of the matrix of the transformation is $p'_{ij} = p_{ji}$. This matrix is the transpose P^T of P.

Second question:

If the matrix P is orthogonal, $P^{-1} = P^T$:

$$\begin{pmatrix} \dfrac{\partial f_\beta}{\partial y_1} \\ \vdots \\ \dfrac{\partial f_\beta}{\partial y_n} \end{pmatrix} = P^{-1} \begin{pmatrix} \dfrac{\partial f_\alpha}{\partial x_1} \\ \vdots \\ \dfrac{\partial f_\alpha}{\partial x_n} \end{pmatrix} \Leftrightarrow \begin{pmatrix} \dfrac{\partial f_\alpha}{\partial x_1} \\ \vdots \\ \dfrac{\partial f_\alpha}{\partial x_n} \end{pmatrix} = P \begin{pmatrix} \dfrac{\partial f_\beta}{\partial y_1} \\ \vdots \\ \dfrac{\partial f_\beta}{\partial y_n} \end{pmatrix}.$$

The $\partial f_\alpha/\partial x_i$ are derived from the $\partial f_\beta/\partial y_j$ by the same associated matrix P as that by which the x_i were derived from the y_i. Therefore, these are the coordinates of the same vector F relative to the bases α and β.

If we consider only orthonormal bases in R^n, the associated matrices P are orthogonal matrices. The vectors F_α are all identical, and the vector F thus defined (by any vector F_α) depends only on the function f and not on the particular basis α chosen.

Third question:

The differential df_α is a linear form of the vector dX whose coordinates relative to the basis α are the dx_i. The coordinates of this vector dX relative to the basis β are the dy_j such that the associated matrix of the dy_j to the dx_i is the matrix P.

We know that the differential $df_\beta = d(f_x \circ p)$ is obtained by replacing the dx_i with their expressions as functions of the dy_j; that is, for a single vector dX, we have $df_\alpha = df_\beta$. The linear form of the vector dX defined by df_α or df_β is the same. It is a linear form df assigned to the function f and is independent of the particular basis chosen.

If α is an orthonormal basis in a Euclidean space, a linear form a is defined by

$$a(X) = \sum_{i=1}^{n} a_i x_i = \langle A, X \rangle,$$

where A is the vector with coordinates a_i relative to the basis α. The equation $a(X) = \langle A, X \rangle$ obtained by using the coordinates relative to the basis α is independent of that basis. The mapping $a \to A$ of the dual into the space is a vector space isomorphism.

If we apply this procedure to the linear form df, we see that the coordinates of the associated vector F relative to the orthonormal basis α are the $\partial f_\alpha / \partial x_i$. Thus, we again find the result of (2). The vector F_α whose coordinates in the orthonormal basis α are the $\partial f_\alpha / \partial x_i$, is a vector independent of the basis α under consideration and depends only on the function f.

607

First question:

Up to a transposition, this is the result obtained in (1) of the preceding exercise (obtained by using the chain rule):

$$\begin{pmatrix} \bar{f}'_u \\ \bar{f}'_v \\ \bar{f}'_w \end{pmatrix} = P^T \begin{pmatrix} f'_x \\ f'_y \\ f'_z \end{pmatrix} \Leftrightarrow (\bar{f}'_u \ \bar{f}'_v \ \bar{f}'_w) = (f'_x \ f'_y \ f'_z) P.$$

By applying the preceding result to the functions X, Y, and Z and to the functions $\bar{X}$, $\bar{Y}$, and $\bar{Z}$, which are composites of X, Y, Z, and p, we obtain

$$\begin{pmatrix} \bar{X}'_u & \bar{X}'_v & \bar{X}'_w \\ \bar{Y}'_u & \bar{Y}'_v & \bar{Y}'_w \\ \bar{Z}'_u & \bar{Z}'_v & \bar{Z}'_w \end{pmatrix} = \begin{pmatrix} X'_x & X'_y & X'_z \\ Y'_x & Y'_y & Y'_z \\ Z'_x & Z'_y & Z'_z \end{pmatrix} P = A(S_M) P.$$

From the definitions of U, V, and W,

$$\begin{pmatrix} X \\ Y \\ Z \end{pmatrix} = P \begin{pmatrix} U \\ V \\ W \end{pmatrix} \Rightarrow \begin{pmatrix} U \\ V \\ W \end{pmatrix} = P^{-1} \begin{pmatrix} \bar{X} \\ \bar{Y} \\ \bar{Z} \end{pmatrix} \Rightarrow \begin{pmatrix} U'_t \\ V'_t \\ W'_t \end{pmatrix} = P^{-1} \begin{pmatrix} \bar{X}'_t \\ \bar{Y}'_t \\ \bar{Z}'_t \end{pmatrix}.$$

The three equations obtained by replacing t with u, v, and w respectively are equivalent to

$$B(S_M) = \begin{pmatrix} U'_u & U'_v & U'_w \\ V'_u & V'_v & V'_w \\ W'_u & W'_v & W'_w \end{pmatrix} = P^{-1} \begin{pmatrix} \bar{X}'_u & \bar{X}'_v & \bar{X}'_w \\ \bar{Y}'_u & \bar{Y}'_v & \bar{Y}'_w \\ \bar{Z}'_u & \bar{Z}'_v & \bar{Z}'_w \end{pmatrix} = P^{-1} A(S_M) P.$$

Second question:

The matrices $B(S_M)$ and $A(S_M)$ are similar matrices. Therefore, they have the same characteristic polynomial. The coefficients of the characteristic polynomials of $B(S_M)$ and $A(S_M)$ are equal. In particular, equating the coefficients of the second-degree terms yields

$$U'_u + V'_v + W'_w = X'_x + Y'_y + Z'_z.$$

We can obtain two other invariants by equating the coefficients of the first- and zeroth-degree terms. In particular, the last equation is

$$\det B(S_M) = \det A(S_M).$$

608

First question:

The height MH relative to the side AB has a maximum when M is on the perpendicular bisector of AB.

Second question:

Let us define the point M_i by the length s_i of the arc AM_i. For the polygonal line to be convex, it is necessary and sufficient that

$$0 \leqslant s_1 \leqslant s_2 \leqslant \dots \leqslant s_n \leqslant l,$$

where l denotes the length of the arc AB. The area of the polygon $AM_1M_2 \dots M_nB$ is a function of the variable $s = (s_1, s_2, \dots, s_n)$ on the region D defined by the above inequalities.

The region D is bounded since it is contained in the block $0 \leqslant s_i \leqslant l$.

The region D is also closed: if σ denotes a point of accumulation of D, then σ is the limit of a sequence $s^{(k)} = (s_1^{(k)}, \dots, s_n^{(k)})$ of elements of D. The coordinate σ_i of σ is the limit of the sequence $\{s_i^{(k)}\}$ and

$$0 \leqslant s_1^{(k)} \leqslant s_2^{(k)} \dots \leqslant s_n^{(k)} \leqslant l \Rightarrow 0 \leqslant \sigma_1 \leqslant \sigma_2 \dots \leqslant \sigma_n \leqslant l.$$

Therefore, σ is an element of D.

The area of the triangle AM_iM_{i+1} is a continuous function of the curvilinear coordinates s_i and s_{i+1} of the vertices M_i and M_{i+1} and hence is a continuous function of s. The area of the polygon $AM_1M_2 \dots M_nB$, being the sum of the areas of the preceding triangles, is a continuous function of s, which we shall denote by f. Since f is defined and continuous in the closed and bounded (hence, compact) region D, it is bounded and it attains its least upper bound S. Therefore, there exists at least one polygon $AP_1 \dots P_nB$ with area equal to S.

Third question:

Consider three consecutive vertices P_{i-1}, P, and P_{i+1} of the above polygon. If M is a point on the arc $P_{i-1}P_{i+1}$, the area of the polygon

$$\Pi_{P_i} = AP_1 \dots P_{i-1}P_iP_{i+1} \dots B$$

must be at least equal to the area of the polygon

$$\Pi_M = AP_1 \dots P_{i-1}MP_{i+1} \dots B.$$

Now, the area of Π_{P_i} is equal to the sum of the area of the polygon

$$AP_1 \dots P_{i-1} P_{i+1} \dots B$$

and the area of the triangle $P_{i-1}P_iP_{i+1}$; the area of Π_M is equal to the sum of the area of this polygon and the area of the triangle $P_{i-1}MP_{i+1}$. The maximum area of the triangle $P_{i-1}MP_i$ is obtained when M is at the point P_i. We know from (1) that P_i is equidistant from P_{i-1} and P_{i+1}.

In the above proof, the index i was arbitrary. Therefore, all sides of the polygonal line $AM_1M_2 \dots M_nB$ are equal.

609

First question:

Let $u_1, ..., u_p$ denote the coordinates of X_0. The mapping

$$f(u_1, ..., u_{i-1}, x_i, u_{i+1}, ...)$$

has an extremum at the value $x_i = u_i$. It is defined in an interval containing u_i and is differentiable at u_i. Therefore, its derivative vanishes at that value (this result is valid for functions of a real variable):

$$\frac{\partial f}{\partial x_i} (u_1, ..., u_{i-1}, u_i, u_{i+1}, ..., u_p) = \frac{\partial f}{\partial x_i} (X_0) = 0.$$

For $X_0 = (0, ..., 0)$, the partial derivatives of the function φ are all 0, but there is neither a maximum nor a minimum at that point because, for any $x_1 > 0$,

$$\varphi(-x_1, 0, ..., 0) < 0 = \varphi(0, ..., 0) < \varphi(x_1, 0, ..., 0).$$

Second question:

The function $\sqrt{6a^2 - x^2}$ is infinitely differentiable if $|x| < a\sqrt{6}$. The functions $x^2/4a$ and $\cos y$ are infinitely differentiable. The same is true therefore for the function f:

$$\frac{\partial f}{\partial x} = x \left[\frac{1}{2a} - (6a^2 - x^2)^{-\frac{1}{2}} \cos y \right], \qquad \frac{\partial f}{\partial y} = - \sin y \sqrt{6a^2 - x^2}.$$

The function $\sqrt{6a^2 - x^2}$ does not vanish in the interior of E. The derivative $\partial f/\partial y$ vanishes if $\sin y$ does. Because of the periodicity of the function f with respect to the variable y, let us consider only the cases $y = 0$ and $y = \pi$. Thus, if we set $\partial f/\partial x$ equal to 0, we obtain the six solutions

$$(1) \begin{cases} x = 0 \\ y = 0 \end{cases} \qquad (2) \begin{cases} x = 0 \\ y = \pi \end{cases} \qquad (3) \begin{cases} x = \pm a\sqrt{2} \\ y = 0 \end{cases} \qquad (4) \begin{cases} x = \pm a\sqrt{2} \\ y = \pi. \end{cases}$$

If the first derivatives of a function f vanish at a point $X_0 = (u, v)$, Taylor's formula for this function giving second-degree terms and a remainder in a neighborhood of X_0 is

$$\Delta f = f(u+h, v+k) - f(u, v)$$

$$= h^2 f''_{x^2}(X_0) + 2hk f''_{xy}(X_0) + k^2 f''_{y^2}(X_0) + (h^2 + k^2) \alpha(h, k),$$

where the function α approaches zero as (h, k) approaches $(0, 0)$. By setting

$$h = \cos\theta\sqrt{h^2 + k^2}, \qquad k = \sin\theta\sqrt{h^2 + k^2},$$

we may write it in the form

$$\Delta f = (h^2+k^2)[f''_{x^2}(X_0)\cos^2\theta + 2f''_{xy}(X_0)\cos\theta\sin\theta + f''_{y^2}(X_0)\sin^2\theta + \alpha(h, k)]$$
$$= (h^2+k^2)[q(\theta)+\alpha(h, k)].$$

If $q(\theta) \geqslant m > 0$, there exists a sphere B with center at X_0 in which $|\alpha(x, k)| < m$. In that sphere, $\Delta f > 0$, so that we have a minimum. Similarly, if $q(\theta) \leqslant M < 0$, we have a maximum.

Finally, if $q(\theta)$ assumes both positive and negative values, there is neither a maximum nor a minimum. These three cases correspond respectively to

$$[f''_{xy}(X_0)]^2 - f''_{x^2}(X_0)f''_{y^2}(X_0) < 0, \qquad f''_{y^2}(X_0) > 0,$$
$$[f''_{xy}(X_0)]^2 - f''_{x^2}(X_0)f''_{y^2}(Y_0) < 0, \qquad f''_{y^2}(X_0) < 0,$$
$$[f''_{xy}(X_0)]^2 - f''_{x^2}(X_0)f''_{y^2}(X_0) > 0.$$

In the present case,

$$f''_{x^2}(X) = \left[\frac{1}{2a} - (6a^2-x^2)^{-\frac{1}{2}}\cos y\right] - x^2(6a^2-x^2)^{-\frac{3}{2}}\cos y,$$

$$f''_{xy}(X) = x(6a^2-x^2)^{-\frac{1}{2}}\sin y,$$

$$f''_{y^2}(X) = -\cos y(6a^2-x^2)^{\frac{1}{2}}.$$

Thus, in the four cases in question, we have

(1) $x = 0, \quad y = 0,$ $\delta = a\sqrt{6}\left(\dfrac{1}{2a} - \dfrac{1}{a\sqrt{6}}\right) > 0,$

no extremum.

(2) $x = 0, \quad y = \pi,$ $\delta = -a\sqrt{6}\left(\dfrac{1}{2a} + \dfrac{1}{a\sqrt{6}}\right) < 0,$

$f''_{y^2}(0, \pi) = a\sqrt{6},$ minimum.

(3) $x = \mp a\sqrt{2}, \quad y = 0,$ $\delta = -\dfrac{1}{2} < 0,$

$f''_{y^2}(X_0) = -2a < 0,$ maximum.

(4) $x = \pm a\sqrt{2}, \quad y = \pi,$ $\delta = -\dfrac{5}{2} < 0,$

$f''_{y^2}(X_0) = 2a > 0,$ mimimum.

Third question:

The above study cannot be applied to points with abscissa $x = +a\sqrt{6}$. At these points f assumes the value $\frac{3}{2}$. It assumes the same value at the points (x, y) and $(-x, y)$. Therefore, we need examine this function only in a neighborhood of $(a\sqrt{6}, y_0)$. Let us set $t = \sqrt{6a^2 - x^2}$ and consider only nonnegative values of t:

$$\Delta f = -\frac{t^2}{4a} + t \cos y = t\left(\cos y - \frac{t}{4a}\right).$$

The factor t is nonnegative. Therefore, Δf is of the same sign as $\cos y - t/4a$ and hence of the same sign as $\cos y_0$ if $\cos y_0$ is nonzero:

$$-\frac{\pi}{2} < y_0 - 2n\pi < \frac{\pi}{2} \Rightarrow \cos y_0 > 0 \quad \text{minimum,}$$

$$\frac{\pi}{2} < y_0 - 2n\pi < \frac{3\pi}{2} \Rightarrow \cos y_0 < 0 \quad \text{maximum.}$$

(In these two cases, the extremum is not strict since the function f is constant if $a = \pm\sqrt{6}$.)

If $y_0 = \pi/2 + n\pi$, then, in every neighborhood of $(a\sqrt{6}, y_0)$, the ratio $(\cos y)/t$ assumes arbitrary values and the expression $\cos y - t/4a$ assumes values of opposite sign. Therefore, there is no extremum at that point.

In the interior of E, there are five points at which the function f has an extremum. At all points on the straight lines serving as boundaries for E with the exception of those whose ordinates are equal to one of the numbers $\pi/2 + n\pi$, the function f has an extremum.

610

First question:

If f is differentiable, there exist, by definition, real constants A and B and real functions φ and ψ of the variable $(\Delta x, \Delta y)$ that approach 0 as (x, y) approaches $(0, 0)$ so that

$$\Delta f = \Delta P + i\Delta Q = (A + iB)\Delta z + [\varphi(\Delta x, \Delta y) + i\psi(\Delta x, \Delta y)]|\Delta z|.$$

Therefore, if we replace Δz by $\Delta x + i\Delta y$, for f to be differentiable it is necessary and sufficient that

$$\Delta P = A\Delta x - B\Delta y + \varphi(\Delta x, \Delta y)\sqrt{\Delta x^2 + \Delta y^2},$$

$$\Delta Q = B\Delta x + A\Delta y + \psi(\Delta x, \Delta y)\sqrt{\Delta x^2 + \Delta y^2}.$$

By the definition of a differential, this means that P and Q must be differentiable and that the coefficients of their differentials (that is, the partial derivatives) satisfy the equations

(1) $A = P'_x(x, y) = Q'_y(x, y), \qquad B = - P'_y(x, y) = Q'_x(x, y)$.

From the definition of the functions P and Q, we know that

$$P(x, y) = U(x, y) \cos [V(x, y)], \qquad Q(x, y) = U(x, y) \sin [V(x, y)].$$

If U and V are differentiable, so are P and Q. Conversely, if P and Q are differentiable, the function U defined by

$$U(x, y) = \sqrt{P^2(x, y) + Q^2(x, y)}$$

is differentiable and hence so are the functions $\cos V$ and $\sin V$, which implies that the function V is differentiable.

At a point (x_0, y_0) at which the functions U and V assume the values U_0 and V_0, we obtain, by using the rules of differential calculus,

$$df = dP + idQ = dUe^{iV_0} + iU_0 e^{iV_0} dV = e^{iV_0}(dU + iU_0 dV).$$

On the other hand, for f to be a differentiable function of z, it is necessary and sufficient that

$$df = (A + iB)(dx + idy) = e^{iV_0}(A' + iB')(dx + idy).$$

(For convenience in calculation, we set $A + iB = e^{iV_0}(A' + iB')$.)

In conclusion, for f to be a differentiable function, it is necessary and sufficient that the functions U and V be differentiable and that

(2) $A' = U'_x(x_0, y_0) = U_0 V'_y(x_0, y_0), \qquad B' = - U'_y(x_0, y_0) = U_0 V'_x(x_0, y_0)$.

If we differentiate the first of eqs. (1) with respect to x and the second with respect to y, we obtain

$$P''_{x^2}(x, y) = Q''_{yx}(x, y), \qquad - P''_{y^2}(x, y) = Q''_{xy}(x, y).$$

The continuity of the derivatives Q''_{yx} and Q''_{xy} implies

$$Q''_{yx}(x, y) = Q''_{xy}(x, y) \Rightarrow P''_{x^2}(x, y) = - P''_{y^2}(x, y).$$

A similar proof holds for the function Q. Regarding the functions U and V, we need note only that the functions $\ln U$ and V satisfy the same equations as do the functions P and Q.

Second question:

In exercise 604, we found the form of all functions φ of the single variable $x^2 + y^2$ that satisfy the equation $\varphi''_{x^2} + \varphi''_{y^2} = 0$. Here, this result is valid for the function

In U:

$$\ln[U(x, y)] = \alpha \ln(\sqrt{x^2+y^2})+\beta \Rightarrow U(x, y) = e^{\beta}(x^2+y^2)^{\alpha/2}.$$

Thus, the function V is determined from its differential:

$$dV = V'_x dx + V'_y dy = -\frac{1}{U}U'_y dx + \frac{1}{U}U'_x dy = \alpha\frac{x\,dy - y\,dx}{x^2+y^2}$$

$$V(x, y) = \alpha \arctan\left(\frac{y}{x}\right) + V_0,$$

or, if we set $z = x + iy = re^{i\theta}$,

$$f(z) = e^{\beta}r^{\alpha}e^{i\alpha\theta}e^{iV_0} = Cr^{\alpha}e^{i\alpha\theta},$$

where C denotes the number $e^{\beta}e^{iV_0}$.

In particular, if α is an integer, $f(z) = z^{\alpha}$.

611

First question:

We know that $|z^{(2^n)}| = |z|^{(2^n)}$ and that $2^n > n$. Therefore,

$$|z| < 1 \Rightarrow |z^{(2^n)}| < |z|^n \quad \text{and} \quad \lim|z|^n = 0,$$

so that

$$\lim z^{(2^n)} = 0;$$

$$|z| > 1 \Rightarrow |z^{(2^n)}| > |z|^n \quad \text{and} \quad \lim|z|^n = +\infty.$$

Therefore,

$$\lim|z^{(2^n)}| = +\infty,$$

so that the sequence does not converge.

Second question:

If $z^{(2^n)}$ approaches l as $n \to \infty$, we know that

$$|l| = \lim|z^{(2^n)}| = 1,$$

$$l = \lim z^{(2^{n+1})} = \lim[z^{(2^n)}]^2 = l^2.$$

These two conditions imply that $l = 1$.

The sequence $\{u_n\}$, where $u_n = z^{(2^n)} - 1$, converges to 0. Therefore, for $n \geqslant$ some integer p, we have $|u_n| \leqslant \frac{1}{2}$. By definition,

$$1 + u_{n+1} = z^{(2^{n+1})} = [z^{(2^n)}]^2 = (1 + u_n)^2 \Rightarrow u_{n+1} = u_n(2 + u_n).$$

Therefore,

$$n \geqslant p \Rightarrow |u_n| \leqslant \tfrac{1}{2} \Rightarrow |2 + u_n| \geqslant 2 - |u_n| \geqslant \tfrac{3}{2}$$

$$\Rightarrow |u_{n+1}| \geqslant \tfrac{3}{2} |u_n| \Rightarrow |u_n| \geqslant (\tfrac{3}{2})^{n-p} |u_p|.$$

Thus, the inequality $|u_n| \leqslant \frac{1}{2}$ implies

$$|u_p| \leqslant (\tfrac{2}{3})^{n-p}(\tfrac{1}{2}) \quad \forall n > p \Rightarrow |u_p| = 0 \Rightarrow |u_n| = 0 \quad \forall n \geqslant p.$$

Thus, z is a (2^n)th root of unity (where n is an integer $\geqslant p$) and, in particular, it is a (2^p)th root of unity:

$$\arg z = \frac{2k\pi}{2^p} = \frac{k}{2^{p-1}} \pi.$$

This necessary condition for convergence of the sequence is also a sufficient condition since it implies that $z^{(2^p)} = 1$ and hence, for $n \geqslant p$, $z^{(2^n)} = 1$.

612

First question:

The image of z_0 under $H(\alpha, z_0)$ is 0. The composite mapping $H(0, Z_0) \circ H$ leaves the set E invariant and maps z_0 into 0. It is a transformation of the type $H(\alpha, z_0)$:

$$H(0, Z_0) \circ H = H(\alpha, z_0) \Rightarrow H = [H(0, Z_0)]^{-1} \circ H(\alpha, z_0).$$

The equality of the images of z under $H(\alpha, z_0)$ and $H(0, Z_0) \circ H$ is written

$$e^{i\alpha} \frac{z - z_0}{1 - \bar{z}_0 z} = \frac{Z - Z_0}{1 - \bar{Z}_0 Z}.$$

Second question:

The rational function Z is differentiable and

$$\frac{dZ}{dz} = e^{i\alpha} \frac{1 - |z_0|^2}{(1 - \bar{z}_0 z)^2}.$$

In particular, if $z = z_0$, we obtain

$$\left(\frac{dZ}{dz}\right)_{z_0} = e^{i\alpha} \frac{1}{1 - |z_0|^2}.$$

If Z is the image of z under $H(\alpha, z_0)$, we saw in (1) that the image U of the number u under $H(\alpha, z_0)$ can be defined by

$$v = \frac{U-Z}{1-\bar{Z}U} = e^{i\beta}\frac{u-z}{1-\bar{z}u}.$$

where z and Z play the role played by z_0 and Z_0 in (1).
By equating the two expressions for the differential of v, we obtain

$$dv = \frac{1-|Z|^2}{(1-\bar{Z}U)^2}\,dU = e^{i\beta}\frac{1-|z|^2}{(1-\bar{z}u)^2}\,du.$$

If we assign to u the value z, we have

$$\frac{dU}{1-|Z|^2} = e^{i\beta}\frac{du}{1-|z|^2} \Rightarrow \frac{|dU|}{1-|Z|^2} = \frac{|du|}{1-|z|^2}.$$

In a different notation (i.e., with dZ and dz replaced by dU and du), this is the result required.

Third question:

The segment in question can be defined by $z = te^{i\theta}$, with $0 \leqslant t \leqslant r$ and $z = x + iy$:

$$l = \int_r \frac{|dz|}{1-|z|^2} = \int_0^r \frac{dt}{1-t^2} = \tfrac{1}{2}\ln\frac{1+r}{1-r}.$$

Another arc Γ is defined by $z = \rho e^{i\omega}$, where ρ and ω are functions of the same parameter t. Then,

$$dz = d\rho\, e^{i\omega} + i\rho\, e^{i\omega}\,d\omega \Rightarrow |dz| = \sqrt{d\rho^2 + \rho^2\,d\omega^2}.$$

The length of this curve can be bounded above by writing

$$\int_\Gamma \frac{\sqrt{d\rho^2 + \rho^2\,d\omega^2}}{1-\rho^2} \geqslant \int_\Gamma \frac{|d\rho|}{1-\rho^2} \geqslant \int_\Gamma \frac{d\rho}{1-\rho^2} = \frac{1}{2}\ln\frac{1+r}{1-r} = l.$$

The arc length is invariant under the transformation defined on these points by

$$Z = H(\alpha, z) = e^{i\alpha}\frac{z-z_0}{1-\bar{z}_0 z},$$

since, as we showed in (2),

$$\frac{|dZ|}{1-|Z|^2} = \frac{|dz|}{1-|z|^2}.$$

Thus, by transforming z_1 into O and z_2 into Z_2 under $H(0, z_1)$, we again obtain the preceding case: the arc Γ' is the minimum length among those curves whose end-points are the images of O, and Z_2 is the segment $Z = tZ_2$. Therefore, the arc Γ of minimum length with end-points at M_1 and M_2 is defined by

$$\frac{z - z_1}{1 - z\bar{z}_1} = t\,\frac{z_2 - z_1}{1 - z_2\bar{z}_2}, \qquad (t \text{ real}, \ 0 \leqslant t \leqslant 1).$$

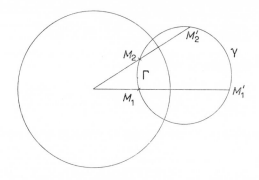

Since t is real and nonnegative, the argument of $\dfrac{z - z_1}{1 - z\bar{z}_1}$ is constant, as is that

of $\dfrac{z - z_1}{z - 1/\bar{z}_1}$.

$$\arg\frac{z - z_1}{z - 1/\bar{z}_1} = \arg(z - z_1) - \arg\left(z - \frac{1}{\bar{z}_1}\right)$$

$$= (\overrightarrow{Ox}, \overrightarrow{M_1 M}) - (\overrightarrow{Ox}, \overrightarrow{M'_1 M}) = (\overrightarrow{M'_1 M}, \overrightarrow{M_1 M}).$$

Thus, the point M lies on the circle γ passing through the point M_1 and through the point M_1 that is the inverse of M_1 with respect to the circle of unit radius centered at O.

Since M_1 and M_2 play the same role, the circle γ passes through the point M_2 and through its inverse M'_2; Γ is the arc $M_1 M_2$ of the circle $M_1 M_2 M'_2 M'_1$.

613

First question:

The functions $s + it$ and e^{s+it} are continuous functions of s and t. The function f, being the quotient of two nonzero continuous functions is a continuous function in the complement of $(0, 0)$ with respect to B.

As (s, t) approaches $(0, 0)$ in $\boldsymbol{R}^2$, the variable $s+it$ tends to 0 in C and (because of the differentiability of the exponential function)

$$\frac{e^{s+it}-1}{s+it} \to 1 .$$

The function f approaches 1 as (s, t) approaches $(0, 0)$. It is continuous at the point $(0, 0)$ and hence continuous at every point of B.

Second question:

The restriction of f to the segment $s=s_0$, $0 \leqslant t \leqslant \pi$, is continuous. Therefore, the integral $g(s_0)$ is defined.

Every closed rectangle defined by

$$s_1 \leqslant s \leqslant s_2, \qquad 0 \leqslant t \leqslant \pi,$$

lies in the interior of the region B. Therefore, the function f is continuous in this rectangle, and g is continuous in $[s_1, s_2]$. (Cf. PZ, Book III, Chapter IV, part 4, § 2. In that book, the theorem is stated for real functions f but it obviously can be extended to complex functions by considering the real and imaginary parts separately.)

Since s_1 and s_2 are arbitrary, we can always associate with a value s_0 two numbers s_1 and s_2 such that $s_1 \leqslant s_0 \leqslant s_2$. The function g is continuous at all values of s. (We note that this result holds in particular for $s=0$.)

If $s>1$, we have

$$|e^{s+it}-1| \geqslant |e^{s+it}|-1 = e^s-1$$

and

$$|g(s)| \leqslant \int_0^\pi \frac{|s+it|}{|e^{s+it}-1|} \, dt \leqslant \int_0^\pi \frac{s+\pi}{e^s-1} \, dt = \pi \frac{s+\pi}{e^s-1}.$$

The majorant

$$\pi \frac{s+\pi}{e^s-1}$$

converges to 0 as s approaches $+\infty$. Therefore, so does the function g.

Third question:

In the complement B' of $\{(0, 0)\}$ with respect to B, the function f is differentiable with respect to s and

$$\frac{\partial f}{\partial s}(s, t) = \frac{e^{s+it}-1-(s+it)e^{s+it}}{(e^{s+it}-1)^2}.$$

The function $\partial f/\partial s$ is also continuous in B' and hence in the rectangles defined by

$$0 \leqslant t \leqslant \pi, \qquad s_1 \leqslant s \leqslant s_2,$$

where the numbers s_1 and s_2 are both positive or both negative.

Every nonzero value s lies in the interior of an interval $[s_1, s_2]$ satisfying the preceding conditions, and we know that the function g is differentiable at s and that

$$g'(s) = \int_0^\pi \frac{\partial f}{\partial s}(s, t)\, dt.$$

Since the function f depends on s and t through the single variable $s+it$, the derivatives $\partial f/\partial s$ and $\partial f/\partial t$ are proportional to each other:

$$\frac{\partial f}{\partial t}(s, t) = i\frac{\partial f}{\partial s}(s, t),$$

$$g'(s) = -i\int_0^\pi \frac{\partial f}{\partial t}(s, t)\, dt = i[f(s, 0) - f(s, \pi)],$$

$$g'(s) = \frac{2\,is\,e^s}{e^{2s}-1} - \frac{\pi}{e^s+1}.$$

Fourth question:

The derivative g' converges to a limit as its argument approaches 0:

$$\frac{e^{2s}-1}{s} \rightarrow \left(\frac{d\,e^{2s}}{ds}\right)_{s=0} = 2 \Rightarrow g'(s) \rightarrow i - \frac{\pi}{2}.$$

The function g is defined and continuous in a neighborhood of 0 and its derivative g' approaches $i - \pi/2$ as s approaches 0. Therefore, the function g is differentiable at 0 and $g'(0) = i - \pi/2$ (cf. exercise 427 of Part 4).

The function g' is defined and continuous for every s. (We have just seen that $g'(0)$ is the limit of $g'(s)$ as s approaches 0.) Therefore, it is integrable over every closed interval $[0, \sigma]$ and

$$\int_0^\sigma g'(s)\, ds = g(\sigma) - g(0).$$

As σ approaches $+\infty$, $g(\sigma)$ approaches 0. By definition, the integral

$$\int_0^{+\infty} g'(s)\, ds$$

Finally, by setting $t/2=u$, we have

$$\int_{+0}^{\pi} \ln\left(\sin\frac{t}{2}\right) dt = 2 \int_{+0}^{\pi/2} \ln(\sin u)\, du = -\pi \ln 2.$$

614

First question:

The functions P and Q are rational functions the denominators of which do not vanish in D. Therefore, they are differentiable. The partial derivatives are also rational functions whose denominators do not vanish in D. They are continuous. A necessary condition for ω to be the differential of a function U is therefore that $\partial P/\partial y = \partial Q/\partial x$:

$$\frac{\partial P}{\partial y} = \frac{4y^2(x^2-y^2)-(3x^2-y^2)(x^2+y^2)}{x^2 y^2} = \frac{-3y^4+2x^2y^2-3x^4}{x^2 y^2}.$$

If we reverse the positions of x and y, we transform P into Q and hence $\partial P/\partial y$ into $\partial Q/\partial x$. Since $\partial P/\partial y$ is symmetric in x and y, the derivative $\partial Q/\partial x$ has the same expression and the two derivatives are equal.

Second question:

The function U is determined by the two equations

$$\frac{\partial U}{\partial x}(x, y) = P(x, y) = \frac{1}{y}\left[3x^2+2y^2 - \frac{y^4}{x^2}\right],$$

$$\frac{\partial U}{\partial y}(x, y) = Q(x, y) = \frac{1}{x}\left[-\frac{x^4}{y^2} + 2x^2+3y^2\right].$$

We shall obtain a solution of the first equation by taking for U a primitive of the function $P(x, y)$, where the variable y is treated as a fixed parameter. The other solutions are derived from this one by adding to it an arbitrary function of the variable y:

$$U(x, y) = \frac{1}{y}\left[x^3+2xy^2 + \frac{y^4}{x}\right] + \varphi(y) = \frac{(x^2+y^2)^2}{xy} + \varphi(y).$$

Thus, we write that the second equation is satisfied, from which we get

$$\varphi'(y)=0 \quad \Rightarrow \quad \varphi(y) \text{ is constant}$$

is convergent and equal to $-g(0)$:

$$-g(0) = \int_0^{+\infty} \frac{2\,\mathrm{i}\,s\,e^s\,ds}{e^{2s}-1} - \pi \int_0^{+\infty} \frac{ds}{e^s+1} = 2\,\mathrm{i} \int_0^{+\infty} \frac{s\,e^s\,ds}{e^{2s}-1} - \pi \ln 2\,.$$

(We do not need to determine the convergence of

$$\int_0^{+\infty} \frac{s\,e^s\,ds}{e^{2s}-1}\,,$$

the imaginary part of

$$\int_0^{+\infty} g'(s)\,ds\,,$$

since we have established the convergence of this last integral.)

By definition,

$$g(0) = \int_0^{\pi} \frac{\mathrm{i}t\,dt}{e^{\mathrm{i}t}-1} = \int_0^{\pi} \frac{\mathrm{i}t\,e^{-\mathrm{i}t/2}}{2\,\mathrm{i}\,\sin t/2}\,dt = \frac{1}{2} \int_0^{\pi} t \cot \frac{t}{2}\,dt - \int_0^{\pi} \frac{\mathrm{i}t\,dt}{2}\,.$$

By equating the two expressions for $g(0)$, we obtain

$$\pi \ln 2 = \frac{1}{2} \int_0^{\pi} t \cot \frac{t}{2}\,dt\,, \qquad \int_0^{+\infty} \frac{s\,e^s\,ds}{e^{2s}-1} = \frac{1}{2} \int_0^{\pi} \frac{t\,dt}{2} = \frac{\pi^2}{8}\,.$$

The function

$$\frac{1}{2} t \cot \frac{t}{2} = f(0,\,t)$$

is continuous at $t=0$ if we take $f(0, 0)=1$ (this is the result of (1)). Therefore, there is no convergence problem for the integral

$$\int_0^{\pi} \frac{1}{2} t \cot \frac{t}{2}\,dt\,.$$

But to transform this integral by integrating by parts, we need to consider a closed interval $[\theta, \pi]$ in which the two functions t and $\cot (t/2)$ are continuous (we take $\theta > 0$):

$$\int_\theta^{\pi} \frac{1}{2} t \cot \frac{t}{2}\,dt = \left[t \ln \left(\sin \frac{t}{2} \right) \right]_\theta^{\pi} - \int_\theta^{\pi} \ln \left(\sin \frac{t}{2} \right) dt\,.$$

As θ approaches 0,

$$\ln \left(\sin \frac{\theta}{2} \right) \simeq \ln \frac{\theta}{2} \qquad \text{and} \qquad \theta \ln \left(\sin \frac{\theta}{2} \right) \to 0\,,$$

$$\int_{+0}^{\pi} \ln \sin \frac{t}{2}\,dt = \lim_{0 \to 0} \int_\theta^{\pi} \ln \left(\sin \frac{t}{2} \right) dt = -\frac{1}{2} \int_0^{\pi} t \cot \frac{t}{2} = -\pi \ln 2\,.$$

(for example, $\varphi(y)=0$) and

$$U(x, y) = \frac{(x^2+y^2)^2}{xy}.$$

Third question:

If A and B are the end-points of an arc Γ in the interior of the region D, we know (cf. PZ, Book III, Chapter VIII, § 2) that

$$\int_\Gamma d\omega = \int_\Gamma dU = U(B)-U(A).$$

We can verify immediately that the particular arc Γ in question is indeed in the interior of D. The end-points of this arc are the points A and B with coordinates

$$A\begin{vmatrix} 1 \\ 1 \end{vmatrix}, \qquad B\begin{vmatrix} \pi/2 \\ 2 \end{vmatrix},$$

$$\int_\Gamma d\omega = U\left(\frac{\pi}{2}, 2\right) - U(1, 1) = \frac{(\pi^2+16)^2}{16\pi} - 4.$$

615

First question:

The function U is defined by

$$\frac{\partial U}{\partial x} = P, \qquad \frac{\partial U}{\partial y} = Q.$$

If U has continuous second partial derivatives, then P and Q have continuous partial derivatives and hence so does the function φ defined by

$$\varphi(x, y) = (x^2+y^2)[P(x, y)-y]/y.$$

Furthermore, we know that continuity of the second derivatives of U implies

$$\frac{\partial^2 U}{\partial x\, \partial y} = \frac{\partial^2 U}{\partial y\, \partial x} \Rightarrow \frac{\partial P}{\partial y} = \frac{\partial Q}{\partial x},$$

$$1 + \frac{x^2-y^2}{(x^2+y^2)^2}\, \varphi(x, y) + \frac{y}{x^2+y^2}\frac{\partial\varphi}{\partial y}(x, y)$$

$$= -1 - \frac{y^2-x^2}{(x^2+y^2)^2}\, \varphi(x, y) - \frac{x}{x^2+y^2}\frac{\partial\varphi}{\partial x}(x, y),$$

from which we obtain equation (E) by elementary calculations.

Second question:

The invariance of the differential under a change of variables enables us to obtain the differential of the function V associated with U, that is, defined by $V(r, \theta) = U(r \cos \theta, r \sin \theta)$:

$$dV = P(r \cos \theta, r \sin \theta) \, dx + Q(r \cos \theta, r \sin \theta) \, dy$$

$$= r \sin \theta \left[1 + \frac{\Phi(r, \theta)}{r^2} \right] (dr \cos \theta - r \sin \theta \, d\theta)$$

$$- r \cos \theta \left[1 + \frac{\Phi(r, \theta)}{r^2} \right] (dr \sin \theta + r \cos \theta \, d\theta)$$

$$= - [\Phi(r, \theta) + r^2] \, d\theta \, .$$

Thus, V is a function of θ alone and the condition under which this is possible is

$$\Phi(r, \theta) + r^2 = f(\theta),$$

where f is an arbitrary differentiable function, and therefore

$$V(\theta) = V_0 - \int_{\theta_0}^{\theta} f(t) \, dt \, .$$

The function φ is determined by

$$\varphi(x, y) + x^2 + y^2 = g \left(\frac{y}{x} \right).$$

If we form the equation E', we find

$$r\Phi'_r + 2r^2 = 0 \, .$$

Third question:

The function φ is indeed of the type indicated:

$$\Phi(r, \theta) = -r^2 + 4 \cos^2 \theta \implies V(\theta) = V_0 - 2\theta - \sin 2\theta.$$

In a neighborhood of the point $(1, 0)$, we can take for θ the value $\arctan (y/x)$ and

$$U(x, y) = - 2 \arctan \frac{y}{x} - \frac{2xy}{x^2 + y^2} \, .$$

If the function U_0 exists, the line integral

$$\int_\Gamma (P(x, y)\,dx + Q(x, y)\,dy) = \int_\Gamma dU_0 = 0,$$

where Γ is a closed curve that does not pass through the origin (the integral is equal to $U_0(B) - U_0(A)$, where A and B are the end-points of the arc Γ, and it is therefore zero if the curve is closed since then A and B are the same point).

If Γ is the circle of radius 1 with center at the origin and if we take the point $(1, 0)$ for A and B, we obtain

$$\int_\Gamma (P(x, y)\,dx + Q(x, y)\,dy) = -\int_0^{2\pi} 4\cos^2\theta\,d\theta = -4\pi.$$

Thus, the problem has no solution.

REMARK: If Γ is a closed rectifiable Jordan curve and if the region $\bar{D}$ consisting of points of Γ together with points enclosed by Γ is contained in Ω, sufficient conditions for the validity of Riemann's formula are satisfied and

$$\int_{\Gamma+} P(x, y)\,dx + Q(x, y)\,dy = \iint_{\bar{D}} \left[\frac{\partial Q}{\partial x}(x, y) - \frac{\partial P}{\partial y}(x, y)\right] dx\,dy = 0.$$

In other words, if the point O does not belong either to Γ or to the region enclosed by Γ, the line integral over the curve Γ is 0. But we have seen that this is no longer the case if O lies in the area enclosed by Γ.

616

The region D can be defined by

$$0 \leqslant y \leqslant b(1 - x/a), \qquad \text{definition of the segment } MN;$$

$$0 \leqslant x < a, \qquad \text{requirement that } M \text{ lie between } O \text{ and } A.$$

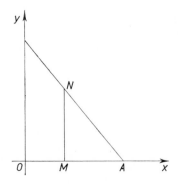

Thus, we know that

$$I = \int_0^a dx \left(\int_0^{b(1-x/a)} (x^2 + y^2) dy \right),$$

$$I = \int_0^a \left[bx^2 \left(1 - \frac{x}{a} \right) + \frac{1}{3} b^3 \left(1 - \frac{x}{a} \right)^3 \right] dx$$

$$= b \left(\frac{a^3}{3} - \frac{a^3}{4} \right) + \frac{1}{3} b^3 \frac{a}{4} = \frac{ab(a^2 + b^2)}{12}.$$

REMARK: We can also integrate first with respect to x. Then,

$$I = \int_0^b dy \left(\int_0^{a(1-y/b)} (x^2 + y^2) dx \right).$$

617

Let us integrate first with respect to y, then with respect to x. For each value of x in $[0, 2]$, we integrate with respect to y over the interval

$$-\sqrt{2x-x^2} \leqslant y \leqslant \sqrt{2x-x^2} \qquad \text{(segment } MN\text{)}.$$

Then we integrate with respect to x from 0 to 2:

$$I = \int_0^2 x\, dx \left(\int_{-\sqrt{2x-x^2}}^{\sqrt{2x-x^2}} dy \right) = \int_0^2 2x \sqrt{2x-x^2}\, dx .$$

We replace the relation

$$u = \sqrt{2x-x^2} = \sqrt{1-(x-1)^2}$$

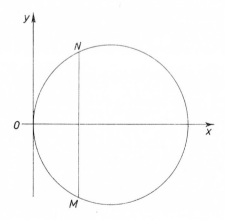

with the parametric equations

$$x = 1 + \cos \varphi, \qquad u = \sin \varphi, \qquad 0 \leqslant \varphi \leqslant \pi,$$

$$i = \int_0^\pi (2 \sin^2 \varphi + 2 \cos \varphi \sin^2 \varphi) \, d\varphi$$

$$= [\varphi - \tfrac{1}{2} \sin 2\varphi + \tfrac{2}{3} \sin^3 \varphi]_0^\pi = \pi.$$

Another way of carrying out the calculation:

We recall (Riemann's formula) that, if Γ denotes the boundary of a region C, then

$$\int_{\Gamma+} (f(x, y) \, dx + g(x, y) \, dy) = \iint_C \left[\frac{\partial g}{\partial x}(x, y) - \frac{\partial f}{\partial y}(x, y) \right] dx \, dy.$$

We can take the present case as an example:

$$f(x, y) = 0, \qquad g(x, y) = \frac{x^2}{2} \qquad \text{or} \qquad f(x, y) = -xy, \qquad g(x, y) = 0$$

and

$$I = \int_\Gamma \frac{x^2}{2} \, dy = \int_0^{2\pi} \frac{1}{2}(1 + \cos \varphi)^2 \cos \varphi \, d\varphi = \pi.$$

618

First question:

We shall integrate first with respect to x because, when we consider y as a parameter, we have

$$x \, dx = \tfrac{1}{2} \, d(x^2) = -\tfrac{1}{2} \, d(1 - x^2 - y^2).$$

The region D can be defined by

$$0 \leqslant x \leqslant \sqrt{1 - y^2}, \qquad 0 \leqslant y \leqslant 1.$$

Then,

$$I = \int_0^1 dy \left(\int_0^{\sqrt{1-y^2}} x \sqrt{1 - x^2 - y^2} \, dx \right)$$

$$= \int_0^1 dy \left[-\tfrac{1}{3}(1 - x^2 - y^2)^{\frac{3}{2}} \right]_0^{\sqrt{1-y^2}} = \frac{1}{3} \int_0^1 (1 - y^2)^{\frac{3}{2}} \, dy.$$

If we set $\varphi = \arcsin y$, we obtain

$$I = \frac{1}{3} \int_0^{\pi/2} \cos^4 \varphi \, d\varphi$$

$$= \frac{1}{3} \int_0^{\pi/2} \left(\frac{\cos 4\varphi}{8} + \frac{\cos 2\varphi}{2} + \frac{3}{8} \right) d\varphi = \frac{\pi}{16}.$$

The transformation of $\cos^4 \varphi$ into a sum is obtained, for example, by writing

$$\cos^4 \varphi = \left(\frac{e^{i\varphi} + e^{-i\varphi}}{2} \right)^4 = \frac{e^{4i\varphi} + e^{-4i\varphi}}{16} + \frac{e^{2i\varphi} + e^{-2i\varphi}}{4} + \tfrac{3}{8}.$$

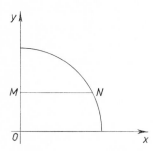

Second question:

In the integral T, we make the change of variables

$$x = r \cos \theta, \qquad y = r \sin \theta.$$

The transformed region $\varDelta$ is defined by

$$0 \leqslant r \leqslant 1, \qquad 0 \leqslant \theta \leqslant \frac{\pi}{2},$$

and the jacobian of the mapping is equal to r. Therefore,

$$I = \iint_\varDelta r \cos \theta \sqrt{1 - r^2} \, r dr \, d\theta.$$

The region $\varDelta$ is a rectangle and the integrand is the product of a function of r and a function of θ. In this case.

$$I = \left(\int_0^{\pi/2} \cos \theta \, d\theta \right) \left(\int_0^1 r^2 \sqrt{1 - r^2} \, dr \right) = \int_0^1 r^2 \sqrt{1 - r^2} \, dr.$$

When we make the change of variable $r = \sin \varphi$, we have

$$I = \int_0^{\pi/2} \sin^2 \varphi \cos^2 \varphi \, d\varphi = \int_0^{\pi/2} \frac{\sin^2 2\varphi}{4} \, d\varphi = \left[\frac{\varphi}{8} - \frac{\sin 4\varphi}{32} \right]_0^{\pi/2} = \frac{\pi}{16}.$$

619

First question:

Let us assume that $x > x_0$. The left-hand side of the equation is the double integral I of the function

$$(u-v)^n f(v)$$

over the region T defined by

$$x_0 \leqslant v \leqslant u \, (\text{segment } MN) \qquad \text{and} \qquad x_0 \leqslant u \leqslant x,$$

that is, in the triangle T the sides of which are segments of the bisector of the angle between the axes and of the lines (parallel to the axes) representing the equations $v = x_0$ and $u = x$.

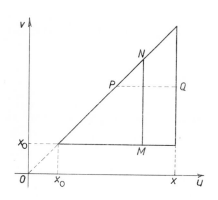
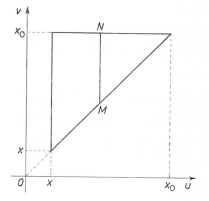

If we evaluate the integral I by integrating first with respect to u and then with respect to v, the region T is defined by

$$v \leqslant u \leqslant x \, (\text{segment } PQ) \qquad \text{and} \qquad x_0 \leqslant v \leqslant x :$$

$$I = \int_{x_0}^x dv \left(\int_v^x (u-v)^n f(v) \, du \right) = \int_{x_0}^x \frac{(x-v)^{n+1}}{n+1} f(v) \, dv .$$

If $x < x_0$, we obtain the same result by replacing the triangle T by the triangle T' defined by

$$u \leqslant v \leqslant x_0 \, (\text{segment } MN) \qquad \text{and} \qquad x \leqslant u \leqslant x_0,$$

that is, the triangle lying above the bisector of the angle between the axes. In fact, the integral J over T' of the function

$$(u-v)^n f(v)$$

has the value

$$J = \int_x^{x_0} du \left(\int_u^{x_0} (u-v)^n f(v)\, dv \right)$$

$$= \int_x^{x_0} dv \left(\int_x^{v} (u-v)^n f(v)\, du \right) = -\int_x^{x_0} \frac{(x-v)^{n+1}}{n+1} f(v)\, dv.$$

Upon reversal of all the limits of integration, the left-hand member is multiplied by $(-1)^2 = 1$ and the right-hand member by -1, and we then obtain the required result.

Second question:

Proof by induction on n

The result is obvious if $n=1$ (we take $(n-1)! = 0! = 1$). Let us suppose it is true that the function F_n defined by

$$F_n(x) = \int_{x_0}^x \frac{(x-t)^{n-1}}{(n-1)!} f(t)\, dt,$$

is the nth-order primitive of f that vanishes at x_0, and suppose also that its first $n-1$ derivatives vanish at x_0.

The result of (1) enables us to write

$$F_{n+1}(x) = \int_{x_0}^x \frac{(x-v)^n}{n!} f(v)\, dv = \int_{x_0}^x du \int_{x_0}^u \frac{(u-v)^{n-1}}{(n-1)!} f(v)\, dv = \int_{x_0}^x F_n(u)\, du.$$

Thus, F_{n+1} is the primitive of F_n that vanishes at x_0: It is the $(n+1)$st-order primitive of f with the property that it and its first n derivatives vanish at x_0.

We can also evaluate the derivative of F_{n+1} by noting that the variable x appears as a limit of integration and in the integrand. Since the derivative of this last function is a continuous function, we know that F_n is differentiable and that

$$F_{n+1}'(x) = \left[\frac{(x-v)^n}{n!} f(v) \right]_{v=x} + \int_{x_0}^x \frac{\partial}{\partial x}\left[\frac{(x-v)^n}{n!} f(v) \right] dv$$

$$= \int_{x_0}^x \frac{(x-v)^{n-1}}{(n-1)!} f(v)\, dv = F_n(x).$$

620

$$\left(\int_0^a e^{-x^2}\,dx\right)^2 = \left(\int_0^a e^{-x^2}\,dx\right)\left(\int_0^a e^{-y^2}\,dy\right) = \iint_D e^{-x^2}e^{-y^2}\,dx\,dy,$$

where the region D is the square $0 \leqslant x \leqslant a, 0 \leqslant y \leqslant a$ (since the integral over a rectangle [in the present case, the square D] of the product of a function of x alone and a function of y alone is the product of the simple integrals of those two functions).

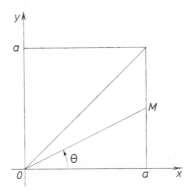

At points symmetric with respect to the line bisecting the angle between the coordinate axes, the function $e^{-(x^2+y^2)}$ assumes the same value. The region D is symmetric with respect to this bisecting line. The integral I in the square D is therefore twice the integral I over the triangle D_0 consisting of points in D lying below the bisecting line, that is defined by

$$0 \leqslant y \leqslant x, \qquad 0 \leqslant x \leqslant a.$$

To evaluate I_0, let us make the change of variables:

$$x = r\cos\theta, \qquad y = r\sin\theta.$$

The region $\varDelta_0$ representing D_0 transformed by this change of variables is defined by

$$0 \leqslant r \leqslant \frac{a}{\cos\theta}, \qquad 0 \leqslant \theta \leqslant \frac{\pi}{4}$$

and the jacobian of the transformation is equal to r:

$$I = 2I_0 = 2\iint_{\varDelta_0} e^{-r^2} r\,dr\,d\theta = \int_0^{\pi/4} d\theta \left(\int_0^{a/\cos\theta} 2e^{-r^2} r\,dr\right),$$

$$I = \int_0^{\pi/4} (1-e^{-a^2/\cos^2\theta})\,d\theta.$$

621

We know (cf. PZ, Book III, Chapter IX, part 2, § 5) that

$$\mu = \iint_d \frac{1}{2}\left(\frac{x^2}{a} + \frac{y^2}{b}\right) dx\, dy,$$

where d is the projection onto the xy-plane of the region D, that is, d is defined by

$$\frac{x^2}{a^2} + \frac{y^2}{b^2} - 1 \leqslant 0.$$

Let us now make the change of variables

$$x = au \cos v, \qquad y = bu \sin v, \qquad 0 \leqslant v < 2\pi.$$

The region δ representing d after this change of variables is defined by

$$0 \leqslant u \leqslant 1, \qquad 0 \leqslant v < 2\pi,$$

and the jacobian is equal to abu. Therefore,

$$\mu = \iint_\delta \tfrac{1}{2} u^2 (a \cos^2 v + b \sin^2 v)\, abu\, du\, dv.$$

The region δ is a rectangle and the integrand is the product of a function of u and a function of v:

$$\mu = \frac{ab}{2}\left(\int_0^1 u^3\, du\right)\left(\int_0^{2\pi} (a \cos^2 v + b \sin^2 v)\, dv\right) = \pi ab\, \frac{a+b}{8}.$$

622

First question:

Referring to exercise 605, we see that we can take for ρ, θ, and φ numbers satisfying the inequalities

$$\rho \geqslant 0, \qquad -\frac{\pi}{2} \leqslant \varphi \leqslant \frac{\pi}{2}, \qquad 0 \leqslant \theta < 2\pi.$$

If these inequalities are satisfied, the region Σ representing the image of S under the transformation is given by

$$0 \leqslant \rho \leqslant 1, \qquad 0 \leqslant \varphi \leqslant \frac{\pi}{2}, \qquad 0 \leqslant \theta \leqslant \frac{\pi}{2}.$$

and the value of the jacobian of the transformation is

$$\frac{D(x, y, z)}{D(\rho, \theta, \varphi)} = \begin{vmatrix} \cos \varphi \cos \theta & -\rho \cos \varphi \sin \theta & -\rho \sin \varphi \cos \theta \\ \cos \varphi \sin \theta & \rho \cos \varphi \cos \theta & -\rho \sin \varphi \sin \theta \\ \sin \varphi & 0 & \rho \cos \varphi \end{vmatrix} = \rho^2 \cos \varphi.$$

Since this jacobian is positive in the region Σ,

$$I = \iiint_\Sigma \rho^4 \cos \varphi \, d\rho \, d\theta \, d\varphi = \left(\int_0^1 \rho^4 \, d\rho \right) \left(\int_0^{\pi/2} d\theta \right) \left(\int_0^{\pi/2} \cos \varphi \, d\varphi \right).$$

(I is the product of three simple integrals since Σ is a parallelepiped and since the integrand is the product of three functions each depending on only one of the variables ρ, θ, φ.)

$$I = \frac{1}{5} \times \frac{\pi}{2} \times 1 = \frac{\pi}{10}.$$

Second question:

Let us make the change of variables

$$x = aX, \qquad y = bY, \qquad z = cZ.$$

The image of E is then given by

$$X \geqslant 0, \qquad Y \geqslant 0, \qquad Z \geqslant 0, \qquad X^2 + Y^2 + Z^2 - 1 \leqslant 0.$$

This is the region S of (1). The jacobian is equal to abc and therefore,

$$J = \iiint_S (a^2 X^2 + b^2 Y^2 + c^2 Z^2) abc \, dX \, dY \, dZ.$$

Evaluation of J is equivalent to evaluating the three integrals

$$\iiint_S X^2 \, dX \, dY \, dZ, \qquad \iiint_S Y^2 \, dX \, dY \, dZ, \qquad \iiint_S Z^2 \, dX \, dY \, dZ.$$

Obviously, the region S remains unchanged if we switch the positions of X and Y or X and Z. Therefore, these three integrals are equal and each is equal to $I/3$:

$$J = (a^2 + b^2 + c^2) abc \, \frac{I}{3} = (a^2 + b^2 + c^2) abc \, \frac{\pi}{30}.$$

623

First question:

Let us make the change of variables

$$x = r \cos \theta, \qquad y = r \sin \theta.$$

The region D' representing the image of D is defined by $0 \leqslant r \leqslant R$ and $0 \leqslant \theta \leqslant \pi/2$. The jacobian is equal to r. Therefore,

$$I_R = \iint_{D'} e^{-r^2} r \, dr \, d\theta = \left(\int_0^R e^{-r^2} r \, dr \right) \left(\int_0^\theta d\theta \right) = \frac{\pi}{4}(1 - e^{-R^2}).$$

When we integrate the function

$$e^{-(x^2+y^2)} = e^{-x^2} e^{-y^2}$$

over a square Δ we obtain

$$J_a = \left(\int_0^a e^{-x^2} \, dx \right) \left(\int_0^a e^{-y^2} \, dy \right) = [K(a)]^2.$$

Second question:

The function $e^{-(x^2+y^2)}$ is positive. Therefore,

$$D_1 \subset D_2 \Rightarrow \iint_{D_1} e^{-(x^2+y^2)} \, dx \, dy \leqslant \iint_{D_2} e^{-(x^2+y^2)} \, dx \, dy.$$

The square Δ is contained in the circle of radius $a\sqrt{2}$ and it contains the circle of radius a. Therefore,

$$\frac{\pi}{4}(1 - e^{-a^2}) = I_a \leqslant J_a = [K_a]^2 \leqslant I_{a\sqrt{2}} = \frac{\pi}{4}(1 - e^{-2a^2}).$$

As a tends to $+\infty$, the functions e^{-a^2} and e^{-2a^2} approach 0. The function $[K_a]^2$ approaches $\pi/4$ and hence K_a approaches $\sqrt{\pi}/2$. By definition, the integral

$$\int_0^{+\infty} e^{-t^2} \, dt$$

is convergent, and

$$\int_0^{+\infty} e^{-t^2} \, dt = \frac{\sqrt{\pi}}{2}.$$

(This result was obtained in exercise 539 of Part 5 by quite a different method.)

624

First question:

The functions f and g defined in R^2 by

$$f(x, y) = e^{-x^2+y^2} \cos 2xy, \qquad g(x, y) = e^{-x^2+y^2} \sin 2xy,$$

are continuous and continuously differentiable throughout the entire plane, and we can easily verify that

$$\frac{\partial f}{\partial y}(x, y) = \frac{\partial g}{\partial x}(x, y) = e^{-x^2+y^2}[2y \cos (2xy) - 2x \sin (2xy)].$$

We can now apply Riemann's formula to the curve Γ and to the region $\bar{D}$ that it defines:

$$\int_{\Gamma^+} (f(x, y)\,dx + g(x, y)\,dy) = \iint_{\bar{D}} \left[\frac{\partial g}{\partial x}(x, y) - \frac{\partial f}{\partial y}(x, y)\right] dx\,dy = 0.$$

Second question:

The line integral around the rectangle in the counterclockwise direction is the sum of the integrals over the four sides: for the four sides,

$$y = 0, \qquad I_1 = \int_0^a e^{-x^2}\,dx,$$

$$x = a, \qquad I_2 = \int_0^b e^{-a^2}\,e^{y^2} \sin (2ay)\,dy,$$

$$y = b, \qquad I_3 = \int_a^0 e^{b^2}\,e^{-x^2} \cos (2bx)\,dx,$$

$$x = 0, \qquad I_4 = 0.$$

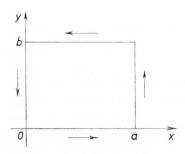

The rectangle is a rectifiable Jordan curve. The result of (1) is therefore applicable to it and

$$I_1 + I_2 + I_3 = 0.$$

Third question:

As a tends to $+\infty$, I_1 tends to

$$\int_0^{+\infty} e^{-x^2} dx = \frac{\sqrt{\pi}}{2};$$

$$0 \leqslant y \leqslant b \Rightarrow |e^{y^2} \sin(2ay)| \leqslant e^{b^2} \Rightarrow |I_2| \leqslant e^{-a^2} e^{b^2} \pi.$$

Therefore, if a tends to $+\infty$, I_2 tends to 0 and

$$\int_0^a e^{b^2} e^{-x^2} \cos(2bx) dx = -I_3 = I_1 + I_2 \rightarrow \frac{\sqrt{\pi}}{2},$$

which proves that the function $x \rightarrow e^{-x^2} \cos(2bx)$ is integrable over the interval $[0, +\infty[$ and that

$$\int_0^{+\infty} e^{-x^2} \cos(2bx) dx = e^{-b^2} \frac{\sqrt{\pi}}{2}.$$

625

First question:

The functions P and Q are continuously differentiable in the complement Ω of $\{(0, 0)\}$ with respect to $\mathbf{R}^2$, and we can verify that

$$\frac{\partial P}{\partial y}(x, y) = \frac{\partial Q}{\partial x}(x, y)$$

$$= e^{-y}\left[\frac{-x \sin x + y \cos x}{x^2 + y^2} - \frac{(x^2 - y^2) \cos x + 2xy \sin x}{(x^2 + y^2)^2}\right].$$

Therefore, if the region $\bar{D}$ consisting of the closed curve Γ and the points enclosed by it is contained in Ω, we can apply Riemann's formula and

$$\int_{\Gamma^+} (P(x, y) dx + \bar{Q}(x, y) dy) = \iint_{\bar{D}} \left[\frac{\partial Q}{\partial y}(x, y) - \frac{\partial P}{\partial x}(x, y)\right] dx \, dy = 0.$$

Second question:

Let us represent the semicircle C by

$$x = R \cos \theta, \qquad y = R \sin \theta, \qquad 0 \leqslant \theta \leqslant \pi.$$

Then,

$$x \, dx + y \, dy = 0, \qquad \frac{x \, dy - y \, dx}{x^2 + y^2} = d\theta,$$

$$I_c = \int_0^\pi e^{-R \sin \theta} \cos (R \cos \theta) \, d\theta.$$

To study the limit of I_C as R tends to $+\infty$, let us note that

$$|I_c| \leqslant \int_0^\pi e^{-R \sin \theta} \, d\theta = 2 \int_0^{\pi/2} e^{-R \sin \theta} \, d\theta.$$

(The last result is obtained by taking $t = \pi - \theta$ in the integral over the interval $[\pi/2, \pi]$.)

We know (cf. exercise 439 of Part 4) that

$$0 \leqslant \theta \leqslant \frac{\pi}{2} \Rightarrow \frac{2}{\pi} \theta \leqslant \sin \theta \leqslant \theta \Rightarrow e^{-R \sin \theta} \leqslant e^{-2R\theta/\pi},$$

$$|I_c| \leqslant 2 \int_0^{\pi/2} e^{-2R\theta/\pi} \, d\theta = \pi \frac{1 - e^{-R}}{R} \leqslant \frac{\pi}{R}.$$

As R tends to $+\infty$, I_C tends to 0. The function

$$e^{-R \sin \theta} \cos (R \cos \theta)$$

is continuous throughout the entire $R\theta$-plane. The integral I_C of this function from 0 to π is therefore a continuous function of R. As R tends to 0, I_C tends to the integral relative to the value 0 of R, namely the integral of 1.

If R tends to 0, I_C tends to π.

Third question:

The integral over the contour Γ is 0 (since Γ is a closed contour excluding 0). This integral is the sum of the integrals $I_{C_2^+}$ and $-I_{C_1^+}$ over the semicircles C_1 and C_2 of radius r and R respectively (where we go around C_2 in the counterclockwise direction and around C_1 in the clockwise direction) and the integrals J and J' over the segments $[r, R]$ and $[-R, -r]$ of the x-axis.

For the integrals J and J' over the x-axis,

$$y = 0, \qquad dy = 0, \qquad P(x, 0) = \frac{\sin x}{x},$$

$$J' = \int_{-R}^{-r} \frac{\sin x}{x}\, dx = \int_{r}^{R} \frac{\sin x}{x}\, dx = J.$$

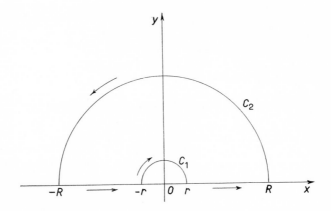

Writing that the integral over Γ is 0, we obtain

$$2 \int_{r}^{R} \frac{\sin x}{x}\, dx + I_{c_2^+} - I_{c_1^+} = 0.$$

As r tends to 0, $I_{c_1^+}$ tends to π. The integral

$$\int_{r}^{R} \frac{\sin x}{x}\, dx$$

thus has a limit (this is obvious since the function $(\sin x)/x$ can be extended as a continuous function to the interval $[0, R]$ by assigning to it the value 1 at $x=0$) and

$$2 \int_{0}^{R} \frac{\sin x}{x}\, dx = \pi - I_{c_2^+}.$$

As r tends to $+\infty$, $I_{c_2^+}$ tends to 0 and the integral

$$\int_{0}^{R} \frac{\sin x}{x}\, dx$$

has a limit. By definition, the function $(\sin x)/x$ is integrable over the interval $[0, +\infty[$ and

$$\int_{0}^{+\infty} \frac{\sin x}{x}\, dx = \lim \frac{1}{2}(\pi - I_{c_2^+}) = \frac{\pi}{2}.$$

Part 7

NUMERICAL SERIES AND SERIES OF FUNCTIONS

Exercises dealing with the results of Chapter III of Book IV
of *Mathématiques générales* by C. Pisot and M. Zamansky

Review

Series of positive terms

We shall frequently have occasion to use the following facts:

If $u_n \leqslant v_n$ and if the series $\sum v_n$ converges, so does the series $\sum u_n$. In particular, if $u_n = O(n^{-\alpha})$, where $\alpha > 1$, the series $\sum u_n$ converges. (This fact is interesting for the use of asymptotic expansions.)

If $u_n \geqslant w_n$ and if the series $\sum w_n$ diverges, so does the series $\sum u_n$.

If $u_n \simeq v_n$ as n tends to $+\infty$, the two series $\sum u_n$ and $\sum v_n$ either both converge or both diverge.

To supplement the tests of Cauchy and d'Alembert and the $n^x u^n$-test, we give the following result: If

$$n\left(1 - \frac{u_{n+1}}{u_n}\right) > k \geqslant 1,$$

the series $\sum u_n$ converges; if

$$n\left(1 - \frac{u_{n+1}}{u_n}\right) \leqslant 1,$$

the series $\sum u_n$ diverges. (Proof of this will be the subject of the first question of exercise 707.)

Series of arbitrary real or complex terms

One should first ascertain whether the series converges absolutely by applying the rules for series of positive terms to test the convergence of the series of absolute values. There are two reasons for this: first, absolutely convergent series have special properties of their own. Second, it is much easier to test a series of real positive terms for convergence than a series of arbitrary real terms or complex terms.

If the series does not converge absolutely, but if the sequence $\{a_n\}$ is positive and nonincreasing and tends to zero, the series whose general terms are

$$(-1)^n a_n, \qquad a_n \cos n\theta, \qquad a_n \sin n\theta, \qquad a_n e^{in\theta},$$

converge.

In other cases, a special study is necessary. Sometimes the problem can be simplified by grouping two or three consecutive terms, which does not change the convergence or divergence of the series if the general term approaches zero as n approaches $+\infty$.

Finally, if the absolute value a_n of the general term of a convergent alternating series $\sum(-1)^n a_n$ is nonincreasing, the limit of the series lies between any two consecutive partial sums.

Absolutely convergent series

The properties of commutativity and associativity of finite sums carry over to absolutely convergent series and only for such series. This means that the sum is independent of the order of the terms and that we may replace an arbitrary set (finite or infinite) of the terms with the sum of those terms (in the case of an infinite set, the limit of the partial sums of those terms).

We have, in particular, the following theorem for denumerable families x_α of real or complex numbers:

If the set of sums of absolute values of finitely many of the numbers in the family in question, that is, if every sum of the form $\sum_\alpha^n |x_\alpha|$, is bounded above (for example, if one of the series obtained by ordering and summing the x_α is absolutely convergent), then all the series obtained by ordering the terms in the family x_α (in any arrangement) are absolutely convergent and they all have the same sum s. If the terms x_α of the family are partitioned into an infinite collection of series $\sum_n u_{k,n}$, with each x_α belonging to one and only one of those series, the series $\sum_n u_{k,n}$ are absolutely convergent, and the series whose general term is

$$U_k = \sum_n u_{k,n}$$

is absolutely convergent and its limit is the number s.

The uniform convergence of series of functions

By definition, the sum S of a series is the limit of the sequence $\{S_n\}$ of the partial sums of the first n terms. Therefore, we need to know how to test whether the sequence $\{S_n\}$ converges uniformly. As I have indicated already, my advice is to use the norm relative to the interval in question:

$$\|f\| = \sup_{a \leqslant x \leqslant b} |f(x)|.$$

If we set

$$U(x) = \sum_0^\infty u_n(x) \quad \text{and} \quad U_p(x) = \sum_0^p u_n(x),$$

we can then say that the series arising from the functions u_n converges uniformly on $[a, b]$ if any one of the following three conditions holds:

(1) The sequence whose general term is

$$\|U - U_p\| = \left\| \sum_{p+1}^{+\infty} u_n \right\|$$

converges to 0,

(2) the double sequence whose general term is

$$\|U_p - U_q\| = \left\| \sum_{p+1}^{q} u_n \right\|$$

converges to 0,

(3) for every $\varepsilon > 0$, there exists an N such that

$$\left| \sum_{p+1}^{+\infty} u_n(x) \right| < \varepsilon \qquad \text{for every } x \in [a, b].$$

The following tests are frequently used:
If the series $\sum \|u_n\|$ converges, the series $\sum u_n$ converges uniformly since

$$\|U_p - U_q\| \leqslant \sum_{p+1}^{q} \|u_n\|.$$

If $|u_n(x)| \leqslant \alpha_n$ and if the numerical series $\sum \alpha_n$ converges, the series $\sum u_n$ converges uniformly. This is a corollary of the preceding since

$$\|u_n\| = \sup_{a \leqslant x \leqslant b} |u_n(x)| \leqslant \alpha_n.$$

The essential theorems on uniformly convergent series

(1) If a series $\sum u_n$ converges uniformly on $[a, b]$ and if the functions $\sum u_n$ are continuous on $[a, b]$, the sum of the series $\sum u_n$ is continuous on $[a, b]$.

(2) If the series $\sum u_n$ is uniformly convergent on $[a, b]$ and if the functions v_n are primitives of the u_n such that, for a particular value x_0 in $[a, b]$, the series $\sum v_n(x_0)$ is convergent, then the series $\sum v_n$ is uniformly convergent on $[a, b]$, and its sum V is a primitive of the sum U of the series $\sum u_n$.

This theorem is often used by taking for v_n the primitive of u_n that vanishes at x_0. Then,

$$\int_{x_0}^{x} \left(\sum_{0}^{\infty} u_n(t) \right) dt = \sum_{0}^{\infty} \int_{x_0}^{x} u_n(t) dt.$$

If we take for x_0 and x two particular values α and β, we thus obtain again the theorem on term-by-term integration of a uniformly convergent series.

(3) If the functions u_n are differentiable and if *the series of the derivatives u'_n converges uniformly on $[a, b]$*, the sum U of the series $\sum u_n$ (assumed to converge) is differentiable and its derivative is given by

$$U'(x) = \sum_0^\infty u'_n(x).$$

Note carefully that it is the uniform convergence of the series $\sum u_n$ that is considered in (1) and (2) but that it is the uniform convergence of the series $\sum u'_n$ that is considered in (3).

Power series

One should never write a power series without at the same time indicating its radius of convergence. We know that the radius of convergence is the same for a given series as for the series of its derivatives and its primitives.

A Maclaurin series converges uniformly in every circle centered at the origin of radius less than the radius of convergence. However, it does not in general converge uniformly in the interior of the circle of convergence itself.

The sum of a power series defines a function that is continuous and infinitely differentiable in the interior in the circle of convergence, and the derivatives are the sums of the series obtained by term-by-term differentiation of the original series.

Exercises

701

Examine the convergence of the series whose general terms are given by

$$\frac{n^2+n+1}{n^3+1}, \qquad \frac{n^2+n+1}{n^4+1}, \qquad \frac{n!\,x^n}{n^n},$$

$$\frac{n^{1/2}\ln n}{n^2+1}\sin n\theta, \qquad \frac{\pi}{2}-\arcsin\frac{1}{n+1},$$

where x is a complex parameter and θ a real parameter.

702

Show that the series whose general term is

$$u_n = \sin\left(\sqrt{n^2+a^2}\,\pi\right),$$

where a is a given positive number, converges.

703

(1) Suppose that $\{u_n\}$ is a nonincreasing sequence that converges to zero. Show that the series $\sum u_n$ and $\sum v_n$, where $v_n = 2^n u_{2^n}$, either both converge or both diverge. Apply this result to Riemann's series, where $u_n = n^{-\beta}$.

(2) Examine the convergence of the series whose general term is

$$t_n = n^{-\alpha}\,(\ln n)^{-\beta}.$$

704

Show that the series whose general term is $(\sin n\alpha)/n$, where α is a given real number, is convergent but not absolutely convergent.

705

Show that the sequence whose general term is

$$u_n = \frac{a(a+1)\cdots(a+n-1)}{n^{a-1}(n!)}$$

converges to a nonzero limit if a is a positive number. Suggestion: Examine the series whose general term is $\ln (u_n/u_{n-1})$. Does this result hold if $a \leqslant 0$?

706

(1) Suppose that the trigonometric series whose general term is

$$u_n(x) = a_n \cos nx + b_n \sin nx$$

converges absolutely at $x=x_0$. Show that this series converges at $x=x_0-h$ if it converges at $x=x_0+h$. Show also that it is absolutely convergent at x_0-h if it is absolutely convergent at x_0+h.

(2) Suppose that a trigonometric series converges absolutely at two values x_0 and x_1 such that the difference x_0-x_1 is not commensurable with π. Show that this series is absolutely convergent for infinitely many values of x that are not congruent to each other modulo 2π.

707

(1) Let $\sum u_n$ be a series of positive numbers. Define

$$v_n = 1 - \frac{u_{n+1}}{u_n}.$$

Show that if, from some integer on,

$$nv_n \geqslant k > 1, \text{ the series } \sum u_n \text{ converges};$$

$$nv_n \leqslant 1, \text{ the series } \sum u_n \text{ diverges}.$$

(Compare the series $\sum u_n$ with Riemann's series $\sum n^{-k}$.) What can we conclude if $\{nv_n\}$ converges to a limit l?

(2) Apply the preceding result to an examination of the convergence of the series

$$u_n = (-1)^n \frac{\alpha(\alpha-1)\cdots(\alpha-n+1)}{n!}.$$

(3) Show that, if $\alpha \neq -1$, the series whose general term is

$$\ln \frac{n - \alpha}{n + 1}$$

is divergent and find the limit of u_n. (Use only values of n greater than α.)
(4) Examine the convergence of the series

$$w_n = \frac{\alpha(\alpha - 1) \cdots (\alpha - n + 1)}{n!}.$$

708

(1) Prove that the series whose general term is

$$u_n = \int_{(n-1)\pi}^{n\pi} \frac{t \sin t}{1 + t^2} dt$$

converges. Use this result to show that the integral

$$I = \int_0^{+\infty} \frac{t \sin t}{1 + t^2} dt$$

converges.
(2) Show that the series whose general term is u_n and the integral I are not absolutely convergent.
(3) Show that we obtain the same results for the integral

$$\int_a^{+\infty} \frac{A(t)}{B(t)} \sin t \, dt,$$

where A and B are two polynomials such that $\deg B = \deg A + 1$ and where a is a number greater than any of the zeros of B. (The convergence of this integral was studied in exercise 535 by a different procedure, though the absolute convergence was not studied then.)

709

(1) Suppose that a sequence $\{u_n\}$ is defined by giving the value of u_0 and the recursion equation

$$u_n = \sin u_{n-1}.$$

Show that the sequence $\{u_n\}$ converges to 0.

Consider the function f_n defined by

$$u_n = f_n(u_0).$$

Show that the sequence of the functions f_n converges uniformly to zero on the real line.

(2) Show that the series whose general term is u_n^3 converges (for example, by using the convergence of the series whose general term is $u_{n+1} - u_n$). Show that the series of the functions f_n^3 converges uniformly on the real line.

(3) Show that the series whose general terms are

$$\ln \frac{\sin u_n}{u_n} \quad \text{and} \quad u_n^2$$

diverge.

(4) Examine the convergence of the series $\sum u_n x^n$ for all values of the complex number x.

710

Suppose that the power series $\sum a_n t^n$ has unit radius of convergence and that the coefficients a_n are all positive. Show that if the sum

$$S(t) = \sum_0^\infty a_n t^n$$

is bounded on an interval of the form $[t_0, 1[$, then the series $\sum a_n$ converges and

$$\lim_{t \to 1} \left(\sum_0^\infty a_n t^n \right) = \sum_0^\infty a_n.$$

If the series $\sum a_n$ diverges, does the sum

$$S(t) = \sum_0^\infty a_n t^n$$

have a left-hand limit as $t \to 1$ from below?

711

(1) Express $\cos^{2p} t$ as a linear combination of exponentials of imaginary quantities and use the result to evaluate the integral

$$I_p = \int_0^{2\pi} \cos^{2p} t \, dt.$$

(2) Show, without using the result of (1), that

$$2nI_n = (2n-1)I_{n-1},$$

and thus find again the value of I_p.

(3) Show that the function F defined by

$$F(x) = \int_0^{2\pi} \ln\,(1 - x\cos t)\,dt$$

is defined in the interval $]-1, 1[$ and that it can be expanded in a Maclaurin series in that interval.

(4) Exhibit the Maclaurin-series expansion of the derivative F' of the function F. Define F', and then F, in terms of elementary functions.

712

For real values of x such that $0 < x < 1$, a function f is defined by

$$f(x) = \int_0^x \frac{t^\alpha}{\sqrt{1-t^4}}\,dt,$$

where α is a real constant greater than -1.

(1) Show that the expression for $f(x)$ is meaningful.

(2) Expand $(1-t^4)^{-\frac{1}{2}}$ and show that one can obtain an expansion of the function f in a series S of terms of the form $a_n t^{k_n}$.

(3) Show that the series S converges uniformly in the interval $[0, 1]$. (Suggestion: Study the convergence of the series for $x = 1$ by using the result of (1), exercise 707.)

Use this result to show that the integral

$$\int_0^1 \frac{t^\alpha dt}{\sqrt{1-t^4}}$$

converges and that it is equal to the sum for $x = 1$ of the series S.

(4) Prove directly that the integral

$$\int_0^1 \frac{t^\alpha dt}{\sqrt{1-t^4}}$$

converges.

713

(1) Show that the integral

$$\int_1^{+\infty} e^{-t} t^{x-1} dt$$

converges for every x and that the integral

$$\int_0^1 e^{-t} t^{x-1} dt$$

converges for $x > 0$.

(2) The gamma function Γ is defined for positive values of x by

$$\Gamma(x) = \int_0^{+\infty} e^{-t} t^{x-1} dt.$$

Show that $\Gamma(x+1) = x\Gamma(x)$ and find an expression for $\Gamma(x)$ for positive integral values of x.

(3) Show that, for $x > 0$,

$$\int_0^1 e^{-t} t^{x-1} dt = \sum_0^\infty \frac{(-1)^n}{n\,!(n+x)}.$$

(4) Show that the series whose general term is

$$\frac{(-1)^n}{n\,!(n+x)},$$

converges for all x not equal to a negative integer. Denote by $S(x)$ the sum of this series.

(5) Define a function G in the complement with respect to $\boldsymbol{R}$ of the set of nonpositive integers by

$$G(x) = S(x) + \int_1^{+\infty} e^{-t} t^{x-1} dt.$$

Show that the restriction of the function G to the set of positive x is the function Γ and that the function G satisfies the relation

$$G(x+1) = xG(x).$$

714

Apply the theorem on the product of two series to show that the function

$$\frac{\ln (1+x)}{1+x}$$

can be expanded in a Maclaurin series if $|x|<1$ and find an expression for the general term of that series. Find directly the radius of convergence of that series.

715

Let x denote a real number such that $-1<x<1$.

(1) Show that every integer n can be expressed in a unique manner as a sum of distinct integral powers of 2:

$$n = \sum_{i=1}^{I} 2^{k_i}.$$

What numbers do we obtain if all the k_i are bounded by an integer P?

(2) Show that the sequence whose general term is

$$u_P = \prod_{k=0}^{P} [1+x^{(2^k)}]$$

converges to $1/(1-x)$.

(3) Show that the series whose general term is $\ln (1+x^n)$ is absolutely convergent and that

$$\sum_{1}^{+\infty} \ln (1+x^n) = - \sum_{1}^{+\infty} \ln (1-x^{2p-1}),$$

$$\lim_{N\infty} \prod_{1}^{N} (1+x^n) = \lim_{N'\infty} \prod_{1}^{N'} \frac{1}{1-x^{2p-1}}.$$

716

Let x and z denote two complex numbers such that

$$2|x||z|+|z|^2 <1.$$

(1) Show that

$$s(x, z) = (1-2xz+z^2)^{-\frac{1}{2}} = \sum_{k=0}^{+\infty} u_k(x, z),$$

where

$$u_0(x, z) = 1, \qquad u_k(x, z) = \frac{1 \cdot 3 \cdots (2k-1)}{2 \cdot 4 \cdots 2k} (2xz - z^2)^k.$$

(2) Let $v_n(x)z^n$ denote the term of degree n in the polynomial in z that is equal to

$$\sum_{k=0}^{n} u_k(x, z).$$

Show that the series whose general term is $v_n(x)z^n$ converges and that its sum is $s(x, z)$.

(3) Show that

$$v_n(x) = \frac{1}{2^n(n\,!)} \frac{d^n}{dx^n} (x^2 - 1)^n.$$

(4) Evaluate the partial derivative $\partial s/\partial z$ and use the expression found to obtain a relation between three consecutive polynomials v_n.

717

(1) Use the Maclaurin-series expansion of e^z to show that, for any complex number z,

$$\left| e^z - \left(1 + \frac{z}{n}\right)^n \right| \leqslant e^{|z|} - \left(1 + \frac{|z|}{n}\right)^n$$

and

$$\left| e^z - \left(1 + \frac{z}{n}\right)^n \right| \leqslant e^{|z|} \frac{|z|^2}{2n}.$$

(2) Show that the sequence of functions f_n defined by

$$f_n(x) = i \frac{\left(1 + \dfrac{ix}{2n}\right)^{2n} + \left(1 - \dfrac{ix}{2n}\right)^{2n}}{\left(1 + \dfrac{ix}{2n}\right)^{2n} - \left(1 - \dfrac{ix}{2n}\right)^{2n}}$$

converges uniformly to $\cot x$ in $]0, \pi/2]$.

718

Show that the series whose general term is

$$\frac{e^{-nx}}{n^2+1},$$

converges if $x \geqslant 0$ and diverges if $x < 0$.

For positive values of x, define a function f by

$$f(x) = \sum_{0}^{+\infty} \frac{e^{-nx}}{n^2+1}.$$

Show that the function f is continuous for $x \geqslant 0$ and differentiable for $x > 0$.

719

Let α denote a given real number and let t and x denote real variables.

(1) Find the Fourier series of the function f defined by

$$f(t) = \cos \alpha t \qquad \text{if} \qquad -\pi < t < \pi.$$

Show why the sum of this series is equal to $f(t)$ for $-\pi < t < \pi$. Find the value of this sum at $t = \pi$.

Use the above result to show that

$$\cot x = \frac{1}{x} + \sum_{k=1}^{+\infty} \frac{2x}{x^2 - k^2\pi^2}.$$

(2) Let M and N be two integers such that $N > M/\pi$. Show that the function

$$\sum_{N}^{+\infty} \frac{1}{x^2 - k^2\pi^2}$$

is continuous for $|x| \leqslant M$. Evaluate

$$\sum_{1}^{+\infty} \frac{1}{k^2}.$$

(3) Show that, for $0 < x < \pi$,

$$\ln \sin x = \ln x + \sum_{1}^{+\infty} \ln\left(1 - \frac{x^2}{k^2\pi^2}\right).$$

(4) Show that, if x is not an integral multiple of π,

$$\frac{1}{\sin^2 x} = \sum_{-\infty}^{+\infty} \frac{1}{(x-k\pi)^2}.$$

720

Define a function f that is periodic with period 2π by

$$f(t) = t, \qquad \text{for } 0 \leqslant t < 2\pi.$$

(1) Find the Fourier series of the function f. What is the sum $S(t)$ of this series for the value t of the variable?

(2) Find the smallest positive value of t, denoted by t_n, at which the sum $S_n(t)$ of the first $n+1$ terms of the preceding series has a relative minimum.

(3) Show that the sequence $\{S_n(t_n)\}$ converges to

$$\pi - 2 \int_0^\pi \frac{\sin t}{t} dt.$$

(4) Find the Maclaurin-series expansion of the function

$$\frac{1}{\pi} \int_0^x \frac{\sin t}{t} dt,$$

and take as an approximate value of

$$I = \frac{1}{\pi} \int_0^\pi \frac{\sin t}{t} dt$$

the sum of the first four nonzero terms of the Maclaurin expansion. Give an indication of the accuracy involved.

(5) Show that the sequence $\{S(t_n) - S_n(t_n)\}$ does not converge to zero. Why do we not have a contradiction with the convergence of $S_n(t)$ to $S(t)$ for every value of t?

Solutions

A series whose terms are of the same sign (from some term on) converges or diverges according as the series obtained by replacing each term with an equivalent term converges or diverges:

$$\frac{n^2+n+1}{n^3+1} > 0 \quad \text{and} \quad \frac{n^2+n+1}{n^3+1} \simeq \frac{1}{n} \quad \Rightarrow \text{ the series diverges;}$$

$$\frac{n^2+n+1}{n^4+1} > 0 \quad \text{and} \quad \frac{n^2+n+1}{n^4+1} \simeq \frac{1}{n^2} \quad \Rightarrow \text{ the series converges.}$$

Let us apply d'Alembert's rule to the series whose general term is $u_n = n!\, x^n/n^n$ (that is, to the series of absolute values):

$$\frac{|u_{n+1}|}{|u_n|} = \frac{(n+1)|x|\, n^n}{(n+1)^{n+1}} = \frac{|x|}{(1+1/n)^n} .$$

The sequence $\{(1+1/n)^n\}$ converges to e (cf. PZ, Book III, Chapter III, part 7, § 6). Therefore, $\{|u_{n+1}|/|u_n|\}$ converges to $|x|/e$. The series converges if $|x| < e$ and diverges if $|x| > e$.

If $|x| = e$, we can verify that $(1+1/n)^n < e$ and hence that $|u_{n+1}|/|u_n| > 1$. The series is divergent since

$$\ln\left[\left(1+\frac{1}{n}\right)^n\right] = n \ln\left(1+\frac{1}{n}\right) = 1 - \frac{1}{2n}\frac{1}{(1+\theta/n)^2} < 1 = \ln e.$$

We know that, for any exponent $\beta > 0$,

$$\ln x = o(x^\beta) \Rightarrow \ln x = O(x^\beta),$$

as x tends to $+\infty$. Since $|\sin n\theta| \leqslant 1$, we have

$$\frac{n^{1/2}\ln n}{n^2+1}\sin n\theta = \frac{n^{1/2}}{n^2+1}O(n^\beta) = O(n^{\beta-3/2}).$$

We need only take $\beta < \frac{1}{2}$ (for example, we can take $\beta = \frac{1}{4}$) to obtain convergence of the series (by the $n^\alpha u_n$-rule).

$$v_n = \frac{\pi}{2} - \arcsin \frac{n}{n+1} \text{ is positive and tends to } 0;$$

$$\cos v_n = \sin\left(\frac{\pi}{2} - v_n\right) = \frac{n}{n+1} \Rightarrow \frac{1}{n+1} = 1 - \cos v_n \simeq \frac{v_n^2}{2}.$$

The positive quantity v_n is asymptotically equal to $\sqrt{2/(n+1)}$, and the series $\sum v_n$ diverges with this last series.

702

Since

$$\sqrt{n^2 + a^2} = n + \frac{a^2}{\sqrt{n^2 + a^2} + n},$$

we have

$$u_n = (-1)^n \sin\left(\frac{a^2}{\sqrt{n^2 + a^2} + n} \pi\right) = (-1)^n \sin v_n.$$

The sequence of the positive terms v_n is a decreasing sequence, and it converges to 0. Therefore, the sequence $\{\sin v_n\}$ has the same properties.

The series $\sum u_n$ is an alternating series; the sequence $\{|u_n|\}$, where $|u_n| = \sin v_n$, is decreasing and its limit is zero. Therefore, the alternating series $\sum u_n$ is convergent (cf. PZ, Book IV, Chapter III, part 1, § 4).

703

First question:

We note first that the u_n are positive since the sequence $\{u_n\}$ is nonincreasing and its limit is zero. Furthermore,

$$2^p < n \leqslant 2^{p+1} \Rightarrow u_{2^p} \geqslant u_n \geqslant u_{2^{p+1}}.$$

If we sum these inequalities over the $2^{p+1} - 2^p = 2^p$ values of n for which they hold, we obtain

$$2^p u_{2^p} \geqslant \sum_{n=2^p+1}^{2^{p+1}} u_n \geqslant 2^p u_{2^{p+1}},$$

or, if we denote by U_k the sum $\sum\limits_{n=1}^{k} u_n$,

$$v_p \geqslant U_{2p+1} - U_{2p} \geqslant \tfrac{1}{2} v_{p+1}.$$

If we sum these inequalities over the values $p \leqslant P$, we obtain

$$V_P = \sum_{p=1}^{P} v_p \geqslant U_{2P+1} - U_2 \geqslant \tfrac{1}{2} \sum_{p=2}^{P+1} v_p = \tfrac{1}{2}(V_{P+1} - V_1).$$

If the series $\sum u_n$ is convergent, the sum U_{2P+1} is majorized by the sum U of the series. Therefore, the sum V_{P+1} is majorized by a number independent of P. The series of the positive terms v_p converges.

If the series $\sum u_n$ diverges, the sum U_{2P+1} tends to $+\infty$ as P tends to $+\infty$, and hence so does the sum V_p. The series $\sum v_n$ diverges:

$$u_n = n^{-\alpha} \Rightarrow v_n = 2^n (2^n)^{-\alpha} = (2^{1-\alpha})^n.$$

v_n is the general term of a geometric series with ratio $2^{1-\alpha}$. This series and hence the series $\sum u_n$ are

convergent	if	$2^{1-\alpha} < 1,$	that is, if $\alpha > 1,$
divergent	if	$2^{1-\alpha} \geqslant 1,$	that is, if $\alpha \leqslant 1.$

Second question:

We know that $\ln n = o(n^h)$ for every $h > 0$. From this, we easily deduce that if $\alpha > 1$, then, for any α' in $(1, \alpha)$,

$$n^{-\alpha}(\ln n)^{-\beta} = O(n^{-\alpha'})$$

and the series $\sum n^{-\alpha} (\ln n)^{-\beta}$ is convergent. On the other hand, if $\alpha < 1$, then, for any α' in $(\alpha, 1)$,

$$n^{-\alpha}(\ln n)^{-\beta} \geqslant n^{-\alpha'}$$

from some term on, and the series $\sum n^{-\alpha} (\ln n)^{-\beta}$ is divergent.

If $\alpha = 1$, we use the result of (1) (since u_n is positive and decreasing for sufficiently large n):

$$2^n t_{2^n} = 2^n 2^{-n} (\ln 2^n)^{-\beta} = n^{-\beta} (\ln 2)^{-\beta}.$$

The series $\sum t_n$ and the series $\sum 2^n t_{2^n}$ are both

convergent	if	$\beta > 1,$
divergent	if	$\beta \leqslant 1.$

704

(1) The series is obviously convergent if α is an integral multiple of π. It either converges for all the values α, $2\pi - \alpha$, and $2n\pi + \alpha$ (where n is an integer) or diverges for all the values. Therefore, we may suppose that $0 < \alpha < \pi$.

To establish the convergence of the series $\sum (\sin n\alpha)/n$, we can use Abel's transformation (cf. PZ, Book IV, Chapter III, part 1, § 4). If we set

$$S_k = \sum_{p=1}^{k} \sin p\alpha,$$

we can write

$$\sum_{k=1}^{n} \frac{\sin k\alpha}{k} = S_1 + \sum_{k=2}^{n} \frac{1}{k}(S_k - S_{k-1}) = \frac{S_n}{n} + \sum_{k=1}^{n-1} \left(\frac{1}{k} - \frac{1}{k+1} \right) S_k.$$

We know that

$$|S_k| = \left| \sum_{p=1}^{k} \sin p\alpha \right| = \left| \frac{\sin (k\alpha/2) \sin ((k+1)\alpha/2)}{\sin (\alpha/2)} \right| \leqslant \frac{1}{\sin (\alpha/2)}.$$

The series

$$\left(\frac{1}{k} - \frac{1}{k+1} \right) S_k$$

is absolutely convergent since the absolute value of the general term is majorized by

$$\left(\frac{1}{k} - \frac{1}{k+1} \right) \frac{1}{\sin (\alpha/2)},$$

which is the general term of a convergent series. The sequence $\{S_n/n\}$ converges to 0. Therefore, the sequence of partial sums

$$\sum_{k=1}^{n} \frac{\sin k\alpha}{k}$$

approaches a limit as n approaches $+\infty$. By definition, the series $\sum(\sin k\alpha)/k$ is convergent.

(2) To examine the series $\sum |\sin n\alpha|/n$, we may assume that $0 < \alpha \leqslant \pi/2$ since the general term assumes the same value for α and $\pi - \alpha$.

Consider the points A and B where the trigonometric circle intersects the x-axis. Let M and N denote the points on the circle whose curvilinear coordinates are $n\alpha$ and $(n+1)\alpha$. We need to consider two cases:

(a) The smaller of the two arcs MN contains either A or B. Since the measure of this arc is α, it is obvious that one of the arcs AM or AN (or BM or BN if it is

B that lies between M and N) is at least equal to $\alpha/2$. Then, either $|\sin n\alpha|$ or $|\sin (n+1)\alpha|$ is at least equal to $\sin (\alpha/2)$.

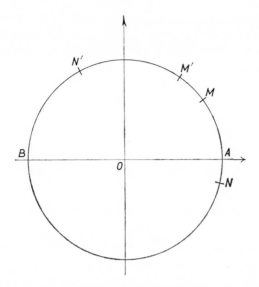

(b) The smaller of the two arcs MN contains neither A nor B. This is the case in the drawing for the points M' and N'. Since the arc $M'N'$ does not exceed $\pi/2$, the sum of the arcs AM' and BN' is at least equal to $\pi/2$. One of these arcs exceeds $\pi/4$ and its sine exceeds $\sin (\pi/4) \geqslant \sin(\alpha/2)$.

In conclusion, at least one of the two numbers $|\sin n\alpha|$, $|\sin (n+1)\alpha|$ is minorized by $\sin (\alpha/2)$. The numerator of at least one of any two consecutive terms

$$\frac{|\sin 2p\alpha|}{2p} \quad \text{and} \quad \frac{|\sin (2p+1)\alpha|}{2p+1}$$

of the series exceeds $\sin (\alpha/2)$. The denominator of this term is at least equal to $(2p+1)$. Therefore,

$$\frac{|\sin 2p\alpha|}{2p}+\frac{|\sin (2p+1)\alpha|}{2p+1} \geqslant \frac{\sin (\alpha/2)}{2p+1}.$$

The series

$$\sum \frac{1}{2p+1} \sin \frac{\alpha}{2}$$

diverges and so does the series

$$\sum \frac{|\sin n\alpha|}{n},$$

the sum of the first $2p+1$ terms of which increases without bound with increasing p.

REMARK: The examination that we have made in the above studies can be carried over to the series $\sum a_n \sin n\alpha$ and $\sum |a_n| \sin n\alpha|$ where the sequence $\{a_n\}$ is decreasing and its limit is 0 and where the series $\sum a_n$ is divergent.

705

We can write

$$u_n = \frac{a}{1}\left(1+\frac{a-1}{2}\right)\cdots\left(1+\frac{a-1}{n}\right)n^{-(a-1)},$$

and, taking the logarithm, replace the product with a sum:

$$\ln u_n = \ln a + \ln\left(1+\frac{a-1}{2}\right)+\cdots+\ln\left(1+\frac{a-1}{n}\right)-(a-1)\ln n.$$

To study the limit of this sequence, let us consider the series $\sum v_n$ such that the sum of the first n terms of that series is $\ln u_n$. Then,

$$v_n = \ln u_n - \ln u_{n-1} = \ln\frac{u_n}{u_{n-1}} = \ln\left(1+\frac{a-1}{n}\right)+(a-1)\ln\left(\frac{n-1}{n}\right)$$

$$= \frac{a-1}{n}+O\left(\frac{1}{n^2}\right)+(a-1)\left[-\frac{1}{n}+O\left(\frac{1}{n^2}\right)\right] = O\left(\frac{1}{n^2}\right).$$

v_n is the general term of a convergent series. The sequence of the partial sums $\ln u_n$ of this series converges to the sum v of the series and the sequence $\{u_n\}$, where $u_n = e^{v_n}$, converges to the positive number e^v.

If a is a negative integer, u_n is 0 from some term on. Otherwise, u_n is never 0 and, since $a+n-1$ is positive from some number n_0 on, the terms u_n all have the same sign from the n_0th term on. Therefore, we need only study $|u_n|$ or $\ln |u_n|$.

The study is identical to that made in the case in which a is positive:

$$\text{if}\quad n \geqslant n_0, \quad v_n = \ln\left|\frac{u_n}{u_{n-1}}\right| = \ln\left(1+\frac{a-1}{n}\right)+(a-1)\ln\left(\frac{n-1}{n}\right).$$

In sum, $u_n = 0$ from some term on if a is a negative integer. For all other values of a, the sequence $\{u_n\}$ converges to a finite nonzero limit.

706

First question:

An elementary trigonometric calculation yields

$$u_n(x_0+h)+u_n(x_0-h) = 2a_n \cos nh \cos nx_0 + 2b_n \sin nx_0 \cos nh$$

$$= 2\,u_n(x_0) \cos nh,$$

so that

$$|u_n(x_0+h)+u_n(x_0-h)| \leqslant 2|u_n(x_0)|.$$

The series $\sum [u_n(x_0+h)+u_n(x_0-h)]$ is absolutely convergent since, by hypothesis, the majorizing series $\sum|u_n(x_0)|$ is convergent.

By writing

$$u_n(x_0-h) = [u_n(x_0+h)+u_n(x_0-h)]-u_n(x_0+h),$$

we see that the series $\sum u_n(x_0-h)$, being the difference between two convergent or absolutely convergent series is itself convergent or absolutely convergent, depending on the hypothesis on $\sum u_n(x_0+h)$.

Second question:

We can apply the above result by setting $x_1 = x_0+h$. Then, the series converges absolutely for the value $x_0-h=x_0'$. The same result applied to the numbers x_0' and $x_0'+h=x_0$ shows that the series converges absolutely for the value $x_0'-h=x_0-2h$. More generally, one can show by induction on n that the series converges absolutely for the values x_0-nh (and also for the values x_0+nh).

If two terms of these series are congruent modulo 2π, there are two distinct integers m and n such that

$$x_0-nh=x_0-mh \bmod 2\pi \implies (m-n)h = 0 \bmod 2\pi,$$

and h is commensurable with π in contradiction with the hypothesis.

707

First question:

By hypothesis,

$$n \geqslant n_0 \implies nv_n \geqslant k \implies \frac{u_{n+1}}{u_n} \leqslant 1-\frac{k}{n}.$$

We know (Maclaurin's formula with two terms plus remainder) that

$$\left(\frac{n+1}{n}\right)^{-k} = \left(1+\frac{1}{n}\right)^{-k} = 1-\frac{k}{n}+\frac{k(k+1)}{2}\left(1+\frac{\theta}{n}\right)^{-k-2} \qquad (0 < \theta < 1),$$

$$\left(\frac{n+1}{n}\right)^{-k} \geq 1-\frac{k}{n} \geq \frac{u_{n+1}}{u_n} > 0.$$

If we multiply all the inequalities corresponding to the values of n beginning with n_0, we obtain

$$\frac{(n+1)^{-k}}{n_0^{-k}} > \frac{u_{n+1}}{u_{n_0}} \Rightarrow u_{n+1} < (n+1)^{-k}\frac{u_{n_0}}{n_0^{-k}}.$$

The series whose general term is

$$\frac{u_{n_0}}{n_0^{-k}}(n+1)^{-k},$$

converges (since u_{n_0}/n_0^{-k} is independent of n), and hence so does the series $\sum u_n$.

In the second case,

$$n \geq n_0 \Rightarrow nv_n \leq 1 \Rightarrow u_{n+1} \geq 1-\frac{1}{n} = \frac{n-1}{n}.$$

If we multiply these equations for the different values of n from n_0 on, we obtain

$$\frac{u_{n+1}}{u_{n_0}} \geq \frac{n_0-1}{n} \qquad \text{or} \qquad u_{n+1} \geq \frac{(n_0-1)u_{n_0}}{n}.$$

The series whose general term is

$$\frac{(n_0-1)u_{n_0}}{n},$$

is divergent and hence so is the series $\sum u_{n+1}$.

Suppose that the sequence $\{nv_n\}$ converges to a limit l. If $l>1$, then, for any $k \in]l, 1[$, we have by the definition of a limit $nv_n \geq k > 1$ for $n \geq$ some integer n_0, so that the series $\sum u_n$ converges.

If $l<1$, then $nv_n \leq 1$ for $n \geq n_0$, and the series $\sum u_n$ diverges.

If $l=1$, we cannot decide about convergence or divergence except when nv_n approaches 1 from below and in that case the series $\sum u_n$ diverges.

Second question:

The ratio

$$\frac{u_{n+1}}{u_n} = \frac{n-\alpha}{n+1}$$

tends to 1 as n tends to $+\infty$. Therefore, the terms u_n all have the same sign from some term on, and we can apply the result of (1):

$$v_n = 1 - \frac{u_{n+1}}{u_n} = \frac{\alpha+1}{n+1} \quad \text{and} \quad \lim_{n\to\infty} nv_n = \alpha+1.$$

If $\alpha+1>1$, that is, if $\alpha>0$, the series $\sum u_n$ converges;
if $\alpha+1<1$, that is, if $\alpha<0$, the series $\sum u_n$ diverges;
if $\alpha+1=1$, that is, if $\alpha=0$, then $u_n=0$ and the series converges.

Third question:

If $\alpha = -1$, we have

$$\ln \frac{n-\alpha}{n+1} = \ln 1 = 0.$$

If $\alpha \neq -1$, the terms have all the same sign. They are positive if $\alpha < -1$ and negative if $\alpha > -1$. To test the series for convergence, we can therefore replace the general term with an equivalent term:

$$\ln \frac{n-\alpha}{n+1} \simeq \frac{n-\alpha}{n+1} - 1 = -\frac{\alpha+1}{n+1}.$$

Therefore, the series diverges and the sequence of partial sums tends to $+\infty$ or $-\infty$ according as $\alpha < -1$ or $\alpha > -1$.

$$\sum_{p>\alpha}^{n} \ln \frac{p-\alpha}{p+1} = \ln \frac{(p_0-\alpha)\cdots(n-\alpha)}{(p_0+1)\cdots(n+1)} = \ln \frac{u_{n+1}}{k},$$

where the coefficient k is the product of the factors of u_{n+1} that correspond to the values of p not exceeding p_0:

$$k = \frac{\alpha(\alpha-1)\cdots(\alpha-p_0+1)}{p_0!} (-1^{p_0}).$$

Here, k does not depend on n. Therefore, the limit of the sequence $\{u_{n+1}\}$ is

$$0 \quad \text{if } \alpha > -1 \text{ since } \lim_{n\to\infty} \ln(u_{n+1}/k) = -\infty,$$

$$\pm\infty \quad \text{if } \alpha < -1 \text{ since } \lim_{n\to\infty} \ln(u_{n+1}/k) = +\infty,$$

$$1 \quad \text{if } \alpha = -1 \text{ since } u_n = n!/n!.$$

Fourth question:

The series $\sum w_n$ is an alternating series, and $|w_n| = |u_n|$. If $\alpha \geqslant 0$, the series $\sum u_n$ converges. All its terms are of the same sign. The series $\sum |w_n| = \sum |u_n|$ converges and the series $\sum w_n$ converges absolutely.

If $-1 < \alpha \leqslant 0$, we know that the sequences $\{u_n\}$ and hence $\{w_n\}$ converge to 0. The ratio

$$\left| \frac{w_{n+1}}{w_n} \right| = \frac{n - \alpha}{n + 1}$$

is always less than 1. Therefore, we know that the alternating series $\sum w_n$ converges (cf. PZ, Book IV, Chapter III, part 1, § 4).

If $\alpha \leqslant -1$, the sequences $\{u_n\}$ and hence $\{w_n\}$ do not converge to 0 as $n \to +\infty$. Therefore, the series $\sum w_n$ diverges.

708

First question:

In the interval $[(n-1)\pi, n\pi]$, the sine function is nonnegative if n is odd and nonpositive if n is even. Therefore, u_n has the same sign as $(-1)^{n-1}$ and

$$|u_n| = (-1)^{n-1} u_n = \int_{(n-1)\pi}^{n\pi} \frac{t|\sin t|}{1 + t^2} \, dt.$$

The function $t/(1 + t^2)$ is a decreasing function for $t \geqslant 1$. Thus, we obtain a majorant and a minorant for $|u_n|$ by replacing $t/(1 + t^2)$ with

$$\frac{(n-1)\pi}{1 + (n-1)^2\pi^2} \quad \text{and} \quad \frac{n\pi}{1 + n^2\pi^2} :$$

$$\frac{(n-1)\pi}{1 + (n-1)^2\pi^2} \int_{(n-1)\pi}^{n\pi} |\sin t| \, dt \geqslant |u_n| \geqslant \frac{n\pi}{1 + n^2\pi^2} \int_{(n-1)\pi}^{n\pi} |\sin t| \, dt.$$

The integral of $|\sin t|$ is also the integral of $(-1)^{n-1} \sin t$, and its value is 2. Then, from the preceding inequalities regarding the indices n and $n+1$, we deduce that

$$\frac{2(n-1)\pi}{1 + (n-1)^2\pi^2} \geqslant |u_n| \geqslant \frac{2n\pi}{1 + n^2\pi^2} \geqslant |u_{n+1}|.$$

These inequalities imply that the sequence $\{|u_n|\}$ is a decreasing sequence and that it converges to 0 since

$$\lim_{n \to \infty} \frac{2(n-1)\pi}{1 + (n-1)^2\pi^2} = 0.$$

The alternating series $\sum u_n$ converges and its sum U is given by

$$U = \lim_{n \to \infty} \sum_{1}^{n} u_n = \lim_{n} \int_0^{n\pi} \frac{t}{1+t^2} \sin t \, dt.$$

To prove the convergence of the integral I, we must ascertain whether

$$\int_0^x \frac{t}{1+t^2} \sin t \, dt$$

has a limit as x tends to $+\infty$. For every x, consider the integer n such that

$$n\pi \leqslant x < (n+1)\pi.$$

This choice of n implies the inequalities

$$\left| \int_0^x \frac{t \sin t}{1+t^2} \, dt - \int_0^{n\pi} \frac{t \sin t}{1+t^2} \, dt \right| = \left| \int_{n\pi}^x \frac{t \sin t}{1+t^2} \, dt \right| < |u_{n+1}|,$$

$$\int_0^{n\pi} \frac{t \sin t}{1+t^2} \, dt - |u_{n+1}| \leqslant \int_0^x \frac{t \sin t}{1+t^2} \, dt \leqslant \int_0^{n\pi} \frac{t \sin t}{1+t^2} \, dt + |u_{n+1}|.$$

As x tends to $+\infty$, so does n. The two extreme quantities in the above expressions approach U. Therefore, the integral I converges, and its value is the sum U of the series.

Second question:

The series $\sum |u_n|$, minorized by the divergent series $\sum 2n\pi/(1+n^2\pi^2)$ (which is bounded by $1/n\pi$), diverges.

The sum of the first n terms tends to $+\infty$ as n tends to $+\infty$. Now,

$$\sum_{1}^{n} |u_n| = \sum_{1}^{n} \int_{(k-1)\pi}^{k\pi} \frac{t|\sin t|}{1+t^2} \, dt = \int_0^{n\pi} \frac{t|\sin t|}{1+t^2} \, dt.$$

The function $t|\sin t|/(1+t^2)$ is not integrable over $[0, +\infty[$. The integral I is not absolutely convergent.

Third question:

The convergence of the integral does not change if we replace, if necessary, the greatest lower bound a of the integral with a number $(n_0-1)\pi$ that exceeds both a and the zeros of A and of the numerator $A'B - AB'$ of the derivative of A/B.

For values equal to or greater than $(n_0-1)\pi$, the function A/B is of constant sign (which we shall assume to be positive) and it is monotonic. Since A/B approaches 0 as x approaches $+\infty$, it is a decreasing function.

The series whose general term is

$$v_n = \int_{(n-1)\pi}^{n\pi} \sin t \, \frac{A(t)}{B(t)} dt,$$

(where $n \geqslant n_0$) is therefore an alternating series. The sequence $\{|v_n|\}$ decreases and converges to 0 (the proof is analogous to that made in (1)). Therefore, this series converges and its sum V is the limit of

$$\sum_{n_0}^{n} v_k = \int_{(n_0-1)\pi}^{n\pi} \sin t \frac{A(t)}{B(t)} dt.$$

Finally, we complete the proof, just as in (1), by using the fact that the double inequality $n\pi \leqslant x < (n+1)\pi$ implies

$$\left| \int_{(n_0-1)\pi}^{x} \sin t \frac{A(t)}{B(t)} dt - \int_{(n_0-1)\pi}^{n\pi} \sin t \frac{A(t)}{B(t)} dt \right| \leqslant |v_{n+1}|.$$

709

Whatever the value of u_0, the term u_1 lies between -1 and $+1$. Changing the sign of u_1 merely changes the sign of u_n. Therefore, we may examine the properties of u_n under the assumption that $0 \leqslant u_1 \leqslant 1$.

First question:

One can verify by induction on n that

$$0 \leqslant u_{n+1} \leqslant u_n \leqslant 1.$$

The sequence $\{u_n\}$, which is decreasing and minorized by 0, converges. The limit of this sequence is a solution of the equation $u = \sin u$, namely, 0.

The sine function is increasing in [0, 1]. Therefore, if we denote by U_n the term of that one of the above sequences $\{u_n\}$ whose first term is $U_1 = 1$, we can show by induction that

$$0 \leqslant u_n \leqslant U_n \Rightarrow U_n = \sup_{0 \leqslant u_1 \leqslant 1} u_n.$$

For negative values of u_1, we obtain the corresponding values of u_n with opposite sign:

$$U_n = \sup_{-1 \leqslant u_1 \leqslant 1} |u_n| = \sup_{u_0} |f_n(u_0)| = \|f_n\|.$$

The sequence $\{U_n\}$ converges to 0 (from (1)). Therefore, the sequence $\{f_n\}$ converges uniformly to 0.

Second question:

The series whose general term is $u_n - u_{n+1}$ converges since

$$\sum_{n=1}^{N} u_n - u_{n+1} = u_1 - u_{N+1}$$

and its limit as N tends to $+\infty$ is u_1. The terms in this series are positive since the sequence $\{u_n\}$ is decreasing and

$$u_n - u_{n+1} = u_n - \sin u_n \simeq \frac{u_n^3}{6}.$$

Therefore, the series $\sum u_n^3/6$ is convergent and so is the series $\sum u_n^3$.
As in (1), let us note that, if $u_1 \geqslant 0$, then $u_n \leqslant U_n$ and hence

$$U_n^3 = \sup_{0 \leqslant u_1 \leqslant 1} u_n^3 = \sup_{-1 \leqslant u_1 \leqslant 1} |u_n|^3 = \sup_{u_0} |f_n^3(u_0)| = \|f_n^3\|.$$

Since the sequence $\{U_n\}$ is a special case of a sequence $\{u_n\}$, the series whose general term is $U_n^3 = \|f_n^3\|$ converges and therefore, the series whose general term is f_n^3 converges uniformly. (Cf. the review at the beginning of Part 7. This is also a consequence of the uniform convergence criterion [cf. PZ, Book IV, Chapter III, part 2, § 1]. One need only take for the value of the term α_n in PZ the number $\|f_n^3\|$.)

Third question:

Proceeding as in the second question, we have

$$\sum_{1}^{N} \ln \frac{\sin u_n}{u_n} = \sum_{1}^{N} \ln \frac{u_{n+1}}{u_n} = \ln \left(\prod_{1}^{N} \frac{u_{n+1}}{u_n} \right) = \ln \frac{u_{N+1}}{u_1}.$$

As N tends to $+\infty$, the ratio u_{N+1}/u_1 approaches 0 and $\ln (u_{N+1}/u_1)$ approaches $-\infty$. The series whose general term is

$$\ln \frac{\sin u_n}{u_n}$$

diverges.
    The terms $(\sin u_n)/u_n$ are less than 1 and hence their logarithms are negative. Furthermore,

$$-\left(\ln \frac{\sin u_n}{u_n} \right) \simeq 1 - \frac{\sin u_n}{u_n} \simeq \frac{u_n^2}{6}.$$

Therefore, the series $\sum u_n^2/6$ diverges, as does the series whose general term is the positive quantity

$$-\ln \frac{\sin u_n}{u_n}.$$

Fourth question:

Let us use d'Alembert's rule to check for absolute convergence:

$$\frac{|u_{n+1}x^{n+1}|}{|u_n x^n|} = \frac{u_{n+1}}{u_n}|x| = \frac{\sin u_n}{u_n}|x|,$$

$$\lim_n u_n = 0 \Rightarrow \lim_n \frac{\sin u_n}{u_n} = 1 \Rightarrow \lim_n \frac{\sin u_n}{u_n}|x| = |x|.$$

Therefore, the series is absolutely convergent for $|x| < 1$ and divergent for $|x| > 1$. (The radius of convergence of the series is 1.)

If $x = 1$, the series $\sum u_n$ diverges since $u_n > u_n^2$ and we know that the series u_n^2 diverges (see (3)).

If $x = -1$, the series $\sum (-1)^n u_n$ is an alternating series. The quantity u_n decreases and the sequence of these terms approaches 0. Thus, the series is convergent.

If $x = e^{i\alpha}$, the two series $\sum u_n \cos n\alpha$ and $\sum u_n \sin n\alpha$ are convergent since the sequence $\{u_n\}$ is decreasing and its limit is zero (cf. PZ, Book IV, Chapter III, part 1, § 4), and hence so is the series whose general term is

$$u_n x^n = u_n e^{in\alpha} = u_n \cos n\alpha + iu_n \sin n\alpha.$$

710

Suppose that

$$\sum_0^{+\infty} a_n t^n \leqslant K$$

for every t in $[t_0, 1[$. Since the terms of the series $\sum a_n t^n$ are positive, the sum of this series is the least upper bound of the set of partial sums of terms of the series. Thus,

$$\sum_0^N a_n t^n \leqslant K, \qquad \forall t \in [t_0, 1[\qquad \text{and} \qquad \forall N.$$

The polynomial

$$\sum_0^N a_n t^n$$

is a continuous function of t. Therefore,

$$\lim_{t \to 1} \sum_0^N a_n t^n = \sum_0^N a_n \leqslant K, \qquad \forall N.$$

K is an upper bound of the set of partial sums of the series with positive terms a_n. Therefore, this series is convergent.

The series $\sum a_n t^n$ is uniformly convergent on [0, 1] since $0 \leqslant a_n t^n \leqslant a_n$ and the numerical series $\sum a_n$ is convergent.

The function

$$\sum_0^{+\infty} a_n t^n$$

is therefore a continuous function of t in [0, 1] and, in particular, it is continuous on the left at 1:

$$\lim_{t \to 1} \sum_0^{+\infty} a_n t^n = \sum_0^{+\infty} a_n.$$

If the series $\sum a_n$ diverges, the sum

$$S(t) = \sum_0^{+\infty} a_n t^n$$

cannot be bounded (since otherwise the series $\sum a_n$ is convergent). The function $S(t)$ is an unbounded increasing function of t in the interval $[t_0, 1[$. Its left-hand limit at $t=1$ is the least upper bound of the set of values that it assumes in $[t_0, 1[$, namely, $+\infty$.

711

First question:

We know that

$$\cos t = \frac{e^{it} + e^{-it}}{2}$$

and hence that

$$\cos^{2p} t = \frac{1}{2^{2p}} (e^{it} + e^{-it})^{2p} = \frac{1}{2^{2p}} \sum_{q=0}^{2p} \binom{2p}{p} e^{2(p-q)it}.$$

The integrals

$$\int_0^{2\pi} e^{2(p-q)it} \, dt$$

are equal to zero if $p - q \neq 0$. Therefore,

$$I_p = \int_0^{2\pi} \frac{1}{2^{2p}} \binom{2p}{p} dt = \frac{2\pi}{2^{2p}} \frac{(2p)!}{(p!)^2} = 2\pi \frac{1 \cdot 3 \cdots (2p-1)}{2 \cdot 4 \cdots 2p}.$$

Second question:

Let us integrate by parts:

$$I_n = \int_0^{2\pi} \cos^{2n-1} t \,(\cos t \, dt)$$

$$= [\sin t \cos^{2n-1} t]_0^{2\pi} + (2n-1) \int_0^{2\pi} \cos^{2n-2} t \sin^2 t \, dt.$$

The first term on the left is equal to zero and

$$I_n = (2n-1) \int_0^{2\pi} \cos^{2n-2} t \,(1-\cos^2 t) dt = (2n-1)(I_{n-1} - I_n),$$

$$2nI_n = (2n-1)I_{n-1}.$$

If we multiply the various equations for $n = 1, 2, ..., p$, we obtain

$$2 \cdot 4 \cdots 2pI_p = 1 \cdot 3 \cdots (2p-1)I_0 = 1 \cdot 3 \cdots (2p-1)2\pi.$$

Third question:

The function that maps t into $1-x \cos t$ assumes all the values of $[1-|x|, 1+|x|]$. For the logarithm to be defined, it is necessary and sufficient that these values be positive, that is, that $|x| < 1$. Then, the function that maps t into $\ln(1-x \cos t)$ is continuous and integrable over $[0, 2\pi]$.

We know that, if $|x \cos t| < 1$, then

$$\ln(1-x \cos t) = -\sum_1^{+\infty} \frac{x^k \cos^k t}{k}.$$

The series converges absolutely and uniformly with respect to t in $[0, 2\pi]$ since

$$\left| \frac{x^k \cos^k t}{k} \right| \leqslant |x^k| = |x|^k$$

and since the geometric series whose general term is $|x|^k$ is convergent.

Therefore, we can integrate term by term:

$$F(x) = \int_0^{2\pi} \ln(1-x \cos t) dt = -\sum_1^{+\infty} \frac{x^k}{k} \int_0^{2\pi} \cos^k t \, dt.$$

If k is odd, the integral of $\cos^k t$ over $[0, 2\pi]$ is 0 as we can see by setting $t = u + \pi$ since then

$$\int_\pi^{2\pi} \cos^k t \, dt = -\int_0^\pi \cos^k u \, du.$$

If k is an even number $2p$, the integral is already calculated. Its value is I_p.

$$F(x) = \int_0^{2\pi} \ln (1-x \cos t)dt = -\sum_{p=1}^{+\infty} \frac{x^{2p}}{2p} 2\pi \frac{1\cdot 3\cdots(2p-1)}{2\cdot 4\cdots(2p)}.$$

The series obtained converges if $|x|<1$. Its radius of convergence is at least equal to 1. By using d'Alembert's rule, we can verify that it is exactly 1. The ratio of two consecutive terms is equal to

$$x^2 \frac{2p}{2p+2} \frac{2p+1}{2p+2},$$

and the limit of this expression is x^2.

Fourth question:

The function represented by a power series is differentiable in the interior of the circle of convergence and its derivative is equal to the series resulting from termwise differentiation:

$$F'(x) = -2\pi \sum_1^{+\infty} x^{2p-1} \frac{1\cdot 3\cdots(2p-1)}{2\cdot 4\cdots 2p}.$$

From the binomial expansion, we can easily verify that

$$(1-x^2)^{-\frac{1}{2}} = 1 + \sum_1^{+\infty} x^{2p}\frac{1\cdot 3\cdots(2p-1)}{2\cdot 4\cdots 2p}.$$

Therefore,

$$F'(x) = \frac{2\pi}{x}[1-(1-x^2)^{-\frac{1}{2}}].$$

Since F vanishes at 0,

$$F(x) = \int_0^x \frac{2\pi}{u}[1-(1-u^2)^{-\frac{1}{2}}]du.$$

This integral is easily evaluated by setting $v=(1-u^2)^{\frac{1}{2}}$:

$$F(x) = 2\pi \int_1^{\sqrt{1-x^2}} \left(1-\frac{1}{v}\right)\frac{-vdv}{1-v^2}$$

$$= 2\pi \int_1^{\sqrt{1-x^2}} \frac{dv}{v+1} = 2\pi[\ln (1+\sqrt{1-x^2})-\ln 2].$$

712

First question:

The function $t^\alpha/(1-t^4)^{\frac{1}{2}}$ is continuous for $t \in [0, x]$ if $\alpha \geqslant 0$. Hence, it is integrable in that case.

If $-1 < \alpha < 0$, this function is continuous in $]0, x]$ and

$$0 < t \leqslant x \Rightarrow 0 < t^\alpha(1-t^4)^{-\frac{1}{2}} \leqslant t^\alpha(1-x^4)^{-\frac{1}{2}}.$$

The majorizing function $t^\alpha(1-x^4)^{-\frac{1}{2}}$ is integrable in $]0, x]$ and hence so is the function $t^\alpha(1-t^4)^{-\frac{1}{2}}$.

Second question:

We know that, if $0 < x < 1$,

$$t^\alpha(1-t^4)^{-\frac{1}{2}} = t^\alpha\left[1 + \sum_1^{+\infty} \frac{1 \cdot 3 \cdots (2n-1)}{2 \cdot 4 \cdots 2n} t^{4n}\right]$$

$$= t^\alpha + \sum_1^{+\infty} \frac{1 \cdot 3 \cdots (2n-1)}{2 \cdot 4 \cdots 2n} t^{4n+\alpha}.$$

The general term of this series

$$\frac{1 \cdot 3 \cdots (2n-1)}{2 \cdot 4 \cdots 2n} t^{4n+\alpha}$$

is positive and majorized by

$$\frac{1 \cdot \cdots 3(2n-1)}{2 \cdot 4 \cdots 2n} t^{4n-1}$$

(since $t^\alpha < t^{-1}$ in $]0, 1[$). This last expression is the general term of a power series whose radius of convergence is 1. This power series converges uniformly in $[0, x]$, and hence so does the series whose general term is

$$\frac{1 \cdot 3 \cdots (2n-1)}{2 \cdot 4 \cdots 2n} t^{4n+\alpha}.$$

We know that we can now integrate this series term by term:

$$f(x) = \int_0^x t^\alpha \, dt + \sum_1^{+\infty} \frac{1 \cdot 3 \cdots (2n-1)}{2 \cdot 4 \cdots 2n} \int_0^x t^{4n+\alpha} \, dt$$

$$= \frac{x^{\alpha+1}}{\alpha+1} + \sum_1^{+\infty} \frac{1 \cdot 3 \cdots (2n-1)}{2 \cdot 4 \cdots (2n)} \frac{x^{4n+\alpha+1}}{4n+\alpha+1}$$

and that the series S obtained thereby is convergent if $0 < x < 1$. Let us denote by $u_n(x)$ the general term of this series.

Third question:

For $x = 1$, we obtain the series whose general term is

$$u_n(1) = \frac{1}{2} \frac{3 \cdots (2n-1)}{4 \cdots 2n} \frac{1}{4n + \alpha + 1},$$

$$\frac{u_{n+1}(1)}{u_n(1)} = \frac{2n+1}{2n+2} \frac{4n+\alpha+1}{4n+\alpha+5} \Rightarrow \lim \frac{u_{n+1}(1)}{u_n(1)} = 1.$$

D'Alembert's rule does not give us in this case the information that we want. However, we can use the result of (1) of exercise 707. Let us define

$$v_n = 1 - \frac{u_{n+1}(1)}{u_n(1)} = \frac{12n + 2\alpha + 9}{(2n+2)(4n+\alpha+5)}.$$

Since the sequence $\{nv_n\}$ converges to $\frac{3}{2} > 1$, the series $\sum u_n(1)$ converges.

The function that maps x into $u_n(x)$ is increasing. Therefore, it is obvious that, in $[0, 1]$,

$$0 \leqslant u_n(x) \leqslant u_n(1),$$

and this double inequality implies the uniform convergence of the series S. (One can apply the uniform-convergence test given in PZ, Book IV, Chapter III, part 2, §1, or one can note that the series whose general term is $\|u_n\| = u_n(1)$ is convergent [cf. review at the beginning of Part 7].)

The sum of the series S, which is uniformly convergent on $[0, 1]$, is continuous on that interval and, in particular, it is continuous on the left at 1; that is,

$$f(x) \to \frac{1}{\alpha+1} + \sum_1^{+\infty} u_n(1) \qquad \text{as } x \to 1.$$

By definition, the function $t^\alpha (1 - t^4)^{-\frac{1}{2}}$ is therefore integrable on $]0, 1[$ and

$$\int_0^1 t^\alpha (1 - t^4)^{-\frac{1}{2}} dt = \lim_{x \to 1} \int_0^x t^\alpha (1 - t^4)^{-\frac{1}{2}} dt = \frac{1}{\alpha+1} + \sum_1^{+\infty} u_n(1).$$

Fourth question:

We know that the function $t^\alpha (1 - t^4)^{-\frac{1}{2}}$ is integrable over every interval of the form $]0, x]$ such that $0 < x < 1$. Therefore, we need only consider the interval $[x, 1[$.

$$t^\alpha (1 - t^4)^{-\frac{1}{2}} = (1 - t)^{-\frac{1}{2}} [t^\alpha (1 + t + t^2 + t^3)^{-\frac{1}{2}}].$$

For $t \in [x, 1]$, the function $t^\alpha(1+t+t^2+t^3)^{-\frac{1}{2}}$ is continuous and hence bounded by a number M:

$$x \leqslant t < 1 \Rightarrow 0 \leqslant t^\alpha(1-t^4)^{-\frac{1}{2}} \leqslant M(1-t)^{-\frac{1}{2}}.$$

The majorizing function $M(1-t)^{-\frac{1}{2}}$ is integrable over $[x, 1]$ and hence so is the function $t^\alpha(1-t^4)^{-\frac{1}{2}}$.

713

First question:

For any number x, we know that

$$\lim_{t \to +\infty} e^{-t}t^{x-1}t^2 = 0,$$

which implies that $e^{-t}t^{x-1} \leqslant t^{-2}$ for all t greater than some t_0. Also,

$$0 < t \leqslant 1 \Rightarrow t^{x-1} > e^{-t}t^{x-1} \geqslant e^{-1}t^{x-1} \geqslant 0.$$

The integrability of the function t^{-2} over the interval $0 \leqslant t < \infty$ therefore implies integrability of the positive function $e^{-t}t^{x-1}$ over that interval. Therefore, the function $e^{-t}t^{x-1}$ is integrable or not over the interval $0 < t \leqslant 1$ according as the function t^{x-1} is integrable or not over that interval and hence is integrable if $x - 1 > -1$, that is, if $x > 0$.

Second question:

The results of (1) imply that, if $x > 0$, the function $e^{-t}t^{x-1}$ is integrable over $0 < t < \infty$. Let us integrate by parts over the closed interval $[u, v]$ and then pass to the limit:

$$\int_u^v e^{-t}t^x \, dt = e^{-u}u^x - e^{-v}v^x + x \int_u^v e^{-t}t^{x-1} \, dt,$$

$$\lim_{u = +0} e^{-u}u^x = 0, \qquad \text{since} \qquad x > 0,$$

$$\lim_{v \to +\infty} e^{-v}v^x = 0$$

(because of the relative speeds of increase of the exponential function and a power).

By definition, the two integrals tend to $\Gamma(x+1)$ and $\Gamma(x)$ as u tends to 0 and v tends to $+\infty$. Therefore,

$$\Gamma(x+1) = x\Gamma(x).$$

Since

$$\Gamma(1) = \int_0^{+\infty} e^{-t}dt = 1,$$

we conclude by induction on n that

$$\Gamma(n+1) = n\Gamma(n) = n[(n-1)!] = n!.$$

Third question:

The function e^{-t} can be expanded in a power series that converges for every t:

$$e^{-t}t^{x-1} = t^{x-1} \sum_0^{+\infty} (-1)^n \frac{t^n}{n!} = \sum_0^{+\infty} (-1)^n \frac{t^{n+x-1}}{n!}.$$

In $[0, 1]$, the function t^{n+x-1} is majorized by 1 if $n+x-1 \geq 0$, as is the case if $n \geq 1$ since $x > 0$. Therefore, if $n \geq 1$,

$$\left| (-1)^n \frac{t^{n+x-1}}{n!} \right| \leq \frac{1}{n!}.$$

Convergence of the series $\sum 1/n!$ implies uniform convergence of the series in question. Therefore, we can integrate term by term:

$$\int_0^1 e^{-t}t^{x-1}dt = \sum_0^{+\infty} \frac{(-1)^n}{n!} \int_0^1 t^{n+x-1}dt = \sum_0^{+\infty} \frac{(-1)^n}{n!(n+x)}.$$

Fourth question:

If x is not a negative integer, $|n+x|$ is minorized by a positive number d that is independent of n, namely, the absolute value of the difference between x and the integer closest to it, and

$$\left| \frac{(-1)^n}{n!(n+x)} \right| \leq \frac{1}{d} \frac{1}{n!}.$$

Since the series $\sum 1/n!$ converges, the series in question converges absolutely.

Fifth question:

We know that the integral

$$\int_1^{+\infty} e^{-t}t^{x-1}dt$$

converges for every x. Therefore, the function G is defined if $S(x)$ is defined, that is, if x is not a nonpositive integer.

We showed in (3) that, if $x > 0$,

$$S(x) = \int_0^1 e^{-t} t^{x-1} \, dt.$$

Therefore,

$$G(x) = \int_0^{+\infty} e^{-t} t^{x-1} \, dt = \Gamma(x).$$

As in (2), we show by integrating by parts over the interval $[1, v]$ and passing to the limit, that

$$\int_1^{+\infty} e^{-t} t^x \, dt = e^{-1} + x \int_1^{+\infty} e^{-t} t^{x-1} \, dt.$$

To compare the term

$$\frac{1}{n! \, (n+x+1)}$$

in $S(x+1)$ with the term in $S(x)$ containing the expression $n+x+1$, namely,

$$\frac{1}{(n+1)! \, (n+x+1)},$$

we write

$$\frac{1}{n! \, (n+x+1)} = \frac{n+x+1-x}{(n+1)! \, (n+x+1)}$$

$$= \frac{1}{(n+1)!} - \frac{x}{(n+1)! \, (n+x+1)},$$

$$S(x+1) = \sum_{n=0}^{+\infty} \frac{(-1)^n}{n! \, (n+x+1)}$$

$$= \sum_{n=0}^{+\infty} \frac{(-1)^n}{(n+1)!} - x \sum_{n=0}^{+\infty} \frac{(-1)^n}{(n+1)! \, (n+x+1)}.$$

To evaluate the series on the right, let us change our index of summation by setting $p = n+1$, so that p assumes the values $1, 2, \ldots$:

$$\sum_{n=0}^{+\infty} \frac{(-1)^n}{(n+1)!} = \sum_{p=1}^{+\infty} \frac{(-1)^{p-1}}{p!} = -\sum_{p=1}^{+\infty} \frac{(-1)^p}{p!} = 1 - e^{-1},$$

$$\sum_{n=0}^{+\infty} \frac{(-1)^n}{(n+1)! \, (n+x+1)} = \sum_{p=1}^{+\infty} \frac{(-1)^{p-1}}{p! \, (p+x)} = -\sum_{p=1}^{+\infty} \frac{(-1)^p}{p! \, (p+x)} = \frac{1}{x} - S(x).$$

Finally,

$$S(x+1) = 1-e^{-1}-x\left(\frac{1}{x}-S(x)\right) = xS(x)-e^{-1}.$$

If we add the integrals studied above, we obtain

$$G(x+1) = xG(x).$$

714

We know that

$$\ln(1+x) = \sum_{n=1}^{+\infty} (-1)^{n-1}\frac{x^n}{n} \qquad \text{if} \qquad |x| < 1,$$

$$(1+x)^{-1} = \sum_{n=0}^{+\infty} (-1)^n x^n \qquad \text{if} \qquad |x| < 1.$$

The two power series have the same radius of convergence, namely, 1. They are absolutely convergent if $|x| < 1$. Therefore, the product series is absolutely convergent if $|x| < 1$ and its sum is the product of the sums of these two series. The general term of the product series is

$$u_n(x) = \sum_{k=1}^{n} (-1)^{k-1}\frac{x^k}{k}(-1)^{n-k}x^{n-k} = (-1)^{n-1}x^n\left(\sum_{k=1}^{n}\frac{1}{k}\right).$$

Therefore, if $|x| < 1$,

$$\frac{\ln(1+x)}{1+x} = \sum_{n=1}^{+\infty} (-1)^{n-1}\left(1+\frac{1}{2}+\cdots+\frac{1}{n}\right)x^n.$$

To determine the radius of convergence of the power series obtained, we apply d'Alembert's rule:

$$\left|\frac{u_{n+1}(x)}{u_n(x)}\right| = |x|\frac{\displaystyle\sum_{k=1}^{n+1}\frac{1}{k}}{\displaystyle\sum_{k=1}^{n}\frac{1}{k}}.$$

The sequence of the sums

$$\sum_{k=1}^{n}\frac{1}{k}$$

converges to $+\infty$ and the sequence $\{1/(n+1)\}$ converges to 0. Therefore,

$$\lim_{n \to +\infty} \frac{\sum\limits_{k=1}^{n+1} \frac{1}{k}}{\sum\limits_{k=1}^{n} \frac{1}{k}} = \lim_{n \to +\infty} \left(1 + \frac{\frac{1}{n+1}}{\sum\limits_{k=1}^{n} \frac{1}{k}}\right) = 1.$$

The ratio

$$\left|\frac{u_{n+1}(x)}{u_n(x)}\right|$$

approaches $|x|$. Therefore, the radius of convergence of the series is equal to 1.

715

First question:

The result is obvious if $n=1$ since $n=2^0$. Let us suppose the result is valid for $n<2^k$. Consider the numbers n satisfying the inequalities $2^k \leqslant n < 2^{k+1}$. There are unique numbers q and r such that

$$n = 2q + r, \quad \text{with} \quad r = 0 \text{ or } 1, \quad \text{and} \quad 2^{k-1} \leqslant q < 2^k.$$

The decomposition of n into $\sum 2^{k_i}$ will or will not include 2^0 according as $r=0$ or 1. Therefore, it is clear that to every decomposition of q there corresponds a decomposition of n and conversely. By hypothesis, the decomposition of q is unique and hence so is that of n.

If none of the k_i exceeds P, we obtain a number n that does not exceed

$$\sum_{k=0}^{P} 2^k = 2^{P+1} - 1.$$

Conversely, in the decomposition of a number not exceeding $2^{P+1} - 1$, the exponents k_i are at most equal to P. All the numbers not exceeding $2^{P+1} - 1$ are therefore obtained when we take the sums

$$\sum_i 2^{k_i},$$

where $k_i \leqslant P$ for every i.

REMARK: Note that the study that we have just made is that of the number system expressed in the base 2.

Second question:

By expanding the product, we obtain

$$u_P = 1 + \sum x^{(2^{k_1})} \times \cdots \times x^{(2^{k_j})} = 1 + \sum x^{2^{k_1}+2^{k_2}+\cdots+2^{k_j}} \quad,$$

where $k_1, k_2, ..., k_j$ are distinct numbers not exceeding P. Then, we know that the numbers

$$2^{k_1} + 2^{k_2} + \cdots + 2^{k_j}$$

are all numbers less than or equal to $2^{P+1} - 1$, each of them obtained once, and hence that

$$u_P = 1 + \sum_{h=1}^{2^{P+1}-1} x^h = \frac{1 - x^{(2^{P+1})}}{1-x}.$$

As P tends to $+\infty$, $x^{(2^{P+1})}$ tends to 0 and u_P tends to $1/(1-x)$.

Third question:

As n tends to $+\infty$, x^n tends to 0. Thus,

$$\ln(1+x^n) \simeq x^n \Rightarrow |\ln(1+x^n)| \simeq |x|^n.$$

The series $\sum \ln(1+x^n)$ is absolutely convergent since the series $\sum |x|^n$ is convergent.

Thus, we know that, if the terms of this series appear each once and only once in an infinite collection of series $\sum u_{p,k}$, the sums U_p of the series $\sum u_{p,k}$ are the terms of an absolutely convergent series the sum of which is equal to the sum of the series

$$\sum_{0}^{+\infty} \ln(1+x^n)$$

(cf. PZ, Book IV, Chapter III, part 1, § 3).

The result of (2) enables us to write

$$\ln \frac{1}{1-x} = \sum_{k=0}^{+\infty} \ln[1+x^{(2^k)}]$$

and, replacing x with x^{2p-1}, we find

$$\ln \frac{1}{1-x^{2p-1}} = \sum_{k=0}^{+\infty} \ln[1+x^{(2p-1)2^k}] = \sum_{k=0}^{+\infty} u_{p,k}.$$

Every integer can be expressed uniquely as the product of an odd number and a power of 2. Therefore, every term of the form $\ln(1+x^n)$ is equal to exactly one of the u_{pk}. By applying the result referrred to above, we therefore obtain

$$\sum_{0}^{+\infty} \ln(1+x^n) = \sum_{p=1}^{+\infty} U_p = \sum_{p=1}^{+\infty} \ln\left(\frac{1}{1-x^{2p-1}}\right) = -\sum_{p=1}^{+\infty} \ln(1-x^{2p-1}).$$

From the definition of the sum of a series,

$$\lim_{N\to\infty} \sum_{1}^{N} \ln(1+x^n) = \lim_{N'\to\infty} \sum_{1}^{N'} \ln\left(\frac{1}{1-x^{2p-1}}\right),$$

or, by using the continuity of the exponential function,

$$\lim_{N\to\infty} \prod_{1}^{N}(1+x^n) = \lim_{N'\to\infty} \prod_{1}^{N'} \frac{1}{1-x^{2p-1}}.$$

716

First question:

The expression $(1-u)^{-\frac{1}{2}}$ can be expanded according to the binomial-series formula as a series in powers of u if $|u| < 1$. Here, $u = 2xz - z^2$ and, by hypothesis,

$$|u| \leqslant 2|xz| + |z|^2 < 1,$$

so that

$$(1-2xz+z^2)^{-\frac{1}{2}} = 1 + \sum_{1}^{+\infty} \frac{1\cdot 3\,\cdots\,(2k-1)}{2\cdot 4\,\cdots\,2k}(2xz-z^2)^k = \sum_{k=0}^{+\infty} u_k(x, z).$$

Second question:

Let us denote by $t_{k,p}$ the coefficient of z^p in the polynomial $u_k(x, z)$ (for $t_{k,p} \neq 0$, it is necessary that $k \leqslant p \leqslant 2k$). We can write

$$s(x, z) = \sum_{k}\left(\sum_{p=k}^{2k} t_{k,p}\right).$$

To obtain $|t_{k,p}|$, we need to replace x and z with $|x|$ and $|z|$ and the coefficient with its absolute value. This means replacing $2xz - z^2$ with $2|x|\,|z| + |z|^2$. Thus, we have proven that the series of the sums

$$\sum_{p=k}^{2k} |t_{k,p}|$$

converges since $2|x|\,|z|+|z|^2<1$:

$$\sum_k\left(\sum_{p=k}^{2k}|t_{k,\,p}|\right)=\sum_k(2|x|\,|z|+|z|^2)^k=(1-2|x|\,|z|-|z|^2)^{-\frac{1}{2}}.$$

If we group the terms $t_{k,p}$ having the same index, that is, having z^p as a factor, we obtain

$$\sum_k t_{k,\,p}.$$

In this sum, we need only assign to k values not exceeding p since the degree in z of the terms of $u_k(x,z)$ is at least equal to k and hence, if $k>p$, the term $t_{k,p}$ is 0. Therefore,

$$\sum_{k=0}^p t_{k,\,p}$$

is the term of degree p in

$$\sum_{k=0}^p u_k(x,\,z).$$

By definition, this is $v_p(x)z^p$.

The series $\sum v_p\,(x)\,z^p$ is absolutely convergent since

$$|v_p(x)z^p|=\left|\sum_{k=0}^p t_{k,\,p}\right|\leqslant\sum_{k=0}^p|t_{k,\,p}|.$$

Therefore, a finite sum of the terms $|v_p(x)\,z^p|$ does not exceed a finite sum of the terms $|t_{k,p}|$. Now, the set of finite sums of the terms $|t_{k,p}|$ is bounded. Specifically, $(1-2|x|\,|z|-|z|^2)^{-\frac{1}{2}}$ is an upper bound of that set. The set of finite sums of the terms $|v_p(x)z^p|$ is bounded. This is a sufficient condition for convergence of the series $\sum|v_p(x)z^p|$.

Finally, we know (cf. PZ, Book IV, Chapter III, § 1 on series of series) that, if $\sum\theta_i$ is a simple series formed by taking the terms of a double sequence $\{t_{k,\,p}\}$ in some order, the convergence of the two series

$$\sum_k\left(\sum_p|t_{k,\,p}|\right)\qquad\text{and}\qquad\sum_p\left(\sum_k|t_{k,\,p}|\right)$$

implies that

$$\sum_i\theta_i=\sum_k\left(\sum_p t_{k,\,p}\right)=\sum_p\left(\sum_k t_{k,\,p}\right).$$

Thus, by taking the $t_{k,p}$ in increasing order of powers of z, we find

$$s(x,\,z)=\sum_p v_p(x)z^p.$$

Third question:

When

$$(2xz - z^2)^k = z^k(2x - z)^k,$$

the term in z^n (which will exist if $n/2 \leqslant k \leqslant n$) is the product obtained by multiplying z^k and the term in z^{n-k} in the binomial expansion of $(2x - z)^k$. Therefore, the coefficient of z^n is

$$(-1)^{n-k}\binom{k}{n-k}(2x)^{2k-n}$$

and

$$v_n(x) = \sum_{n/2 \leqslant k \leqslant n} (-1)^{n-k}\binom{k}{n-k}(2x)^{2k-n}\frac{1 \cdot 3 \cdots (2k-1)}{2 \cdot 4 \cdots 2k}$$

$$= \sum_{n/2 \leqslant k \leqslant n} (-1)^{n-k}2^{2k-n}\frac{k!}{[(2k-n)!][(n-k)!]}\frac{2k!}{2^{2k}(k!)^2}x^{2k-n}$$

$$= \sum_{n/2 \leqslant k \leqslant n} \frac{(-1)^{n-k}}{2^n}\frac{(2k)!}{[(2k-n)!][(n-k)!][k!]}x^{2k-n}.$$

We now make the calculations

$$\frac{d^n}{dx^n}(x^2 - 1)^n = \frac{d^n}{dx^n}\left[\sum_{k=0}^{n}\binom{n}{k}x^{2k}(-1)^{n-k}\right]$$

$$= \sum_{n/2 \leqslant k \leqslant n}\frac{n!}{[k!][(n-k)!]}\frac{(2k)!}{(2k-n)!}x^{2k-n}(-1)^{n-k}$$

$$= (n!)2^n v_n(x).$$

Fourth question:

$$\frac{\partial s}{\partial z} = (x - z)(1 - 2xz + z^2)^{-\frac{3}{2}} = s\frac{x - z}{1 - 2xz + z^2}.$$

The power series

$$\sum_{n=0}^{+\infty} v_n(x)z^n$$

can be differentiated term by term in the interior of the circle of convergence. This circle contains, in particular, those z such that

$$|z|^2 + 2|x||z| < 1.$$

Thus, we know that, for these values of z,

$$\frac{\partial s}{\partial z} = \sum_{n=0}^{+\infty} n v_n(x) z^{n-1},$$

and, consequently, we have

$$(1 - 2xz + z^2)\left(\sum_{n=0}^{+\infty} n v_n(x) z^{n-1}\right) = \left(\sum_{n=0}^{+\infty} v_n(x) z^n\right)(x - z).$$

Two power series that assume the same values in a neighborhood of z have the same coefficients. In particular, equating the coefficients of z^{n-1} yields

$$(n-2)v_{n-2}(x) - 2(n-1)xv_{n-1}(x) + nv_n(x) = xv_{n-1}(x) - v_{n-2}(x),$$

$$(n-1)v_{n-2}(x) - (2n-1)xv_{n-1}(x) + nv_n(x) = 0.$$

717

First question:

We know that

$$e^z - \left(1 + \frac{z}{n}\right)^n = \sum_{k=0}^{+\infty} \frac{z^k}{k!} - \sum_{k=0}^{n} \frac{n(n-1)\cdots(n-k+1)}{n^k} \frac{z^k}{k!}$$

$$= \sum_{k=2}^{n}\left[1 - \left(1 - \frac{1}{n}\right)\cdots\left(1 - \frac{k-1}{n}\right)\right]\frac{z^k}{k!} + \sum_{n+1}^{+\infty} \frac{z^k}{k!}.$$

The absolute value of the finite summation and the absolute value of the sum of the series are both majorized when we replace each term with its absolute value.

Since the coefficients of z^k on the right side of the above equation are nonnegative real numbers, this means

$$\left|e^z - \left(1 + \frac{z}{n}\right)^n\right| \leq \sum_{k=2}^{n}\left[1 - \left(1 - \frac{1}{n}\right)\cdots\left(1 - \frac{k-1}{n}\right)\right]\frac{|z|^k}{k!} + \sum_{n+1}^{+\infty} \frac{|z|^k}{k!} = e^{|z|} - \left(1 + \frac{|z|}{n}\right)^n.$$

Let us write Maclaurin's formula for the function $\ln(1+u)$:

$$\ln(1+u) = u - \frac{u^2}{2}\frac{1}{(1+\theta u)^2}, \qquad 0 < \theta < 1,$$

from which we get, for positive values of u,

$$u - \frac{u^2}{2} \leq \ln(1+u) \leq u.$$

Then, taking for u the value $|z|/n$ and multiplying by n,

$$|z| - \frac{|z|^2}{2n} \leqslant n \ln\left(1 + \frac{|z|}{n}\right) \leqslant |z|,$$

$$e^{|z|} e^{-|z|^2/2n} \leqslant \left(1 + \frac{|z|}{n}\right)^n \leqslant e^{|z|}.$$

One can easily show (for example, by using Maclaurin's formula) that $e^{-u} \geqslant 1 - u$ if $u \geqslant 0$. Therefore,

$$e^{|z|}\left(1 - \frac{|z|^2}{2n}\right) \leqslant \left(1 + \frac{|z|}{n}\right)^n \leqslant e^{|z|}.$$

Subtracting $(1 + |z|/n)^n$ from this double inequality yields

$$e^{|z|}\left(1 - \frac{|z|^2}{2n}\right) - \left(1 + \frac{|z|}{n}\right)^n \leqslant 0 \leqslant e^{|z|} - \left(1 + \frac{|z|}{n}\right)^n.$$

Since the first of these expressions is nonpositive, subtracting it from the third can only increase the latter, and we have

$$0 \leqslant e^{|z|} - \left(1 + \frac{|z|}{n}\right)^n \leqslant e^{|z|} \frac{|z|^2}{2n}.$$

Since this result is established for a *majorant* of the quantity

$$\left| e^z - \left(1 + \frac{z}{n}\right)^n \right|,$$

it holds *a fortiori* for that quantity itself.

Second question:

We set

$$\left(1 + \frac{ix}{2n}\right)^{2n} = e^{ix} + r_n, \qquad \left(1 - \frac{ix}{2n}\right)^{2n} = e^{-ix} + s_n$$

and we know that $|r_n|$ and $|s_n|$ are majorized by

$$e^{|x|} \frac{|x|^2}{4n} \leqslant \frac{e^{\pi/2}}{4} \frac{x^2}{n} = C \frac{x^2}{n},$$

since $0 < x \leqslant \pi/2$ in the present case.

$$f_n(x) - \cot x = i\frac{e^{ix}+e^{-ix}+r_n+s_n}{e^{ix}-e^{-ix}+r_n-s_n} - i\frac{e^{ix}+e^{-ix}}{e^{ix}-e^{-ix}}$$

$$= \frac{i(r_n+s_n)\sin x - (r_n-s_n)\cos x}{\sin x(2i\sin x + r_n - s_n)},$$

$$|f_n(x) - \cot x| \leqslant \frac{|i(r_n+s_n)\sin x - (r_n-s_n)\cos x|}{\sin x\,|2i\sin x + r_n - s_n|}.$$

Let us use the inequalities

$$|\sin x| \leqslant 1, \qquad |\cos x| \leqslant 1, \qquad |\sin x| \geqslant \frac{2}{\pi}x,$$

to majorize the numerator and minorize the denominator (in connection with the last, cf. exercise 439 of Part IV).

Since the absolute value of the sum of two quantities is equal to or greater than the difference between the absolute values of those quantities, we have

$$|f_n(x) - \cot x| \leqslant \frac{|r_n+s_n|+|r_n-s_n|}{\frac{2}{\pi}x\left(\frac{4}{\pi}x-|r_n|-|s_n|\right)} \leqslant \frac{2(|r_n|+|s_n|)}{\frac{2}{\pi}x\left(\frac{4}{\pi}x-|r_n|-|s_n|\right)}.$$

Let us now replace $|r_n|$ and $|s_n|$ with their majorant Cx^2/n:

$$|f_n(x) - \cot x| \leqslant \frac{4Cx^2/n}{\frac{2}{\pi}x\left(\frac{4}{\pi}x-2C\frac{x^2}{n}\right)} = \frac{C\pi^2}{n}\cdot\frac{1}{\left(2-\frac{C\pi x}{n}\right)}.$$

Finally, we note that, if $n \geqslant 5$,

$$2 - \frac{C\pi x}{n} \geqslant 2 - \frac{C\pi^2}{2n} \geqslant 2 - \frac{5\,\pi^2}{4\,10} \geqslant \frac{3}{4}.$$

Finally, for $n \geqslant 5$,

$$|f_n(x) - \cot x| \leqslant \frac{\pi^2 e^{\pi/2}}{4n}\cdot\frac{4}{3} = \frac{\pi^2 e^{\pi/2}}{3n}.$$

The convergence of the sequence $\{f_n\}$ to the cotangent function is uniform in the interval $]0, \pi/2]$.

718

The terms of the series are positive. Therefore,

$$x \geqslant 0 \Rightarrow \frac{e^{-nx}}{n^2+1} \leqslant \frac{1}{n^2+1} \Rightarrow \text{convergence},$$

$$x < 0 \Rightarrow \lim_{n \to \infty} \frac{e^{-nx}}{n^2+1} = +\infty \Rightarrow \text{divergence}.$$

(The ratio of the exponential to any power approaches $+\infty$.)

For $x \geqslant 0$, we have seen that the general term of the series is majorized by $1/(n^2+1)$, which is the general term of a convergent numerical series. This implies that the series converges uniformly in the interval $[0, +\infty[$. Therefore, the sum of the series is a continuous function of x.

The general term of the series of derivatives is

$$\frac{-ne^{-nx}}{n^2+1},$$

and this series does not converge for $x=0$. It does converge, however, for every positive value of x. The sequence $\{ne^{-nx}\}$ converges to 0 and hence, from some value of n on,

$$\left| \frac{ne^{-nx}}{n^2+1} \right| < \frac{1}{n^2+1}.$$

Since the function e^{-nx} is a decreasing function of x, it is obvious that

$$x \geqslant a > 0 \Rightarrow e^{-nx} \leqslant e^{-na} \Rightarrow \left| \frac{-ne^{-nx}}{n^2+1} \right| \leqslant \frac{ne^{-na}}{n^2+1}.$$

The series whose general term is

$$\frac{ne^{-na}}{n^2+1}$$

converges. Therefore, the series of derivatives converges uniformly in the interval $[a, +\infty[$. Thus, we know that the function f is differentiable for $x \geqslant a$ and that

$$f'(x) = -\sum_0^{+\infty} \frac{ne^{-nx}}{n^2+1}.$$

If $x_0 > 0$, we can take a number a such that $0 < a < x_0$. The above result is then valid and the function f is differentiable at x_0. Therefore, the function f is differentiable at every value $x_0 > 0$.

719

First question:

The function f is even and the coefficients b_k are 0. For the coefficients a_k, we can integrate over $[0, \pi]$ and double the result obtained:

$$a_0 = \frac{2}{\pi} \int_0^\pi \cos \alpha t \, dt = 2 \frac{\sin \alpha \pi}{\alpha \pi},$$

$$a_k = \frac{2}{\pi} \int_0^\pi \cos \alpha t \cos kt \, dt = \frac{(-1)^k}{\pi} \frac{2\alpha \sin \alpha \pi}{\alpha^2 - k^2}.$$

In the theory of Fourier series, the functions in question are assumed to be periodic. Therefore, we need to extend the function f by taking

$$f(-\pi) = f(\pi) = \cos \alpha \pi, \qquad f(t + 2\pi) = f(t).$$

Then, the function is continuous on $[-\pi, \pi]$. It is also differentiable in that interval; that is, it has a derivative in the interior of that interval, a right-hand derivative at $-\pi$, and a left-hand derivative at π. From this, we conclude that the periodic function f is continuous and has a right- and a left-hand derivative at every value of x. However, it does not have a derivative at the value π since

$$f'(\pi - 0) = -\alpha \sin \alpha \pi, \qquad f'(\pi + 0) = f'(-\pi + 0) = \alpha \sin \alpha \pi.$$

Thus, we know that the Fourier series converges to $f(t)$ for every value of t and hence it converges to $\cos \alpha t$ for every value of t in $[-\pi, \pi]$, including the value π since we have set $f(\pi) = \cos \alpha \pi$.

In particular, if $t = \pi$,

$$\cos \alpha \pi = \sin \alpha \pi \left[\frac{1}{\alpha \pi} + \sum_1^{+\infty} \frac{(-1)^k 2\alpha \cos k\pi}{\pi(\alpha^2 - k^2)} \right]$$

$$= \sin \alpha \pi \left[\frac{1}{\alpha \pi} + \sum_1^{+\infty} \frac{2\alpha \pi}{\pi^2(\alpha^2 - k^2)} \right].$$

Let us set $\alpha \pi = x$. Then,

$$\cot x = \frac{1}{x} + \sum_1^{+\infty} \frac{2x}{x^2 - k^2 \pi^2}.$$

Second question:

If $|x| < M$, the series converges uniformly since

$$\left| \frac{1}{x^2 - k^2 \pi^2} \right| \leqslant \frac{1}{k^2 \pi^2 - M^2} \qquad \text{if} \qquad k \geqslant \frac{M}{\pi}$$

and

$$\frac{1}{k^2\pi^2 - M^2}$$

is the general term of a convergent numerical series.

If $k \geqslant N > M/\pi$, the functions

$$\frac{1}{x^2 - k^2\pi^2}$$

are continuous since their denominator is nonzero. The sum of the series therefore defines a continuous function of x. The function

$$\sum_N^{+\infty} \frac{1}{x^2 - k^2\pi^2}$$

is continuous for $|x| \leqslant M$.

In particular, if $|x| \leqslant 1$, the function

$$\sum_1^{+\infty} \frac{1}{x^2 - k^2\pi^2}$$

is a continuous function of x and therefore

$$\sum_1^{+\infty} \frac{1}{-k^2\pi^2} = \lim_{x \to 0} \sum_1^{+\infty} \frac{1}{x^2 - k^2\pi^2} = \lim_{x \to 0} \frac{1}{2x}\left(\cot x - \frac{1}{x}\right).$$

Let us use the asymptotic expansions:

$$\frac{1}{2x}\left(\cot x - \frac{1}{x}\right) = \frac{x\cos x - \sin x}{2x^2 \sin x} = \frac{x - x^3/2 - x + x^3/6 + o(x^3)}{2x^3 + o(x^3)}$$

$$= \frac{-x^3/3 + o(x^3)}{2x^3 + o(x^3)} = -\frac{1}{6} + o(1),$$

$$\sum_1^{+\infty} \frac{1}{k^2} = -\pi^2 \lim_{x \to 0} \frac{1}{2x}\left(\cot x - \frac{1}{x}\right) = \frac{\pi^2}{6}.$$

Third question:

If $0 \leqslant x \leqslant \pi$, the series

$$\sum_2^{+\infty} \frac{2x}{x^2 - k^2\pi^2}$$

converges uniformly since, as we saw in (2), the series

$$\sum_{2}^{+\infty} \frac{1}{x^2 - k^2\pi^2}$$

converges uniformly on $[0, \pi]$ and the factor $2x$ is bounded on $[0, \pi]$.

Thus, we know that

$$\int_0^x \left(\sum_2^{+\infty} \frac{2u}{u^2 - k^2\pi^2} \right) du = \sum_2^{+\infty} \int_0^x \frac{2u}{u^2 - k^2\pi^2} du = \sum_2^{+\infty} \ln\left(1 - \frac{x^2}{k^2\pi^2}\right).$$

Another primitive of

$$\sum_2^{+\infty} \frac{2x}{x^2 - k^2\pi^2} = \cot x - \frac{1}{x} - \frac{2x}{x^2 - \pi^2}$$

is

$$\ln \frac{\sin x}{x} - \ln\left(1 - \frac{x^2}{\pi^2}\right).$$

As x tends to 0, this function tends to 0. Therefore, it is the primitive that vanishes at 0, and

$$\sum_2^{+\infty} \ln\left(1 - \frac{x^2}{k^2\pi^2}\right) = \ln \frac{\sin x}{x} - \ln\left(1 - \frac{x^2}{\pi^2}\right),$$

$$\ln \sin x = \ln x + \sum_1^{+\infty} \ln\left(1 - \frac{x^2}{k^2\pi^2}\right).$$

Fourth question:

Let us study the series of the derivatives. The general term is

$$\left(\frac{2x}{x^2 - k^2\pi^2}\right)' = \frac{-1}{(x - k\pi)^2} + \frac{-1}{(x + k\pi)^2}.$$

If $|x| \leqslant M$ and if $k > M/\pi$,

$$|x \pm k\pi| \geqslant k\pi - M \implies \left|\left(\frac{2x}{x^2 - k^2\pi^2}\right)'\right| \leqslant \frac{2}{(k\pi - M)^2}.$$

The series of the derivatives

$$\left(\frac{2x}{x^2 - k^2\pi^2}\right)'$$

is uniformly convergent on the interval $[-M, M]$. Each of the terms of index $k \geqslant N > M/\pi$ is continuous on $[-M, M]$. In this interval, the function

$$\sum_{N}^{+\infty} \frac{2x}{x^2 - k^2\pi^2}$$

is differentiable and its derivative is the sum of the series of the derivatives:

$$\sum_{N}^{+\infty} \left(\frac{2x}{x^2 - k^2\pi^2} \right)' = -\sum_{N}^{+\infty} \left[\frac{1}{(x-k\pi)^2} + \frac{1}{(x+k\pi)^2} \right].$$

If in addition x is an integral multiple of π, each of the first N terms of the series is differentiable and

$$(\cot x)' = -\frac{1}{x^2} + \sum_{1}^{N-1} \left(\frac{2x}{x^2 - k^2\pi^2} \right)' + \left(\sum_{N}^{+\infty} \frac{2x}{x^2 - k^2\pi^2} \right)',$$

or, by changing the signs of all of the terms,

$$\frac{1}{\sin^2 x} = \frac{1}{x^2} + \sum_{1}^{+\infty} \left[\frac{1}{(x-k\pi)^2} + \frac{1}{(x+k\pi)^2} \right] = \sum_{-\infty}^{+\infty} \frac{1}{(x-k\pi)^2}$$

The above result was proven for the numbers x in $[-M, M]$, but, since M is arbitrary, it is valid for every number x.

720

First question:

Let us evaluate a_k and b_k simultaneously by observing that

$$a_k + ib_k = \frac{1}{\pi} \int_0^{2\pi} t e^{ikt} dt = \left[\frac{t e^{ikt}}{ik\pi} \right]_0^{2\pi} - \int_0^{2\pi} \frac{e^{ikt}}{ik\pi} dt = -\frac{2i}{k},$$

$$a_k = 0, \qquad b_k = -\frac{2}{k}, \qquad a_0 = \frac{1}{\pi} \int_0^{2\pi} t dt = 2\pi,$$

$$S(t) = \pi - \sum_{1}^{+\infty} \frac{2}{k} \sin kt.$$

The function f is continuous and differentiable everywhere in the interval $]0, 2\pi[$. For these values, the series converges and its sum is $f(t) = t$. For $t = 0$, we see that $S(0) = \pi$, and we note that

$$\frac{f(+0) + f(-0)}{2} = \frac{f(+0) + f(2\pi - 0)}{2} = \frac{0 + 2\pi}{2} = S(0).$$

Second question:

Let us evaluate the derivative of the function $S_n(t)$:

$$S_n(t) = \pi - 2 \sum_1^n \frac{\sin kt}{k} \Rightarrow S_n'(t) = -2 \sum_1^n \cos kt = -2 \operatorname{Re}\left(\sum_1^n e^{ikt}\right),$$

$$\sum_1^n e^{ikt} = e^{it} \frac{e^{int} - 1}{e^{it} - 1} = e^{i(n+1)t/2} \frac{e^{int/2} - e^{-int/2}}{e^{it/2} - e^{-it/2}}$$

$$= e^{i(n+1)t/2} \frac{\sin nt/2}{\sin t/2},$$

$$S_n'(t) = -2 \operatorname{Re}\left(\sum_1^n e^{ikt}\right) = -2 \frac{\cos (n+1)t/2 \, \sin nt/2}{\sin t/2}.$$

The derivative $S_n'(t)$ is negative as t tends to 0 from the right. The function S_n has a relative minimum at the smallest positive zero of $S_n'(t)$. Therefore, the value t_n is defined by

$$\frac{n+1}{2} t_n = \frac{\pi}{2} \Leftrightarrow t_n = \frac{\pi}{n+1}.$$

Third question:

The corresponding minimum of S_n is $\pi - 2\sigma_n(t_n)$, where we define

$$\sigma_n(t_n) = \sum_1^n \frac{1}{k} \sin kt_n = t_n \sum_1^n \frac{\sin kt_n}{kt_n} = \frac{\pi}{n+1} \sum_1^n \frac{\sin kt_n}{kt_n}.$$

The values kt_n define a partition of $[0, \pi]$ into $n+1$ equal subintervals, and $\sigma_n(t_n)/\pi$ is the mean value of the function $x^{-1} \sin x$ at the end-points of these subintervals. (The $(n+1)$st value seems to be missing, but it is $\pi^{-1} \sin \pi = 0$.)

Thus, we know that

$$\lim_{n \to \infty} \sigma_n(t_n) = \int_0^\pi \frac{\sin x}{x} dx,$$

$$\lim_{n \to \infty} S_n(t_n) = \pi - 2 \int_0^\pi \frac{\sin x}{x} dx.$$

Fourth question:

From the power-series expansion of the sine, we deduce that

$$\frac{\sin t}{t} = \sum_0^{+\infty} (-1)^n \frac{t^{2n}}{(2n+1)!}$$

for every value of t. A power series can be integrated term by term in the interior of the circle of convergence. This means in the present case that, for any value of x

$$\frac{1}{\pi}\int_0^x \frac{\sin t}{t}dt = \frac{1}{\pi}\sum_0^{+\infty}(-1)^n \frac{x^{2n+1}}{[(2n+1)!][2n+1]},$$

$$I = \frac{1}{\pi}\int_0^\pi \frac{\sin t}{t}dt = \sum_0^{+\infty}(-1)^n \frac{\pi^{2n}}{[(2n+1)!][2n+1]} = \sum_0^{+\infty} u_n.$$

I is the sum of an alternating series. The sequence of absolute values of the general term is decreasing since

$$\left|\frac{u_{n+1}}{u_n}\right| = \frac{(2n+1)\pi^2}{(2n+2)(2n+3)^2} < \frac{\pi^2}{(2n+3)^2} < \frac{10}{25} \qquad \text{if } n \geqslant 1.$$

The sum I of the series thus lies between two consecutive sums:

$$\sum_0^3 u_n \qquad \text{and} \qquad \sum_0^4 u_n$$

for example. Since u_n is positive, the quantity

$$\sum_0^3 u_n$$

is less than I and we shall find a decimal approximation to it, less than the true value, by minorizing the positive terms and majorizing the negative ones:

$$\sum_0^3 u_n = 1 - \frac{\pi^2}{18} + \frac{\pi^4}{600} - \frac{\pi^6}{35\,280} = 1 - 0.549 + 0.162 - 0.028,$$

which is less than the true value with an error not exceeding 0.003;

$$|u_4| = u_3 \times \frac{7\pi^2}{8 \times 81} < \frac{0.028 \times 10}{81} < 0.004.$$

The majorant $\sum_0^4$ should be approximated from above:

$$\sum_0^4 u_n < (0.585 + 0.003) + 0.004 = 0.592.$$

Thus, we have

$$0.585 < I < 0.592.$$

Fifth question:

For $t \in \,]0, 2\pi[$ and, in particular, for the value t_n, the series converges to $f(t) = t$.

The sequence whose general term is $S(t_n) = t_n$ thus converges to 0 and the sequence $\{S(t_n) - S_n(t_n)\}$ converges to $\pi - 2\pi I = \pi(1 - 2I)$. The calculation in (4) shows that $\pi(1 - 2I) < -0.17\pi$. Therefore, the sequence $\{S(t_n) - S_n(t_n)\}$ does not converge to 0.

There is no contradiction with convergence at every point of $S_n(t)$ to $S(t)$. This result can be expressed as follows: For fixed t_n, the limit of the sequence $\{S(t_n) - S_p(t_n)\}$ as p tends to $+\infty$ is zero. On the other hand, in the sequence $\{S(t_n) - S_n(t_n)\}$, the value t_n of the variable depends on the index n.

We have just shown that the convergence of S_n to S is not uniform on $[0, 2\pi]$: for no matter how high a number N, there exist values of t and numbers n greater than N such that

$$|S(t) - S_n(t)| > 0.17\pi.$$

(We need only take for t any number t_n of sufficiently great index n.) Thus, the sequence of the numbers

$$\|S - S_n\| = \sup_{0 \leqslant t \leqslant 2\pi} |S(t) - S_n(t)|$$

does not converge to 0.

Part 8

DIFFERENTIAL EQUATIONS, LINEAR DIFFERENTIAL SYSTEMS

Exercises dealing with the material in Chapter VI of Book IV of
Mathématiques Générales by C. Pisot and M. Zamansky

Introduction

In these exercises, calculating techniques play a certain role. Therefore, they should provide the student with an opportunity to test his ability to carry a calculation to its completion. It would be quite harmful for him to satisfy himself with having an idea of the method to be followed and not seeking to obtain explicit results.

Review

Linear equations and systems

The general solution is always obtained by adding to a particular solution the general solution of the corresponding homogeneous equation or system.

Solution of first-order equations

Three essential results: (1) For equations with separable variables

$$f(x)\,dx+g(y)\,dy = 0,$$

the solution is given by

$$\int_{x_0}^{x} f(u)\,du + \int_{y_0}^{y} g(v)\,dv = C.$$

(2) For homogeneous equations

$$y' = f\left(\frac{x}{y}\right),$$

we seek linear solutions ($y=mx$). For other solutions, we set $y=tx$ and solve the resulting equation (in t and x). Another method is to shift to polar coordinates.

(3) For linear equations

$$y' = a(x)y+b(x),$$

we find a solution y_1 of the corresponding homogeneous equation ("homogeneous" here means that $b(x)=0$) and then we define a new unknown z by setting $y=y_1 z$.

The existence and uniqueness of solutions

If a function $f(x,y)$ is continuous in a region D of the xy-plane and if its partial derivative with respect to y exists and is bounded in D, then the equation

$$y' = f(x,y)$$

has exactly one solution y such that $y(x_0) = y_0$ for every point (x_0, y_0) of the region D.

Second-order equations

For incomplete equations, that is, equations in which the independent variable x does not appear (except in the differential dx), we set $y' = z$ and convert the equation into one in which z plays the role of the unknown dependent variable. This will be a first-order equation. If we know how to solve it for z (in terms of y), we then re-replace z by y' to solve for y in terms of x.

For homogeneous linear equations, the set of solutions constitutes a two-dimensional vector space over the field of real or complex numbers according as the solutions assume real or complex values. If we know a nonzero solution y_1, we can obtain the remaining solutions by setting $y = y_1 z$ in the original equation and solving for z.

First-order linear differential systems with constant coefficients

Let us write symbolically

$$\frac{dX}{dt} = AX.$$

If the matrix A has distinct characteristic values r_i and if V_i is an eigenvector relative to the value r_i, the general solution of the system is

$$X = \sum \lambda_i V_i e^{r_i t},$$

where the λ_i are arbitrary constants.

If the eigenvalues are not all distinct, we simplify the matrix A as much as possible by making a change of basis (though we cannot always obtain a diagonal matrix) and then study the system transformed by the corresponding change of variables.

Linear equations with constant coefficients

The equation

$$y^{(n)} + a_1 y^{(n-1)} + \cdots + a_n y = 0,$$

is equivalent to the system

$$\begin{cases} u_1 = y', \qquad u_2 = u'_1, \ldots, u_{n-1} = u'_{n-2}, \\ \quad u'_{n-1} + a_1 u_{n-1} + \cdots + a_n y = 0. \end{cases}$$

Therefore, the methods of the preceding paragraph are applicable and, in particular, the numbers r, such that the exponential function e^{rt} is a solution of the equation, are given by the characteristic equation

$$r^n + a_1 r^{n-1} + \cdots + a_n = 0.$$

If the roots of this equation are distinct, the corresponding exponential functions e^{rt} constitute a basis for the vector space of the solutions.

Exercises

801

Solve the differential equation

$$y' = \sqrt{\frac{1+y^2}{1+x^2}},$$

and show that the solutions can be written in the form of an equation stating that a certain polynomial in x and y is equal to 0.

802

Find the solutions of the differential equation

$$x' \sin t - x \cos t = e^t \sin^4 t.$$

Show that for certain values of t there exist either no solutions or infinitely many solutions that assume a given value x_0.

803

(1) Find the solutions of the differential equation

$$x(1+x^2)y' - y(x^2-1) + 2x = 0.$$

Do there exist solutions that are defined for all real values of x?
 (2) Consider the tangents to the different integral curves at points of given abscissa $x_0 \neq 0$. Show that all these tangents pass through a single point and express the coordinates of that point as a function of x_0.

804

(1) Find the solutions of the differential equation

$$(E) \qquad y'(y^2 - x^2 - 2xy) + y^2 - x^2 + 2xy = 0.$$

(2) Find a function φ such that the product

$$\varphi(x^2 + y^2)\left[(y^2 - x^2 + 2xy)\,dx + (y^2 - x^2 - 2xy)\,dy\right]$$

is the differential of a function U of the two variables x and y. Choose one of the possible functions φ and find the corresponding function U. Use the result to find again the solutions of equation (E).

805

Find the solutions of the equation

$$xy' + v = xy^3.$$

806

Consider the differential equation

$$(E) \qquad\qquad x(x+2)y' + (x+1)y - 1 = 0.$$

(1) Find a power series of the form $\sum a_n x^n$ representing the solution of the equation (E) and determine its radius of convergence.

(2) Express the function represented by the above series in terms of elementary functions.

807

Consider the differential equation

$$(E) \qquad\qquad y' + xy'^2 - y = 0.$$

(1) Use the theorem on the existence of solutions of a first-order differential equation to show that, at every point of the region D defined by $4xy + 1 > 0$, there passes at least one integral curve (consider, in particular, points on the y-axis).

(2) Find the integral curves of equation (E) that are straight lines.

(3) Set $y' = t$ and find parametric expressions for the other integral curves of (E). That is, find parametric equations for the solutions with x and y expressed as functions of t. Do not, however, make a study of these integral curves.

(4) Find the locus of points on the integral curves at which the tangent has a given slope m.

808

Find the solutions of the differential equation

$$y - xy' + y'^3 = 0.$$

809

Find the solutions of the differential equation

(E) $$y^3 + y'^3 - yy' = 0.$$

810

Find the solutions of the differential equation

$$3(x^2 - 1)y^2 y' + xy^3 = x^3 + x^2 - x - 1$$

in the two intervals $-1 < x < 1$ and $x > 1$.

811

Find the solutions of the following differential equations (belonging to standard types):

(A) $$(x^2 + y^2)(1 + y'^2) = 4(y - xy')^2,$$

(B) $$2x(x+1)y' - y = y^3 \arcsin x,$$

(C) $$x^6 y'^2 = 4y(x-2)^2,$$

(D) $$x + yy' = y'^3.$$

812

Consider the differential equation

(E) $$4x^2 y'^2 - y^2 = xy^3.$$

(1) Show that the curve that is the symmetric image about the origin of an integral curve of (E) is also an integral curve of (E).

(2) Transform equation (E) by making the change of variables

$$x = aX, \qquad y = bY,$$

where a and b are constants. What relation (R) must exist between a and b for the transformed equation to be the same as (E)? Find a function f of the two variables x and y that is invariant under all changes of variables of the form indicated (that is, such that $f(x,y) = f(X,Y)$) that verify the relation (R).

(3) For positive values of y, we set

$$u = f(x,y), \qquad v = \ln y.$$

Show without calculation that the equation (E') resulting from this change of variables in (E) can be solved in terms of an integral.

(4) Write the equation (E'). Find the solutions of (E') and give the equations of the integral curves of equation (E).

813

Consider the differential equation

$$(y-x)y''+(1+y')(1+y'^2) = 0.$$

(1) Find a function f such that the set of solutions of (E) is identical to the set of solutions of the equations

(e_a) $\qquad\qquad (y-x)f(y') = a,$

where a is an arbitrary constant.

(2) Find the solutions of (E).

814

Find the solutions of the differential equation

(E) $\qquad\qquad 1+y'^2 - yy'' = 0.$

815

Find the solutions of the equation

$$2y'y''+y'^2 = 1.$$

816

(1) Consider the differential equation

$$(1+x^2)y''+xy'-\tfrac{1}{4}y = 0.$$

Verify that the function

$$x \rightarrow \sqrt{x+\sqrt{1+x^2}}$$

is a solution of (E) and find the other solutions.

(2) Find the series expansion in powers of x of the function

$$\sqrt{x+\sqrt{1+x^2}}.$$

817

(1) Let p and q denote two continuous and continuously differentiable complex functions of a real variable t. How should these two functions be chosen so that the differential equation

$$(E) \qquad\qquad x'' + px' + qx = 0$$

will have two solutions u and v whose product is equal to 1?

(2) Suppose now that the functions p and q are real. Under what condition will (E) have two real solutions u and v whose product is equal to 1?

Show that, if such solutions do not exist, there exist real solutions y and z such that $y^2 + z^2 = 1$.

818

Find the solution of the differential equation

$$y'' - 2y' + y = 2 \sinh x,$$

such that $y(0) = y'(0) = 0$.

819

Find the solutions of the differential equation

$$L(x) = x''' + x'' + x' + x = \sin t + t^2 e^t.$$

820

Find the general solution of the differential equation

$$x^2 y'' - 2y = x^4 \cos x.$$

821

Consider the differential equation

$$\frac{d^2 y}{dx^2} - 2x \frac{dy}{dx} + (x^2 - p)y = 0,$$

where p is a given real number.

(1) Denote by T_k the transformation defined by

$$(E_1) \qquad\qquad x = x_1+k, \qquad y = \varphi(k)e^{kx}y_1,$$

where k is an arbitrary constant and φ is a given function. Show that the equation (E_1) that results from the transformation T_k on (E) is identical to (E).

(2) Show that the function φ can be chosen so that the transformations T_k constitute a group (G) isomorphic to the additive group of real numbers. (Suggestion: show that it is possible to take a polynomial in k for the function $\ln \varphi$.) Find a function f of the two variables x and y that is invariant under the transformations T_k of the group G.

(3) Set

$$z(x) = f[x, y(x)].$$

What is *a priori* the form of the differential equation (E') satisfied by the function z? Write out (E') explicitly. Find the solutions of (E') and (E).

822

(1) Find the solutions of the differential system

(1)
$$\begin{cases} \dfrac{dx}{dt} = x+ y, \\[2mm] \dfrac{dy}{dt} = -x+2y+z, \\[2mm] \dfrac{dz}{dt} = x \quad +z. \end{cases}$$

(2) Find the solution of the differential system

(2)
$$\begin{cases} \dfrac{dx}{dt} = x+ y \quad -e^{2t}, \\[2mm] \dfrac{dy}{dt} = -x+2y+z+e^{2t}, \\[2mm] \dfrac{dz}{dt} = x \quad +z+e^{2t}, \end{cases}$$

such that $x(0)=1$, $y(0)=0$, and $z(0)=0$.

823

Find the solution of the differential equation

$$
\begin{cases}
\dfrac{dx}{dt} = -x+y, \\[2mm]
\dfrac{dy}{dt} = -y+z, \\[2mm]
\dfrac{dz}{dt} = x-z,
\end{cases}
$$

such that $x(0)=1$, $y(0)=j$, and $z(0)=j^2$ (cf. problem 216). Show that the triangle MNP, where M, N, and P are the images in the complex plane of $x(t)$, $y(t)$, and $z(t)$ in the complex plane, is equilateral.

824

Find the solutions of the differential system

$$
(S) \quad
\begin{cases}
\dfrac{d^2x}{dt^2} + 3\dfrac{dy}{dt} - 4x + 6y = 0, \\[3mm]
\dfrac{d^2y}{dt^2} + \dfrac{dx}{dt} - 2x + 4y = 0.
\end{cases}
$$

825

Consider the differential system

$$
(S) \quad
\begin{cases}
x' = x+y, \\
y' = -x+3y.
\end{cases}
$$

(1) Show that, by a suitable change of variables, the system (S) can be transformed into the system

$$
S_1
\begin{cases}
u' = 2u+kv, \\
v' = 2v.
\end{cases}
$$

Solve the systems (S_1) and (S).

(2) Show that the functions x representing the first member of an ordered pair (x, y) representing a solution of the system (S) are solutions of a second-

order linear differential equation. Use this fact to find again the solutions of the system (S).

(3) Find the solutions of the system

$$\Sigma \begin{cases} x' = x+y+\sin t, \\ y' = -x+3y. \end{cases}$$

826

Find the solutions of the differential system

$$(S) \begin{cases} \dfrac{dx}{dt} = 8x - y - 5z, \\[2mm] \dfrac{dy}{dt} = -2x+3y+z. \\[2mm] \dfrac{dz}{dt} = 4x - y - z. \end{cases}$$

The matrix A of this system was studied in exercise 346 of Part 3. It is suggested that the reader use the results of that exercise.

Solutions

801

The equation is one with separable variables:

$$\frac{dy}{\sqrt{1+y^2}} = \frac{dx}{\sqrt{1+x^2}}.$$

The solutions are given by

$$\int_0^x \frac{du}{\sqrt{1+u^2}} - \int_0^y \frac{du}{\sqrt{1+u^2}} = \ln(x+\sqrt{1+x^2}) - \ln(y+\sqrt{1+y^2}) = C,$$

or, if we set $C = \ln \lambda$,

$$x + \sqrt{1+x^2} = \lambda(y+\sqrt{1+y^2}),$$

Then, elementary calculations yield

$$4\lambda^2(x^2+y^2) - 4\lambda(1+\lambda^2)xy - (\lambda^2-1)^2 = 0.$$

802

The equation is linear. Therefore, let us first determine the solution of the homogeneous equation. We also assume that $n\pi < t < (n+1)\pi$, where n is an integer, so that the interval under consideration will contain no zeros of the sine function.

$$\frac{x'}{x} = \frac{\cos t}{\sin t} \Rightarrow x = C \sin t.$$

To find the solutions of the given nonhomogeneous equation, we take as our unknown function y, defined by $x = y \sin t$,

$$y' \sin^2 t = e^t \sin^4 t \Rightarrow y(t) - y_0 = \int_{t_0}^t e^u \sin^2 u \, du.$$

Then, writing the sine in its exponential form

$$\sin u = \frac{e^{iu} - e^{-iu}}{2i},$$

we get

$$y(t) - y_0 = \int_{t_0}^{t} e^u \frac{2 - e^{2iu} - e^{-2iu}}{4} du = \left[\frac{e^u}{2} - \frac{e^{u(1+2i)}}{4(1+2i)} - \frac{e^{u(1-2i)}}{4(1-2i)} \right]_{t_0}^{t}$$

$$= \left[e^u \left(\frac{1}{2} - \frac{e^{2iu}(1-2i)}{20} - \frac{e^{-2iu}(1+2i)}{20} \right) \right]_{t_0}^{t},$$

$$y(t) = e^t \left(\frac{1}{2} - \frac{\cos 2t + 2\sin 2t}{10} \right) + \lambda,$$

$$x(t) = e^t \sin t \left(\frac{1}{2} - \frac{\cos 2t + 2\sin 2t}{10} \right) + \lambda \sin t.$$

As t tends to $n\pi$, $x(t)$ tends to 0. Therefore, there exist infinitely many solutions that vanish at $n\pi$, and *no* solution has any value other than zero at $n\pi$.

803

First question:

The equation is linear. The solution of the corresponding homogeneous equation is

$$\frac{y'}{y} = \frac{x^2 - 1}{x(1+x^2)} = -\frac{1}{x} + \frac{2x}{x^2 + 1},$$

$$y = C \frac{x^2 + 1}{x}.$$

Therefore, we take z, defined by $y = z(x^2 + 1)/x$, as our unknown:

$$(1+x^2)^2 z' + 2x = 0 \Rightarrow z = \frac{1}{1+x^2} + \lambda.$$

The general solution of the equation, for $x \neq 0$, is

$$y = \frac{1}{x} + \lambda \frac{x^2 + 1}{x} = \frac{\lambda x^2 + (1+\lambda)}{x}.$$

As x tends to 0, y tends to $+\infty$ unless $\lambda = -1$. If $\lambda = -1$, the above expression for y is equal to $-x$ for all $x \neq 0$. If we set $y(0) = 0$ (so that $y = -x$ everywhere), we have a solution defined for all x, and this is the only such solution.

The theorem on the existence of solutions of a differential equation can be applied to an equation put in the form $y'=f(x, y)$. In the present case, we cannot put the differential equation in this form for $x=0$ since the coefficient of y' is 0.

Second question:

A linear equation solved for y' can be written in the form

$$y' = p(x) \, y + q(x).$$

The tangent to the integral curve at the point (x_0, y_0) is given by the equation

$$Y - y_0 = y_0'(X - x_0) = [p(x_0)y_0 + q(x_0)](X - x_0),$$

or

$$[Y - q(x_0) \, (X - x_0)] - y_0[1 + p(x_0)(X - x_0)] = 0.$$

This equation is satisfied, whatever the value of y_0, by the solution (X, Y) of the system

$$\begin{cases} Y - q(x_0)(X - x_0) = 0 \\ \\ 1 + p(x_0)(X - x_0) = 0 \end{cases} \Leftrightarrow \begin{cases} X = x_0 - \dfrac{1}{p(x_0)} \\ \\ Y = -\dfrac{q(x_0)}{p(x_0)}. \end{cases}$$

Here,

$$p(x) = \frac{x^2 - 1}{x(x^2 + 1)}, \qquad q(x) = \frac{-2}{1 + x^2},$$

$$X = \frac{-2x_0}{x_0^2 - 1}, \qquad Y = \frac{2x_0}{x_0^2 - 1}.$$

804

First question:

The equation is homogeneous. The standard method of solving such an equation is to make the substitution $y = tx$, solve the resulting equation, and then substitute y/x for t. However, this method does not yield solutions for which y/x is a constant m. Let us therefore first look for particular solutions of this type. If m is a solution of the equation,

$$m(m^2 - 1 - 2m) + (m^2 - 1 + 2m) = (m - 1)(m^2 + 1) = 0.$$

Therefore, the only (real) solution of this type is $y = x$.

For the other solutions, we write

$$y = tx, \qquad dy = d(tx) = x\,dt + t\,dx,$$

$$(x\,dt + t\,dx)(t^2 - 1 - 2t) + (t^2 - 1 + 2t)dx = 0,$$

$$\frac{dx}{x} = -dt\,\frac{t^2 - 1 - 2t}{(t-1)(t^2+1)} = \left(\frac{1}{t-1} - \frac{2t}{t^2+1}\right)dt,$$

$$x = C\frac{t-1}{t^2+1}, \qquad y = \frac{Ct(t-1)}{t^2+1}.$$

The integral curves are circles that are tangent at the origin to the line $y=x$. The Cartesian equation of such a circle is obtained, for example, by replacing t with y/x in the expression for x (assuming $x \neq 0$):

$$x = C\frac{x(y-x)}{y^2+x^2} \Rightarrow x^2 + y^2 = C(y-x).$$

Second question:

For $A(x, y)dx + B(x, y)dy$, where A and B are continuously differentiable, to be the differential of a function U, it is necessary that

$$\frac{\partial A}{\partial y} = \frac{\partial B}{\partial x}.$$

In the present case, this means that

$$2y\varphi'(x^2+y^2)[y^2-x^2+2xy]+2\varphi(x^2+y^2)[x+y]$$
$$= 2x\varphi'(x^2+y^2)[y^2-x^2-2xy]-2\varphi(x^2+y^2)[x+y],$$

or

$$2\varphi'(x^2+y^2)[x^3+y^3+x^2y+xy^2]+4\varphi(x^2+y^2)[x+y] = 0,$$

and, when we divide by $x+y$,

$$2\varphi'(x^2+y^2)[x^2+y^2]+4\varphi(x^2+y^2) = 0;$$

$$u\varphi'(u)+2\varphi = 0 \Rightarrow \varphi(u) = \frac{\lambda}{u^2} \Rightarrow \varphi(x^2+y^2) = \frac{\lambda}{(x^2+y^2)^2}.$$

If we take $\lambda = 1$, the function U is defined by

$$\frac{\partial U}{\partial x} = \frac{y^2-x^2+2xy}{(x^2+y^2)^2}, \qquad \frac{\partial U}{\partial y} = \frac{y^2-x^2-2xy}{(x^2+y^2)^2},$$

and we obtain

$$U = \frac{x-y}{(x^2+y^2)} + U_0.$$

The equation (E) can be replaced with the equation $dU=0$. Thus, the solutions are defined by $U=$const.:

$$\frac{x-y}{(x^2+y^2)} = K.$$

For $K=0$, we have $y=x$; for other values of K,

$$x^2+y^2 = -\frac{1}{K}(y-x).$$

If we set $-1/K=C$, we obtain the equations of the circles found in (1).

805

This is a Bernoulli equation, which we can rewrite

$$xy'y^{-3}+y^{-2} = x.$$

Then, substituting $y^{-2}=u$, we find

$$-x\frac{u'}{2}+u = x.$$

The corresponding homogeneous equation is

$$\frac{u'}{u} = \frac{2}{x},$$

which implies that $u=cx^2$. Therefore, we take for our unknown function v, defined by $u=vx^2$:

$$-v'\frac{x^3}{2} = x \Rightarrow v = \frac{2}{x}+\lambda \Rightarrow u = 2x+\lambda x^2,$$

$$y = (2x+\lambda x^2)^{-1/2}.$$

806

First question:

Suppose that the radius of convergence ρ of the series is not zero. We know that, in the interior of the interval of convergence,

$$y = \sum_0^{+\infty} a_n x^n \Rightarrow y' = \sum_0^{+\infty} na_n x^{n-1},$$

so that the left-hand member of the equation is

$$a_0 - 1 + (3a_1 + a_0)x + \sum_{2}^{+\infty} [na_{n-1} + (2n+1)a_n]x^n = 0.$$

If a power series is identically equal to 0, all its coefficients must be zero (these are the derivatives, evaluated at $x=0$, of its sum divided by $n!$):

$$a_0 = 1, \qquad a_1 = -\frac{1}{3}, \cdots, a_n = -\frac{n}{2n+1}a_{n-1}.$$

D'Alembert's rule enables us to determine immediately the radius of convergence ρ of the series:

$$\lim_{n\to\infty} \left| \frac{a_n x^n}{a_{n-1} x^{n-1}} \right| = \lim_{n\to\infty} \frac{n}{2n+1} |x| = \frac{|x|}{2} \Rightarrow \rho = 2.$$

Therefore, the equation has a power-series solution. Its radius of convergence is 2. This is the series

$$S(x) = \sum_{0}^{+\infty} (-1)^n \frac{n!}{1 \cdot 3 \cdots (2n+1)} x^n.$$

Second question:

The equation (E) is a linear equation. Let us seek its solutions by the usual method. The corresponding homogeneous equation is

$$\frac{y'}{y} = -\frac{x+1}{x(x+2)},$$

and its solution is

$$y = \frac{C}{\sqrt{|x(x+2)|}}.$$

Therefore, we take for our unknown the function z defined by

$$y = \frac{z}{\sqrt{|x(x+2)|}}.$$

When we make this substitution, our equation becomes

$$\frac{x(x+2)z'}{\sqrt{|x(x+2)|}} = 1,$$

from which we get

$$z(x) = \lambda + \int_{x_0}^{x} \frac{\sqrt{|t(t+2)|}}{t(t+2)} dt.$$

To evaluate this integral, we treat two cases separately according to the sign of $t(t+2)$:

(a)
$$-2 < t < 0 \Rightarrow |t(t+2)| = -t(t+2),$$

$$z(x) = \lambda + \int_{-1}^{x} \frac{-dt}{\sqrt{-t(t+2)}} = \lambda - \arcsin(x+1);$$

(b)
$$t > 0 \Rightarrow |t(t+2)| = t(t+2),$$

$$z(x) = \lambda + \int_{x_0}^{x} \frac{dt}{\sqrt{t(t+2)}} = \mu + \ln[(x+1) + \sqrt{x(x+2)}].$$

(The study for $t < -2$ is analogous, but it does not interest us since the series converges in $]-2, 2[$.)

The solutions of the equation (E) are given by

$$y(x) = \frac{\lambda - \arcsin(x+1)}{\sqrt{-x(x+2)}}$$

for $-2 < x < 0$ and by

$$y(x) = \frac{\mu + \ln[(x+1) + \sqrt{x(x+2)}]}{\sqrt{x(x+2)}}$$

for $0 < x < 2$.

As x tends to 0 the denominator tends to 0. Therefore, these solutions tend to $\pm \infty$ except possibly for values of λ and μ such that the numerators vanish at $x=0$. These are the values

$$\lambda = \arcsin 1 = \frac{\pi}{2}, \qquad \mu = 0.$$

We know that there does exist a solution that remains bounded as x tends to 0, namely, the function $S(x)$ defined by the infinite series of (1). We have just seen that there can exist only one such solution in each of the intervals $]-2, 0[$ and $]0, 2[$. Therefore, we see that

$$S(x) = \frac{\pi/2 - \arcsin(x+1)}{\sqrt{-x(x+2)}} = \frac{\arccos(x+1)}{\sqrt{-x(x+2)}}$$

for $-2 < x < 0$ and

$$S(x) = \frac{\ln[(x+1) + \sqrt{x(x+2)}]}{\sqrt{x(x+2)}} = \frac{\cosh^{-1}(x+1)}{\sqrt{x(x+2)}}$$

for $0 < x < 2$.

807

First question:

The equation in y' has two solutions if $4xy+1>0$ and $x\neq0$. As x tends to 0, one of these solutions tends to $\pm\infty$ and the other to y. It is this second solution that interests us. We have

$$y' = f(x, y),$$

where

$$f(x, y) = \frac{-1+\sqrt{4xy+1}}{2x} \quad \text{if} \quad x \neq 0, \qquad f(0, y) = y.$$

The function f is defined and continuous in the region D since

$$\sqrt{1+4xy} = 1+2xy+2xy\varepsilon(x, y) = 1+2xy_0+x\varepsilon_1(x, y),$$

where $\varepsilon(x, y)$ and $\varepsilon_1(x, y)$ tend to 0 as (x, y) tends to $(0, y_0)$, and

$$f(x, y) = y_0+\tfrac{1}{2}\varepsilon_1(x, y) \to y_0 = f(0, y_0) \qquad \text{as} \qquad (x, y) \to (0, y_0).$$

The function f is differentiable with respect to y in the region D, and

$$f'_y(x, y) = \frac{1}{\sqrt{4xy+1}} \quad \text{if} \quad x \neq 0, \qquad f'_y(0, y) = 1,$$

so that

$$f'_y(x, y) = \frac{1}{\sqrt{4xy+1}}$$

for all x.

The derivative f'_y is bounded in the region D_a defined by $4xy+1\geqslant a>0$. Through every point in the region D_a there passes exactly one integral curve of the equation $y'=f(x, y)$ (cf. PZ, Book IV, Chapter VI, part 1, § 3). Every point (x_0, y_0) of the region D belongs to a region D_a (we need only take $4x_0y_0+1>a>0$). Therefore, through every point of the region D there passes exactly one integral curve of the equation $y'=f(x, y)$.

REMARKS: (a) A similar study of the equation

$$y' = \frac{-1-\sqrt{4xy+1}}{2x}$$

shows that through every point of the region D not lying on the y-axis there passes exactly one integral curve of this equation.

(b) If we seek the integral curves instead of the functions that satisfy (E), x and y play the same role and we can write (E) in the form

$$dx\,dy + x\,dy^2 - y\,dx^2 = 0.$$

We now note that the y-axis, defined by $x=0$ (and on which therefore $dx=0$) is an integral curve.

In sum, through every point of the region D defined by $4xy+1>0$ there pass exactly two integral curves of (E).

To every point (x_0, y_0) of D there correspond two functions y representing solutions of (E) such that $y(x_0)=y_0$ if $x_0 \neq 0$ but only one function if $x_0=0$.

Second question:

For the line representing the equation $y=mx+p$ to be an integral curve of (E), it is necessary and sufficient, since $y'=m$, that

$$y = mx+p = xm^2 + m,$$

which means that

$$m = 0 \text{ or } 1 \qquad \text{and} \qquad p = m,$$

from which we get the two solutions

$$y = 0 \qquad \text{and} \qquad y = x+1.$$

Third question:

For the other integral curves, y' is variable and can be used as the parameter for the curve. Thus, we replace equation (E) with the system

$$\begin{cases} y = t+xt^2 \\ dy = t\,dx = d(t+xt^2) \end{cases} \Leftrightarrow \begin{cases} y = t+xt^2 \\ \dfrac{dx}{dt}t(1-t) = 2xt+1. \end{cases}$$

The last equation is linear, and we use standard procedures for solving it. The corresponding homogeneous equation

$$\frac{dx}{x} = \frac{2}{1-t} \Rightarrow x = \frac{\lambda}{(t-1)^2}.$$

Making the change of unknown function

$$x = \frac{u}{(t-1)^2},$$

we get

$$u'\frac{t}{1-t} = 1 \Rightarrow u' = \frac{1-t}{t} \Rightarrow u = \ln(Ct)-t.$$

Resubstituting,

$$\begin{cases} x = \dfrac{\ln{(Ct)}-t}{(t-1)^2} \\[2mm] y = t+xt^2 = \dfrac{t^2\ln{(Ct)}-2t^2+t}{(t-1)^2}. \end{cases}$$

To check our calculations, let us find the expressions for dx/dt, dy/dt, and y':

$$\frac{dx}{dt} = \frac{(1/t-1)(t-1)-2[\ln{(Ct)}-t]}{(t-1)^2} = \frac{-2t\ln{(Ct)}+t^2+2t-1}{t(t-1)^2},$$

$$\frac{dy}{dt} = \frac{(2t\ln{(Ct)}-3t+1)(t-1)-2[t^2\ln{(Ct)}-2t^2+t]}{(t-1)^2}$$

$$= \frac{-2t\ln{(Ct)}+t^2+2t-1}{(t-1)^2}.$$

From these expressions, we see that we do indeed have

$$y' = \frac{dy}{dt} \div \frac{dx}{dt} = t.$$

Fourth question:

The slopes μ of the tangents to the integral curves passing through the point (x, y) are solutions of the equation

$$x\mu^2 + \mu - y = 0.$$

If this equation has two distinct roots, the point (x, y) belongs to the region D and we know that there then exist two integral curves passing through that point.

Therefore, the desired locus is the straight line Δ_m representing the equation

$$xm^2 + m - y = 0,$$

except possibly for the point whose coordinates are $x = -\frac{1}{2}m$, $y = m/2$, which does not belong to D since $4xy+1=0$.

A more precise study enables us to establish that the points of the hyperbola representing the equation $4xy+1=0$ are the cusps of the integral curves (at these points, $dx/dt=dy/dt=0$) and that the slope of the tangent to the integral curve is again a double root of the equation $xt^2+t-y=0$. Therefore, the entire line Δ_m is the locus sought.

808

This equation is a particular case of Lagrange's equation. It is called *Clairaut's equation*. The general solution consists of linear functions of the form

$$y(x) = mx - m^3,$$

where m is an arbitrary constant.

There is also a special solution for which y' is variable and will be taken as parameter t. Then,

$$\begin{cases} y = xt - t^3 \\ dy = t\,dx = d(xt - t^3) \end{cases} \Leftrightarrow \begin{cases} y = xt - t^3 \\ (x - 3t^2)\,dt = 0 \end{cases}$$

$$\begin{cases} x = 3t^2 \\ y = 2t^3 \end{cases} \Leftrightarrow y = \frac{2\varepsilon}{3\sqrt{3}} x^{3/2}, \qquad (\varepsilon = \pm 1).$$

We know that the tangents of the integral curve obtained in this way are the straight lines representing the linear functions found above.

809

The variables in this equation are separable since x does not appear in it explicitly. However, it is difficult to solve the equation for y'. Therefore, we use a parametric representation by setting

$$y = ty'.$$

Then,

$$y' = \frac{t}{1 + t^3}, \qquad y = \frac{t^2}{1 + t^3}.$$

The function that expresses x in terms of t is given by

$$dy = \frac{t(2 - t^3)}{(1 + t^3)^2}\,dt = y'\,dx = \frac{t}{1 + t^3}\,dx.$$

This equation is satisfied if $t = 0$ (corresponding to which, the equation (E) is satisfied by the solution $y = 0$). If $t \neq 0$, we have

$$dx = \frac{2 - t^3}{1 + t^3}\,dt = \left(-1 + \frac{3}{1 + t^3}\right)dt \Rightarrow x = x_0 - t + \int_{t_0}^{t} \frac{3\,du}{1 + u^3},$$

$$\int_{t_0}^{t} \frac{3\,du}{1 + u^3} = \int_{t_0}^{t} \left(\frac{1}{u + 1} + \frac{-u + 2}{u^2 - u + 1}\right)du$$

$$= \frac{1}{2} \ln\left[\frac{(t + 1)^2}{t^2 - t + 1}\right] + \sqrt{3} \arctan \frac{\sqrt{3}}{3}(2t - 1) + C.$$

Thus, finally, for the two intervals $t < -1$ and $t > -1$,

$$\begin{cases} x = \lambda - t + \dfrac{1}{2} \ln\left[\dfrac{(t+1)^2}{t^2 - t + 1}\right] + \sqrt{3}\arctan\left[\dfrac{\sqrt{3}}{3}(2t-1)\right] \\[3mm] y = \dfrac{t^2}{1 + t^3}. \end{cases}$$

810

The equation given is not one of the standard types. However, if we make the substitution $Y = y^3$, we obtain a linear equation

$$(x^2 - 1)Y' + xY = x^3 + x^2 - x - 1.$$

The corresponding homogeneous equation

$$\frac{Y'}{Y} = \frac{-x}{x^2 - 1} \Rightarrow Y = \frac{C}{\sqrt{|x^2 - 1|}}.$$

If we make the substitution

$$Y = \frac{z}{\sqrt{|x^2 - 1|}},$$

this equation becomes

$$(x^2 - 1)\frac{z'}{\sqrt{|x^2 - 1|}} = x^3 + x^2 - x - 1 = (x+1)(x^2 - 1),$$

or, since $x^2 - 1 \neq 0$,

$$z' = (x+1)\sqrt{|x^2 - 1|} \Rightarrow z - z_0 = \int_{x_0}^{x} (t+1)\sqrt{|t^2 - 1|}\, dt.$$

If $x > 1$, we make the substitution $t = \cosh u$:

$$z - z_0 = \int_{u_0}^{U} (1 + \cosh u)\sinh^2 u\, du = \left[-\frac{u}{2} + \frac{\sinh 2u}{4} + \frac{\sinh^3 u}{3}\right]_{u_0}^{U},$$

$$z = -\frac{1}{2}\ln(x + \sqrt{x^2 - 1}) + \frac{x\sqrt{x^2 - 1}}{2} + \frac{(x^2 - 1)\sqrt{x^2 - 1}}{3} + \lambda,$$

and

$$Y = y^3 = \frac{z}{\sqrt{x^2 - 1}} \Rightarrow y = \frac{\sqrt[3]{z}}{\sqrt[6]{x^2 - 1}}.$$

If $|x| < 1$, we set $t = \cos u$:

$$z - z_0 = -\int_U^{U_0} (1 + \cos u) \sin^2 u \, du = -\left[\frac{u}{2} - \frac{\sin 2u}{4} + \frac{\sin^3 u}{3} \right]_U^{U_0},$$

$$z = -\frac{1}{2} \arccos x + \frac{x\sqrt{1 - x^2}}{2} - \frac{(1 - x^2)\sqrt{1 - x^2}}{3} + \lambda.$$

Proceeding as above, we get

$$y = \frac{\sqrt[3]{z}}{\sqrt[6]{1 - x^2}}.$$

811

The solutions will be given in an abridged form since analogous examples have been treated in the preceding exercises.

(A)

The equation is homogeneous (see exercise 804, (1)). We may either set $t = y/x$ or change to polar coordinates $x = r \cos \theta$, $y = r \sin \theta$:

$$(x^2 + y^2)(dx^2 + dy^2) = r^2 \left[r^2 + \left(\frac{dr}{d\theta} \right)^2 \right] d\theta^2,$$

$$x \, dy - y \, dx = r^2 \, d\theta,$$

$$r^2(r^2 + r'^2) = 4r^4 \Leftrightarrow r^2 + r'^2 = 4r^2,$$

$$r' = \varepsilon r \sqrt{3} \quad (\varepsilon = \pm 1) \Rightarrow r = r_0 e^{\varepsilon \sqrt{3} \theta}.$$

(B)

The equation is a Bernoulli equation (see exercise 805), applicable for $-1 < x < 1$. The equation becomes linear if we set $u = y^{-2}$:

$$-x(x + 1)u' - u = \arcsin x.$$

The corresponding homogeneous equation

$$\frac{u'}{u} = -\frac{1}{x(x + 1)} \Rightarrow u = C \frac{x + 1}{x}.$$

If we now take a new unknown v defined by

$$u = v \frac{x+1}{x},$$

we have

$$-v'(x+1)^2 = \arcsin x \Rightarrow v - v_0 = -\int_{x_0}^{x} \frac{\arcsin t}{(t+1)^2} \, dt.$$

Integration by parts yields

$$v - v_0 = \left(\frac{\arcsin t}{t+1}\right)_{x_0}^{x} - \int_{x_0}^{x} \frac{dt}{(t+1)\sqrt{1-t^2}} = \frac{\arcsin x}{x+1} + \sqrt{\frac{1-x}{1+x}} + C,$$

$$y^{-2} = u = \frac{\arcsin x}{x} + \frac{\sqrt{1-x^2}}{x} + \lambda \frac{x+1}{x}.$$

(C)

The variables are separable (see exercise 801):

$$\frac{dy}{2\sqrt{y}} = \varepsilon \frac{x-2}{x^3} \, dx \Rightarrow \varepsilon \sqrt{y} = -\frac{1}{x} + \frac{1}{x^2} + \lambda,$$

$$y = \frac{(\lambda x^2 - x + 1)^2}{x^4}.$$

(D)

The equation is a Lagrange equation (see exercise 807). No linear function $y = mx + p$ can be a solution since it is necessary that $m = -1/m$, that is, that $m^2 + 1 = 0$. We write the equation parametrically and set $y' = t$:

$$\begin{cases} x + yt = t^3 \\ dx = \frac{1}{t} dy = d(t^3 - yt) \end{cases} \Leftrightarrow \begin{cases} x + yt = t^3 \\ \frac{t^2+1}{t} \frac{dy}{dt} + y = 3t^2. \end{cases}$$

The last equation is linear. The corresponding homogeneous equation

$$\frac{dy}{y} = \frac{-t \, dt}{t^2 + 1} \Rightarrow y = \frac{C}{\sqrt{t^2 + 1}}.$$

We introduce a new unknown z defined by $y=z/\sqrt{t^2+1}$:

$$z'\,\frac{t^2+1}{t\sqrt{t^2+1}} = 3t^2 \Rightarrow z = z_0 + \int_{x_0}^{x} \frac{3t^3\,dt}{\sqrt{t^2+1}}.$$

If we then take $u=\sqrt{t^2+1}$, we obtain

$$z = u^3 - 3u + \lambda \Rightarrow y = t^2 - 2 + \frac{\lambda}{\sqrt{t^2+1}},$$

$$x = t^3 - yt \qquad \Rightarrow x = 2t - \frac{\lambda t}{\sqrt{t^2+1}}.$$

812

First question:

The symmetric image of a given curve about the origin has the same equation as the original curve except that x is replaced by $-x$ and y by $-y$; y', if it appears in the original equation, is left unchanged. Thus, equation (E) is still satisfied.

Second question:

Equation (e), when operated upon by T_{ab}, becomes

$$4b^2X^2Y'^2 - b^2Y^2 = ab^3XY^3.$$

It is identical to equation (E) if $ab=1$ (relation (R)). The function f is easily obtained if we note that the relation (R) can be written

$$ab = \frac{x}{X}\frac{y}{Y} = 1 \Rightarrow xy = XY.$$

Therefore, we can define f by $f(x,y)=xy$. (Of course, every function of xy is also a solution.)

Third question:

The equation (E'), resulting from the substitutions $u=xy$ and $v=\ln y$ in (E) is invariant under the transformations $\theta_{a,\,1/a}$ derived from the transformations $T_{a,\,1/a}$ by the change of variables and hence defined by

$$\begin{cases} x = aX \\ y = \dfrac{1}{a}Y \end{cases} \quad \text{or} \quad \begin{cases} u = U \\ v = V - \ln a. \end{cases}$$

Thus, if (E') is the equation $dv/du = \varphi(u, v)$, we must have

$$\frac{dV}{dU} = \varphi(U, V) \Rightarrow \frac{dv}{du} = \varphi(u, v + \ln a) = \varphi(u, v).$$

Since $\ln a$ is arbitrary, this relation implies that the function φ is a function only of u and hence that

$$\frac{dv}{du} = \varphi(u) \Rightarrow v = v_0 + \int_{u_0}^{u} \varphi(t)\,dt.$$

∎

Fourth question:

By the definition of u and v,

$$\begin{cases} x = u\,e^{-v} \\ y = e^v \end{cases} \Rightarrow \begin{cases} dx = e^{-v}(du - u\,dv) \\ dy = e^v\,dv. \end{cases}$$

The equation (E'), representing (E) transformed, is therefore

$$4u^2\,dv^2 = (1+u)(du - u\,dv)^2.$$

It follows from this that $1+u$, hence $1+xy$, must be nonnegative. (Turning back to equation (E), we see that this condition is necessary for the existence of a value of y' satisfying the equation.) Therefore,

$$2\,\varepsilon u\,dv = \sqrt{1+u}\,(du - u\,dv), \qquad (\varepsilon = \mp 1),$$

$$u\,dv(\sqrt{1+u} + 2\,\varepsilon) = \sqrt{1+u}\,du.$$

The equation (E') is verified if $u = 0$ (that is, if either x or y is 0) or if $\sqrt{1+u} = 2$ (that is, if $xy = 3$).

Otherwise, if we take as our parameter $t = \sqrt{1+u}$, we obtain

$$dv = \frac{2t^2\,dt}{(t^2-1)(t+2\,\varepsilon)} = \left[\frac{1}{1+2\varepsilon}\frac{1}{t-1} + \frac{1}{1-2\varepsilon}\frac{1}{t+1} + \frac{8}{3}\frac{1}{t+2\varepsilon}\right] dt.$$

Then,

$$v = v_0 + \tfrac{1}{3}\ln|(t-1)(t+1)^{-3}(t+2)^8|$$

if $\varepsilon = 1$, and

$$v = v_0 + \tfrac{1}{3}\ln|(t+1)(t-1)^{-3}(t-2)^8|$$

if $\varepsilon = -1$.

From the expression for v, we derive immediately the expression for y and then that for x since $xy = u = t^2 - 1$. The results obtained are valid for $y > 0$.

We now pass to the general case by using the result of (1). Corresponding to every integral curve is its symmetric image about the origin.

The parametric equations of the two families of integral curves (except for the coordinate axes and the hyperbola $xy=3$) are therefore

$$\begin{cases} y = y_0(t-1)^{1/3}(t+2)^{8/3}(t+1)^{-1}, \\ x = y_0^{-1}(t+1)^2(t-1)^{2/3}(t+2)^{-8/3}; \end{cases}$$

$$\begin{cases} y = y_0(t+1)^{1/3}(t-2)^{8/3}(t-1)^{-1}, \\ x = y_0^{-1}(t-1)^2(t+1)^{2/3}(t-2)^{-8/3}. \end{cases}$$

But we note that if $t'=-t$, the two families of curves are interchanged. Thus, we obtain all the integral curves of (E) by considering just one of the two families.

<h3 style="text-align:center">813</h3>

First question:

The solutions of the equation (e_a) satisfy the equation obtained by differentiating the original one:

$$(y'-1)f(y')+(y-x)f'(y')y''=0.$$

Conversely, every solution of this equation is a solution of one of the equations (e_a).

For a function f to meet the requirements, it is necessary and sufficient that (E') be identical to (E), that is, that

$$\frac{(y'-1)f(y')}{f'(y')} = (1+y')(1+y'^2).$$

Therefore, the function f is defined by

$$\frac{f'(u)}{f(u)} = \frac{u-1}{(u+1)(u^2+1)} = \frac{-1}{u+1} + \frac{u}{u^2+1} \Rightarrow f(u) = C\frac{\sqrt{u^2+1}}{u+1}.$$

We seek one such function f. Therefore, we assign C an arbitrary value, for example, the value 1.

Second question:

The equation (e_a) is a Lagrange equation. The only linear function that satisfies (e_a) is the function

$$y = x + \frac{a}{f(1)} = x+a\sqrt{2}.$$

We find the other solutions by taking for our parameter $t = y'$:

$$\begin{cases} y = x + a \dfrac{t+1}{\sqrt{t^2+1}} \\[2mm] dy = t\,dx = dx + a\,\dfrac{(1-t)\,dt}{(t^2+1)^{3/2}} \end{cases} \Leftrightarrow \begin{cases} y = x + a \dfrac{t+1}{\sqrt{t^2+1}}, \\[2mm] dx = \dfrac{-a\,dt}{(t^2+1)^{3/2}}. \end{cases}$$

Then,

$$\begin{cases} x = \dfrac{-at}{\sqrt{t^2+1}} + x_0, \\[4mm] y = \dfrac{a}{\sqrt{t^2+1}} + x_0. \end{cases}$$

If we set $t = \tan \varphi$, we see that the integral curves are circles.

814

The variable x does not appear in the equation. Therefore, we begin by determining the relationship between y and y'. If we set $y' = z(y)$, we have

$$y'' = \frac{dy'}{dx} = \frac{dz}{dy}\frac{dy}{dx} = z\frac{dz}{dy}.$$

If we now set $u = z^2$, we have

$$1 + z^2 - yz\frac{dz}{dy} = 1 + u - \frac{1}{2}y\frac{du}{dy} = 0.$$

This is a linear equation in u, and it has the particular solution $u = -1$. Therefore, its general solution is

$$u = -1 + \lambda y^2.$$

The solutions of equation (E) are therefore defined by

$$y'^2 = -1 + \lambda y^2 = \frac{1}{k^2}(y^2 - k^2),$$

where the positive constant λ is denoted by $1/k^2$. Then,

$$\frac{k\,dy}{\sqrt{y^2-k^2}} = \pm\,dx \Rightarrow \ln\left|\frac{y+\sqrt{y^2-k^2}}{k}\right| = \pm\frac{x-x_0}{k}.$$

Taking the hyperbolic cosine and simplifying the left-hand member, we have

$$\left|\frac{y}{k}\right| = \cosh\left(\pm\frac{x-x_0}{k}\right) = \cosh\frac{x-x_0}{k} \Rightarrow y = k\cosh\frac{x-x_0}{k}.$$

815

As in the preceding exercise we let $z(y)=y'$. As we have seen, $y''=z\,dz/dy$. Then,

$$2z^2 \frac{dz}{dy} + z^2 = 1,$$

$$dy = \frac{2z^2}{1-z^2}\,dz \implies y-y_0 = -2z+\ln\left|\frac{1+z}{1-z}\right|.$$

Instead of solving this equation for $z=y'$, let us find x as a function of z:

$$dx = \frac{1}{z}\,dy = \frac{2z}{1-z^2}\,dz \implies x-x_0 = -\ln|1-z^2|.$$

We might also note that the equation may be written

$$\frac{d(y'^2)}{dx} + y'^2 = 1.$$

From this, we can easily obtain an expression for y'^2 as a function of x and then solve for y in terms of an integral.

816

First question:

Let us set

$$y_1(x) = \sqrt{x + \sqrt{1+x^2}},$$

$$y_1'(x) = \frac{1}{2}\frac{1+x/\sqrt{1+x^2}}{\sqrt{x + \sqrt{1+x^2}}} = \frac{\sqrt{x + \sqrt{1+x^2}}}{2\sqrt{1+x^2}} = \frac{y_1(x)}{2\sqrt{1+x^2}}.$$

We calculate $y_1''(x)$ by differentiating the equation

$$2\sqrt{1+x^2}\,y_1'(x) = y_1(x).$$

This yields

$$2\sqrt{1+x^2}\,y_1''(x) + \frac{2x}{\sqrt{1+x^2}}y_1'(x) = y_1'(x) = \frac{y_1(x)}{2\sqrt{1+x^2}}.$$

If we multiply both sides by $\frac{1}{2}\sqrt{1+x^2}$, we obtain equation (E), so that y_1 is indeed a solution of this equation.

To find a solution of (E) that is independent of y_1, let us substitute $y_1 z$ for y in (E). This gives us

$$y_1 z'' + z'\left(2y_1' + \frac{x}{1+x^2} y_1\right) = 0,$$

$$\frac{z''}{z'} = -\frac{2y_1'}{y_1} - \frac{x}{1+x^2} \Rightarrow z' = \frac{C}{y_1^2 \sqrt{1+x^2}},$$

$$z' = \frac{C}{(x+\sqrt{1+x^2})\sqrt{1+x^2}} = C\frac{\sqrt{1+x^2}-x}{\sqrt{1+x^2}} = C\left(1 - \frac{x}{\sqrt{1+x^2}}\right),$$

$$z = C(x - \sqrt{1+x^2}) + D.$$

Thus, the general solution of (E) is

$$y(x) = C\sqrt{x+\sqrt{1+x^2}}(x-\sqrt{1+x^2}) + D\sqrt{x+\sqrt{1+x^2}}$$

$$= -C\sqrt{\sqrt{1+x^2}-x} + D\sqrt{x+\sqrt{1+x^2}}.$$

Second question:

The function y_1 is the only solution of equation (E) such that

$$y_1(0) = 1, \qquad y_1'(0) = \tfrac{1}{2}.$$

Let us find a power series solution of (E) that satisfies these two conditions. If such a series exists, it must represent the function y_1.
    We set

$$S(x) = 1 + \frac{x}{2} + \sum_{2}^{+\infty} a_n x^n,$$

and we assume that the radius of convergence is nonzero. Then, in the interval of convergence,

$$S'(x) = \frac{1}{2} + \sum_{2}^{+\infty} na_n x^{n-1}, \qquad S''(x) = \sum_{2}^{+\infty} n(n-1)a_n x^{n-2}.$$

The left-hand member of equation (E) is a power series if we replace y, y', and y'', with $S(x)$, $S'(x)$, and $S''(x)$. Equation (E) will be verified by $S(x)$ if all the coefficients of the power series so obtained will vanish. Thus, for the first term,

$$2a_2 - \tfrac{1}{4} = 0 \Rightarrow a_2 = \tfrac{1}{8};$$

for the second term,

$$6a_3 + \tfrac{1}{2} - \tfrac{1}{8} = 0 \Rightarrow a_3 = -\tfrac{1}{16};$$

....

for the nth term,

$$(n+2)(n+1)a_{n+2}+n(n-1)a_n+na_n-\tfrac{1}{4}a_n = 0$$

$$\Rightarrow a_{n+2} = -\frac{n^2-\tfrac{1}{4}}{(n+1)(n+2)}a_n.$$

The ratio $|a_{n+2}/a_n|$ approaches 1 as $n\to\infty$. Therefore, the radius of convergence of the series of even-degree terms is 1 (by virtue of d'Alembert's rule), and the same is true of the series of odd-degree terms. Thus, the radius of convergence of the series of both even and odd terms is 1. The calculations made are valid in the interval $]-1, 1[$. In this interval, the function $S(x)$ defined by the power series is therefore equal to

$$\sqrt{x+\sqrt{1+x^2}}.$$

The coefficients a_{2p} and a_{2p+1} can be calculated by induction, and we see that

$$a_{2p} = (-1)^{p-1}\,2^{\frac{\prod\limits_{k=1}^{p-1}(4k^2-\tfrac{1}{4})}{(2p)!}}\,a_2 = \frac{(-1)^{p-1}}{4}\frac{\prod\limits_{k=1}^{p-1}(4k^2-\tfrac{1}{4})}{(2p)!},$$

$$a_{2p+1} = (-1)^p\frac{\prod\limits_{k=1}^{p}[(2k-1)^2-\tfrac{1}{4}]}{(2p+1)!}\,a_1 = \frac{(-1)^p}{2}\frac{\prod\limits_{k=1}^{p}[(2k-1)^2-\tfrac{1}{4}]}{(2p+1)!}.$$

817

First question:

Let us find v and its derivatives as functions of u and its derivatives:

$$v = \frac{1}{u}, \qquad v' = -\frac{u'}{u^2}, \qquad v'' = \frac{2u'^2-uu''}{u^3}.$$

If v is a solution of equation (E), we see that u is a solution of the system

$$(1) \qquad \begin{cases} 2u'^2-uu''-puu'+qu^2 = 0, \\ u''+pu'+qu = 0. \end{cases}$$

We obtain an equivalent system by multiplying the second equation by u and then adding it to the first:

(2)
$$\begin{cases} 2u'^2 + 2qu^2 = 0 \\ u'' + pu' + qu = 0 \end{cases} \Leftrightarrow \begin{cases} u' = \varepsilon Qu, \\ u'' + pu' + qu = 0, \end{cases}$$

where Q denotes one of the square roots of $-q$ and ε denotes one of the two numbers ± 1.

Then, the first equation implies

$$u(t) = u_0 \exp\left(\varepsilon \int_{t_0}^{t} Q(\theta)\, d\theta\right).$$

From the second equation, we obtain

$$q' + 2pq = 0.$$

This condition is satisfied in particular if $q=0$. We shall disregard this solution in what follows since the corresponding functions u and v are constants.

Second question:

For the solution u found in (1) to be real, it is necessary and sufficient that

$$\int_{0}^{t} Q(\theta)\, d\theta$$

be real (we then take for u_0 a real number), that is, that Q be a real function and hence that q be a nonpositive function. Therefore, the equation (E) has solutions u and v the product of which is equal to 1 in the intervals in which the function q is nonpositive.

If the function q is nonnegative in the interval $[\alpha, \beta]$, the function Q and its primitive

$$\int_{t_0}^{t} Q(\theta)\, d\theta$$

then assume purely imaginary values and

$$u = \exp\left(\int_{t_0}^{t} Q(\theta)\, d\theta\right)$$

assumes values the absolute value of which is equal to 1. The solution $v=1/u$ is equal to $\bar{u}$. The two solutions $y=(u+\bar{u})/2$ and $z=(u-\bar{u})/2i$ are real and

$$1 = uv = (y+iz)(y-iz) = y^2 + z^2.$$

818

We shall begin by seeking the general solution of this linear equation with constant coefficients. The coefficients r in the arguments of the exponentials that are solutions of the corresponding homogeneous equation are roots of the characteristic equation

$$r^2 - 2r + 1 = (r-1)^2 = 0.$$

The characteristic equation has a double root equal to 1. In this case, the general solution of the homogeneous equation is

$$y(x) = Ae^x + Bxe^x.$$

We now need to find a particular solution of the nonhomogeneous equation. We know that $2 \sinh x = e^x - e^{-x}$. Therefore, let us find particular solutions of the nonhomogeneous equations with right-hand members e^x and $-e^{-x}$ and then take the sum of the results.

Since 1 is a double root of the characteristic equation, we seek a solution of the form $\lambda x^2 e^x$ that will satisfy the equation with right-hand member e^x. When we substitute this expression into the equation, we obtain $\lambda = \frac{1}{2}$.

On the other hand, -1 is not a root of the characteristic equation. Therefore, for the nonhomogeneous equation whose right-hand member is $-e^{-x}$, we seek a solution of the form μe^{-x} and, proceeding as above, we find that $\mu = -\frac{1}{4}$.

Thus, the general solution of the equation is

$$y(x) = Ae^x + Bxe^x + \tfrac{1}{2}x^2 e^x - \tfrac{1}{4}e^{-x}.$$

Expressing the fact that this solution satisfies the given conditions, we have

$$y(0) = A - \tfrac{1}{4} = 0,$$
$$y'(0) = A + B + \tfrac{1}{4} = 0.$$

Therefore, the desired solution is obtained for

$$A = \tfrac{1}{4} \quad \text{and} \quad B = -\tfrac{1}{2}.$$

819

The method is the same as for a second-order equation. We seek the general solution of the corresponding homogeneous equation and a particular solution of the nonhomogeneous equation.

(1) *Solutions of L(x)=0.* The exponential function e^{rt} is a solution if r is a root of the characteristic equation

$$r^3 + r^2 + r + 1 = 0,$$

that is, if

$$r = -1, i, \text{ or } -i.$$

The general solution is a linear combination of these three particular solutions.

(2) *The right-hand member* $\sin t + t^2 e^t$ is the sum of the three functions

$$\frac{e^{it}}{2i}, \qquad -\frac{e^{-it}}{2i} \qquad \text{and} \qquad t^2 e^t.$$

The particular solution x_0 of the nonhomogeneous equation is that sum of the three functions x_1, x_2, and x_3 such that

$$L(x_1) = \frac{e^{it}}{2i}, \qquad L(x_2) = -\frac{e^{-it}}{2i}, \qquad L(x_3) = t^2 e^t.$$

(3) *Particular solutions x_1 and x_2.* The number i is a simple root of the characteristic equation. Therefore, let us seek a solution x_1 of the form $\lambda t e^{it}$:

$$L(\lambda t e^{it}) = 2\lambda(i-1)e^{it} = \frac{e^{it}}{2i},$$

from which we get

$$\lambda = \frac{1}{4i(i-1)} = -\frac{1}{8} + \frac{i}{8} \Rightarrow x_1 = \frac{-1+i}{8} t e^{it}.$$

In $L(x)$, the coefficients of the derivatives are real numbers. Therefore,

$$L(\bar{x}_1) = \overline{L(x_1)} = \overline{\left(\frac{e^{it}}{2i}\right)} = \frac{e^{-it}}{-2i}.$$

The function $\bar{x}_1$ is a function x_2.

(4) *A particular solution x_3.* We may take for x_3 the product of e^t multiplied by a second-degree polynomial A. Then,

$$L(x_3) = L(e^t A) = e^t(4A + 6A' + 4A'') = t^2 e^t,$$

$$4A + 6A' + 4A'' = t^2 \Rightarrow A(t) = \frac{2t^2 - 6t + 5}{8}.$$

A particular solution of the nonhomogeneous equation is then

$$x_0(t) = x_1(t) + x_2(t) + x_3(t) = -\frac{t}{4}(\sin t + \cos t) + \frac{2t^2 - 6t + 5}{8}e^t,$$

and the general solution of the nonhomogeneous equation is

$$x(t) = -\frac{t}{4}(\sin t + \cos t) + \frac{2t^2 - 6t + 5}{8}e^t + u\sin t + v\cos t + we^{-t},$$

where u, v, and w are arbitrary constants.

820

The corresponding homogeneous equation is an Euler equation (cf. PZ, Book IV, Chapter VI, part 2, § 3). It is transformed into an equation with constant coefficients when we set $t = \ln|x|$. It has solutions of the form $y = x^r$, where r is a solution of the equation

$$r(r-1) - 2 = (r+1)(r-2) = 0.$$

The general solution of the corresponding homogeneous equation is therefore

$$y(x) = \frac{\lambda}{x} + \mu x^2.$$

We now need to find a particular solution of the nonhomogeneous equation. We use the so-called method of variation of parameters. Let u and v be two functions such that

$$y(x) = \frac{u(x)}{x} + v(x)x^2,$$

$$y'(x) = u(x)\left(\frac{1}{x}\right)' + v(x)(x^2)' = -\frac{u(x)}{x^2} + 2xv(x).$$

The functions u' and v' then satisfy the relation

$$\frac{u'(x)}{x} + v'(x)x^2 = 0.$$

We obtain a second equation by writing that the function y is a solution of our original equation:

$$y''(x) = \frac{2u(x)}{x^3} + 2v(x) - \frac{u'(x)}{x^2} + 2xv'(x),$$

$$x^2 y''(x) - 2y(x) = -u'(x) + 2x^3 v'(x) = x^4 \cos x.$$

Then, the functions u and v are determined by

$$\begin{cases} \dfrac{u'(x)}{x} + v'(x)x^2 = 0 \\[2mm] -u'(x)+2x^3v'(x) = x^4 \cos x \end{cases} \Leftrightarrow \begin{cases} u'(x) = -\dfrac{x^4}{3} \cos x \\[2mm] v'(x) = \dfrac{x}{3} \cos x. \end{cases}$$

A particular solution of this system is

$$\begin{cases} u(x) = \tfrac{1}{3}(-x^4 \sin x - 4x^3 \cos x + 12x^2 \sin x + 24x \cos x - 24 \sin x) \\[2mm] v(x) = \tfrac{1}{3}(x \sin x + \cos x), \end{cases}$$

from which we get

$$y(x) = (8-x^2) \cos x + 4\left(x - \frac{2}{x}\right) \sin x.$$

Therefore, the general solution of the equation is

$$y(x) = (8-x^2) \cos x + 4\left(x - \frac{2}{x}\right) \sin x + \frac{\lambda}{x} + \mu x^2.$$

821

First question:

The derivatives with respect to x and x_1 are equal since $dx = dx_1$:

$$\frac{dy}{dx} = \frac{dy}{dx_1} = \varphi(k)e^{kx}\left(\frac{dy_1}{dx_1} + ky_1\right),$$

$$\frac{d^2y}{dx^2} = \frac{d^2y}{dx_1^2} = \varphi(k)e^{kx}\left(\frac{d^2y_1}{dx_1^2} + 2k\frac{dy_1}{dx_1} + k^2y_1\right).$$

In the equation (E_1), the function $\varphi(k)\,e^{kx}$ appears as a factor, and we obtain

$$\frac{d^2y_1}{dx_1^2} + 2k\frac{dy_1}{dx_1} + k^2 y_1 - 2(x_1+k)\left(\frac{dy_1}{dx_1} + ky_1\right) + [(x_1+k)^2 - p]\,y_1 = 0,$$

which, on simplification, becomes an equation identical to (E).

Second question:

The product $T_k T_{k'}$ of the two transformations T_k and $T_{k'}$ is defined by

$$\begin{cases} x = x_2 + k + k', \\[2mm] y = y_2\,\varphi(k)\,\varphi(k')\,e^{(k+k')x - kk'}. \end{cases}$$

This operation is commutative. If $T_k T_{k'}$ is a transformation of the family, it can only be $T_{k+k'}$. For this to be the case, it is necessary and sufficient that

$$\varphi(k)\varphi(k')e^{-kk'} = \varphi(k+k').$$

Therefore, the mapping $k \to T_k$ is a group isomorphism; that is, the T_k constitute a group isomorphic to the additive group of real numbers.

If we set $\psi(k) = \ln \varphi(k)$, the relation verified by φ is transformed into

$$\psi(k) + \psi(k') - kk' = \psi(k+k').$$

Let us find solutions ψ that are differentiable, and let us differentiate with respect to k':

$$\psi'(k') - k = \psi'(k+k') \implies \psi'(0) - k = \psi'(k) \qquad (\text{if } k' = 0),$$

$$\psi(k) = \psi(0) + k\psi'(0) - \frac{k^2}{2}.$$

$\psi(0)$ is 0 (as a consequence of the fundamental relation obtained by assigning to k' the value 0), and we take $\psi'(0) = 0$.

$$\psi(k) = -\frac{k^2}{2} \implies \varphi(k) = e^{-k^2/2}.$$

The transformation T_k is then defined by

$$x = x_1 + k, \qquad y = e^{kx - k^2/2} y_1.$$

When we eliminate k from these two equations, we obtain

$$y = y_1 e^{(x-x_1)x - (x-x_1)^2/2} = y_1 e^{x^2/2 - x_1^2/2},$$

$$y e^{-x^2/2} = y_1 e^{-x_1^2/2}.$$

Therefore, we take

$$f(x, y) = y e^{-x^2/2}.$$

Third question:

The equation (E'), which is a transformation of (E), is also a linear equation. Like the equation (E), it is invariant under the transformation T_k defined for the variables x and z by

$$x = x_1 + k, \qquad z = z_1.$$

Thus, if we take the coefficient of d^2z/dx^2 equal to 1 in (E'), the coefficients of z and dz/dx are invariant whereas the image of x under T_k is arbitrary. These coefficients therefore are independent of x and the equation (E') is an equation with constant coefficients.

The determination of (E') is elementary:

$$y = z\,e^{x^2/2}, \qquad \frac{dy}{dx} = e^{x^2/2}\left(\frac{dz}{dx} + xz\right),$$

$$\frac{d^2y}{dx^2} = e^{x^2/2}\left[\frac{d^2z}{dx^2} + 2x\frac{dz}{dx} + (x^2+1)z\right].$$

Substituting these values into equation (E), we obtain

$$\frac{d^2z}{dx^2} + (1-p)z = 0.$$

If $p > 1$, let us set $1-p = -q^2$:

$$z = A'\,e^{qx} + B'\,e^{-qx} \implies y = A'\,e^{qx+x^2/2} + B'\,e^{-qx+x^2/2}.$$

Then, if we set

$$A'\,e^{-q^2/2} = A, \qquad B'\,e^{-q^2/2} = B,$$

we have

$$y = A\,e^{(x+q)^2/2} + B\,e^{(x-q)^2/2}.$$

If $p = 1$,

$$z = A + Bx \implies y = e^{x^2/2}(A + Bx).$$

If $p < 1$, we set $1 - p = q^2$:

$$z = A\cos qx + B\sin qx \implies y = e^{x^2/2}(A\cos qx + B\sin qx).$$

822

First question:

For simplicity in writing, we set

$$X = \begin{pmatrix} x \\ y \\ z \end{pmatrix}, \qquad A = \begin{pmatrix} 1 & 1 & 0 \\ -1 & 2 & 1 \\ 1 & 0 & 1 \end{pmatrix}.$$

We know that if V is an eigenvector of A relative to the eigenvalue r, the vector $X = V e^{rt}$ is a solution of the system since

$$\frac{dX}{dt} = rV e^{rt} = AV e^{rt} = AX.$$

In the case in which the eigenvalues are distinct, the general solution is a linear combination of the preceding solutions.

The coordinates x, y, z of an eigenvector relative to the eigenvalue r are solutions of the system

$$\begin{cases} x(1-r)+y & = 0 \\ -x & +y(2-r)+z = 0 \\ x & +z(1-r) = 0 \end{cases} \Leftrightarrow \begin{cases} x = z(r-1), \\ y = x(r-1) = z(r-1)^2, \\ z[-(r-1)+(2-r)(r-1)^2+1] = 0. \end{cases}$$

The characteristic equation is therefore

$$(r-1)^2(2-r)+2-r = (2-r)[(r-1)^2+1] = 0$$

and the eigenvalues are $r_1 = 2$, $r_2 = 1+i$, $r_3 = 1-i$.

The first two equations of the second system immediately yield the coordinates of the eigenvectors relative to each of these values: by taking for these vectors $z = 1$, we have

$$V_1 \begin{array}{c} 1 \\ 1 \\ 1 \end{array} \qquad V_2 \begin{array}{c} i \\ -1 \\ 1 \end{array} \qquad V_3 \begin{array}{c} -i \\ -1 \\ 1. \end{array}$$
$$(r_1 = 2) \qquad (r_2 = 1+i) \qquad (r_3 = 1-i)$$

The general solution of the system (1) is therefore

$$X = aV_1 e^{r_1 t} + bV_2 e^{r_2 t} + cV_3 e^{r_3 t},$$

where a, b, and c are arbitrary constants. In coordinate (parametric) form,

$$\begin{cases} x = a e^{2t} + i e^t (b e^{it} - c e^{-it}), \\ y = a e^{2t} - e^t (b e^{it} + c e^{-it}), \\ z = a e^{2t} + e^t (b e^{it} + c e^{-it}). \end{cases}$$

REMARK: If we seek real solutions of the system, it is necessary and sufficient that a be real and that b and c be conjugates. We can therefore give an expression containing only real terms by setting $b=(\beta+i\gamma)/2$ and $c=(\beta-i\gamma)/2$:

$$x = a\,e^{2t}-e^{t}(\gamma\cos t+\beta\sin t),$$
$$y = a\,e^{2t}-e^{t}(\beta\cos t-\gamma\sin t),$$
$$z = a\,e^{2t}+e^{t}(\beta\cos t-\gamma\sin t).$$

However, it is usually more convenient to keep the expressions with imaginary exponentials.

Second question:

For the two problems confronting us (the search for a particular solution of the nonhomogeneous system and the determination of the solution that satisfies the given conditions), it will be convenient to use the basis V_1, V_2, V_3 of the eigenvectors of A. Therefore, let us determine the matrices P and P^{-1}.

Let X and U denote the column matrices of the coordinates of a vector in the canonical basis and the basis of the eigenvectors respectively. We set

$$X = PU, \qquad U = P^{-1}X.$$

We know that P is the matrix in which the column vectors are the vectors V_1, V_2, V_3:

$$P = \begin{pmatrix} 1 & i & -i \\ 1 & -1 & -1 \\ 1 & 1 & 1 \end{pmatrix}.$$

We can determine P^{-1} either by calculating the determinant of the matrix P and its minors or by solving the system

$$\begin{cases} x = u+iv-iw \\ y = u-v-w \\ z = u+v+w \end{cases} \Rightarrow \begin{cases} u = \dfrac{y+z}{2}, \\[2mm] v+w = \dfrac{z-y}{2}, \\[2mm] v-w = -i\dfrac{2x-y-z}{2}. \end{cases}$$

$$U = P^{-1}X \Leftrightarrow \begin{cases} u = \dfrac{1}{2}y+\dfrac{1}{2}z, \\[2mm] v = -\dfrac{i}{2}x-\dfrac{1-i}{4}y+\dfrac{1+i}{4}z, \\[2mm] w = \dfrac{ix}{2}-\dfrac{1+i}{4}y+\dfrac{1-i}{4}z. \end{cases}$$

The system (2′) representing the transformation of (2) under the change of basis is then obtained immediately:

$$\frac{dX}{dt} = AX + e^{2t}\begin{pmatrix} -1 \\ 1 \\ 1 \end{pmatrix} \Leftrightarrow \frac{dU}{dt} = P^{-1}APU + e^{2t}P^{-1}\begin{pmatrix} -1 \\ 1 \\ 1 \end{pmatrix},$$

(2′)
$$\begin{cases} \dfrac{du}{dt} = 2u + e^{2t}, \\[2mm] \dfrac{dv}{dt} = (1+i)v + ie^{2t}, \\[2mm] \dfrac{dw}{dt} = (1-i)w - ie^{2t}. \end{cases}$$

We need to determine the solution of this system such that

$$U_0 = \begin{pmatrix} u(0) \\ v(0) \\ w(0) \end{pmatrix} = P^{-1}X_0 = P^{-1}\begin{pmatrix} 1 \\ 0 \\ 0 \end{pmatrix} = \begin{pmatrix} 0 \\ -\dfrac{i}{2} \\ \dfrac{i}{2} \end{pmatrix}.$$

To do this, let us determine the general solution of each of the three equations in the system and then the value of the constants:

$$\begin{cases} u = ae^{2t} + te^{2t} \\[2mm] v = be^{t(1+i)} - \dfrac{1-i}{2}e^{2t} \\[2mm] w = ce^{t(1-i)} - \dfrac{1+i}{2}e^{2t} \end{cases} \qquad \begin{cases} u(0) = 0 \quad \Rightarrow a = 0, \\[2mm] v(0) = -\dfrac{i}{2} \Rightarrow b = \dfrac{1}{2} - i, \\[2mm] w(0) = \dfrac{i}{2} \Rightarrow c = \dfrac{1}{2} + i. \end{cases}$$

Therefore, the solution of the system (2) is obtained by determining the vector $X = PU$. We obtain

$$x = e^t(2\cos t - \sin t) + te^{2t} - e^{2t},$$
$$y = -e^t(\cos t + 2\sin t) + te^{2t} + e^{2t},$$
$$z = e^t(\cos t + 2\sin t) + te^{2t} - e^{2t}.$$

823

Let us use the method of (1) of the preceding exercise to obtain the general solution of the system and let us determine the constants by writing the given conditions for $t=0$. (In the present case, we need not determine the matrix P^{-1} since the problem can be treated without explicitly writing the equations in the basis of the eigenvectors.)

The coordinates of an eigenvector of the matrix of the system are solutions of the equations

$$\begin{cases} -(1+r)x+y & = 0 \\ -(1+r)y+z = 0 \\ x \qquad -(1+r)z = 0 \end{cases} \Leftrightarrow \begin{cases} y = (1+r)x, \\ z = (1+r)y = (1+r)^2x, \\ x[1-(1+r)^3] = 0. \end{cases}$$

The characteristic values are the zeros of the polynomial $1-(1+r)^3$, namely, 0, $j-1$, and j^2-1.

The corresponding eigenvectors have the coordinates

$$V_1 \begin{cases} 1 \\ 1 \\ 1 \end{cases} \qquad V_2 \begin{cases} 1 \\ j \\ j^2 \end{cases} \qquad V_3 \begin{cases} 1 \\ j^2 \\ j \end{cases}$$
$$(r_1 = 0) \qquad\qquad (r_2 = j-1) \qquad\qquad (r_3 = j^2-1)$$

Therefore, the general solution of the system is

$$X = aV_1 + be^{(j-1)t}V_2 + ce^{(j^2-1)t}V_3$$

(where a, b, and c are arbitrary constants); that is,

$$\begin{cases} x = a+be^{(j-1)t}+ce^{(j^2-1)t}, \\ y = a+bje^{(j-1)t}+cj^2e^{(j^2-1)t}, \\ z = a+bj^2e^{(j-1)t}+cje^{(j^2-1)t}. \end{cases}$$

The particular solution desired is obtained by writing

$$\begin{cases} x(0) = a+b+c & = 1 \\ y(0) = a+bj+cj^2 = j \\ z(0) = a+bj^2+cj = j^2 \end{cases} \Rightarrow \begin{cases} a = 0, \\ b = 1, \\ c = 0. \end{cases}$$

Therefore, the points MNP correspond to

$$x(t) = e^{(j-1)t}, \qquad y(t) = je^{(j-1)t}, \qquad z(t) = j^2e^{(j-1)t}.$$

These values are the products of 1, j, and j^2 respectively and a single complex number $e^{(j-1)t}$. Under these conditions, we know that the triangle MNP is similar to the triangle $M_0N_0P_0$ corresponding to 1, j, and j^2 (cf. PZ, Book IV, Chapter IV, part 2), and the triangle $M_0N_0P_0$ is equilateral.

824

By introducing the two functions $u = dx/dt$ and $v = dy/dt$, we obtain a first-order system equivalent to the given system:

$$(S') \quad \begin{cases} \dfrac{dx}{dt} = u \\[2mm] \dfrac{dy}{dt} = v \\[2mm] \dfrac{du}{dt} = 4x - 6y \quad -3v \\[2mm] \dfrac{dv}{dt} = 2x - 4y - u. \end{cases}$$

The method used in the preceding two exercises can also be used for this one. We determine the characteristic values and eigenvectors of the matrix of the system. We can then immediately derive the general expression for the solution.

However, in the present problem, the functions u and v are auxiliary unknowns and we need not, for example, seek the four coordinates of the eigenvectors of the matrix. Since we know the form of the solutions of a linear differential system with constant coefficients, we can seek directly solutions defined by

$$x = \lambda e^{rt}, \qquad y = \mu e^{rt}.$$

In this way, we obtain a solution of the system (S) if

$$\begin{cases} \lambda(r^2 - 4) + 3\mu(r + 2) = 0, \\ \lambda(r - 2) + \mu(r^2 + 4) = 0. \end{cases}$$

Here, r must be a zero of the polynomial

$$(r^2 - 4)(r^2 + 4) - 3(r^2 - 4) = (r^2 - 4)(r^2 + 1).$$

This yields the values 2, -2, i, and $-i$ for r.

The corresponding values of λ and μ will then be

$$\begin{aligned} &\text{if } r = 2, & \lambda &= 1, & \mu &= 0; \\ &\text{if } r = -2, & \lambda &= 2, & \mu &= 1; \\ &\text{if } r = i, & \lambda &= 3, & \mu &= 2 - i; \\ &\text{if } r = -i, & \lambda &= 3, & \mu &= 2 + i. \end{aligned}$$

The general solution of the system, being a linear combination of the four independent solutions, is

$$\begin{cases} x = ae^{2t}+2be^{-2t}+3ce^{it}+3de^{-it}, \\ y = \qquad be^{-2t}+(2-i)ce^{it}+(2+i)de^{-it}, \end{cases}$$

where a, b, c, and d are arbitrary constants.

For a solution to be real, it is necessary and sufficient that c and d be complex conjugates.

825

First question:

The matrix A of the system (S) has a double characteristic value equal to 2, and the corresponding eigenvectors constitute a one-dimensional vector space. One of these vectors has coordinates $x=y=1$.

In the present case, it is impossible to take a basis consisting of eigenvectors. For new basis vectors, we take the vectors $(1, 1)$ and $(0, 1)$. The new coordinates u and v and the matrix A' representing the transformation of A are then

$$\begin{cases} x = u \\ y = u+v \end{cases} \qquad \begin{cases} u = x \\ v = y-x \end{cases} \qquad A' = \begin{pmatrix} 2 & 1 \\ 0 & 2 \end{pmatrix}.$$

Therefore, the system (S_1), representing the transformation of (S) is

$$\begin{cases} u' = 2u+v \\ v' = \qquad 2v \end{cases} \Rightarrow \begin{cases} u' = 2u+ae^{2t}, \\ v = ae^{2t}. \end{cases}$$

The first of these equations is a first-order linear differential equation, which can be solved immediately:

$$\begin{cases} u = (at+b)e^{2t}, \\ v = ae^{2t}, \end{cases}$$

so that the general solution of the system (S) is

$$\begin{cases} x = (at+b)e^{2t}, \\ y = (at+a+b)e^{2t}. \end{cases}$$

Second question:

The system (S) is equivalent to the system

$$\begin{cases} x' = x+y, \\ x'' = x'-x+3y. \end{cases}$$

(By differentiating the first equation and subtracting it from the second, we see that $y' = -x + 3y$.)

We again obtain an equivalent system by writing

$$\begin{cases} x' = x + y \\ x'' = x' - x + 3(x' - x) = 4x' - 4x \end{cases} \quad \text{or} \quad \begin{cases} x' = x + y \\ x'' - 4x' + 4x = 0. \end{cases}$$

The last equation is a second-order linear equation with constant coefficients, the characteristic equation of which is

$$r^2 - 4r + 4 = 0.$$

This equation has a double root equal to 2. Therefore, the general solution of the linear equation is

$$x = (at + b)e^{2t},$$

from which we get

$$y = x' - x = (at + a + b)e^{2t}.$$

Third question:

The change of variables indicated in (1) transforms the system $\sum$ into the system

$$(\Sigma_1) \begin{cases} u' = 2u + v + \sin t \\ v' = 2v - \sin t. \end{cases}$$

Instead of decomposing $\sin t$ into

$$(e^{it} - e^{-it})/2i,$$

we note that, since the coefficients in the system are real, the real solutions u and v can be defined as the imaginary parts of the solutions of the system:

$$\begin{cases} u' = 2u + v + e^{it}, \\ v' = 2v - e^{it}. \end{cases}$$

A particular solution of the second equation is

$$v = e^{it}\frac{2 + i}{5}.$$

The corresponding function u is a solution of the system

$$u' = 2u + e^{it}\frac{7 + i}{5},$$

and a particular solution is

$$u = -\frac{13+9i}{25} e^{it}.$$

The general solution of the system $(\sum_1)$ is obtained by adding to the imaginary part of the preceding solution the general solution of the corresponding homogeneous system:

$$\begin{cases} u = (at+b)e^{2t} - \dfrac{9 \cos t + 13 \sin t}{25}, \\[2mm] v = \qquad a e^{2t} + \dfrac{5 \cos t + 10 \sin t}{25}, \end{cases}$$

so that the general solution of the system $(\sum)$ is

$$\begin{cases} x = \qquad (at+b)e^{2t} - \dfrac{9 \cos t + 13 \sin t}{25}, \\[2mm] y = (at+a+b)e^{2t} - \dfrac{4 \cos t + 3 \sin t}{25}. \end{cases}$$

826

We showed in exercise 346 of Part 3 that the matrix A is transformed into

$$B = \begin{pmatrix} 2 & 0 & 0 \\ 0 & 4 & 1 \\ 0 & 0 & 4 \end{pmatrix}$$

if the basis vectors are

$$V_1 \begin{vmatrix} 1 \\ 1 \\ 1 \end{vmatrix} \qquad V_2 \begin{vmatrix} 1 \\ -1 \\ 1 \end{vmatrix} \qquad \varepsilon_3 \begin{vmatrix} \frac{1}{3} \\ \frac{1}{3} \\ 0. \end{vmatrix}$$

Let u, v, and w denote the coordinates relative to the basis V_1, V_2, V_3. Then, the system (S) is transformed into

$$(S_1) \begin{cases} u' = 2u, \\ v' = \qquad 4v+w, \\ w' = \qquad 4w. \end{cases}$$

The system consisting of the last two equations is of the same type as that studied in the preceding exercise. We obtain without difficulty the general solution

$$\begin{cases} u = & a\,e^{2t}, \\ v = & (bt+c)e^{4t}, \\ w = & b\,e^{4t}. \end{cases}$$

The solution of the system (S) can be derived from it immediately if we note that the vector X whose coordinates are x, y, z, is

$$X = uV_1 + vV_2 + w\varepsilon_3.$$

Therefore,

$$\begin{cases} x = a\,e^{2t} + bt\,e^{4t} + \left(\dfrac{b}{3}+c\right)e^{4t}, \\[2mm] y = a\,e^{2t} - bt\,e^{4t} + \left(\dfrac{b}{3}-c\right)e^{4t}, \\[2mm] z = a\,e^{2t} + bt\,e^{4t} + c\,e^{4t}. \end{cases}$$

Glossary

Abelian group (or commutative group):   A group in which the group operation is commutative, that is, $a \circ b = b \circ a$ for every a and b in the group.

Asymptotic expansion: A formal summation of the form $\sum\limits_{n=0}^{\infty} a_n x^{-n}$ is an asymptotic expansion of a function $f(x)$ if

$$\lim_{x \to \infty} \left[f(x) - \sum_{k=0}^{n} a_k x^{-k} \right] = 0 \quad \text{for each } n.$$

Bijective mapping: A mapping that is both injective and surjective.

Bilinear form: A mapping F defined on the Cartesian product of two vector spaces V and W into the space of real or complex numbers is a bilinear form if, for every v, v_1, $v_2 \in V$, every w, w_1, $w_2 \in W$ and every real or complex λ

$$F(v_1 + v_2, w) = F(v_1, w) + F(v_2, w),$$
$$F(v, w_1 + w_2) = F(v, w_1) + F(v, w_2),$$
$$F(\lambda v, w) = F(v, \lambda w) = \lambda F(v, w).$$

Binary operation: A mapping into S of the Cartesian product of a set S with itself: $S \times S \to S$.

Bolzano–Weierstrass theorem: A bounded infinite subset of R^p has at least one point of accumulation.

Borel–Lebesgue theorem: Every closed bounded set in R^p that is covered by a collection of open sets (that is, is contained in their union) is also covered by some finite subcollection of that collection. (This theorem is often called the Heine–Borel or Heine–Borel–Lebesgue theorem.)

Closed interval: see *Interval*.

Commutative group: see *Abelian group*.

Commutative ring: A ring in which both ring operations are commutative, that is, for every a and b in the ring, not only

$$a + b = b + a.$$

but also

$$a \times b = b \times a$$

Compact interval: An interval is said to be compact if it is closed and bounded.

Complement: A set E is the complement of a set F in a set A if E is the set of all members of A that are not members of F.

Continuous function: A function f is continuous at a point x_0 if the two quantities

$$\lim_{x \to x_0} f(x) \quad \text{and} \quad f(x_0)$$

both exist and are equal to each other. A function f is continuous on an interval if it is continuous at every point of that interval (in the case of a closed interval, the limits at the end points are taken from the interior of the interval).

Convex function (of a real variable): A function f is convex on an interval $[\alpha, \beta]$ if, for every subinterval $[a, b]$ and every $x \in [a, b]$

$$f(a) + \frac{f(b) - f(a)}{b - a}(x - a) \geq f(x).$$

Graphically, this means that every point on the segment connecting two points of the graph of the function lies on or above the graph.

Determinant: Let A denote an $n \times n$ matrix with elements belonging to a ring. Consider all products of the form $(-1)^p x_1 x_2 \ldots x_n$, where each x_i is an element of the ith row of A but no two are from the same column and where p is 0 or 1 according as the number of transpositions, necessary to make the column numbers of the x's read 1, 2, 3, ..., $n-1$, n in that order, is even or odd. The sum of all such products is called the determinant of A. It is denoted by det A or by exhibiting the elements of A (in the usual $n \times n$ array) with a single vertical bar on each side. For example

$$\det \begin{pmatrix} a & b \\ c & d \end{pmatrix} = \begin{vmatrix} a & b \\ c & d \end{vmatrix} = ad - bc.$$

Eigenvector: An eigenvector of an $n \times n$ matrix A with elements belonging to a ring is an $n \times 1$ matrix (i.e., a column vector) $X \neq 0$ such that $AX = \lambda X$ for some element λ of the ring.

Equivalence class: Let $\mathcal{R}$ denote an equivalence relation (q.v.) on a set E. Then A is an equivalence class (with respect to $\mathcal{R}$ if

(1) A is nonempty,
(2) $a, b \in A$ implies $a \mathcal{R} b$,
(3) $\left. \begin{array}{l} a \in A \\ b \notin A \end{array} \right\}$ imply a not $\mathcal{R} b$.

Equivalence relation: $\mathcal{R}$ is an equivalence relation on a set E if for every $a, b, c, \in E$,

$$a \mathcal{R} a,$$
$$a \mathcal{R} b \quad \text{implies} \quad b \mathcal{R} a,$$
$$\left. \begin{array}{l} a \mathcal{R} b \\ b \mathcal{R} c \end{array} \right\} \text{imply} \quad a \mathcal{R} c.$$

Field: Suppose that two binary operations $+$ and $\times$ are defined on a set S. The set S is said to be a field with respect to these two operations if
(1) S is a commutative group with respect to the operation $+$;
(2) S, exclusive of the zero element (q.v.) of that group, is a commutative group with respect to the operation $\times$;
(3) for every x, y, z in S,

$$x \times (y+z) = x \times y + x \times z.$$

Group: Let S be a set on which a closed binary operation (denoted in what follows by $\circ$) is defined. Suppose that
(1) for any $x, y, z \in S$,

$$(x \circ y) \circ z = x \circ (y \circ z);$$

(2) there exists an $a \in S$ such that for any $x \in S$

$$a \circ x = x \circ a = x;$$

(3) for every $x \in S$ there exists an $y \in S$ such that

$$x \circ y = y \circ x = a,$$

where a is the number of S referred to in requirement (2). Then S is a group with respect to the operation $\circ$.

Homomorphism: $h : G \to H$ is a homomorphism of the group G into the group H if $h(xy) = h(x)h(y)$ for all x, y in G.

Homothetic transformation: A mapping $x \to y$ defined by $y - a = b(x-a)$, where a and b are constants, is a homothetic transformation with center a and ratio b.

Ideal: A subset $\mathscr{I}$ of a commutative ring with ring operations $+$ and $\times$ is an ideal if
(1) $\mathscr{I}$ is a group with respect to $+$,
(2) $a, b \in \mathscr{I} \Rightarrow a \times b \in \mathscr{I}$ for all a in the ring.

Injective function: A function f is injective if, for a and b in its domain of definition, $a \neq b$ implies $f(a) \neq f(b)$.

Integral curve: An integral curve of an ordinary differential equation is a curve representing a particular solution of that equation.

Interval: A set I of real numbers is called an interval if

$$\left. \begin{array}{l} a \in I \\ a < b < c \\ c \in I \end{array} \right\} \quad \text{imply } b \in I.$$

An interval is said to be closed if it has a minimum α and a maximum β; it is denoted by $[\alpha, \beta]$. An interval is said to be open if it has neither a maximum nor a minimum. If an open interval has greatest lower bound α and least upper bound β, it is denoted by $]\alpha, \beta[$ or, in many books, by (α, β). An interval with minimum but no maximum is denoted by $[\alpha, \beta[$ or $[\alpha, \beta)$; an interval with maximum but no minimum is denoted by $]\alpha, \beta]$ or $(\alpha, \beta]$.

Isomorphism: A bijective mapping f of a group G onto a group G' if, for every $x, y \in G$,

$$f(x \circ y) = f(x) \circ f(y)$$

where $\circ$ is the group operation. With a ring or field, this equation must hold for both operations. A vector-space isomorphism is a bijective linear mapping, of a vector space V onto a vector space V'.

Jacobian determinant: Let φ_1, φ_2, ..., φ_n be n functions of n real variables $x_1, x_2, ..., x_n$. The determinant

$$\begin{vmatrix} \dfrac{\partial \varphi_1}{\partial x_1} & \dfrac{\partial \varphi_1}{\partial x_2} & \cdots & \dfrac{\partial \varphi_1}{\partial x_n} \\[2ex] \dfrac{\partial \varphi_2}{\partial x_1} & \dfrac{\partial \varphi_2}{\partial x_2} & \cdots & \dfrac{\partial \varphi_2}{\partial x_n} \\[2ex] \vdots & \vdots & & \vdots \\[2ex] \dfrac{\partial \varphi_n}{\partial x_1} & \dfrac{\partial \varphi_n}{\partial x_2} & \cdots & \dfrac{\partial \varphi_n}{\partial x_n} \end{vmatrix}$$

is the Jacobian determinant (or simply the Jacobian) of the φ_i with respect to the x_i.

Jordan curve: Let a curve in two-dimensional space have the parametric representation $x = f(t)$, $y = g(t)$ where t ranges over a *closed* interval $[a, b]$. Suppose the only pair of values t_1 and t_2 where $t_1 < t_2$ such that $f(t_1) = f(t_2)$ and $g(t_1) = g(t_2)$ is the pair $t_1 = a$, $t_2 = b$. (Graphically, the fact that this is true of a and b means that the curve is closed, and the fact that it is true *only* of a and b means that it does not intersect itself.) Such a curve is called a Jordan curve. Jordan curves in higher-dimensional spaces are defined analogously.

Length of a curve (in xy-space): Let $x = f(t)$, $y = g(t)$ be the parametric representation of a curve C. Then the length of C is the least upper bound of the set of all sums

$$\sum_{i=0}^{n-1} [(f(t_{i+1}) - f(t_i))^2 + (g(t_{i+1}) - g(t_i))^2]^{\frac{1}{2}}$$

where $a \leqq t < b$, $a = t_0 < t_1 < t_2 < ... < t_n - b$, and all possible values of n, $t_0, t_1, ..., t_n$ are considered.

Linear form: A mapping f defined on a vector space V into the space of real or complex numbers is a linear form if, for every $v, w \in V$ and every real or complex λ

$$f(v + w) = f(v) + f(w)$$
$$f(\lambda v) = \lambda f(v).$$

Lower bound: Let E denote a set, $\mathscr{R}$ an ordering of E, and E' a subset of E. Then $x \in E$ is a lower bound of E' if $y \in E'$ implies $x \mathscr{R} y$.

Maclaurin series: A series of the form $\sum\limits_{n=0}^{\infty} a_n z^n$, where the a_n are constants.

Maximal element: x is a maximal element of an ordered set E if it is an upper bound (q.v.) of E that belongs to E.

Minimal element: x is a minimal element of an ordered set E if it is a lower bound (q.v.) of E that belongs to E.

Monotonic function: A function is monotonic if it is either nonincreasing or nondecreasing (q.q.v.).

Nondecreasing function: A function f is nondecreasing if, for a and b in the domain of f, the inequality $a < b$ implies $f(a) \leq f(b)$.

Nonincreasing function: A function f is nonincreasing if, for a and b in the domain of f the inequality $a < b$ implies $f(a) \geq f(b)$.

Nonsingular: An $n \times n$ matrix A is nonsingular if there exists a matrix B such that the matrix products AB and BA exist, are equal and have as common value the n-dimensional unit matrix.

Norm: A function f on a vector space V into $\mathbf{R}$ is a norm if, for every x, y in V and every real or complex number λ,

$$f(x) = \begin{cases} 0 \text{ if } x \text{ is the zero element of } V, \\ \text{positive otherwise}, \end{cases}$$
$$f(x+y) \leq f(x) + f(y),$$
$$f(\lambda x) = |\lambda| f(x).$$

Open interval: see *Interval*

Ordering: A relation $\mathscr{R}$ is an ordering of a set E if, for every a, b, c in E,

(1) $a \mathscr{R} a$,

(2) $\left.\begin{array}{l} a \mathscr{R} b \\ b \mathscr{R} c \end{array}\right\}$ imply $a \mathscr{R} c$,

(3) $a \mathscr{R} b$ or $b \mathscr{R} a$,

(4) $\left.\begin{array}{l} a \mathscr{R} b \\ b \mathscr{R} a \end{array}\right\}$ imply $a = b$.

Partition: A collection of disjoint subsets of A is called a partition of A if the union of the subsets in the collection is equal to A. The word is used in a more restricted sense in the study of Riemann integration. There, a partition of an interval is a finite disjoint collection of subintervals the union of which is the original interval.

Radius of convergence: A complex power series $\sum\limits_{n=0}^{\infty} a_n(z - z_0)^n$ where the a_n are constants does one of the following:

(1) converges only for $z = z_0$,

(2) converges for all z inside a circle of radius R with center at z_0 but diverges for all z outside that circle,

(3) converges for all z in the complex plane.

The number R in (2) is called the radius of convergence of the series. If situation (3) obtains, one says that the radius of convergence is ∞. In the case of a real power series, instead of a circle, there is an interval of convergence, half the length of which is called the radius of convergence.

Rectifiable curve: A curve with finite length (q.v.).

Regulated function: f is a regulated function on $[a, b]$ if f is the uniform limit on $[a, b]$ of a sequence of step-functions.

Relatively prime polynomials: Two polynomials are relatively prime if no single nonconstant polynomial divides both.

Riemann's formula: Let Γ denote a piecewise-smooth Jordan curve serving as boundary of a closed plane region C. Let $f(x, y)$ and $g(x, y)$ denote two functions with continuous partial derivatives defined on C. Then

$$\int_{+} f(x, y)\,\mathrm{d}x + g(x, y)\,\mathrm{d}y = \iint_{C}\left[\frac{\partial g(x, y)}{\partial x} - \frac{\partial f(x, y)}{\partial y}\right]\mathrm{d}x\,\mathrm{d}y$$

where the plus sign means that the line integral over Γ is in the conventional positive direction.

Ring: Let S be a set on which two closed binary operations, $+$ and $\times$, are defined. Suppose that S is a commutative group with respect to the operation $+$. Suppose that, for every $x, y, z \in S$,

$$(x \times y) \times z = x \times (y \times z),$$
$$x \times (y+z) = (x \times y) + (x \times z),$$
$$(x+y) \times z = x \times z + y \times z.$$

Then S is a ring with respect to the operations $+$ and $\times$.

Step function: f is a step function on $[a, b[$ if there exist numbers

$$a = \alpha_0 < \alpha_1 < \ldots < \alpha_n = b$$

such that f is constant on each $[\alpha_i, \alpha_{i+1}[$ for $i = 0, 1, \ldots, n-1$.

Subgroup: S is a subgroup of a group G if S is a subset of G and a group with respect to the binary operation that makes G a group.

Subring: S is a subring of a ring R if S is a subset of R and a ring with respect to the binary operations that make R a ring.

Surjective mapping: f is a surjective mapping of a set A onto a set B if, for every b in B, there exists an a in A such that $f(a) = b$.

Symmetric matrix: A matrix that is equal to its transpose.

Transpose: Let $A = (a_{ij})$ be an $m \times n$ matrix. Then the $n \times m$ matrix $B = (b_{ij})$ is the transpose of A if $a_{ij} = b_{ji}$ for every $i = 1, \ldots, m$ and $j = 1, \ldots, n$. (That is, the rows of A become the columns of B and vice versa.)

Uniform convergence: Let $\{f_n\}$ denote a sequence of functions defined on an interval I. Suppose that there exists a function f such that, for every x_1 in I the sequence $\{f_n(x)\}$ converges to $f(x)$. This convergence of the sequence to f is said to be uniform if, for every $\varepsilon > 0$, there exists an integer N (independent of x) such that the inequality $n > N$ implies $|f_n(x) - f(x)| < \varepsilon$ for every x in I.

Upper bound: Let E denote a set, $\mathscr{R}$ an ordering of E, and E' a subset of E. Then $x_1 \in E$ is an upper bound of E' if $y \in E'$ implies $y \mathscr{R} x$.

Vector space: Let V denote a commutative group with group operation denoted by $+$. Let F denote a field with field operations $+$ and $\times$. Suppose that an operation, denoted here by $\circ$, on F and V together (with result in V) is defined such that, for every a, b in F and x, y in V

$$
\begin{aligned}
a \circ (x+y) &= a \circ x + a \circ y, \\
(a+b) \circ x &= a \circ x + b \circ x, \\
a \circ (b \circ x) &= (a \times b) \circ x, \\
i \circ x &= x,
\end{aligned}
$$

where i is the unit element of F. Then V is a vector space over F.

Vector subspace: U is a vector subspace of a vector space V if U is a subset of V and a vector space relative to the operations that make V a vector space.

Zero element of a group: The element a of a group such that, for every x in the group, $a \circ x = x \circ a = x$.

X^T: A symbol for the transpose (q.v.) of the matrix X.

Subject Index